FUNDAMENTALS OF DATA COMMUNICATION

By
K.K. Sharma
Assistant Professor
Deptt. of Electronics & Communication Engg.
LINGAYA's University
Faridabad

S.K. KATARIA & SONS®
Publishers of Engineering Books
4885/109, Prakash Mahal, Dr. Subhash Bhargav Lane,
Daryaganj, New Delhi-110002 (India)
Tel.: +91-11-23243489, 43551243, Mobile: +91-9871775858
E-mail: katariabook@gmail.com, katariabooks@yahoo.com
website: www.skkatariaandsons.com
(e-books, subscriptions and bulk buying also available on our website)

Published by :
S.K. KATARIA & SONS®
4885/109, Prakash Mahal, Dr. Subhash Bhargav Lane,
Daryaganj, New Delhi-110002 (India)
Tel.: +91-11-23243489, 43551243, Mobile: +91-9871775858
E-mail: katariabook@gmail.com, katariabooks@yahoo.com

First Edition: 2012, Reprint: 2023

Price: ₹ 450/-

Laser Typesetting by :
AISA ENTERPRISES, Delhi

Printed at :
Fine Offset Printers, Delhi

Preface to Second Edition

This text book, '**Fundamentals of Data Communication Engineering**' is intended to serve as a one semester text for undergraduate course of Electronics and Communication Engineering, Electrical Engineering, and Electrical and Electronic Engineering students. The purpose of this book is to extend and update the knowledge of the reader to the basic concepts of modern data communication systems and networks. In this book, the fundamental concepts of data communication engineering are explained in a simple manner.

This revised second edition includes few topic such as RFID Technology, Application of RFID's etc.

Although there are lots of standard books available on the subject of Data communication system for the students but author has felt that there is no such book which cater the needs for the students of ECE. Author has made efforts to include all the topics from the basic to into sufficient depth in a single volume.

The book has been written in a very simple language keeping in focus the syllabi prescribed for the subject. All the topics of this book have been illustrated with the photograph and solved problems.

This book is an outgrowth of this need and hope it will be very helpful for the students.

Inspite of sincere efforts there may be some misprints, omission and errors. Any suggestion towards the improvement of this text will be highly appreciated.

FARIDABAD **K. K. SHARMA**
JULY, 2011

Acnowledgement

I take this opportunity to express my gratitude to the respected Late Prof G.V.K Sinha, founder Chairman of Lingaya's University, Dr. K. K. Aggarwal, Chancellor, Lingaya's University, Dr. P. Gadde, CEO and Pro Chancellor, Lingaya's University and Dr. G. S. Yadava, Pro Vice-Chancellor, Lingaya's University, Faridabad for providing innovative and encouraging environment to work on this project.

I would like to express my sincere gratitude to Cdr Rajendera Kumar, Professor, Dean and Head of ECE, Echelon Institute of Technology, Faridabad, for constant inspiration and motivation.

I would also like to thank to the Management, teachers, students and supporting staff of Lingaya's University. I would like to express my gratitude to all the authors whose work has been freely consulted in the preparation of this text.

I am grateful to my wife Sunita, my daughters Aparna and Vaishali, and son Aman for supporting me to complete this project.

A special thanks to Mr. Sanjeev Kataria, S. K. Kataria & Sons, New Delhi and Mr. Sachin for publishing this book in such a beautiful get-up and well in time.

Lastly, I am grateful to my father and mother without their blessing this book would have not come into existence.

AUTHOR

This Book is Dedicated

to

My Beloved Wife Sunita

Contents

7. NETWORK CONFIGURATIONS 200–258

8. INTEGRATED SERVICE DIGITAL NETWORK (ISDN) 259–274

1

CHAPTER

Digital Communication

1.1. INTRODUCTION TO COMMUNICATION

INSIDE THIS CHAPTER

People can communicate with other people. This is one unmistakable characteristic that points out the difference between humans and other animals (*i.e.,* the ability to communicate). During the early days, communication among humans was limited by distance. People communicated by natural senses of hearing and sight or by written words. Actual words could be conveyed although the process was very slow. As the distance increased beyond the normal range of vision and hearing it became very difficult or near to impossible to communicate using the above methods. People searched for better ways of communicating faster and over long distances. Smoke signals and flag signaling were used to send messages. These were used for centuries until the nineteenth century when electrical signals came in use. **In 1869, the telephone was invented.** This marked a real breakthrough. At that time, it seemed like a miracle to be able to talk to someone so many miles away as easily and quickly as they were standing next to you. The arrival of wireless communication nearly three decades after the Telephone, completed the breakthrough.

Communication is the process of conveying information at a distance or it is the basic process of exchanging information. In other words, communication is the science of transmitting, receiving and processing of information. Electrical communication is a process in which the message or information is transmitted from one point to another point or from one person to another in the form of electrical signal, through some communication link.

The electronic equipments or devices which are used for the purpose of communication are known as communication equipments and integrating of different communication equipments for the purpose of communication form a **communication system**.

DO YOU KNOW

Communication is the process of conveying information at a distance or it is the basic process of exchanging information.

1.2. ELECTRICAL COMMUNICATION SYSTEM

In communication, the physical information, such as sounds, pictures, words etc., is converted into an electrical message called signal and this electrical signal is being conveyed at the far place, and there, these electrical signals are reconverted into physical message.

Figure 1.1 shows the simple block diagram of electrical communication system and it has three main basic components,

1. Transmitter
2. Communication channel or transmission media
3. Receiver

Transmitter

Transmitter is the sub-system that takes the information signal and processes it prior to transmission. The transmitter modulates the information onto a carrier signal, amplifies the signal and broadcasts it over the channel.

Communication Channel

Communication Channel is the medium which transports the modulated signal to the receiver. Air acts as the channel for broadcasts like radio may also be a wiring system like cable TV or the Internet.

Receiver

Receiver is the sub-system that takes in the transmitted signal from the channel and processes it to retrieve the information signal. The receiver must be able to discriminate the signal from other signals which may use the same channel (called tuning), amplify the signal for processing and demodulate (remove the carrier) to retrieve the information. It also, then processes the information for reception (for example, broadcast on a loudspeaker).

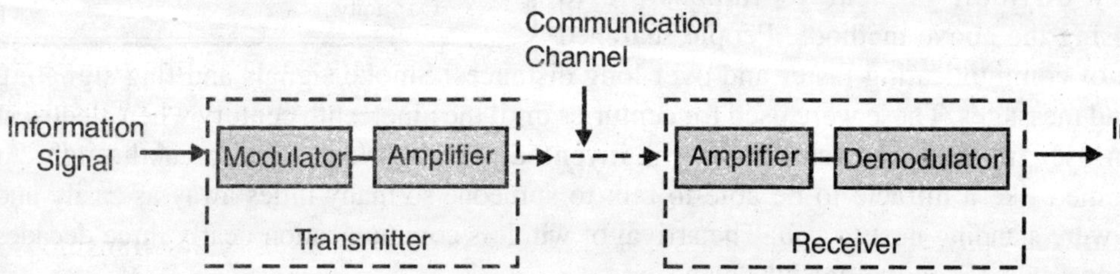

Fig. 1.1. *Simple block diagram of communication system*

1.3. TYPES OF COMMUNICATION SYSTEMS

Broadly we classify the communication system in two types depending upon the type of the signal which is used in the communication system for processing, namely

1. Analog Communication System
2. Digital Communication System

1.3.1. Analog Communication System

We have described an electrical communication system in rather broad terms based on the implicit assumption that the message signal is a continuous time-varying waveform. We refer to such continuous-time signal waveforms as ***analog signals*** and to the corresponding information sources that produce such signals as ***analog sources***.

Analog signals can be transmitted directly via carrier modulation over the communication channel and demodulated accordingly at the receiver. We call such a communication system an ***analog communication system.***

In general, an analog communication system can be represented by the functional block diagram shown in Fig. 1.2. The information generated by the source may be of the form of voice (speech source), a picture (image source), or plain text.

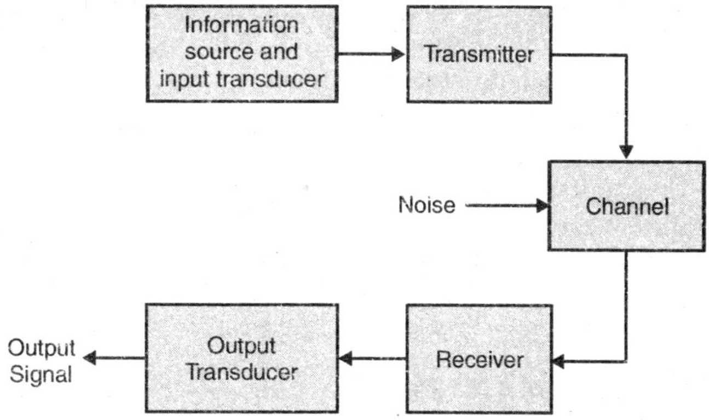

Fig. 1.2. *Functional block diagram of an analog communication system*

Information Source

The communication system exists to convey a message. This message comes from the **information source**, which originates it, in the sense of selecting one message from a group of messages. In other words we can say that the function of information source is to provide required message that is to be transmitted. An essential feature of any source that generates information is that its output is described in probabilistic terms; *i.e.,* the output of a source is not deterministic. Otherwise, there would be no need to transmit the message.

Transducer

A **transducer** is usually required to convert the output of a source into an electrical signal that is suitable for transmission. For example, a microphone serves as the transducer that converts an acoustic speech signal into an electrical signal, and a video camera converts an image into an electrical signal. At the destination, a similar transducer is required to convert the electrical signals that are received into a form that is suitable for the user; *e.g.*, acoustic signals, images, etc.

The heart of the communication system consists of three basic parts, namely, the *transmitter*, the *communication channel,* and the *receiver.* The functions performed by these three elements are described as follows:

Transmitter

The **transmitter** converts the electrical signal into a form that is suitable for transmission through the physical channel or transmission medium. For example, in radio and TV broadcast, the Federal Communications Commission (FCC) specifies the frequency range for each transmitting station. Hence, the transmitter must translate the information signal to be transmitted into the appropriate frequency range that matches the frequency allocation assigned to the transmitter. Thus, signals transmitted by multiple radio stations do not interfere with one another. Similar, functions are performed in telephone communication systems where the electrical speech signals from many users are transmitted over the same wire.

In general, the transmitter performs the matching of the message signal to the channel by a process called *modulation.* Usually, modulation involves the use of the information signal to systematically vary either the amplitude, frequency, or phase of a sinusoidal carrier. For example, in AM radio broadcast, the information signal that is transmitted is contained in the amplitude variations of the sinusoidal carrier, which is the center frequency in the frequency band allocated to the radio transmitting station. This is an example of *amplitude modulation.* In FM radio broadcast, the information signal that is transmitted is contained in the frequency variations of the sinusoidal carrier. This is an example of *frequency modulation. Phase modulation* (*PM*) is yet a third method for impressing the information signal on a sinusoidal carrier.

In another words, carrier modulation such as AM, FM, and PM is performed at the transmitter, as indicated above, to convert the information signal to a form that matches the characteristics of the channel.

In addition to modulation, other functions that are usually performed at the transmitter are filtering of the information-bearing signal, amplification of the modulated signal, and in the case of wireless transmission, radiation of the signal by means of a transmitting antenna.

At last it is concluded that the transmitter is required to make the suitable for conduction over the communication channel.

Communication Channel

The **communications channel** is the physical medium that is used to send the signal from the transmitter to the receiver. There are two types of channels, namely point to point channel and broadcast channels. In point to point channel usually employ a variety of physical media, including wire lines, optical fiber cables, and wireless (microwave radio) such as telephone channels. On the other hand, in broadcast channels provide a capability where many receiving stations can be reached simultaneously from a single transmitter.

In wireless transmission, the channel is usually the atmosphere (free space). Examples of broadcast channels are satellite communications, television broadcasting, radio broadcasting, etc.

The words "*tele*", "*phon*", and "*graph*" are derived from Greek.

Tele – means 'at a distance'

Phon – means sound or speech

Graph - means writing or drawing

Therefore, *telecommunication* means communication at a distance. This can be done through wires called transmission lines or through atmosphere by a radio link. Other examples include:

Telephone – speaking at a distance

Television – seeing at a distance

Telegraph – writing at a distance

Complex telephone networks are built using a lot of devices and methods, some of which will be discussed later in this book.

Whatever the physical medium for signal transmission, the essential feature is that the transmitted signal is distorted in a random manner by a variety of possible mechanisms. During the transmission and reception process the signal get corrupted due to noise inducted in the system.

Noise may be defined as unwanted signals which tend to interfere with the required signals. The most common form of signal degradation comes in the form of additive noise, which is generated at the front end of the receiver, where signal amplification is performed. This noise is often called *thermal noise*.

In wireless transmission, additional additive disturbances are man-made noise, and atmospheric noise picked up by a receiving antenna. Automobile ignition noise is an example of man-made noise, and electrical lightning discharges from thunderstorms is an example of atmospheric noise. Interference from other users of the channel is another form of additive noise that often arises in both wireless and wire line communication systems.

Receiver

The function of the **receiver** is to recover the message signal contained in the received signal. If the message signal is transmitted by carrier modulation, the receiver performs *carrier demodulation* in order to extract the message from the sinusoidal carrier. Since, the signal demodulation is performed in the presence of additive noise and possibly other signal distortion, the demodulated message signal is generally degraded to some extent by the presence of these distortions in the received signal. Other than performing the primary function of signal demodulation, the receiver also performs a number of peripheral functions, including signal filtering and noise suppression.

Destination is the final stage which is used to convert an electrical signal in to its original form by using the output transducer.

> **DO YOU KNOW**
>
> Analog signals can be transmitted directly via carrier modulation over the communication channel and demodulated accordingly at the receiver. We call such a communication system an ***analog communication system.***

1.3.2. Digital Communication System

An analog source output may be converted into a digital form and the message can be transmitted via digital modulation and demodulated as a digital signal at the receiver. The communication system which uses the digital signals as input is known as **digital communication system**.

There are some main advantages to transmitting an analog signal by means of digital modulation.

1. The most important reason is that signal fidelity is better controlled through digital transmission than analog transmission.
2. In particular, digital transmission allows us to regenerate the digital signal in long-distance transmission, thus eliminating effects of noise at each regeneration point.
3. In contrast, the noise added in analog transmission is amplified along with the signal when amplifiers are used periodically to boost the signal level in long-distance transmission.
4. Another reason for choosing digital transmission over analog is that the analog message signal may be highly redundant. With digital processing, redundancy may be removed prior to modulation, thus conserving channel bandwidth.
5. Yet another reason may be that digital communication systems are often cheaper to implement.

In some applications, the information to be transmitted is inherently digital; *e.g.*, in the form of computer data, etc. In such cases, the information source that generates the data is called a ***discrete (digital) source***.

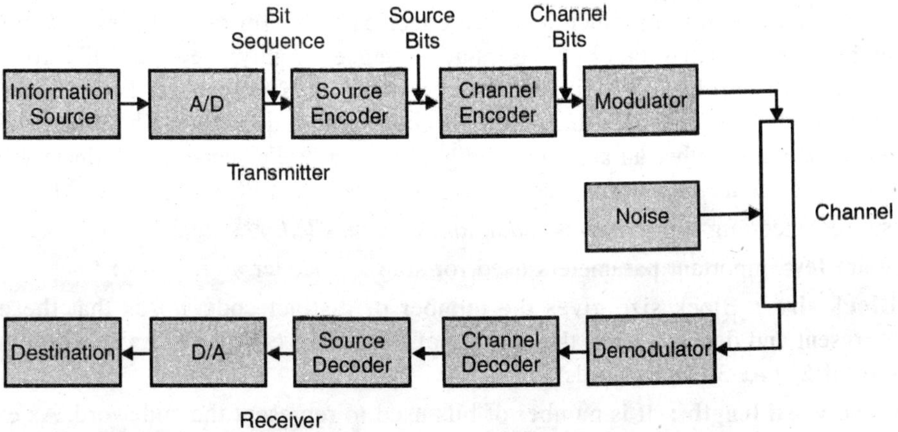

Fig. 1.3. *Typical block diagram of a digital communication system*

In a digital communication system, the functional operations performed at the transmitter and receiver must be expanded to include message signal discretization at the transmitter and message signal synthesis or interpolation at the receiver. Additional functions include redundancy removal, and channel coding and decoding.

Figure 1.3 shows a typical block diagram of a digital communication system.

Information Source

The source output may be either an analog signal, such as audio or video signal, or a digital signal, such as the output of a computer which is discrete in time and has a finite number of output characters. The analog information is converted to an equivalent digital signal by sampling and quantization that process is known as **Analog to Digital conversion (ADC).** The A/D produces analog-to-digital conversion (for analog source) including sampling and quantizing.

Sampling : Pulse amplitude modulation (PAM) : Converting the continuous input to discrete sequence.

Quantization : Pulse code modulation (PCM) : Encoding each quantized sample into a *digital word* including uniform and nonuniform quantization.

The discrete information sources are characterized by source alphabet, symbol rate, source alphabet probabilities and probabilistic dependence of symbols in a sequence.

The *source alphabet* will have letters, digits, and special characters available from the information source.

The *symbol rate* is the rate at which the information sources thc symbols/characters and its unit is symbol/sec.

Source alphabet occurs with its own probabilities. Each source alphabet from a source has independent reoccurrence rate in the sequence. A parameter called source entropy may be defined as the average information contents per symbol in a long message. It is expressed in bits/symbol. The source information rate is given by the product of source entropy and symbol rate expressed in bits per second. It represents the minimum number of bits per sec on an average to represent the information from the source. That is to say, source information rate representations the minimum average data rate needed to convey the information from source to destinations.

Source Encoding/Decoding

In a digital communication system, the messages produced by the source are usually converted into a sequence of binary digits (0's & 1's) by assigning codeword to the symbols in the input sequence. The simplest way in which a source coder can perform this process is to assign a fixed length binary code to each symbol of the input sequence. Ideally, we would like to represent the source output (message) by as few binary digits as possible. In other words, we seek an efficient representation of the source output that results in little or no redundancy. The process of efficiently converting the output of either an analog or a digital source into a sequence of binary digits is called ***source encoding* or *data compression*,** such as Huffman coding, JPEG (for picture), MPEG (for video).

The source-encoding step removes *redundant* or *unneeded* messages.

There are few important parameters used for source encoder as follows :

- **Block size :** Block size gives the number of distinct code-words that the encoder can represent and depends upon the number of bits in the codeword. As an example, the block size of 2^5 *i.e.,* 32 code-words.
- **Code-word length :** It is number of bits used to represent the codeword. As example, if 5 bits are assigned to each code-word, then code-word length will be 5 bits.
- **Average data rate :** It is the output bits per second from the source encoder. It may be the long term data rate that can be achieved by the encoder. Actually, the source encoder

assigns many numbers of bits to each input symbol and that's why the data rate is higher than the symbol rate. For example, if we consider that the symbols are given to the source encoder and the length of code-word is 5 bits, then the output data rate from the source encoder would be,

$$\text{Data rate} = \text{Codeword length} \times \text{Symbol rate}$$

$$\text{Data rate} = 10 \times 5 = 50 \text{ bits/sec}$$

The coding efficiency is the rate of actual data rate to minimum achievable data rate.

The **source decoder** at the receiver converts the binary output of the channel decoder in to a symbol sequence.

Channel Encoder/Decoder

The sequence of binary digits from the source encoder, which we call the *information sequence,* is passed to the *channel encoder.* The purpose of the channel encoder is to introduce, in a controlled manner, some redundancy in the binary information sequence which can be used at the receiver to overcome the effects of noise and interference encountered in the transmission of the signal through the channel. Thus, the added redundancy serves to increase the reliability of the received data and improves the fidelity of the received signal. In effect, redundancy in the information sequence aids the receiver in decoding the desired information sequence. Channel coding is used to counteract channel impairments such as noise, fading and jamming by selectively introducing redundancies in the transmitted data. Through channel coding, communication system can obtain coding gain which is the decrease of signal to noise ratio (SNR) to get the same Bit error rate (BER) performance.

The channel encoder provides a different type of communication security than that provided by the encryptor. By adding some known redundancy to the message. Error control is achieved by adding extra bits to the output of the source encoded data. The extra redundancy bits added do not convey any information but provides detection and correction of errors in the message bits at the receiver. The coding and decoding process at the encoder and decoder needs the memory for processing of binary data. Typical channel codes: *block codes, convolutional codes.*

There are some important parameters that a channel encoder must possess.

(*i*) The coding rate which is depend upon the extra bits added by the channel encoder.

(*ii*) Coding techniques used.

(*iii*) Coding efficiency.

(*iv*) Error detection/correction capability.

(*v*) Complexity of encoder and decoder.

Digital Modulator/Demodulator

As we have already discussed about the need of modulation in communication system. Analog modulation methods are employed in analog communication system such as amplitude, frequency and phase modulation. A bit sequence cannot be transmitted from channel (fiber, cable, atmosphere and other transmission medium) directly. We must use the bit sequence to generate appropriate digital waveform, which is suitable to transmission medium. For example *Pulse waveform* (baseband signal) can directly transmitted through *cable*. For wireless, the baseband signals must be shift to a frequency band (RF signal), which is suitable to transmit through atmosphere.

In digital communication system, digital modulation techniques are used such as amplitude shift keying (ASK), frequency shift keying (FSK), phase shift keying (PSK) and minimum shift keying (MSK).

The output of channel encoder is a digital signal composed of symbols. These digital symbols can be transmitted directly on a physical channel after passing through a line coding system (such as NRZ, RZ, Manchester, Differential Manchester). This type of transmission is referred to as base band modulation because the base band signal has been coded and transmitted at the base band frequency only. However, for wireless communication this digital signal is carrier modulated using sinusoidal as carrier signal. Carrier modulation enables efficient transmission of the signal by minimizing the effect of channel noise.

At the receiving end, we use digital demodulator to converts modulated signals in to binary bits.

Communication Channel

The **communications channel** is the physical medium that is used to send the signal from the transmitter to the receiver. There are two types of channels, namely point to point channel and broadcast channels. In point to point channel usually employ a variety of physical media, including wire lines, optical fiber cables, and wireless (microwave radio) such as telephone channels. On the other hand, in broadcast channels provide a capability where many receiving stations can be reached simultaneously from a single transmitter. In wireless transmission, the channel is usually the atmosphere (free space). Examples of broadcast channels are satellite communications, television broadcasting, radio broadcasting etc.

Whatever the physical medium for signal transmission, the essential feature is that the transmitted signal is distorted in a random manner by a variety of possible mechanisms. During the transmission and reception process the signal get corrupted due to noise inducted in the system.

Noise may be defined as unwanted signals which tend to interfere with the required signals. The most common form of signal degradation comes in the form of additive noise, which is generated at the front end of the receiver, where signal amplification is performed. This noise is often called *thermal noise*.

In wireless transmission, additional additive disturbances are man-made noise, and atmospheric noise picked up by a receiving antenna. Automobile ignition noise is an example of man-made noise, and electrical lightning discharges from thunderstorms is an example of atmospheric noise. Interference from other users of the channel is another form of additive noise that often arises in both wireless and wire line communication systems.

DO YOU KNOW

In a **digital communication system**, the functional operations performed at the transmitter and receiver must be expanded to include message signal discretization at the transmitter and message signal synthesis or interpolation at the receiver.

1.4. COMPARISON OF DIGITAL AND ANALOG COMMUNICATION SYSTEMS

Analog communication involves transmission of a continuous of values over a defined interval of time. Processing of analog information is generally not easy and it is more prominent to the noise. Further, there is no facility of error detection/correction in the analog system. Digital communications overcome these problems very effectively. We now discuss the advantages and few disadvantages of digital communication systems.

1.4.1. Advantages of Digital Communication

(*i*) Reduction of noise is possible in digital communication system.

(*ii*) Reduction of distortion and other impairments is possible in digital communication system.

(*iii*) Error detection and correction is possible in digital communication system that will reduce the probabilities of error by employing channel encoder.

(*iv*) Regeneration of signal is easier than analog communication system.

(*v*) Signal processing and storage is easier due to use of digital signal processing techniques and digital computer.

(*vi*) Security of data is possible due to encryption and decryption and spread spectrum coding provides antijamming facility.

(*vii*) Data compression and image processing is possible in digital communication system.

(*viii*) Large amount of noise interference may be tolerated.

(*ix*) Use of VLSI technology may be possible.

(*x*) Handling of multimedia information such as integration of voice, video and data is easier.

(*xi*) There are many signal processing algorithms can be used to increase transmission rate and reliability of digital communication system.

(*xii*) Digital circuits are more reliable and can be produced at lower cost than analog circuits, such as DSP, FPGA and ASIC.

(*xiii*) Digital signal is less subject to distortion and interference than analog signal.

(*xiv*) The digital signal is more easer to "reborn" or regenerated, compared to analog signals because:

- Digital signal has only two states (0 or 1)
- On the contrary, analog signal are not two-state signal.

It can take an infinite variety of amplitude.

1.4.2. Disadvantages of Digital communication

1. Digital communication requires more bandwidth than analog communication.
2. It needs synchronization.

1.5. LIMITATIONS OF COMMUNICATION SYSTEM

The primary limitations of information transmission by electrical means are bandwidth and noise.

1. Bandwidth Limitation
2. Noise Limitation

1.5.1. Bandwidth Limitation

The **bandwidth** can be defined as the frequency range or the band of frequency needed for a particular transmission. The concept of bandwidth applies to the both systems and signals. If the signal changes frequently with time, its frequency contents extends over a wide range and we may say that the signal has large bandwidth. In the same way, the ability of a system to follow signal variations is reflected in its usable frequency response or transmission bandwidth. Every communication system has a finite bandwidth that restricts the rate of signal variations. The bandwidth is allocated by regulatory agencies. Such type of regulation is required to avoid interference among the signals of the same frequency. However, that allocated bandwidth for a given transmission may not be adequate to convey the entire information.

According to information theory, greater the bandwidth of a transmission of the system, the more is the information that can be transmitted. Now we know that to transmit video signal in Television broadcasting needs bandwidth in MHz (India it is 7 MHz), while the audio signal has slow variations and it requires less bandwidth (approximately 150 kHz). For example, when we listen music in AM and FM radio, the performance of FM radio is better than AM radio, why? The reason is related to the bandwidth used in AM and FM radio. The full amount of information available to the human ear is contained in the range of frequency from 0Hz to 15 kHz. The maximum modulating frequency is only 5 kHz in AM transmission whereas it is 15 kHz in FM. Therefore, an AM radio receiver cannot reproduce all the information contained in the music because this would be needed bandwidth of 30 kHz. On the other hand, the bandwidth allocated to FM transmission is 200 kHz. So that FM receiver can easily reproduce the transmitted information. This is why FM has a better fidelity than AM and it is preferred to listen to a FM receiver.

Thus bandwidth is a fundamental limitation of any communication system.

DO YOU KNOW

The **bandwidth** can be defined as the frequency range or the band of frequency needed for a particular transmission.

1.5.2. Noise Limitation

Noise is another limitation of any communication system. It can be defined as an unwanted form of energy which tends to interfere with the transmission and reception of the desired signal in a communication system. Some sort of noise is always present in the communication system. Noise cannot be removed completely but it can be minimized by using some noise limitation circuits.

DO YOU KNOW

Noise is defined as an unwanted form of energy which tends to interfere with the transmission and reception of the desired signal in a communication system.

1.6. DIGITAL RADIO

In a digital radio, the digital pulses modulate an analog carrier between two or more points. The digital transmission system needs a transmission channel may a guided media such as metallic parallel lines, coaxial cables, optical fiber, or unguided media *i.e.*, wireless transmission.

Digital modulation is sometimes called **digital radio** because digitally modulated signals can be propagated through earth's atmosphere and used in wireless communication systems.

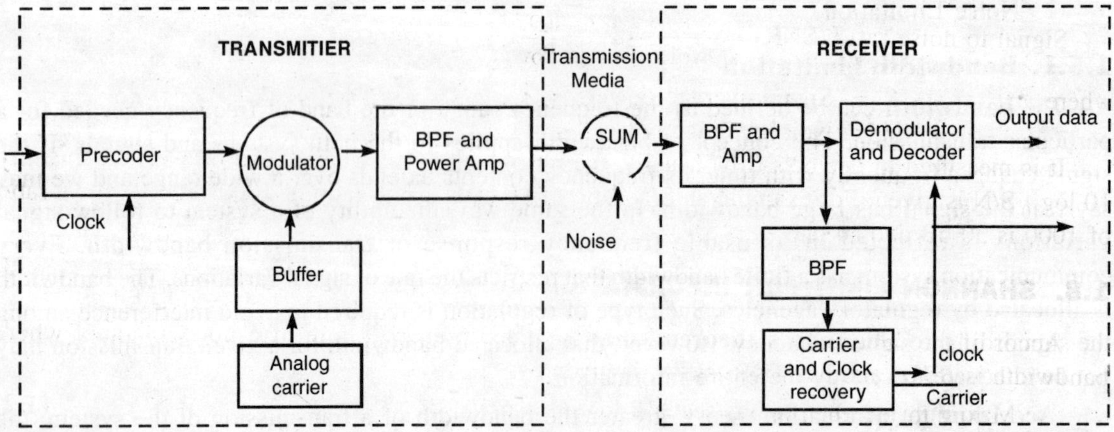

Fig. 1.4. *A simple block diagram of Digital Radio.*

1.7. THE MAXIMUM DATA RATE OF A CHANNEL (IN NOISELESS AND NOISY CHANNEL)

In 1924, Henry Nyquist, has realized that even a perfect channel (*i.e.,* noiseless channel) has a finite transmission capacity. He derived as expression for the maximum data rate for a finite bandwidth noiseless channel.

Sampling is a process by which an analog signal is converted in to discrete time signal. The number of samples to be taken depends upon maximum signal frequency present; sampling theorem gives the complete idea about the sampling of the signal.

Sampling theorem states that :

(*i*) A bandlimited signal of finite energy, which has no frequency components higher that B Hz, is completely recovered from the knowledge of its samples taken at the rate of 2B samples per second, where B is the bandwidth of a signal. 2B is normally referred as sampling rate or Nyquist data rate.

(*ii*) A bandlimited signal of finite energy, which has no frequency components higher that B Hz, is completely described by specifying the values of the signal at instants of time separated by 1/2B seconds.

Nyquist proved that any arbitrary signal has been passed through a low pass filter of bandwidth B, the filtered signal can be reconstructed completely by making only 2B (exact) sample per second. Sampling the line faster than 2B times per second is of no use because the higher frequency components such sampling could recover have already been filtered out.

According to **Nyquist's theorem**, if the signal consists of V discrete levels, the maximum data rate can be given as

$$\text{Maximum data rate} = 2B \log_2 V \text{ bits/sec} \quad \ldots(1.1)$$

$$\boxed{\text{Maximum data rate} = 2B \log_2 V \text{ bits/sec}}$$

Let us consider an example; a noiseless channel of kHz cannot transmit binary (*i.e.,* two levels) signals at a rate encoding 6000 bits per sec.

In 1948, Claude Shannon carried Nyquist's work further and extended it to the practical care of a channel subject to random noise. All practical channels are noisy channels and random noise deteriorates the situation rapidly. It is always that random or thermal noise present due to the motion of the molecules in the system. The amount of thermal noise present is determined by the ratio of the signal power to the noise power. This ratio is called **signal to noise ratio.**

$$\text{Signal to noise ratio (SNR)} = \frac{\text{Signal average power}}{\text{Noise average power}}$$

where S = Signal average power

N = Noise average power

It is measured in decibels (dB) usually, the ratio itself is not quoted, and instead, the quantity $10 \log_{10}$ S/N is given. For example an S/N ratio of 10 is 10 dB, a ratio of 100 is 20 dB, and a ratio of 1000 is 30 dB and so on.

1.8. SHANNON LIMIT FOR INFORMATION CAPACITY

According to **Shannon's theorem**, that the maximum data rate of a noisy channel where bandwidth is B Hz, and has SNR is S/N, is given by.

$$\text{Maximum number of bits/sec} = B \log_2 (1 + S/N) \quad \ldots(1.2)$$

$$\boxed{\text{Maximum number of bits/sec} = B \log_2 (1 + S/N)}$$

Shannon's result was derived from information-theory arguments and applies to any channel subject to thermal noise. Now, for examples, a channel of 3 kHz bandwidth with a signal to noise ratio of 30 dB can never transmit much more than 30,000 bps. This is this maximum data rate in noisy channel.

This theorem is concerned with the rate of transmission of information over a transmission media or communication channel. The term communication channel includes all the features and components parts of the transmission system which introduce noise, or limit the bandwidth.

Shannon's theorem defines that it is possible, in principle, to device a means whereby a communication system will transmit information with an arbitrarily small probability of error provided that the information rate '*R*' is less than or equal to a '*C*' the channel capacity.

$$R = C \qquad \text{...(1.3)}$$

where R = Information rate

C = Channel capacity

It is commonly known as Shannon-Hartley capacity theorem. According to this the system capacity (C) is a function of the average received signal power 'S', the average noise power 'N', and bandwidth 'B'. They are related as:

$$\boxed{C = B \log_2 (1 + S/N)} \qquad \text{...(1.4)}$$

If bandwidth (B) is in hertz (Hz) and base is 2 of logarithmic then unit of channel capacity, is bits/sec.

Theoretically as we have discussed in Eq. (1.2), it is possible to transmit information over such a communication channel at any rate R, where $R = C$, with an arbitrarily small error.

The important feature of the theorem is that it indicates that for $R = C$, error free transmission is possible in the presence of noise.

There is another statement associated with this theorem, that is given a source of M equally likely messages, with $M >> 1$, which is generating information at a rate R and given a channel with a channel capacity C. if, now $R > C$, the probability of error is close to unity for every possible set of M transmitter signals.

If the information rate R exceeds a specific value of channel capacity C, the error probability approaches unity, as M increases.

The channel capacity 'C' decides the maximum permissible rate at which an error free transmission of information is possible through it.

Example 1. Consider a noiseless channel with a bandwidth of 3000 Hz transmitting a signal with two signal levels.

Solution. The maximum bit rate can be calculated as

$$\text{Bit Rate} = 2 \times 3000 \times \log_2 2 = 6000 \text{ bps}$$

Example 2. Consider the same noiseless channel, transmitting a signal with four signal levels (for each level, we send two bits).

Solution. The maximum bit rate can be calculated as:

$$\text{Bit Rate} = 2 \times 3000 \times \log_2 4 = 12{,}000 \text{ bps}$$

Example 3. Consider an extremely noisy channel in which the value of the signal-to-noise ratio is almost zero. In other words, the noise is so strong that the signal is faint. For this channel the capacity is calculated as

$$C = B \log_2 (1 + SNR) = B \log_2 (1 + 0)$$

$$C = B \log_2 (1) = \text{B} \times 0 = 0$$

Example 4. We have a channel with a 1 MHz bandwidth. The SNR for this channel is 63; what is the appropriate bit rate and signal level?

Solution. First, we use the Shannon formula to find our upper limit.

$$\begin{aligned} C &= B \log_2 (1 + SNR) \\ &= 10^6 \log_2 (1 + 63) \\ &= 10^6 \log_2 (64) = 6 \text{ Mbps} \end{aligned}$$

Then we use the Nyquist formula to find the number of signal levels.

$$\mathbf{4\ Mbps = 2 \times 1\ MHz \times \log_2 \mathit{V} \rightarrow \mathit{V} = 4}$$

Summary

- Communication is the process of conveying information at a distance or it is the basic process of exchanging information. In other words, communication is the science of transmitting, receiving and processing of information.
- The electronic equipments or devices which are used for the purpose of communication are known as communication equipments and integrating of different communication equipments for the purpose of communication form a communication system.
- Electrical communication system and it has three main basic components,
  1. Transmitter
  2. Communication channel or transmission media
  3. Receiver
- An electrical communication system in rather broad terms based on the implicit assumption that the message signal is a continuous time-varying waveform. We refer to such continuous-time signal waveforms as ***analog signals*** and to the corresponding information sources that produce such signals as ***analog sources***.
- **Information Source :** The communication system exists to convey a message. This message comes from the information source,
- **Transducer :** A microphone serves as the transducer that converts an acoustic speech signal into an electrical signal
- The **transmitter** converts the electrical signal into a form that is suitable for transmission through the physical channel or transmission medium.
- The **communications channel** is the physical medium that is used to send the signal from the transmitter to the receiver. There are two types of channels, namely point to point channel and broadcast channels.
- **Noise** may be defined as unwanted signals which tend to interfere with the required signals.
- The function of the **receiver** is to recover the message signal contained in the received signal.
- **Destination** is the final stage which is used to convert an electrical signal in to its original form by using the output transducer.
- The communication system which uses the digital signals as input is known as **digital communication system**.
- The information source that generates the data is called a ***discrete (digital) source***.
- **Information Source :** The source output may be either an analog signal, such as audio or video signal, or a digital signal, such as the output of a computer which is discrete in time
- The A/D produces analog-to-digital conversion (for analog source) including sampling and quantizing.

Sampling : Pulse amplitude modulation (PAM) : Converting the continuous input to discrete sequence.

Quantization : Pulse code modulation (PCM) : Encoding each quantized sample into a *digital word* including uniform and nonuniform quantization.

- The *source alphabet* will have letters, digits, and special characters available from the information source.
- The *symbol rate* is the rate at which the information sources the symbols/characters and its unit is symbol/sec.
- The process of efficiently converting the output of either an analog or a digital source into a sequence of binary digits is called ***source encoding* or *data compression.***
- There are few important parameters used for source encoder as follows :
  - **Block size :** Block size gives the number of distinct code-words that the encoder can represent and depends upon the number of bits in the codeword. As an example, the block size of 2^5 *i.e.,* 32 code-words.
  - **Code-word length :** It is number of bits used to represent the codeword. As example, if 5 bits are assigned to each code-word, then code-word length will be 5 bits.
  - **Average data rate :** It is the output bits per second from the source encoder.
- The **source decoder** at the receiver converts the binary output of the channel decoder in to a symbol sequence.
- The purpose of the channel encoder is to introduce, in a controlled manner, some redundancy in the binary information sequence which can be used at the receiver to overcome the effects of noise and interference encountered in the transmission of the signal through the channel.
- Analog modulation methods are employed in analog communication system such as amplitude, frequency and phase modulation.
- In digital communication system, digital modulation techniques are used such as amplitude shift keying (ASK), frequency shift keying (FSK), phase shift keying (PSK) and minimum shift keying (MSK).
- Analog communication involves transmission of a continuous of values over a defined interval of time. Processing of analog information is generally not easy and it is more prominent to the noise. Further, there is no facility of error detection/correction in the analog system.
- The primary limitations of information transmission by electrical means are bandwidth and noise.
  1. Bandwidth Limitation
  2. Noise Limitation
- The **bandwidth** can be defined as the frequency range or the band of frequency needed for a particular transmission.
- Noise is another limitation of any communication system. It can be defined as an unwanted form of energy which tends to interfere with the transmission and reception of the desired signal in a communication system.
- Digital modulation is sometimes called **digital radio** because digitally modulated signals can be propagated through earth's atmosphere and used in wireless communication systems.
- Nyquist proved that any arbitrary signal has been passed through a low pass filter of bandwidth B, the filtered signal can be reconstructed completely by making only 2B (exact) sample per second.
- According to **Nyquist's theorem,** if the signal consists of V discrete levels, the maximum data rate can be given as

$$\text{Maximum data rate} = 2B \log_2 V \text{ bits/sec}$$

- According to **Shannon's theorem**, that the maximum data rate of a noisy channel where bandwidth is B Hz, and has SNR is *S/N*, is given by

$$\text{Maximum number of bits/sec} = B \log_2 (1 + S/N).$$

Review Questions

1. What do you understand by term communication? Explain electronic communication systems in brief.
2. Draw and explain the basic block diagram of communication system.
3. Compare the digital communication system with the analog communication system.
4. Describe the term channel in communication system.
5. Describes the fundamental limitations of communication system.
6. Draw the digital communication block diagram and explain the working of each block.
7. Explain the difference between digital and analog signals.
8. Define signal and explain how they are classified.
9. Write the advantages and disadvantages of digital communication system.
10. The SNR for a channel having 1 MHz bandwidth is 63. Calculate the following for this channel :
    (*i*) Bit rate (*ii*) Signal level.
    [**Ans.** Bit rate = 6 Mbps and signal level = 35.99 dB] **[MDU Dec. 2009]**
11. A telephone line has a frequency range of 300 to 3300 Hz. The signal to noise ratio is 3162 (35 dB). For this channel calculate the channel capacity. **[MDU Dec. 2008]**
12. The SNR for a channel having 1-MHz bandwidth is 63. Calculate the following for this channel
    (*i*) Bit Rate (*ii*) Signal Level **[MDU Dec. 2009]**
13. A telephone line has a frequency range of 300 to 3300 kHz. The signal to noise ratio is 3162 (35 dB) for the channel calculate channel capacity. **[MDU May 2010]**
14. What is the Shannon limit for information capacity for a standard telephone circuit with a signal to noise ratio fo 1000 and a bandwidth of 2.7 kHz. **[MDU May 2011]**
15. Given a channel with an intended capacity of 20 Mbps and bandwidth of this is 3 MHz. Determine the SNR required in order to achieve this capacity. [**Ans.** 100 or 20 dB]
16. Determine the maximum bit rate for a channel having bandwidth of 3100 Hz and SNR is 20 dB. And also calculate the number of levels required to transmit at the maximum bit rate.
    [**Ans.** Maximum bit rate = 20.640 kbps, No. of levels = 16]
17. What does the Shannon capacity have to do with communication? Calculate the maximum bit rate for the noiseless channel with a bandwidth 3000 Hz transmitting a signal with two signal levels.
    [**Hints :** Capacity = $BW \times \log_2 (1 + SNR)$, bit rate = 6 kbps]
18. What is Nyquist theorem ? What is the bit rate, if we want to digitize the human voice, assuming 8 bits per sample ?
19. Write the short notes on following :
    (*i*) Digital Radio
    (*ii*) Shannon's limit for information capacity
    (*iii*) The advantages and disadvantages of digital communication system

□□□

2
CHAPTER

Digital Modulation

INSIDE THIS CHAPTER

2.1. MEANING AND DEFINITION OF MODULATION

For the transmission of message to distant place, sound wave is first converted into electrical signal with the help of microphone. In order to radiate this audio signal in space, we require an antenna as the radiating system, for effective radiation antenna should be fed with high frequency currents (electrical signal). But as the frequency of audio signal remains quite low, it cannot be fed as such to the antenna for radiation. Therefore before feeding audio signal to the antenna, a process is performed in which the audio signal is superimposed on a high frequency wave, called **carrier wave**. The amplitude, or frequency or phase of the high frequency wave (carrier wave) is modified in accordance with the audio signal (called **modulating signal**), so that the resultant wave (called **modulated wave**) inherits the signal.

The process of superimposition of modulating signal (low frequency) over carrier wave (high frequency) is called **modulation.**

In telecommunications, *modulation* is the process of varying a periodic waveform, *i.e.*, a tone, in order to use that signal to convey a message.

Modulation is the act of translating some low frequency or baseband signal (voice, music, data) to a higher frequency.

2.2. WHY DO WE MODULATE SIGNALS?

There are at least two reasons :

(*a*) To allow the simultaneous transmission of two or more baseband signals by translating them to different frequencies, (*b*) To take advantage of the greater efficiency and smaller size of higher-frequency antennas. The size of antenna depends upon the wavelength of the frequency used. To radiate energy the size of antenna should be at least one-fourth of the wavelength ($\lambda/4$) used.

$$\lambda = \frac{c}{f}$$

where, λ = wavelength

c = velocity of electromagnetic wave (existence of electromagnetic waves that travel at the speed of light *i.e.*, $\mathbf{3 \times 10^{-8}}$ **meter/sec**. In Germany, Heinrich Hertz proved Maxwell's theory and in 1888 was the first to create electro-magnetic radio waves.)

f = operating frequency

DEFINITION

The process of superimposition of modulating signal (low frequency) over carrier wave (high frequency) is called **modulation.**

Few more reasons to use modulation are follows :

1. Modulation for Ease of Radiation

Consider again transmission of good quality audio. Assume we choose the carrier frequency to be 1 MHz. The linear modulation schemes that would be discussed shortly give rise to a maximum frequency spread (of the modulated signal) of 40 kHz, the spectrum of the modulated signal extending from (1000 – 20) = 980 kHz to (1000 + 20) = 1020 kHz. If the antenna is designed for 1000 kHz, it can easily take care of the entire range of frequencies involved because modulation process has rendered the signal into a NBBP signal.

2. Modulation for Efficient Transmission

Quite a few wireless channels have their own appropriate passbands. For efficient transmission, it would be necessary to shift the message spectrum into the passband of the channel intended. Ground wave propagation (from the lower atmosphere) is possible only up to about 2 MHz. Long distance ionospheric propagation is possible for frequencies in the range 2 to 30 MHz. Beyond 30 MHz, the propagation is line of sight. Preferred frequencies for satellite communication are around 3 to 6 GHz. By choosing an appropriate carrier frequency and modulation technique, it is possible for us to translate the baseband message spectrum into a suitable slot in the passband of the channel intended. That is, **modulation results in frequency translation.**

3. Modulation for Multiplexing

Several message signals can be transmitted on a given channel, by assigning to each message signal an appropriate slot in the passband of the channel. Take the example of AM broadcast, used for voice and medium quality music broadcast. The passband of the channel used to 550 kHz to 1650 kHz. That is, the width of the passband of the channel that is being used is 1100 kHz. If the required transmission bandwidth is taken as 10 kHz, then it is possible for us to multiplex, atleast theoretically, 110 distinct message signals on the channel and still be able to separate them individually as and when we desire because the identity of each message is preserved in the frequency domain.

4. Modulation for Frequency Assignment

Continuing on the broadcast situation, let us assume that each one of the message signals is being broadcast by a different station. Each station can be assigned a suitable carrier so that the corresponding program material can be received by tuning to the station desired.

5. Modulation to Improve the Signal-to-noise Ratio

Certain modulation schemes (notably frequency modulation and phase modulation) have the feature that they will permit improved signal-to-noise ratio at the receiver output, provided we are willing to pay the price in terms of increased transmission bandwidth (Note that the transmitted power need not be increased). This feature can be taken advantage of when the quality of the receiver output is very important.

2.3. CLASSIFICATION OF MODULATION

Mainly we can classify the modulation as per the type of modulating signal. It may be analog or digital signal in nature, if the modulating signal is analog (such as audio signal from microphone) than it is known as **Analog Modulation** and if that signal is digital (such as output from computer in the form of 1 and 0) in nature than, we may call that type of modulation as **Digital Modulation.**

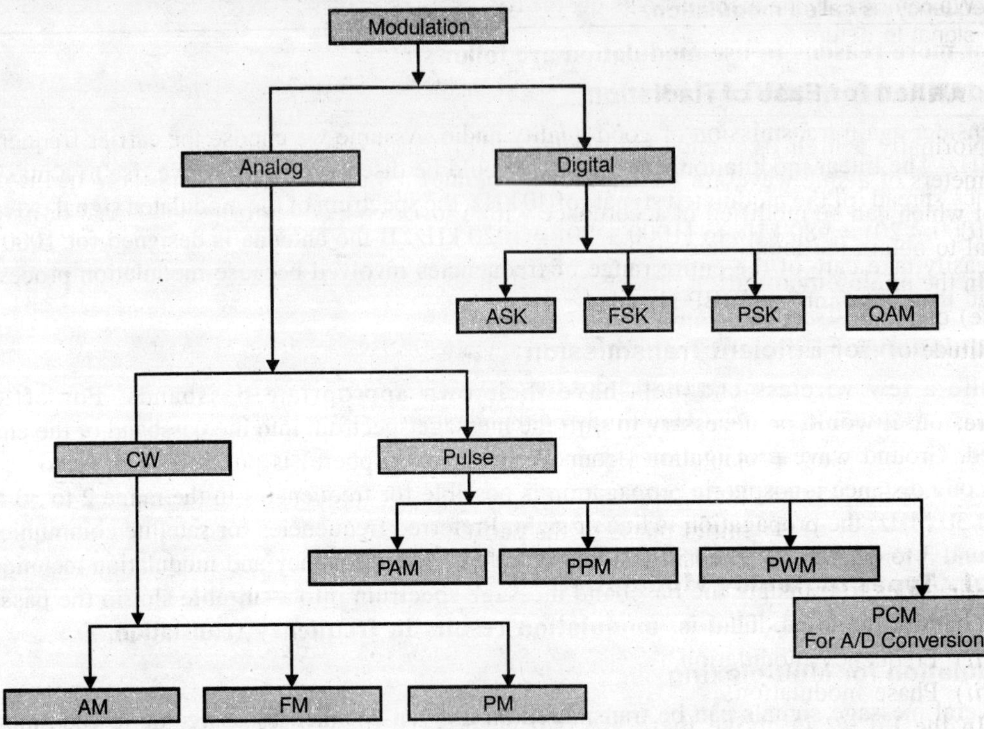

CW—Continuous Wave
FM—Frequency Modulation
PAM—Pulse Amplitude Modulation
PWM—Pulse Width Modulation
ASK—Amplitude Shift Keying
PSK—Phase Shift Keying
AM—Amplitude Modulation
PM—Phase Modulation
PPM—Pulse Position Modulation
PCM—Pulse Code Modulation (Analog to Digital conversion)
FSK—Frequency Shift Keying
QAM—Quadrature Amplitude Modulation

Fig. 2.1. *Classification of modulation.*

Modulation may be classified as per the type of carrier signal; it may be high frequency sinusoidal waveform or may be pulse wave. Now if carrier wave is a sinusoidal wave form than this type of modulation is known as **Continuous wave (CW) modulation** and if it is pulse wave than it is referred as **Pulse modulation**. The examples of continuous wave modulation are amplitude modulation, frequency modulation and phase modulation. On the other hand, if the carrier waveform is a pulse type of waveform, then the modulation process is known as pulse modulation and examples of this are pulse amplitude modulation (PAM), pulse width modulation (PWM), pulse position modulation (PPM), pulse code modulation (PCM) etc. Figure 2.1 illustrated the simple classification of modulation.

The aim of **analog modulation** (AM, FM and PM) is to transfer an analog low pass signal, for example an audio signal or TV signal, over an analog band pass channel, for example a limited radio frequency band or a cable TV network channel.

The aim of **digital modulation** (ASK, FSK and PSK) is to transfer a digital bit stream over an analog band pass channel, for example over the public switched telephone network (where a filter limits the frequency range to between 300 and 3400 Hz) or a limited radio frequency band.

DO YOU KNOW

We can classify the modulation as per the type of modulating signal. It may be analog or digital signal in nature.

2.4. ANALOG MODULATION

Normally a high-frequency sinusoidal waveform is used as carrier signal. The three main parameters of a sine wave are its amplitude (volume), its phase (timing) and its frequency (pitch), all of which can be modified in accordance with a low frequency information signal or modulating signal to obtain the modulated signal.

In the **analog modulation** process, one of the characteristic (such as Amplitude, frequency and phase) of a high-frequency sinusoidal carrier is changed in direct proportion to the instantaneous amplitude of the baseband signal. The carrier itself can be described by the equation

$$e(t) = A \cos(\omega_c t + \theta) \quad ...(2.1)$$

where, A = Peak amplitude of the carrier,

ω_c = Angular frequency of the carrier in radians per second,

t = Time, and

θ = Initial phase of the carrier at time $t = 0$.

2.4.1. Types of Analog Modulation

(*i*) Amplitude modulation

(*ii*) Frequency modulation

(*iii*) Phase modulation.

In the Eq. (2.1) above there are two properties of the carrier that can be changed, the amplitude (A) and the angular position (argument of the cosine function).

DO YOU KNOW

In the **analog modulation** process, some characteristic (such as Amplitude, frequency and phase) of a high-frequency sinusoidal carrier are changed in direct proportion to the instantaneous amplitude of the baseband signal.

Three cases may arise :

(*i*) When amplitude (A) is varied in accordance with the modulating wave, the process is known as **Amplitude modulation**;

(*ii*) When ω_c is varied in accordance with the modulating wave, while E_c and θ remain constant, the process is known as **Frequency modulation**.

(*iii*) When θ is varied in accordance with modulating wave, while ω_c and E_c remain constant, the process is known as **Phase modulation**.

Thus we have amplitude modulation and angular modulation. Angular modulation can be further characterized as either frequency modulation or phase modulation.

2.5. DIGITAL MODULATION

The transmission of digital signals is increasing at a rapid rate. Low-frequency analogue signals are often converted to digital format (PAM) before transmission. The source signals are generally referred to as ***baseband*** signals. Of course, we can send analogue and digital signals directly over a medium. From electro-magnetic theory, for efficient radiation of electrical energy from an antenna it must be at least in the order of magnitude of a quarter-wavelength (λ/4) in size; $c = f\lambda$, where c is the velocity of light, f is the signal frequency and l is the wavelength. For a 1 kHz audio signal, the wavelength is 300 km. An antenna of this size is not practical for efficient transmission. The low-frequency signal is often frequency-translated to a higher frequency range for efficient transmission. The process is called *modulation*. The use of a higher frequency range reduces antenna size. In the modulation process, the baseband signals constitute the *modulating signal* and the high-frequency *carrier signal* is a sinusoidal waveform.

There are three basic methods of modulating a sine wave carrier. For binary digital modulation, they are as follows :

1. Amplitude-shift keying (ASK),
2. Frequency-shift keying (FSK) and
3. Phase shift keying (PSK).

Important Terms Used in Data Communication

1. Signal element : A **signal element** carries data element and it is the shortest unit (time wise) of a digital signal.

2. Data element (bit) : Data element is the smallest entity that can represent a piece of information known a **bit**.

"Data elements are what we need to send;
Signal elements are what we can send.
Data elements are being carried and signal
Element are the carriers"

3. Data rate or bit rate : The **data rate** is defined the number of data element (or bits) sent in one second. Its unit is bps (Bits per second).

4. Signal rate : The **signal rate** is the number of signal element sent in one second. Its unit is baud. It is prefer in data communication to increase the data rate while decreasing the signal rate. Speed of transmission can be increase by increasing the data rate whereas decreasing the signal rate, decrease the bandwidth requirement.

2.6. AMPLITUDE-SHIFT KEYING (ASK)

Amplitude shift keying (ASK) in the context of digital communications is a modulation process that imparts to a sinusoid two or more discrete amplitude levels. These are related to the number of levels adopted by the digital message.

For a binary message sequence, there are two levels, one of which is typically zero.

Thus, the modulated waveform consists of bursts of a sinusoid. The data rate is often made a sub-multiple of the carrier frequency. This has been done in the waveform of Figure 2.2.

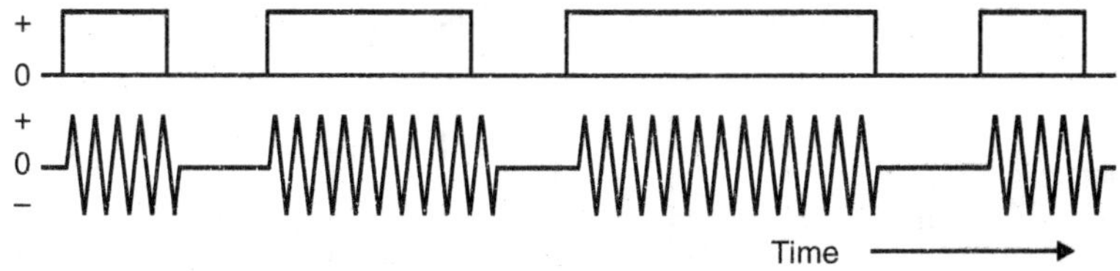

Fig. 2.2. *ASK waveform (below) and message signal (above).*

There are sharp discontinuities shown at the transition points. These result in the signal having an unnecessarily wide bandwidth. Band limiting is generally introduced before transmission, in which case these discontinuities would be 'rounded off'. The band limiting may be applied to the digital message, or the modulated signal itself.

A *amplitude-shift keying* (BASK) signal can be defined by

$$s(t) = A\ m(t) \cos 2\pi fct,\ 0 = t = T \qquad ...(2.2)$$

where A is a constant, $m(t) = 1$ or 0, fc is the carrier frequency, and T is the bit duration.

It has a power $P = A^2/2$, so that $A = 2P$. Thus Eq. (2.2) can be written as

$$s(t) = \sqrt{2}P \cos 2\pi fct,\ 0 \le t \le T$$

$$= \frac{\sqrt{PT}\sqrt{2}}{T \cos 2\pi fct},\ 0 \le t \le T$$

$$= \frac{\sqrt{E}\sqrt{2}}{T \cos 2\pi fct},\ 0 \le t \le T \qquad ...(2.3)$$

where $E = PT$ is the energy contained in a bit duration. If we take

$$\phi_1(t) = \frac{\sqrt{2}}{T} \cos 2\pi fct \qquad ...(2.4)$$

as the orthonormal basis function, the applicable signal space or constellation diagram of the BASK signals is shown in Figure 2.3. The ASK waveform of above Eq. (2.3) for symbol '1' may be represented as

$$s(t) = \sqrt{P_s\ T_b} \cdot \sqrt{2/T_b} \cos(2\pi f_c t) = \sqrt{P_s\ T_b}\ Q_1(t) \qquad ...(2.5)$$

This means that there is only one carrier function ϕ_1(t). The signal space diagram will have two points on ϕ_1(t). One will be at $\sqrt{P_s\ T_b}$ and other will be at zero as shown in Fig. 2.3.

The distance between the two signal points is,

$$d = \sqrt{P_s\ T_b} = \sqrt{E_b} \qquad ...(2.6)$$

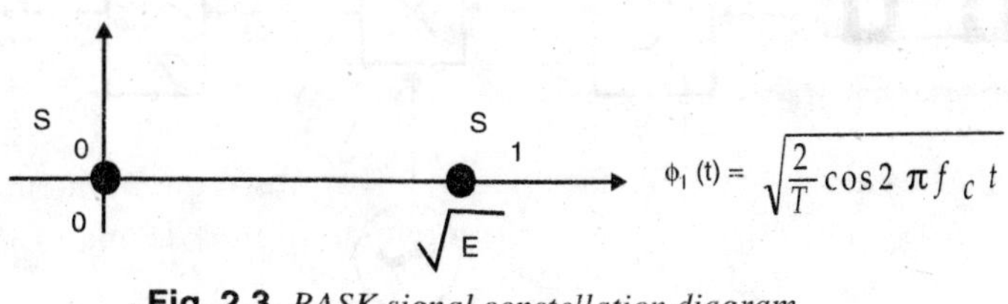

Fig. 2.3. *BASK signal constellation diagram.*

To transmit symbol '0', the $s(t) = 0$ *i.e.,* no signal is transmitted. Signal $s(t)$ contains some complete cycles of carrier frequency 'f_c'.

Hence the ASK waveform look like an ON-OFF of the signal. Therefore, it is also known as the ON-OFF keying (OOK).

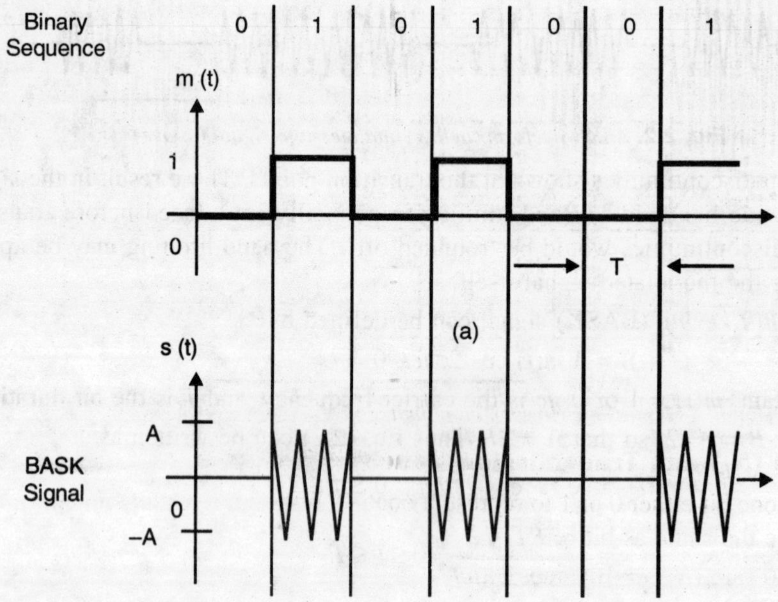

Fig. 2.4. *(a) Binary modulating signal and (b) BASK signal.*

Figure 2.4 shows the BASK signal sequence generated by the binary sequence 0 1 0 1 0 0 1. The amplitude of a carrier is switched or keyed by the binary signal $m(t)$. This is sometimes called **on-off keying (OOK)**.

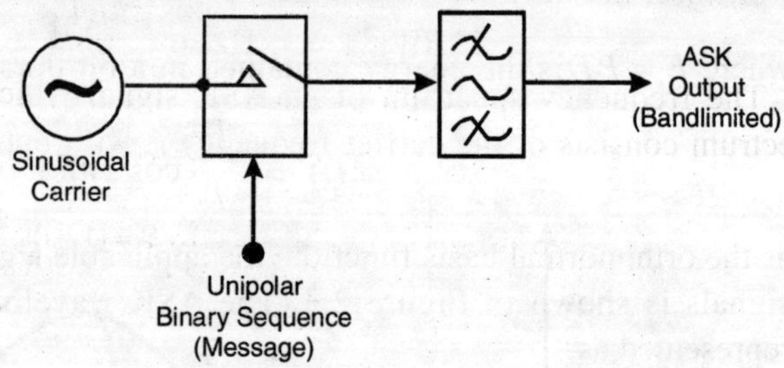

Fig. 2.5. *The principle of ASK generation.*

A block diagram of a basic ASK generator is shown in Fig. 2.5. This show band limiting following modulation.

The switch is opened and closed by the unipolar binary sequence.

As already indicated, the sharp discontinuities in the ASK waveform of Figure 2.6 simply a wide bandwidth. A significant reduction can be accepted before errors at the receiver increase unacceptably. This can be brought about by bandlimiting (pulse shaping) the message *before* modulation, or bandlimiting the ASK signal itself *after* generation.

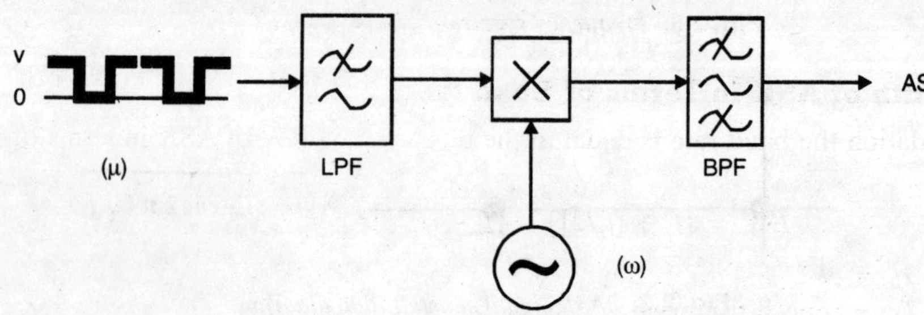

Fig. 2.6. *ASK bandlimiting, with a LPF or a BPF.*

Figure 2.7 shows the signals present in a model of Fig. 2.5, where the message has been bandlimited. The shape, after bandlimiting, depends naturally enough upon the amplitude and phase characteristics of the bandlimiting filter.

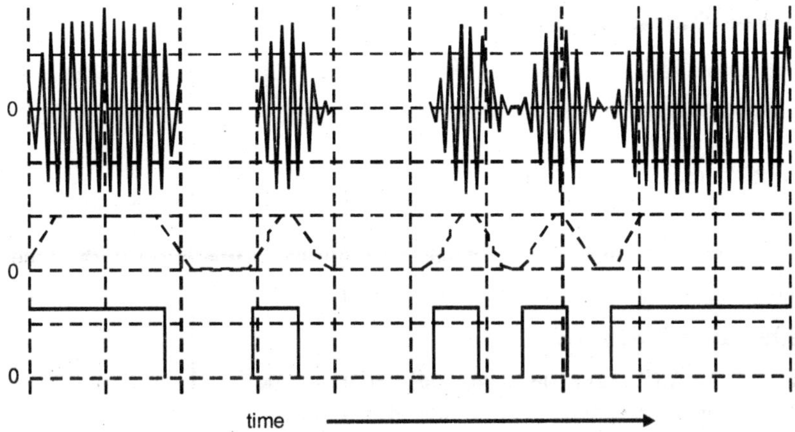

Fig. 2.7. *Original TTL message (lower), bandlimited message (centre), and ASK (above).*

2.6.1. Baud Rate (N_b) and Transmission Bandwidth of ASK

For ASK we use one bit either 0 or 1 to represent one symbol, so the rate of symbol transmission (baud rate) N_b will be the same as bit rate.

$$\text{Baud rate } (N_b) = \text{Bit rate}$$

The BW of ASK signal is dependent on the bit rate (f_b), where bit rate $f_b = 1/T_b$. for a bit rate of (f_b) bits/sec, the maximum bandwidth required for an ASK signal is;

$$BW_{max} = f_b \text{ Hz} \quad ...(2.7)$$

The frequency spectrum of an ASK signal which is shown in Fig. 2.8, represents that the spectrum consists of the carrier frequency f_c with upper and lower sidebands.

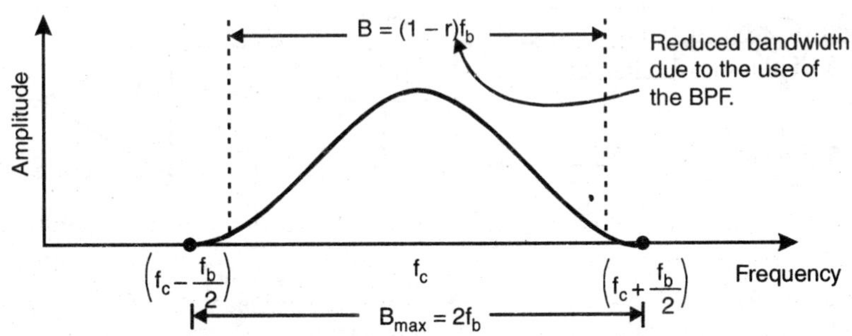

Fig. 2.8. *Frequency spectrum of ASK signal.*

2.6.2. Bandwidth of ASK in Terms of Baud Rate

In ASK modulation the baud rate is equal to the bit rate. The BW of ASK in terms of bit rate is given by,

$$BW = \{f_c + [f_b/2]\} - \{f_c - [f_b/2]\} = f_b \quad ...(2.8)$$

Where $f_b = 1/T_b$.

T_b = one bit interval

Since in ASK bit rate is the same as baud rate, the expression for bandwidth is given by,

$$BW = \{f_c + [N_b/2]\} - \{f_c - [N_b/2]\} = N_b \quad ...(2.9)$$

where, N_b baud rate = f_b bit rate

2.6.3. Merits and Demerits of ASK

Merits

1. ASK is quite simple.
2. It is easy to generate.
3. It is easy to detect.

Demerits

1. It is very sensitive to noise therefore it finds limited application in data transmission.
2. It is used at very low bit rates, typically less than 100 bps.

2.6.4. Demodulation Methods

It is apparent that the ASK signal has a well defined envelope. Thus it is amenable to demodulation by an envelope detector. A synchronous demodulator would also be appropriate.

Note that :

- Envelope detection circuitry is simple.
- Synchronous demodulation requires a phase-locked local carrier and therefore carrier acquisition circuitry.

With band limiting of the transmitted ASK neither of these demodulation methods would recover the original binary sequence; instead, their outputs would be a band limited version. Thus, further processing-by some sort of decision-making circuitry for example, would be necessary. Thus, demodulation is a two-stage process:

1. Recovery of the band limited bit stream
2. Regeneration of the binary bit stream

A simple block diagram of coherent ASK demodulator is shown in Fig. 2.9 ((*a*) and (*b*))

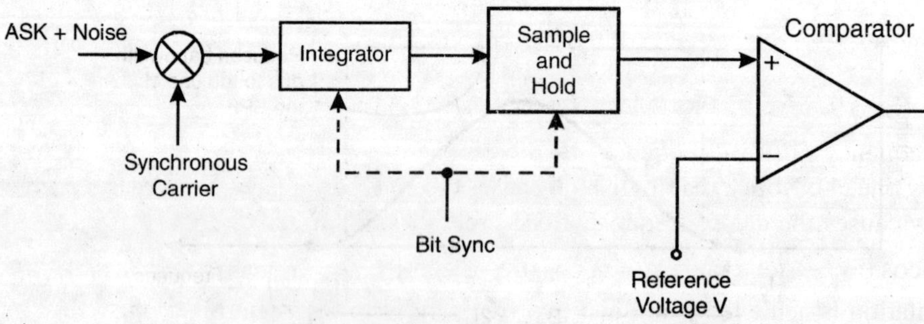

Fig. 2.9 *(a). A coherent ASK Demodulator.*

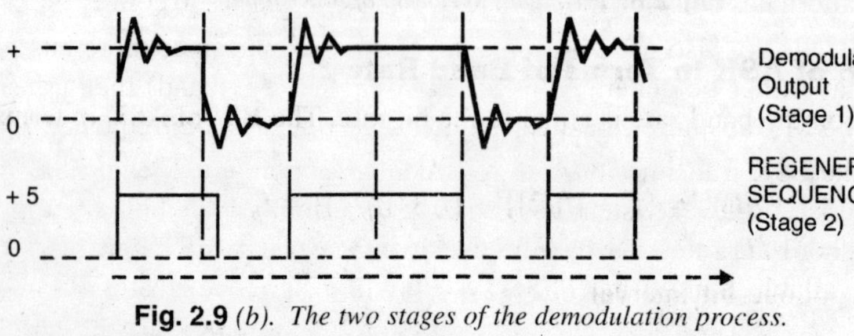

Fig. 2.9 *(b). The two stages of the demodulation process.*

2.7. FREQUENCY SHIFT KEYING (FSK)

One of the simplest, and widest used system, is frequency modulation. It involves making a change to the frequency of the carrier to represent a different level. The generic name for this family of modulation is **Frequency Shift Keying** (FSK) shown in Fig. 2.10.

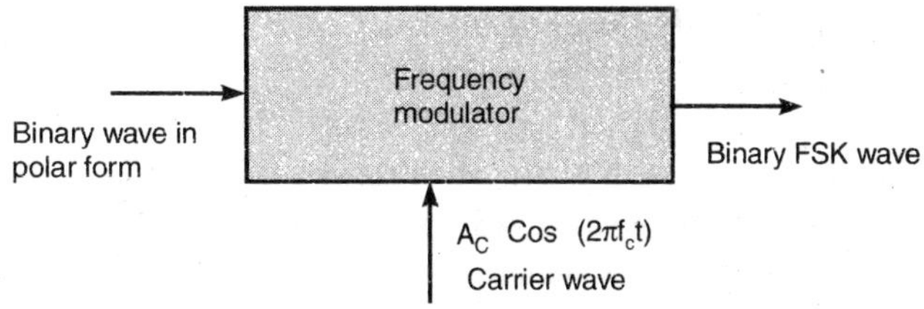

Fig. 2.10. *Block diagram of FSK*

In this method the frequency of the carrier is changed to two different frequencies depending on the logic state of the input bit stream. The typical output waveform of an FSK is shown in Fig. 2.11.

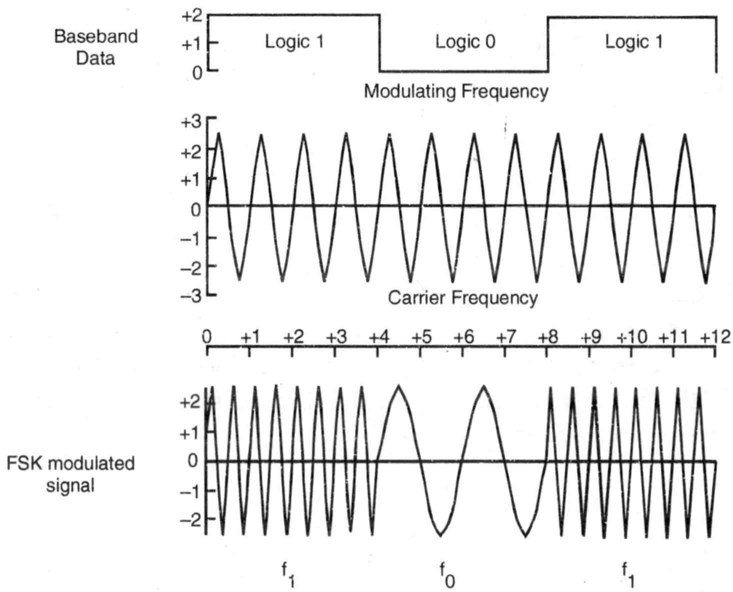

Fig. 2.11. *Frequency Shift Keying (FSK).*

Assume frequency f_0 (low frequency) to represent binary zero (0) and f_1 (high frequency) to represent binary one (1). Notice that logic high causes the centre frequency to increase to a maximum and a logic low causes the centre frequency to decrease to a minimum.

where $f_0 = A\cos(\omega_c - \Delta\omega)t$ and $f_1 = A\cos(\omega_c + \Delta\omega)t$

FSK modulation is achieved using just two frequencies to represent binary data. For example, a frequency of 2000 Hz may represent a binary 1 (mark) and 1000 Hz a binary 0 (space). The energy of the signal alternates between the mark and space frequencies to convey the digital message.

More than two levels are also used by some advanced systems since more bits per tone can be sent. FSK signals may either be generated as an audio signal (baseband) then transmitted using a conventional SSB or FM transmitter, or a transmitter's carrier may be frequency keyed directly.

If the tones of the FSK transmissions are regarded as symbols, a 2-level FSK signal has 1 bit per symbol. By contrast a 4-level system has 2 bits per transmitted symbol. A multi-level signal is capable of much higher data rates but has the disadvantage that more bandwidth is required. If a fixed bandwidth is available, a multi-level FSK system will experience greater difficulty in resolving the symbols.

The rate at which symbols are sent is the baud rate; this need not be the same as the bit rate supported by the modem. Typical baud rates range from 45 baud up to 300 baud on HF with SSB modulation being used. Data rates of 1200 baud or more are commonly used on VHF narrowband FM as the modulation method.

FSK is classified as wide-band if the separation between the two carrier frequencies is larger than the bandwidth of the spectrums of f_0 and f_1. In this case the spectrum of the modulated signal appears as two separate ASK signals. Narrow-band FSK is the term used to describe an FSK signal whose carrier frequencies are separated by less than the width of the spectrum than ASK for the same modulation.

2.7.1. Generation of FSK

A simplified binary FSK modulator is shown on Fig. 2.12. The center frequency f_c is chosen such that it falls halfway between the mark and space frequencies. A logic '0' input shift the Voltage Controlled Oscillator (VCO) output to the space frequency and a logic '1' input shifts the VCO output to mark frequency.

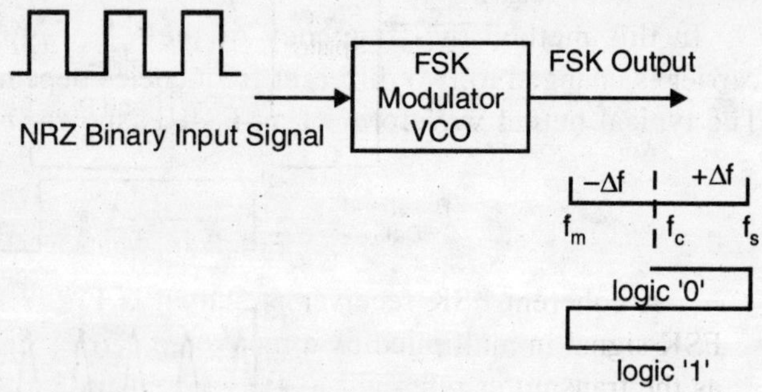

Fig. 2.12. *FSK modulator.*

Now as the binary signal changes from '0' to '1' or '1' or '0' conditions, the VCO output shifts or deviates back and forth between the mark and space frequencies.

In binary FSK modulator, Δf is the peak frequency deviation of the carrier and is equal to the difference between the carrier rest frequency and either the space or the mark frequency (or half the difference between the carrier rest frequency) and either the space or mark frequency (or half the difference between the space and mark frequencies).

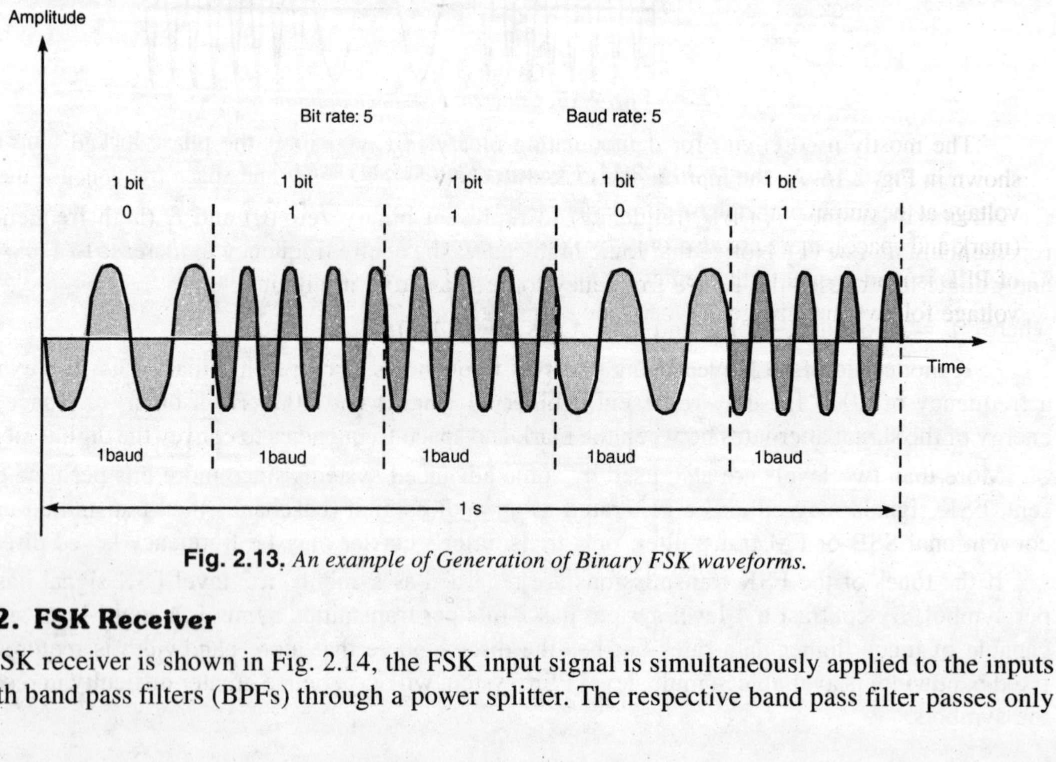

Fig. 2.13. *An example of Generation of Binary FSK waveforms.*

2.7.2. FSK Receiver

FSK receiver is shown in Fig. 2.14, the FSK input signal is simultaneously applied to the inputs of both band pass filters (BPFs) through a power splitter. The respective band pass filter passes only the space or only the mark frequency on to its respective envelope detector. The envelope detector indicates the total power in each pass band and the comparator responds to the larger of the two powers. This type of FSK detection is known as *non-coherent detection*; there is no frequency involved in the demodulation process. That means there is no need of synchronization either in phase, frequency or both with the incoming FSK signal.

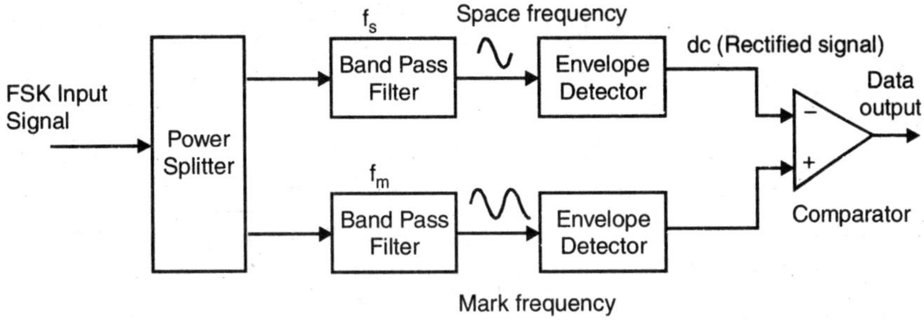

Fig. 2.14. *Non-coherent FSK demodulator.*

A coherent FSK receiver is shown is Fig. 2.15. In coherent FSK demodulator the incoming FSK signal in multiplied by a recovered carrier signal that has the exact same frequency and phase as the transmitter reference. The two transmitted frequencies (mark and space frequencies) are not continuous, so it is not practical to reproduce a local reference that is coherent with both of them. So due to this problem, coherent detection is seldom used.

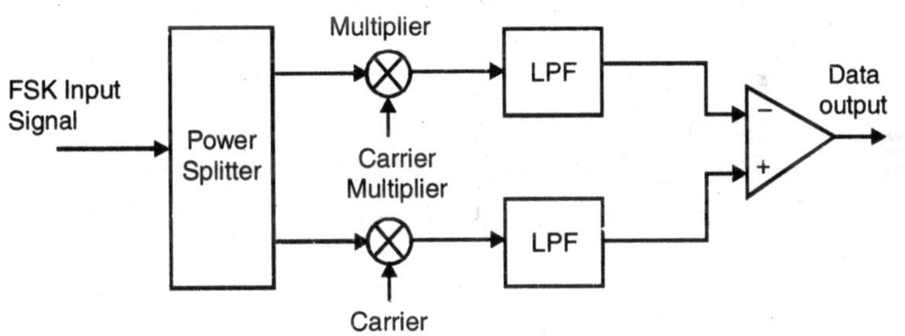

Fig. 2.15. *Coherent FSK demodulator.*

The mostly used circuit for demodulating binary FSK signals is the phase locked loop (PLL) as shown in Fig. 2.16. As the input to the PLL shifts between the mark and space frequencies, the dc error voltage at the output of the phase comparator follows the frequency shift. As we have two input frequencies (mark and space), there are also only two output error voltage for logic '0' and logic '1'. The frequency of PLL is made equal to the center frequency of the FSK modulator so that the changes on the dc error voltage follows the changes in the analog input frequency and are symmetrical around 0 V.

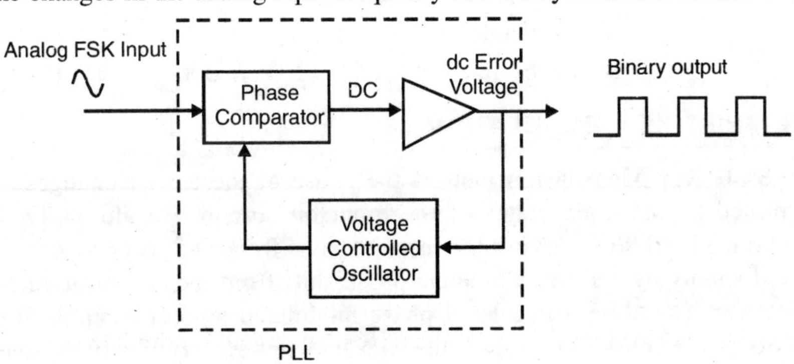

Fig. 2.16. *PLL-FSK demodulator.*

2.7.3. BFSK Merits and Demerits

FSK has the advantage of being very simple to generate, simple to demodulate and due to the constant amplitude can utilize a non-linear PA.

BFSK is normally used in asynchronous data modems that are used for data communication over analog, voice band telephone lines.

BFSK has a poorer error performance than PSK or QAM and consequently, is rarely used for high performance digital radio systems.

BFSK is relatively easy to implement.

It has better noise immunity than ASK. The major drawback is its high bandwidth requirement. Therefore, FSK is extensively used in low speed modems having bit rates lower than 1200 bits/sec.

Significant disadvantages, however, are the poor spectral efficiency and BER performance. This precludes its use in this basic form from cellular and even cordless systems.

2.7.4. FSK Baud and BW

The baud for binary FSK is determined by making N = 1,

$$\text{Baud} = \frac{f_b}{N} \qquad \text{...(2.10)}$$

$$\text{Baud} = \frac{f_b}{1} = f_b \qquad \text{...(2.11)}$$

where, f_b = input Bit rate

N = number of bits encoded in to each signaling unit

$\ddot{A}_f$ = frequency deviation $|(f_m - f_s)|$ in Hertz

The minimum BW for FSK is determined by,

$$B = |(f_s - f_b) - (f_m - f_b)|$$

$$B = |(f_s - f_m)| + 2f_b$$

$$B = 2(\Delta f + f_b) \qquad \text{...(2.12)}$$

where, B = minimum Nyquist bandwidth

f_s = frequency space for logic '0'

f_m = frequency mark for logic '1'

Example 1. Find the minimum bandwidth for an FSK signal transmitting at 2000 bps. Transmission is in half-duplex mode, and the carriers are separated by 3000 Hz.

Solution. For FSK

$$\text{BW} = \text{baud rate} + f_{c1} - f_{c0}$$

$$\text{BW} = \text{bit rate} + f_{c1} - f_{c0} = 2000 + 3000 = 5000 \text{ Hz}$$

2.8. PHASE SHIFT KEY MODULATION

In **Phase Shift Key Modulation** method the phase of the carrier changes between different phases determined by the logic states of the input bit stream. An alternative to imposing the modulation onto the carrier by varying the instantaneous frequency is to modulate the phase. This can be achieved simply by defining a relative phase shift from the carrier, usually equi-distant for each required state. Therefore a two level phase modulated system, such as Binary Phase Shift Keying, has two relative phase shifts from the carrier, + or –90°. Typically this technique will lead to an improved BER performance compared to MSK.

PSK describes the modulation technique that alters the phase of the carrier. Mathematically:

$$s(t) = \sin(2\pi f_C + \phi(t)) \quad ...(2.13)$$

There are several different types of phase shift key (PSK) modulators.

- Binary-phase (BPSK)
- Quadrature-phase (QPSK)
- Eight-phase (8 PSK)
- Sixteen-phase (16 PSK)

2.8.1. Binary-Phase Shift Key Modulation

The simplest form of phase modulation is binary (two level) phase modulation. With theoretical BPSK the carrier phase has only two states, $\pm \pi/2$. Obviously the transition from a one to a zero, or vice versa, will result in the modulated signal crossing the origin of the constellation diagram resulting in 100% AM.

In this modulator the carrier assumes one of two phases. Logic 1 produces no phase change and logic 0 produces a 180° phase change. The output waveform for this modulator is shown in Fig. 2.17.

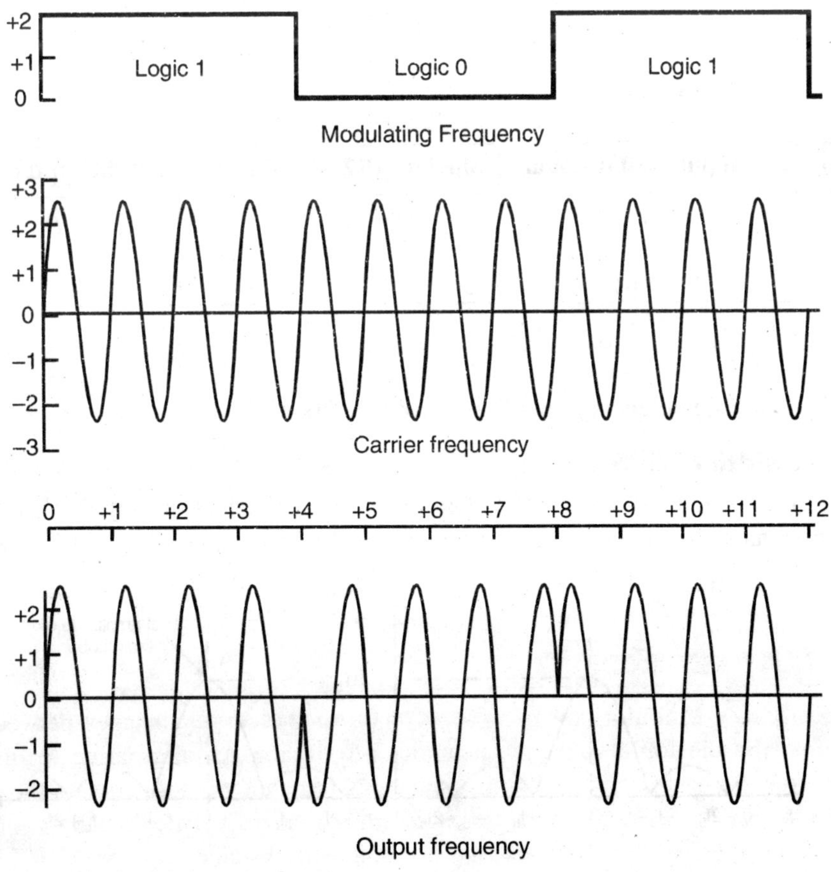

Fig. 2.17. *Phase shift key modulation.*

2.8.2. Generation of BPSK

A simplified block diagram of a BPSK transmitter is shown in Fig. 2.18. BFSK signal may be generated by applying carrier to a balanced modulator. The binary data signal (0's and 1's) is converted into a NRZ bipolar signal by an MRF encoder. Here, the bipolar signal $b(t)$, is applied as a modulating signal to the balanced modulator.

A NRZ level encoder converts the binary data sequence into bipolar NRZ signal.

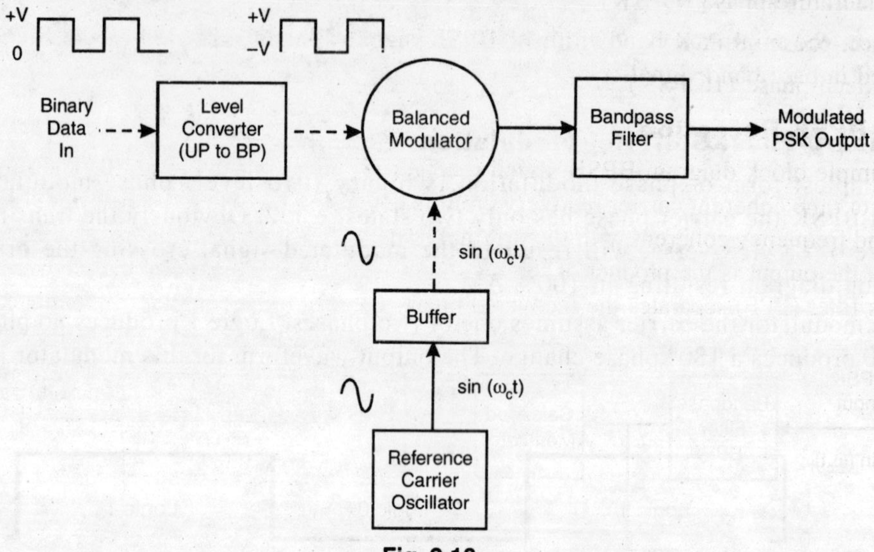

Fig. 2.18.

S. No.	Input digital signal	Bipolar NRZ signal $b(t)$	BPSK output signal
1.	Binary 0	$b(t) = -1$	$-\sqrt{2P}\cos\omega_c t$
2.	Binary 1	$b(t) = +1$	$+\sqrt{2P}\cos\omega_c t$

Here ; $$P = \frac{E_b}{T_b},$$

where E_b is the signal energy and T_b is the bit duration.

2.8.3. Bandwidth of BPSK

The spectrum of the BPSK signal is centered on the carrier frequency f_c. If $f_b = 1/T_b$ then for BPSK, the maximum frequency in the baseband signal will be f_b as shown in Fig. 2.19.

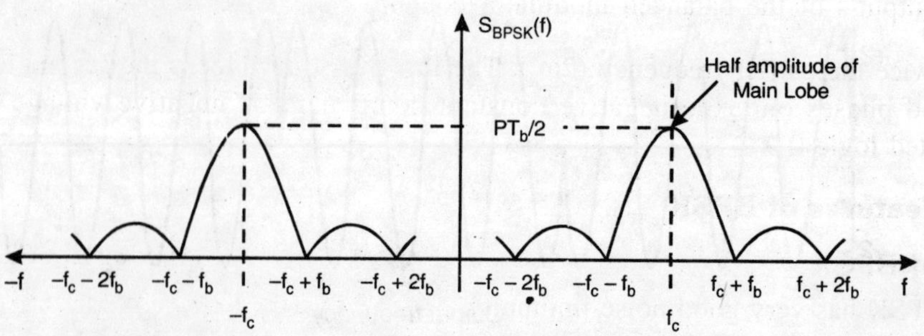

Fig. 2.19. *Plot of power spectral density of BPSK signal.*

In this figure, the main lobe is centered around carrier frequency f_c and extends from $f_c - f_b$ to $f_c + f_b$.

Therefore BW of BPSK signal will be,

$$BW = \text{Highest frequency} - \text{Lower} + \text{Frequency in the main lobe}$$

$$BW = f_c + f_b - (f_c - f_b)$$

or
$$BW = 2f_b \qquad \text{...(2.14)}$$

Hence, the minimum bandwidth of BPSK signal is equal to twice of the highest frequency contained in base band signal.

2.8.4. BPSK Reception

A simple block diagram BPSK signal receiver is shown in the Fig. 2.20. The input signal may be ± sin $\omega_c t$ the coherent carrier recovery circuits detects and regenerates a carrier signal that is both phase and frequency coherent with the original transmit carrier. The balanced modulator is a product detector; the output is the product of the two inputs (the BPSK signal and the recovered carrier). The low pass filter (LPF) separates the recovered binary data from the complex demodulated signal.

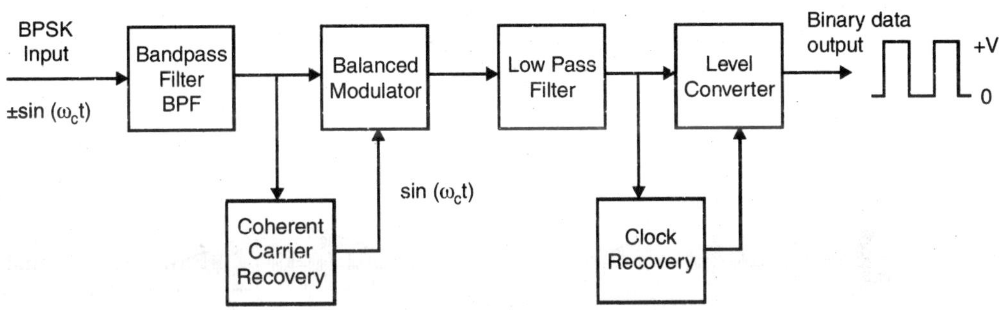

Fig. 2.20. *Block diagram of BPSK receiver.*

If the BPSK input signal is +sin $\omega_c t$ (for logic 1), the output of the balanced modulator is

$$\text{Output} = +\frac{1}{2}V = \text{logic '1'}$$

If the BPSK input signal is – sin $\omega_c t$ (for logic '0'), the output of the balanced modulator is

$$\text{Output} = -\frac{1}{2}V = \text{logic '0'}$$

The output's of the balanced modulator contains a negative voltage $\left(\left[-\frac{1}{2}\right]V\right)$ and a cosine wave at twice the carrier frequency ($2\omega_c$). The low pass filter blocks the second harmonics of the carrier and phases only the negative constant component. A negative voltage represents of a demodulated logic 0.

2.8.5. Features of BPSK

Merits of BPSK

(*i*) BPSK has very good noise immunity.

(*ii*) It has minimum value of probability of error.

(*iii*) BPSK bandwidth is lower than that of a BFSK signal.

Demerits of BPSK

(*i*) The generation and detection is not easy.

(*ii*) The recovered carrier is unchanged even if the input signal has changed its sign. This results in ambiguity in the output signal.

Example 2. Find the minimum bandwidth for an ASK signal transmitting at 2000 bps. The transmission mode is half-duplex.

Solution. In ASK the baud rate and bit rate are the same. The baud rate is therefore 2000. An ASK signal requires a minimum bandwidth equal to its baud rate. Therefore, the minimum bandwidth is 2000 Hz.

Example 3. Given a bandwidth of 10,000 Hz (1000 to 11,000 Hz), draw the full-duplex ASK diagram of the system. Find the carriers and the bandwidths in each direction. Assume there is no gap between the bands in the two directions.

Solution. For full-duplex ASK, the bandwidth for each direction is

$$BW = 10000/2 = 5000 \text{ Hz}$$

The carrier frequencies can be chosen at the middle of each band (see Fig. 2.21).

$$f_c \text{ (forward)} = 1000 + 5000/2 = 3500 \text{ Hz}$$

$$f_c \text{ (backward)} = 11000 - 5000/2 = 8500 \text{ Hz}$$

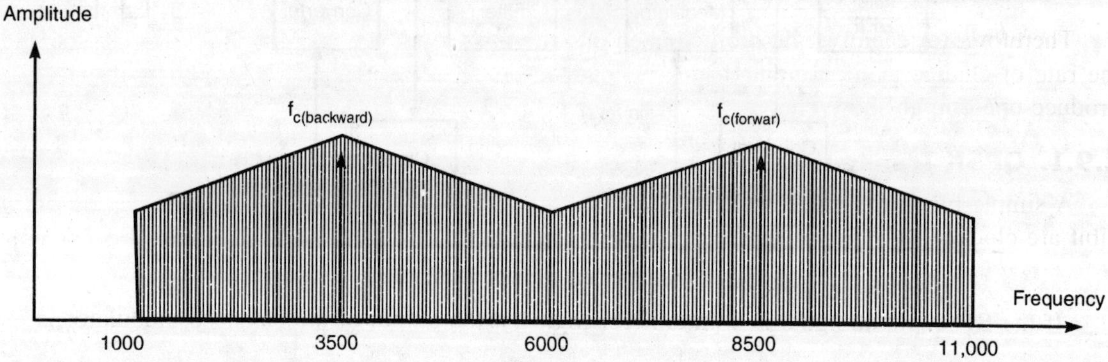

Fig. 2.21.

2.9. QUADRATURE PHASE SHIFT KEYED (QPSK)

Higher order modulation schemes, such as QPSK, are often used in preference to BPSK when improved spectral efficiency is required. In case of BPSK, we know that when symbol changes the level, the phase of the carrier is changed by 180°. Because, there were only two symbols in BPSK, the phase shift occurs in two levels only. However, in QPSK, two successive bits are combined. In fact, this combination of two bits forms four distinct symbols. When the symbol is changed to next symbol, then the phase of the carrier is changed by 45°.

QPSK is an M-ary encoding scheme where N = 2 and M = 4 (hence, the name "quaternary" means '4'). With the QPSK, four output phases are possible for a single carrier frequency. Because there are four output phases, there must be four different input conditions. Because the digital input to a QPSK modulator is a binary signal, to produce four different input combinations, the modulator requires more than a single input bit to determine the output condition. There are four possible conditions present with two numbers of bits. The process of combining two successive bits is explained in Fig. 2.22.

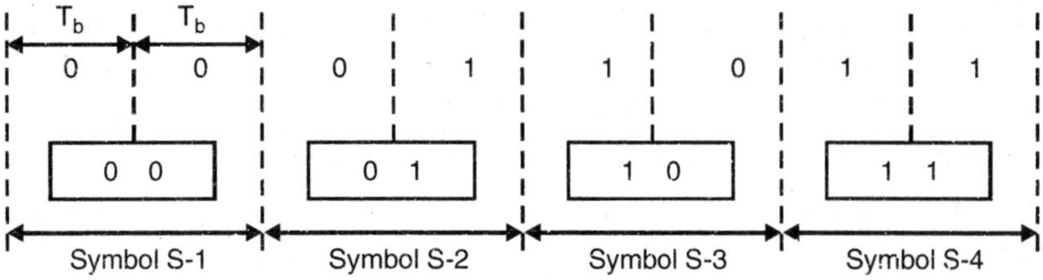

Fig. 2.22. *Grouping of bits in QPSK.*

As each symbol contain two bits so the symbol duration $T_s = 2T_b$. Therefore, with QPSK, the binary input data are combined into groups of two bits, called dibits. These symbols and their phase shifts are shown in Table 2.2.

Table 2.2. Symbols and Corresponding Phase Shift

Symbol	Phase Shift in Carrier
00 (S-1)	–135°
01 (S-2)	–45°
10 (S-3)	+135°
11 (S-4)	+45°

Therefore, for each two bit dibit clocked into the modulator, a single output change occurs, and the rate of change at the output (band) is equal to one half the input bit rate *i.e.*, two input bits produce one output phase charge.

2.9.1. QPSK Transmitter

A simple block diagram of a QPSK modulation is illustrated in Fig. 2.23. Here, two bits or a dibit are clocked into the bit splitter. In the bit splitter we have serial to parallel convertor, which

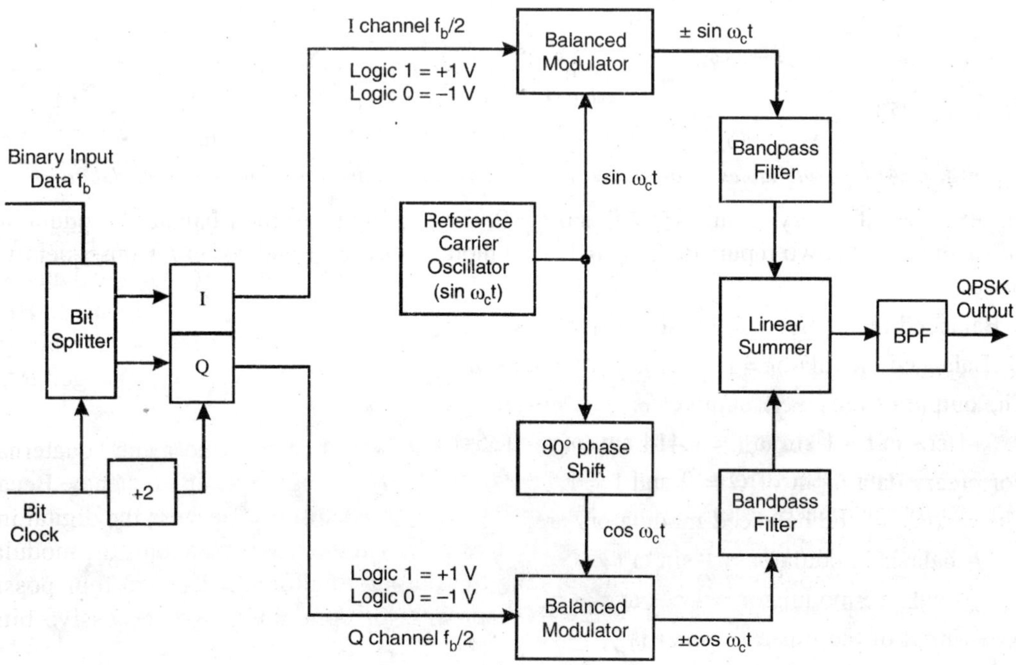

Fig. 2.23. *QPSK modulator.*

outputs the both bits simultaneously parallely. Both the bits are directed forwards the I and Q channels, one bit is given to I-channel and other to the Q-channel. In I-channel, or in-phase channel modulates a carrier that is in phase with the reference oscillator and the Q bit modulates a carrier that is 90° out of phase or in quadrature with the reference carrier.

The operation of the QPSK modulation is same once a digit has been split into I and Q channels as in a BPSK modulator. Essentially a QPSK modulator is the combination of two BPSK modulators combined in parallel.

Further, again for a logic 1 = +1V and logic '0' = -1V, two phases are possible at the output of the I-balanced modulator (+ sin $\omega_c t$ and –sin $\omega_c t$), and similarly two phases all at the output of the Q-channel balanced modulator (+ cos $\omega_c t$ and –cos $\omega_c t$). The linear summer combines the two quadrature (90° out of phase) signals and produces four possible resultant phasors expressed as follows:

$+\sin \omega_c t + \cos \omega_c t$, $+ \sin \omega_c t - \cos \omega_c t$; $-\sin \omega_c t + \cos \omega_c t$ and $-\sin \omega_c t - \cos \omega_c t$

Binary Input		QPSK Output Phase
Q	I	
0	0	–135°
0	1	–45°
1	0	+135°
1	1	+45°

(a)

For the QPSK modulator shown in Fig. 2.23, the truth tables, phasor diagram and constellation diagram are shown in Fig. 2.24.

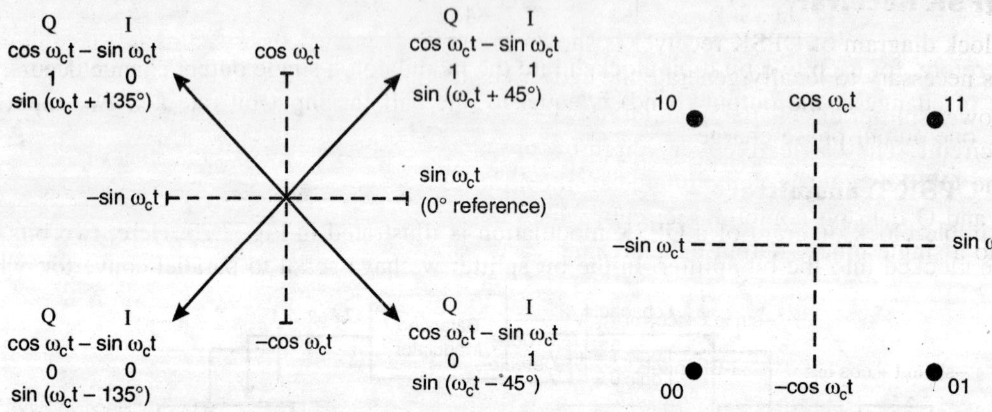

(b) (c)

Fig. 2.24. *The truth tables, phasor diagram and constellation diagram of the QPSK modulator.*

For example, if binary input of $Q = 0$ and $I = 0$, the two inputs to the I balanced modulator are –1 and sin ω_ct and the two inputs to the Q balanced modulator are –1 and cos $\omega_c t$. Consequently, the outputs are

I- Balanced modulator = $(-1)(\sin \omega_c t) = -1 \sin \omega_c t$ ...(2.15)

Q- Balanced modulator = $(-1)(\cos \omega_c t) = -\cos \omega_c t$ ...(2.16)

The output of the linear summer is

$$-1 \cos \omega_c t - 1 \sin \omega_c t = 1.414 \sin (\omega_c t - 135°)$$

For binary data input of $Q = 0$ and I = 1,

The outputs of the balanced modulators are;

I- balance modulator = $+ \sin \omega_c t$

Q- balance modulator = $- \cos \omega_c t$

And output of the linear summer is

$$-\cos \omega_c t + \sin \omega_c t = \sin (\omega_c t - 45°).$$

It can be seen that with QPSK each of the four possible output phasors has exactly the same amplitude, refer to Fig. 2.24(*c*).

Therefore, the binary information must be encoded entirely in the phase of the output signal. Because of the constant amplitude it is differ from QAM (Quadrature Amplitude Modulation) where phase and amplifiers both are varying.

From Fig. 2.24(*b*), it is seen that the angular separation between any two adjacent phasor in QPSK is 90°.

Figure 2.25 shows the output phase-version-time relationship for a QPSK modulator.

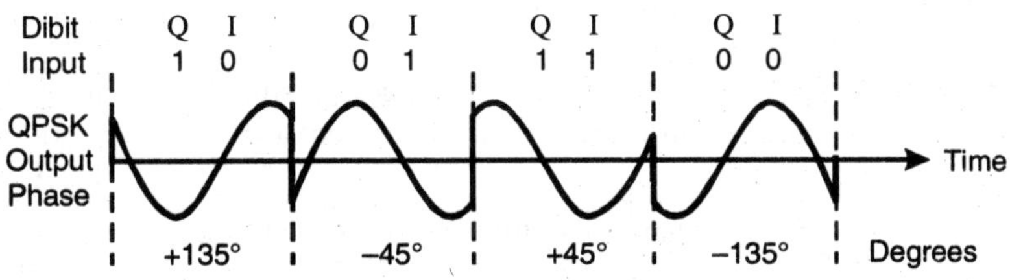

Fig. 2.25. *Output phase-versus-time relationship for a QPSK modulator.*

2.9.2. QPSK Receiver

The block diagram of QPSK receiver is shown in Fig. 2.26 we use the synchronous technique. Hence it is necessary to locally generate the carrier cos $\omega_c t$ and sin $\omega_c t$.

The power splitter directs the input QPSK signal to the I and Q product detector and the carrier recovery circuit. The carrier recovery circuit reproduces the original transmit carrier oscillator signal. The QPSK signal is demodulated in the I and Q product detectors, which generate the original I and Q data bit combining circuit, where they are converted from parallel I and Q data channels to a single binary output data stream.

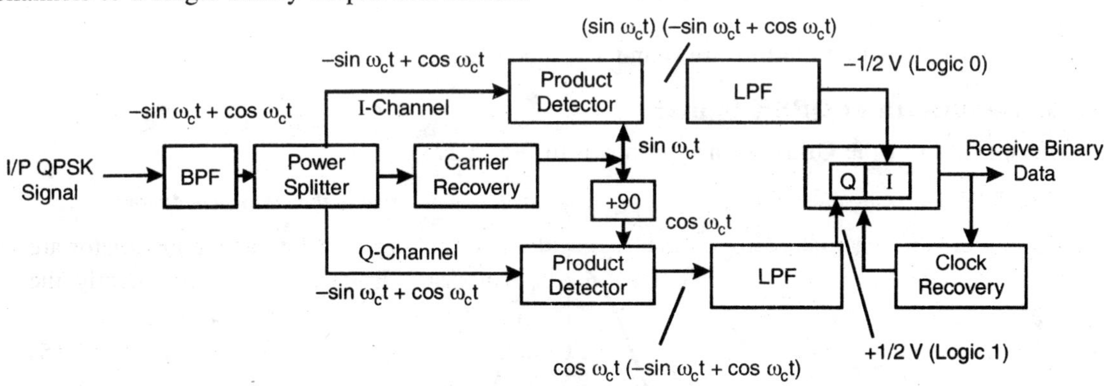

Fig. 2.26. *QPSK receiver.*

The input signal may be any out of the four possible output phases shown in Fig. 2.25.

To understand the demodulation process, let us assume that incoming QPSK signal is $-\sin \omega_c t + \cos \omega_c t$.

The receive QPSK signal ($-\sin \omega_c t + \cos \omega_c t$) is one of the inputs to the I product detector. The other input is the recovered carrier ($\sin \omega_c t$). The output of the I product detector is,

$$I = \underbrace{(-\sin \omega_c t + \cos \omega_c t)}_{\text{(QPSK I/P Signal)}} \underbrace{(\sin \omega_c t)}_{\text{(Carrier)}} \qquad \ldots(2.17)$$

$$I = (-\sin \omega_c t)(\sin \omega_c t) + (\cos \omega_c t)(\sin \omega_c t)$$

$$= -\sin^2 \omega_c t + (\cos \omega_c t)(\sin \omega_c t)$$

$$= -\frac{1}{2}(1 - \cos 2\omega_c t) + \frac{1}{2}\sin(\omega_c + \omega_c)t + \frac{1}{2}\sin(\omega_c - \omega_c)t$$

$$I = -\frac{1}{2} + \underbrace{\frac{1}{2}\cos 2\omega_c t}_{\text{Filtered out}} + \underbrace{\frac{1}{2}\sin 2\omega_c t + \frac{1}{2}\sin 0}_{\text{Equals to zero}}$$

$$I = -\frac{1}{2} V \text{ (logic '0')} \qquad \text{...(2.18)}$$

Again, at the input of Q-product detector, the input receive signal is $(-\sin \omega_c t + \cos \omega_c t)$. The other input is the recovered carrier shifted by 90° in phase $(\cos \omega_c t)$. The output of the Q product detector is

$$Q = \underbrace{(-\sin \omega_c t + \cos \omega_c t)}_{\text{QPSK I/P signal}} = \underbrace{(\cos \omega_c t)}_{\text{Carrier}} \qquad \text{...(2.19)}$$

$$Q = \cos^2 \omega_c t - (\sin \omega_c t)(\cos \omega_c t)$$

$$= \frac{1}{2}(1 + \cos^2 \omega_c t) - \frac{1}{2}\sin(\omega_c + \omega_c)t - \frac{1}{2}\sin(\omega_c - \omega_c)t$$

$$Q = \frac{1}{2} + \underbrace{\frac{1}{2}\cos^2 \omega_c t}_{\text{Filtered out}} - \underbrace{\frac{1}{2}\sin^2 \omega_c t - \frac{1}{2}\sin 0}_{\text{Equals to zero}}$$

$$Q = \frac{1}{2} V \text{ (logic '1')} \qquad \text{...(2.20)}$$

The demodulated I and Q bits are 0 and 1 respectively.

2.9.3. Bandwidth of QPSK Signal

The spectrum of the QPSK signal is shown in Fig. 2.27.

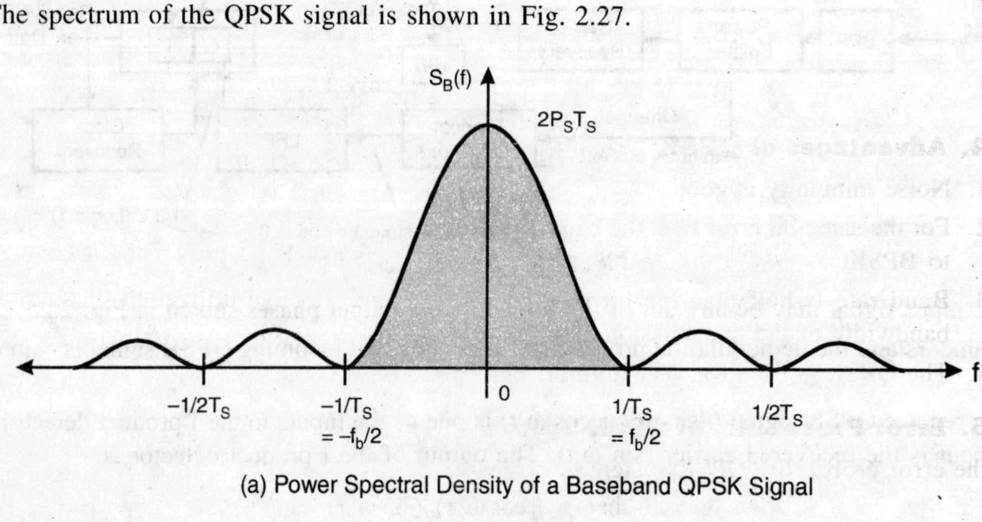

(a) Power Spectral Density of a Baseband QPSK Signal

Fig. 2.27.

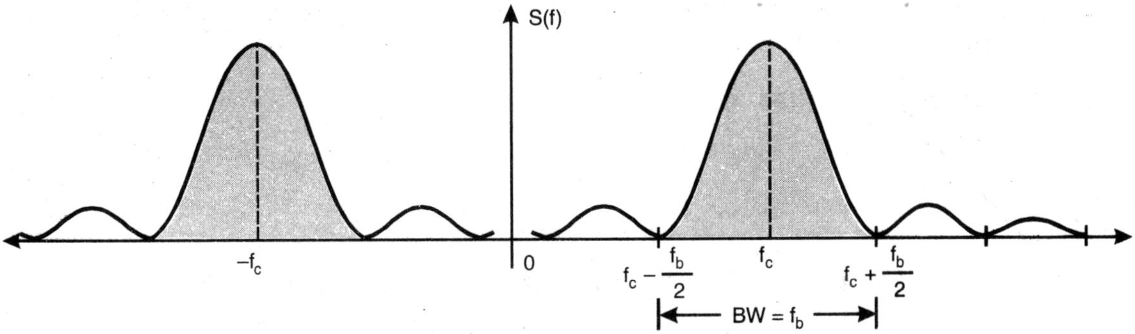

(b) Power Spectral Density of a QPSK Signal

Fig. 2.27. *The spectrum of the QPSK signal.*

We have seen that the bandwidth of BPSK signal is equal of $2\,f_b$. Here $T_b = \frac{1}{f_b}$ is the one bit period. In QPSK, the two waveform forms the baseband signals. One bit period for both of these signals is equals to $2T_b$. Therefore the bandwidth of QPSK signal will be

$$\text{BW} = 2 \times \frac{1}{2T_b} = f_b \qquad \text{...(2.21)}$$

Hence the bandwidth of QPSK signal is half of the bandwidth of BPSK signal.

$$\text{BW} = \text{Highest frequency} - \text{Lowest frequency in main lobe}$$

$$\text{BW} = \frac{1}{T_s} - \left[-\frac{1}{T_s}\right]$$

since carrier frequency f_c cancel out.

$$\text{BW} = \frac{2}{T_s}$$

We know that

$$T_S = 2T_b$$

$$\text{BW} = \frac{2}{2T_b} = \frac{1}{T_b} = f_b$$

$$\text{BW} = f_b \qquad \text{...(2.22)}$$

2.9.4. Advantages of QPSK

1. Noise immunity is good.
2. For the same bit error rate, the bandwidth required by QPSK is reduced to half as compared to BPSK.
3. Baud rate is half than the bit rate, therefore more effective utilization of the available bandwidth of the transmission channel.
4. The QPSK is used for very high bit rate data transmission.

2.9.5. Error Probability of QPSK System

The error probability QPSK system is expressed by

$$P_e = \frac{1}{2}\, efr_c \sqrt{\frac{E_b}{N_0}} \qquad \text{...(2.23)}$$

The P_e of BPSK is same as QPSK, thus the performance of QPSK in presence of noise is equivalent to that of a BPSK systems.

2.10. EIGHT-PHASE SHIFT KEY MODULATION

With this modulator 3 bits are processed to produce a single phase change. This means that each symbol consists of 3 bits.

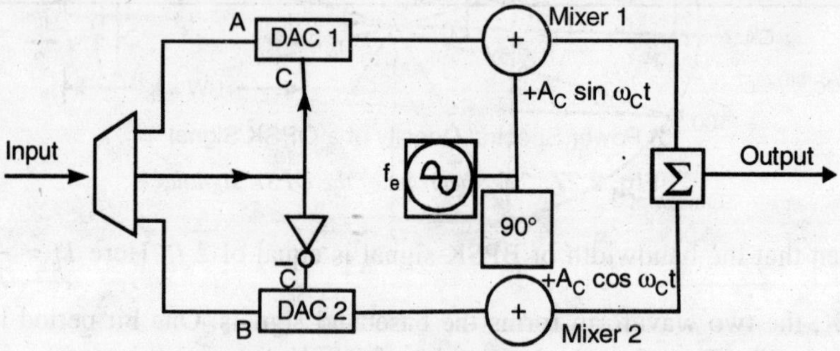

Fig. 2.28. *8-PSK Modulator.*

Figure 2.28 above shows a typical circuit for the *8-PSK modulator*. With this modulator bit A controls the output polarity of the first digital-to-analogue converter (DAC1). Bit B is used to control the output polarity of the second DAC 2 and bit C is used to control the output amplitude of both DACs.

A	Polarity	B	Polarity	C	Amplitude	$\overline{C}$	Amplitude
0	–	0	–	0	0.5	1	1.21
1	+	1	+	1	1.21	0	0.5

Fig. 2.29. *Digital to Analogue Conversion Condition for 8 PSK modulator*

The conditions shown in the table above (Figure 2.29) produce the positions shown in table below (Fig. 2.30) for all the different permutations.

Table 2.3. Input permutations and Positions

$\overline{C}$	C	B	A	C	A	Polarity	Amplitude	$\overline{C}$	B	Polarity	Amplitude	Quad
1	0	0	0	0	0	–	0.5	1	0	–	1.21	4
1	0	1	0	0	0	–	0.5	1	0	–	1.21	4
1	0	1	0	0	0	–	0.5	1	1	+	1.21	2
1	0	1	1	0	1	+	0.5	1	1	+	1.21	1
0	1	0	0	1	0	–	1.21	0	0	–	0.5	3
0	1	0	1	1	1	+	1.21	0	0	–	0.2	4
0	1	1	0	1	0	–	1.21	0	1	+	0.5	2
0	1	1	1	1	1	+	1.21	0	1	+	0.5	1

The constellation diagram can be drawn according to the above Table 2.3 and is shown below in Fig. 2.30.

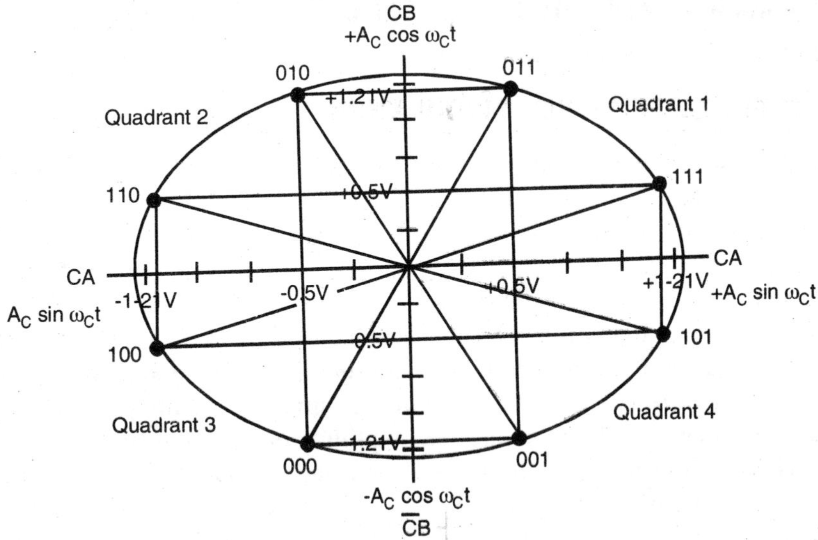

Fig. 2.30. *PSK Constellation.*

2.11. SIXTEEN-PHASE SHIFT KEY MODULATION

With this modulator 4 bits are processed to produce a single phase change. This means that each symbol consists of 4 bits. The constellation for this modulator scheme is shown below Fig. 2.31.

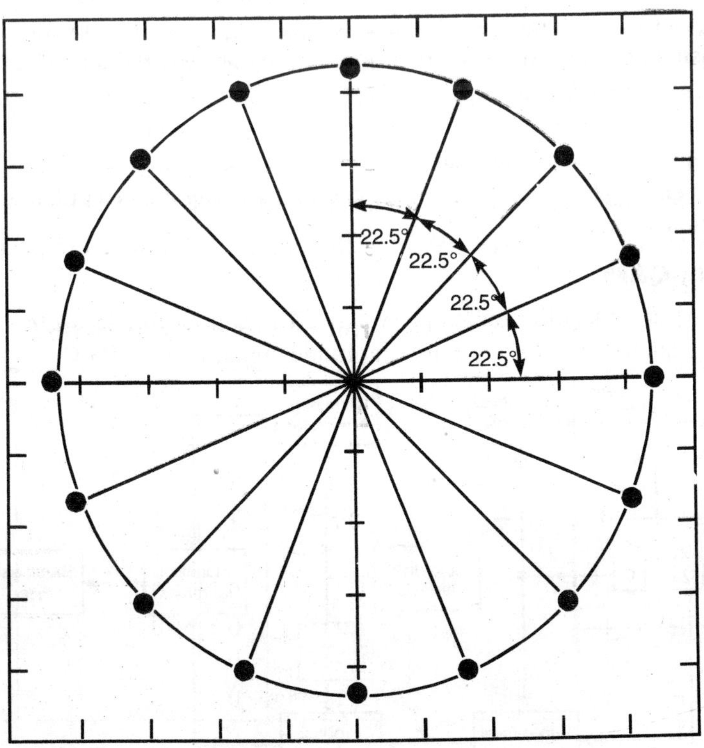

Fig. 2.31. *PSK Modulation Constellation.*

2.12. QUADRATURE AMPLITUDE MODULATION

Quadrature Amplitude Modulation or QAM is a form of modulation which is widely used for modulating data signals onto a carrier used for radio communications. It is widely used because it offers advantages over other forms of data modulation such as PSK, although many forms of data modulation operate alongside each other.

Quadrature Amplitude Modulation, QAM is a signal in which two carriers shifted in phase by 90 degrees are modulated and the resultant output consists of both amplitude and phase variations. In view of the fact that both amplitude and phase variations are present it may also be considered as a mixture of amplitude and phase modulation.

Quadrature Amplitude Modulation (QAM) is a complicated name for a simple technique. In the simplest of terms, Quadrature amplitude modulation is the combination of amplitude modulation and phase shift keying. More technically, quadrature amplitude modulation is a system of modulation in which data is transferred by modulating the amplitude of two separate carrier waves, mostly sinusoidal, which are out of phase by 90 degrees (sine and cosine). Due to their phase difference, they are called quadrature carriers.

QAM, Quadrature amplitude modulation is widely used in many digital data radio communications and data communications applications. A variety of forms of QAM are available and some of the more common forms include 8 QAM, 16 QAM, 32 QAM, 64 QAM, 128 QAM, and 256 QAM. Here the figures refer to the number of points on the constellation, *i.e.,* the number of distinct states that can exist.

The various flavours of QAM may be used when data-rates beyond those offered by 8-PSK are required by a radio communications system. This is because QAM achieves a greater distance between adjacent points in the I-Q plane by distributing the points more evenly. And in this way the points on the constellation are more distinct and data errors are reduced. While it is possible to transmit more bits per symbol, if the energy of the constellation is to remain the same, the points on the constellation must be closer together and the transmission becomes more susceptible to noise. This results in a higher bit error rate than for the lower order QAM variants. In this way there is a balance between obtaining the higher data rates and maintaining an acceptable bit error rate for any radio communications system.

2.12.1. Eight (8)-QAM

8-QAM is an M-ary encoding technique where M is equal to 8. In 8-QAM, the output signal from an 8QAM modulator is not a constant-amplitude signal as in 8 PSK.

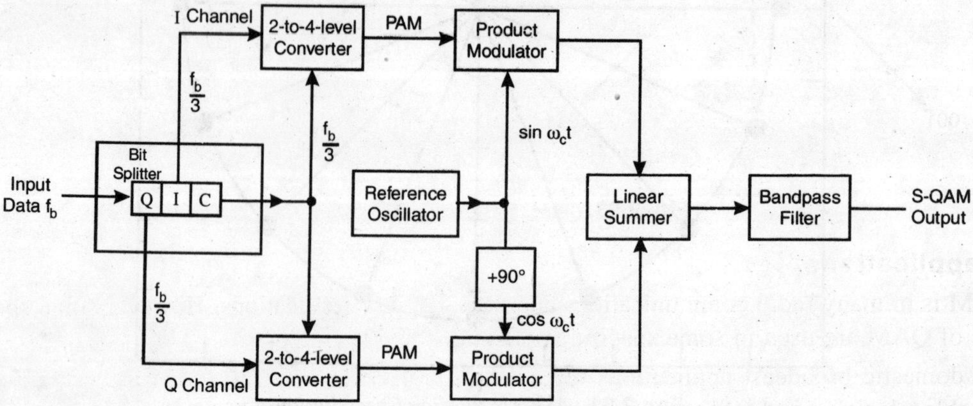

(a)

Fig. 2.32. *The simple block diagram of an 8-QAM transmitter.*

The simple block diagram of an 8-QAM transmitter is shown in Fig. 2.32. The incoming data are divided into groups of three bits (tribits); the I, Q and C bit streams, each with a bit rate equal to one-third of the incoming data rate. The I and Q bits determine the polarity of the PAM signal at the output of the 2-to-4-level convertors, and the C channel determines the magnitude. The magnitude of the I and Q PAM signals are always equal because C bit is fed there uninverted.

I/Q	C	Output
0	0	–0.541 V
0	1	–1.307 V
1	0	+0.541 V
1	1	+1.307 V

(b)

Fig. 2.32.

Their polarities depends on the logic condition of the I and Q bits and, therefore, may be different.

The minimum Bandwidth requires in 8QAM is same as in 8 PSK, *i.e.*, $f_b/3$. Because the bit rate in I and Q channel is one-third of the input binary rate.

Binary Input			8-QAM Output	
Q	I	C	Amplitude	Phase
0	0	0	0.755 V	–135°
0	0	1	0.848 V	–135°
0	1	0	0.765 V	–45°
0	1	0	0.648 V	–45°
1	0	0	0.765 V	+135°
1	0	1	1.848 V	+135°
1	1	0	0.765 V	+45°
1	1	1	1.848 V	+45°

(a)

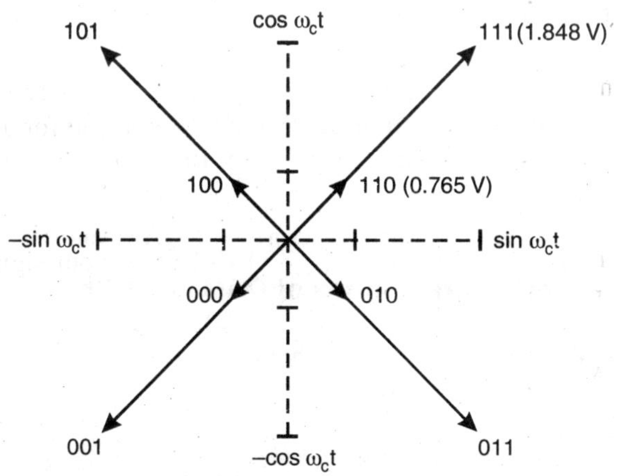

(b)

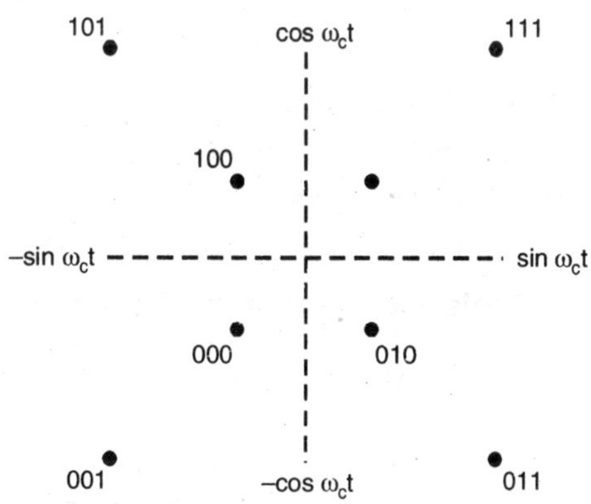

(c)

Fig. 2.33. *QAM constellation diagram.*

QAM applications

QAM is in many radio communications and data delivery applications. However some specific variants of QAM are used in some specific applications and standards.

For domestic broadcast applications for example, 64 QAM and 256 QAM are often used in digital cable television and cable modem applications. In the UK, 16 QAM and 64 QAM are currently used for digital terrestrial television using DVB - Digital Video Broadcasting. In the US, 64 QAM and 256 QAM are the mandated modulation schemes for digital cable as standardized by the SCTE in the standard ANSI/SCTE 07 2000.

In addition to this, variants of QAM are also used for many wireless and cellular technology applications.

2.12.2. Constellation Diagrams for QAM

The constellation diagrams show the different positions for the states within different forms of QAM, quadrature amplitude modulation. As the order of the modulation increases, so does the number of points on the QAM constellation diagram increases.

The Fig. 2.34 below show constellation diagrams for a variety of formats of modulation:

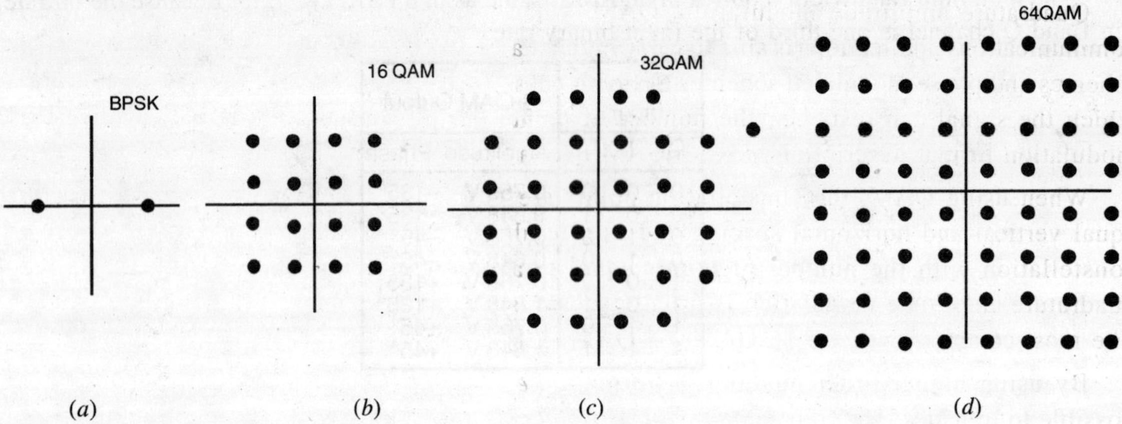

Fig. 2.34.

2.12.3. QAM Bits Per Symbol

The advantage of using QAM is that it is a higher order form of modulation and as a result it is able to carry more bits of information per symbol. By selecting a higher order format of QAM, the data rate of a link can be increased.

The Table 2.4 below gives a summary of the bit rates of different forms of QAM and PSK.

Table 2.4. A summary of the bit rates of different forms of QAM and PSK.

Modulation	Bits per symbol	Symbol Rate
BPSK	1	1 x bit rate
QPSK	2	1/2 bit rate
8PSK	3	1/3 bit rate
16QAM	4	1/4 bit rate
32QAM	5	1/5 bit rate
64QAM	6	1/6 bit rate

2.12.4. Analog and Digital QAM

Quadrature amplitude modulation, QAM may exist in what may be termed either analogue or digital formats. The analogue versions of QAM are typically used to allow multiple analogue signals to be carried on a single carrier. For example it is used in PAL and NTSC television systems, where the different channels provided by QAM enable it to carry the components of chrome or colour information. In radio applications a system known as C-QUAM is used for AM stereo radio. Here the different channels enable the two channels required for stereo to be carried on the single carrier.

Digital formats of QAM are often referred to as "Quantized QAM" and they are being increasingly used for data communications often within radio communications systems. Radio communications systems ranging from cellular technology through wireless systems including WiMAX, and Wi-Fi 802.11 use a variety of forms of QAM, and the use of QAM will only increase within the field of radio communications.

2.12.5. Digital/Quantized QAM Basics

Quadrature amplitude modulation, QAM, when used for digital transmission for radio communications applications is able to carry higher data rates than ordinary amplitude modulated schemes and phase modulated schemes. As with phase shift keying, etc. the number of points at which the signal can rest, *i.e.,* the number of points on the constellation is indicated in the modulation format description, *e.g.,* 16QAM uses a 16 point constellation.

When using QAM, the constellation points are normally arranged in a square grid with equal vertical and horizontal spacing and as a result the most common forms of QAM use a constellation with the number of points equal to a power of 2 *i.e.,* 2, 4, 8, 16 ... . As the quadrature amplitude modulation is usually square, some of these may be missing or atypical. The most common ones are 16-QAM, 64-QAM, 128-QAM and 256-QAM.

By using higher order modulation formats, *i.e.,* more points on the constellation, it is possible to transmit more bits per symbol. However the points are closer together and they are therefore more susceptible to noise and data errors.

Phase modulation can be considered as a special form of QAM where the amplitude remains constant and only the phase is changed. By doing this the number of possible combinations is halved.

2.12.6. QAM Advantages and Disadvantages

QAM appears to increase the efficiency of transmission for radio communications systems by utilizing both amplitude and phase variations.

But it has a number of drawbacks:

1. The first is that it is more susceptible to noise because the states are closer together so that a lower level of noise is needed to move the signal to a different decision point. Receivers for use with phase or frequency modulation are both able to use limiting amplifiers that are able to remove any amplitude noise and thereby improve the noise reliance. This is not the case with QAM.
2. The second limitation is also associated with the amplitude component of the signal. When a phase or frequency modulated signal is amplified in a radio transmitter, there is no need to use linear amplifiers, whereas when using QAM that contains an amplitude component, linearity must be maintained. Unfortunately linear amplifiers are less efficient and consume more power, and this makes them less attractive for mobile applications.

2.13. Sixteen-Quadrature Amplitude Modulation

With this modulator, 4 bits are processed to produce a single vector. The resultant constellation consists of three different amplitudes distributed in 12 different phases as shown below in Fig. 2.35.

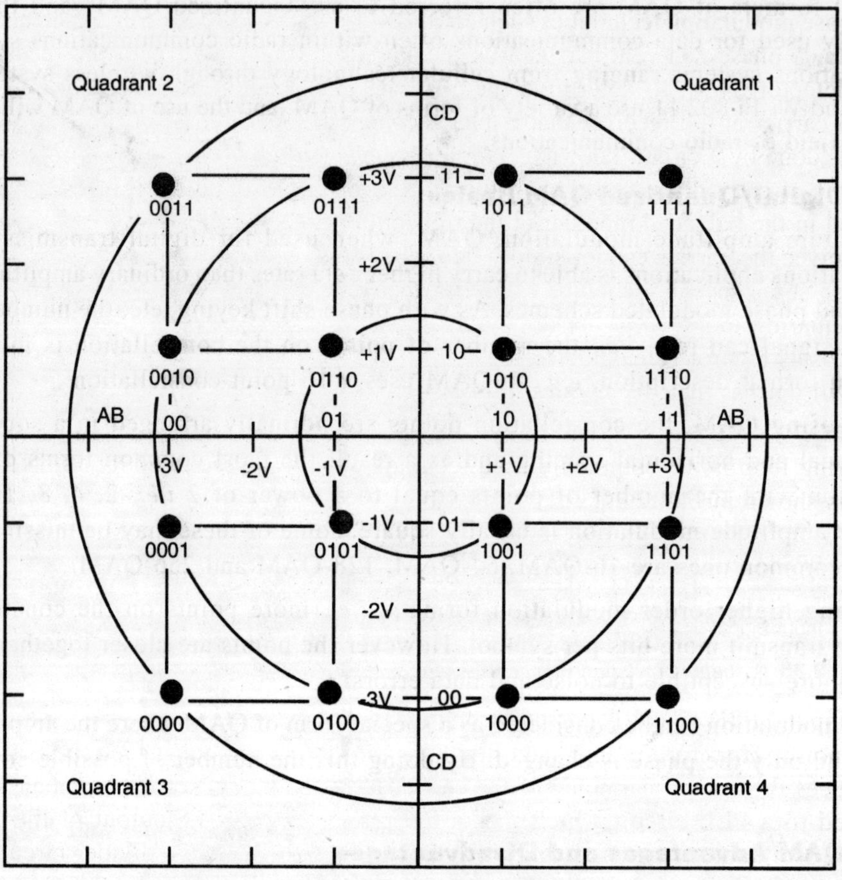

Fig. 2.35. *QAM Constellation.*

2.14. BANDWIDTH EFFICIENCY

Bandwidth Efficiency is the ratio of the transmission bit rate to the minimum bandwidth required for a particular modulation scheme.

$$B\eta = \text{Transmission bit rate (Bps)/ minimum bandwidth (Hz)}$$

(Bits/sec) / Hz

Or

(Bits/sec) /(cycle/sec)

Or

Bits/cycle

Bandwidth efficiency can be given in percentage by simply multiplying by 100.

2.15. CARRIER RECOVERY

Carrier recovery is the process of extracting a phase coherent reference carrier from a receiver signal and sometimes it is referred as *phase referencing*. So for we have seen that in phase modulation techniques the binary data (*i.e.,* 0 and 1) were encoded as a precise phase of the transmitted carrier. The angular separation between adjacent phases varied between 30° and 180°, depending upon the type of phase modulation techniques such as BPSK, QPSK, 8-PSK etc. For the precise demodulation of the data, a phase-coherent carrier was recovered and compared with the received carrier in a product detector. Now, to determine the absolute phase of the received carrier, it is needed to produce a carrier at the receiver that is phase coherent with the transmit reference oscillator. This function is done by a circuit known as *carrier recovery circuit.*

As it is already discussed that in the PSK and QAM modulation techniques, the carrier is suppressed in the balanced modular at the transmitter itself, and therefore it is not transmitted. So, at the receiver the carrier cannot simply be locked with a standard phase-locked loop (PLL).

We use some special technique of carrier recovery namely;

(*i*) Squaring loop

(*ii*) Costas loop or Remodulator

We will discuss their techniques in brief here.

(i) Squaring Loop

Squaring loop is the commonly used technique of carrier recovery for BPSK. A simple block diagram is shown in Fig. 2.36.

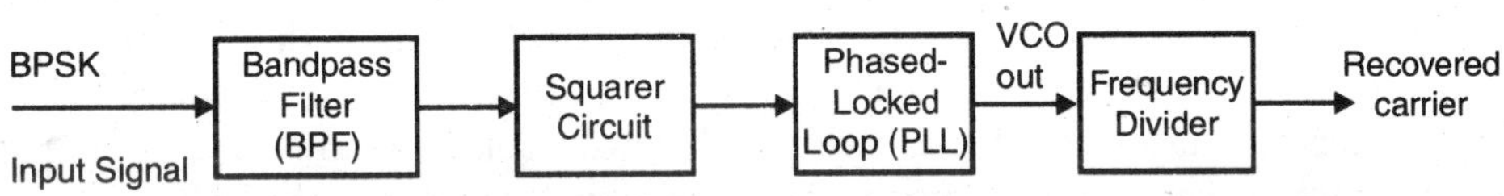

Fig. 2.36. *A simple block diagram of Squaring loop for carrier recovery circuit for a BPSK receiver.*

The input BPSK signal is passed through a BPF to achieve reduction in spectral width of the received noise. Then further this band limited signal is fed to a squarer circuit which demodulates and produces the second harmonic of the carrier frequency. This second harmonic of the carrier is further fed to a PLL circuit which tracks the phase of this. The output of the PLL is voltage controlled oscillator output frequency which is given to frequency divider circuit to divide this frequency by two and used as the reference for the product detectors.

In the BPSK, there are only two output phases are possible such as $+\sin \omega_c t$ and $-\sin \omega_c t$.

The function of squaring circuit may be written mathematically as if they receive signal is $+\sin \omega_c t$;

Output of squaring circuit= $(+\sin \omega_c t)\,(+\sin \omega_c t) = +\sin^2 \omega_c t$

$$= \frac{1}{2}(1 - \cos 2\omega_c t) = \underbrace{\frac{1}{2}}_{\text{Filtered Out}} - \frac{1}{2}\underbrace{\cos 2\omega_c t}_{\text{Second harmonic carrier}}$$

The constant voltage is removed by filtering and leaving only $\cos 2\omega_c t$.

Let us see, it we have received signal is $-\sin \omega_c t$;

Output of squaring circuit = $(-\sin \omega_c t)\,(-\sin \omega_c t) = +\sin^2 \omega_c t$

$$= \frac{1}{2}(1 - \cos 2\omega_c t) = \underbrace{\frac{1}{2}}_{\text{Filtered Out}} - \frac{1}{2}\underbrace{\cos 2\omega_c t}_{\text{Second harmonic carrier}}$$

Similarly, the constant voltage is filtered out and remains only the $\cos 2\omega_c t$.

Carrier recovery for M-ary PSK : Similarly we can the squaring loop method to recover the carrier for the M-ary PSK. A simple block diagram is shown in Fig. 2.37. This circuit may be used in the receiver of BPSK, QPSK and M-ary PSK. This circuit is referred as the Mth power loop and for BPSK it is known as squaring loop because in BPSK the value of M is 2.

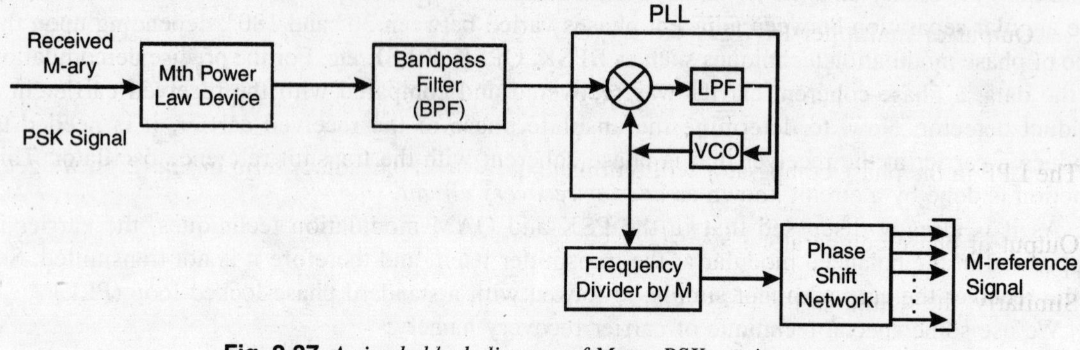

Fig. 2.37. *A simple block diagram of M-ary PSK carrier recover.*

The only disadvantage of this method is a problem of phase ambiguity. There is another method for carrier synchronization is *Costas loop.*

(ii) Costas Loop

The Costas loop method is also used for carrier synchronization in the BPSK receiver. The method of carrier recovery in the Costas, or quadrature loop, shown in Fig. 2.38. Figure shows that the Costas loop consist of two phase comparators and a VCO along with a 90° phase shifter. The Costas loop produces the same result as a squaring circuit followed by an ordinary PLL in place of BPF. This carrier recovery method uses two parallel two tracking loop (I & Q) simultaneously to derive the product of the I and Q components of the signal that drives the VCO.

The VCO generates a carrier frequency f_c with an arbitrary phase angle 'Q'. The carrier signal is applied to the phase comparator-I, *i.e.,* in-phase loop uses the VCO as in PLL and the quadrature (Q) loop uses a 90° shifted VCO signal, phase comparator-Q. Both the phase comparators I and Q

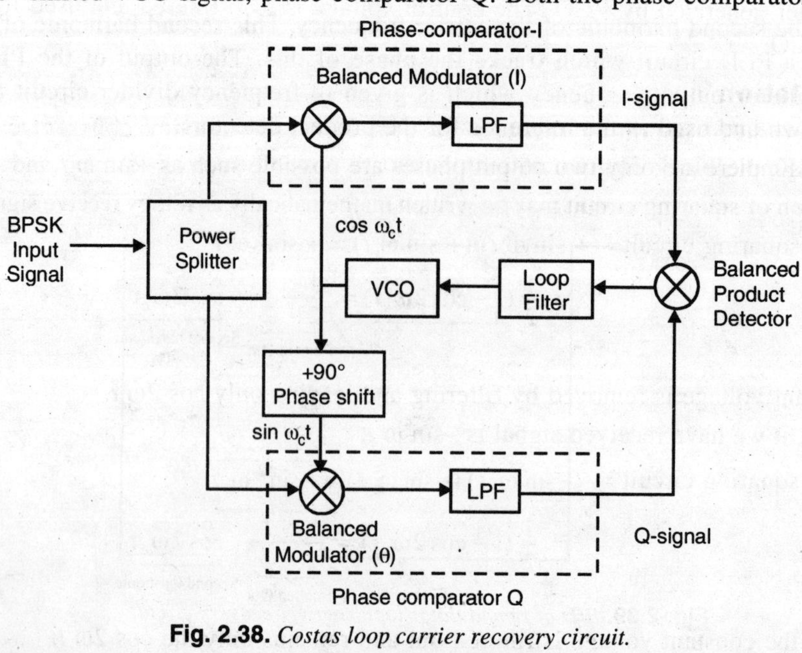

Fig. 2.38. *Costas loop carrier recovery circuit.*

Once the frequency of the VCO is equal to the suppressed carrier frequency, the product of the I and Q signals will produce an error voltage proportional to any phase error in the VCO. The error voltage controls the phase and, thus the frequency of the VCO.

The output of multiplier in-phase comparator I is given by

$$\text{Output of multiplier-I} = \sqrt{2P_s}\,\cos\omega_c t\,(\omega_c t + Q)$$

$$= \frac{1}{2} \times \sqrt{2P_s}\left[\cos(2\omega_c t + \theta + \phi) + \cos(\phi - \theta)\right] \quad ...(2.24)$$

The LPF in the phase comparator will eliminate the second harmonics term in Eq. (2.1), we get

$$\text{Output of phase comparator-I} = \frac{1}{2}\sqrt{2P_s}\,\cos(\phi - \theta) \quad ...(2.25)$$

Similarly, the output of the phase comparator-Q is given by

$$\text{Output of phase comparator-Q} = \frac{1}{2} \times \sqrt{2P_s}\,\sin(\phi - \theta) \quad ...(2.26)$$

The outputs of both the phase comparator are applied to a multiplier produces control voltage V_C. The control voltage is given by

$$V_C = \frac{1}{4} 2P_s \sin(\phi - \theta)\cos(\phi - \theta)$$

$$= \frac{P_s}{2} \sin(\phi - \theta)\cos(\phi - \theta)$$

$$V_C = \frac{P_s}{4} \sin 2(\phi - \theta) \quad ...(2.27)$$

Thus the control voltage V_C is proportional to the phase difference $(\phi - \theta)$ between the received carrier and carrier generated by VCO. The control voltage is then filtered and used for correcting frequency of VCO.

(iii) Remodulation

Another method of obtaining recovery of a phase and frequency coherent carrier is the remodulator. A simple block diagram of remodulator is shown in Fig. 2.39.

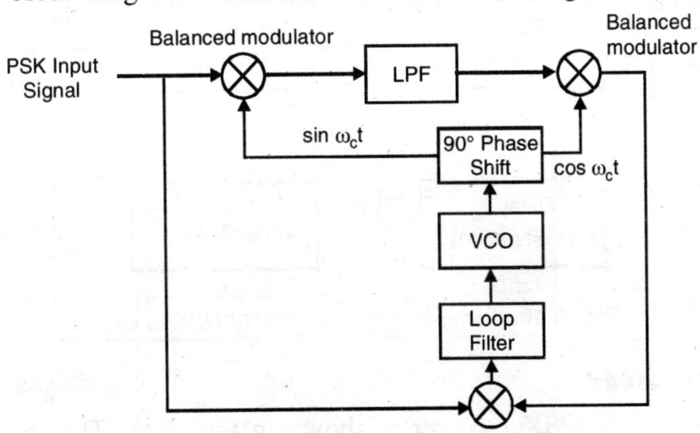

Fig. 2.39. *B/D of remodulator loop carrier recovery circuit.*

The remodulator produces a lop error voltage that proportional to device the phase error between the incoming signal and the VCO signal. The remodulator method has a faster acquisition time than either the squaring or the Costes loops.

2.16. DIFFERENTIAL PHASE-SHIFT KEYING (DPSK)

Differential phase shift keying (DPSK) is an alternative form of digital modulation where the binary input information is contained in the difference between two successive signaling elements rather than the absolute phase. We can view DPSK as the non-coherent version of the PSK. It is differentially coherent modulation method and it also does not need a synchronous carrier at the demodulator. The input sequence of binary bits is modified such that the next bit depends upon the previous received bits. Therefore, in the receiver, the previous received bits are used to detect the present bit.

2.16.1. Generation of DPSK

Encoding Technique : Thus, in order to eliminate the need for phase synchronization of coherent receiver with PSK, a differential encoding system can be used with PSK. The digital information content of the binary data is encoded in terms of signal transitions. As an example, the symbol '1' to indicate no transition in a given binary sequence (with respect to the previous encoded bit) and symbol '0' to indicate transition. This new signaling technique when combines differential encoding with PSK is known as **differential phase shift keying (DPSK).**

DPSK Transmitter : A simple block diagram of DPSK generator is shown in Fig. 2.40. The DPSK is a modification of BPSK. The data stream (bit) is applied to the input of the encoder. The output of the encoder is applied to one input of the product modulator. A sinusoidal carrier of fixed amplitude and frequency is applied to the product modulator as second input signal.

The relationship between the binary sequence and its differentially encoded version is illustrated in Table 2.5 for an assumed data sequence;

00100100111

In this example it has been assumed that the encoding has been done in such a way that transition in the given binary sequence with respect to the previous encoded bit is represented by a symbol 0 and no transition by symbol '1'. It is to be noted that an extra bit (symbol '1') has been arbitrarily added as an initial bit. This is essential to determine the encoded sequence.

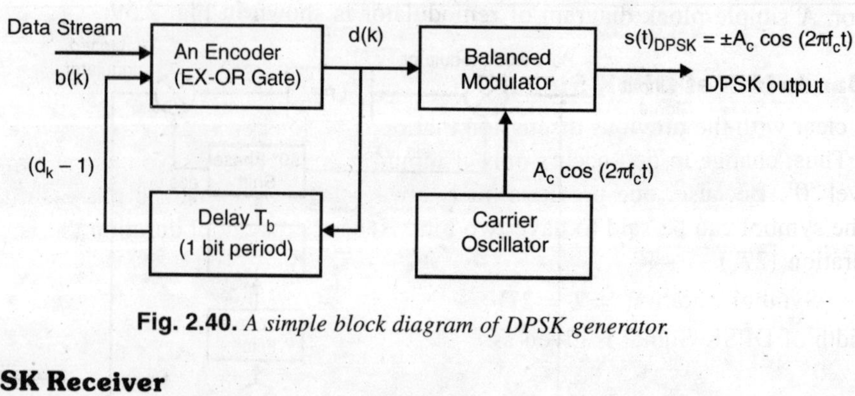

Fig. 2.40. *A simple block diagram of DPSK generator.*

2.16.2. DPSK Receiver

The block diagram of DPSK receiver is shown in Fig. 2.41. The received DPSK signal in denoted $s(t)_{DPSK}$ and it is applied through a band pass filter and correlator. Another input of the correlator is a delayed version of the received DPSK by the time interval T_b is applied. The delayed version of the receiver DPSK signal (in the absence of channel noise) has been shown in the 4th row of the table. The output of difference is proportional to cos (ϕ), here (ϕ) is the difference between the carrier phase angle of the receiver DPSK signal and its delayed version, measured in the same bit internal. The phase angle of the DPSK signal and its delayed version has been shown in 3rd and 5th rows respectively.

The phase difference between the two sequences for each bit internal is used to determine the sign of the phase comparator O/P. When $\phi = 0$, the integrator output is positive whereas when $\phi = \pi$, the integrator output is negative. By comparing the integrator output with a decision level of zero volt, the decision device can reconstruct the binary sequence by assigning a symbol '0' for negative output and a symbol '1' for positive output. The reconstructed data is shown in the last row of the Table 2.5.

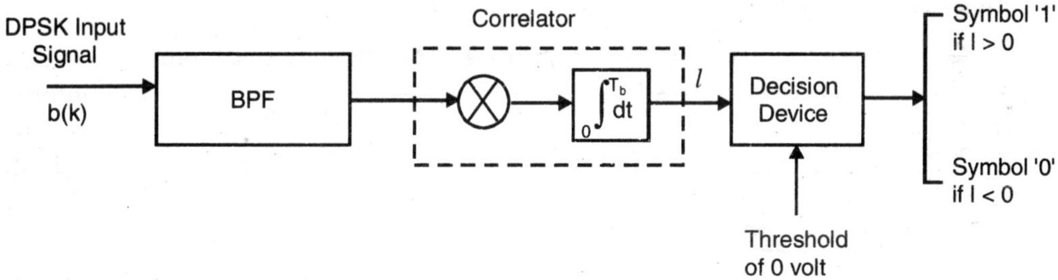

Fig. 2.41. *B/D of DPSK receiver.*

Table 2.5

Binary data {b(k)}		0	0	1	0	0	1	0	0	1	1
Differentially encoded data {d(k)}	1*	0	1	1	0	1	1	0	1	1	1
Phase of DPSK	0	π	0	0	π	0	0	π	0	0	0
Shifted Differentially encoded Data $\{d_{k-1}\}$		1	0	1	1	0	1	1	0	1	1
Phase of shifted DPSK		0	π	0	0	π	0	0	π	0	0
Phase comparison output		–	–	+	–	–	+	–	–	+	+
Detected Binary Sequence		0	0	1	0	0	1	0	0	1	1

* Arbitrary starting reference bit.

2.16.3. Bandwidth of DPSK Signal

As it is clear with the previous discussion that one previous bit is used to decide the phase shift of next bit. Thus, change in $b(t)$ occurs only if input bit is at level '1'. No change happens if input bit is at level '0'. Because, one previous bit is always used to define the phase shift in next bit, therefore, the symbol can be said to have two bits. Hence and symbol duration (T) is equivalent to two bits duration ($2T_b$).

i.e., Symbol duration $= T = 2T_b$

Bandwidth of DPSK Signal is given as

$$BW = \frac{2}{T} = \frac{1}{T_b} = f_b \qquad ...(2.28)$$

Hence the minimum bandwidth in DPSK is equal to f_b, *i.e.*, maximum base band signal frequency.

2.16.4. Comparison of DPSK and BPSK

Advantages of DPSK

(*i*) DPSK does not need carrier at the receiver end; this means that the complicated circuitry for generation of local carrier is not needed.

(*ii*) The bandwidth required by DPSK is lower than the BPSK.

Disadvantages of DPSK

(*i*) The probability of error (*i.e.*, bit error rate) of DPSK is higher than that of BPSK.

(*ii*) Noise interference is more in DPSK.

(*iii*) Because DPSK uses two successive bits for its reception, error in the first bit creates error in the second bit. Therefore, error propagation in DPSK is more.

Performance comparison of BPSK and DPSK is given in the Table 2.6.

Table 2.6. Differentially encoded sequences with phase

Binary data {b(k)}		0	0	1	0	0	1	0	0	1	1
Differentially encoded data {d(k)}	0*	1	0	0	1	0	0	1	0	0	0
Phase of DPSK	π	0	π	π	0	π	π	0	π	π	π
Shifted Differentially encoded Data $\{d_{k-1}\}$		0	1	0	0	1	0	0	1	0	0
Phase of shifted DPSK		π	0	π	π	0	π	π	0	π	π
Phase comparison output		–	–	+	–	–	+	–	–	+	+
Detected Binary Sequence		0	0	1	0	0	1	0	0	1	1

* Starting reference bit.

2.17. CLOCK RECOVERY

Clock recovery in the process of regenerating the timing or synchronization clock, at the receiver, so that transmitter and receiver are in synchronous in the digital radio system.

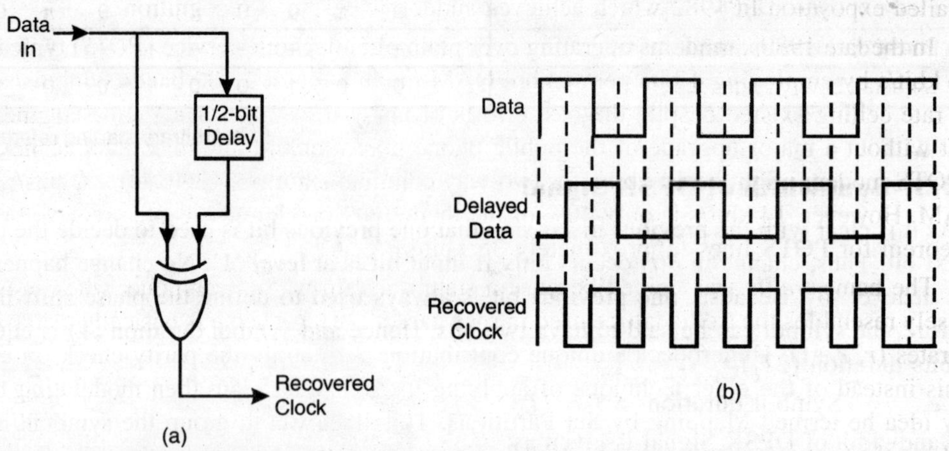

Fig. 2.42. *Clock recovery circuit with timing diagram.*

A simple circuit for the clock recovery is shown in Fig. 2.42. (*a*). In this circuit, the recovered data are delayed by one-half a bit time and then compared with the original data in an Ex-OR circuit.

The frequency of the clock which is recovered with this method is equal to the received data rate (f_b). The relationship between the data and the recovered clock timing is shown in Fig. 2.42 (*b*). As long as they receive data contains a substantial number of transition (I/O sequence), the recovered is maintained but if we have the receive data a long string of either '1' or '0' then the recovered clock would be lost.

So to prevent this lost of recovered clock, we use a method known as *scrambling*, in which enough transitions are introduced in the binary signal using a prescribed algorithm at the transmitter and descrambler uses the same algorithm to remove the transition.

2.18. PROBABILITY OF ERROR AND BIT ERROR RATE

They are often used interchangeably although have slight different meanings.

Probability of error : Probability of error is a theoretical expectation of the bit error rate for a given system.

Bit Error Rate (BER) : BER is an empirical record of a system's actual bit error performance.

For example, if a system has a Pe of 10^{-5}, this means that mathematically you can expect one bit error in every 100000 bits transmitted.

If a system has a BER of 10^{-5}, that means in past performance there was one bit error for every 100000 bits transmitted.

A bit error is measured and then compared with the expected probability of error to evaluate system's performance.

2.19. TRELLIS CODED MODULATION

In telecommunication, **trellis modulation** (also known as **trellis coded modulation**, or simply **TCM**) is a modulation scheme which allows highly efficient transmission of information over band-limited channels such as telephone lines.

Trellis modulation was invented by Gottfried Ungerboeck working for IBM in the 1970s, and first described in a conference paper in 1976; but it went largely unnoticed until he published a new detailed exposition in 1982 which achieved sudden widespread recognition.

In the late 1980s, modems operating over plain old telephone service (*POTS*) typically achieved 9.6 kbit/s by employing 4 bits per symbol QAM modulation at 2,400 baud (symbols/second). This bit rate ceiling existed despite the best efforts of many researchers, and some engineers predicted that without a major upgrade of the public phone infrastructure, the maximum achievable rate for a POTS modem might be 14 kbit/s for two-way communication (3,429 baud × 4 bits/symbol, using QAM. However, 14 kbit/s is only 40% of the theoretical maximum bit rate predicted by Shannon's Theorem for POTS lines (approximately 35 kbit/s).

The name *trellis* was coined because a state diagram of the technique, when drawn on paper closely resembles the trellis lattice used in rose gardens. The scheme is basically a convolution code of rates (r, r +1). Ungerboeck's unique contribution is to apply the parity check on a per symbol basis instead of the older technique of applying it to the bit stream then modulating the bits. The key idea he termed Mapping by Set Partitions. This idea was to group the symbols in a tree like fashion then separate them into two limbs of equal size. At each limb of the tree, the symbols were further apart.

A flood of research activity ensued, and by 1990 the International Telecommunication Union had published modem standards for the first trellis-modulated modem at 14.4 kbit/s (2,400 baud and 6 bits per symbol). Over the next several years further advances in encoding, plus a corresponding symbol rate increase from 2,400 to 3,429 baud, allowed modems to achieve rates up to 34.3 kbit/s (limited by maximum power regulations to 33.8 kbit/s). Today, the most common trellis-modulated V.34 modems use a 4-dimensional set partition which is achieved by treating two 2-dimensional symbols as a single lattice. This set uses 8, 16, or 32 state convol :tional codes to squeeze the equivalent of 6 to 10 bits into each symbol sent by the modem (for example, 2,400 baud × 8 bits/symbol = 19,200 bit/s).

Once manufacturers introduced modems with trellis modulation, transmission rates increased to the point where interactive transfer of multimedia over the telephone became feasible (a 200 kilobyte image and a 5 megabyte song could be downloaded in less than 1 minute and 30 minutes, respectively). Sharing a floppy disk via a BBS could be done in just a few minutes, instead of an hour.

Summary

- The carrier voltage component transmits no information, while each of the two side bands carries complete information.
- **Modulation index** is the depth of the modulation or in other words extent to which carrier is modulated.
- A **signal element** carries data element and it is the shortest unit (time wire) of a digital signal.
- **Data element** is the smallest entity that can represent a piece of information known a **bit**.
- The **data rate** is defined the number of data element (or bits) sent in one second. Its unit is bps (Bits per second).
- The **signal rate** is the number of signal element sent in one second. Its unit is baud.
- The amount of deviation in frequency is proportional to the amplitude of the signal, while the rate is determined by the frequency of modulation source.
- $$e = E_c \sin\{\omega_c t + m_f \sin \omega_m t\}$$
- $$\Delta f = \frac{CS}{2}$$
- The bandwidth of FM signal depends upon the value of modulation index m_f.
- In the process of **phase modulation**, the instantaneous phase of the carrier is varied by an amount that is proportion to the instantaneous amplitude the signal and at a rate that is proportional to modulating frequency.
- $$e = E_c \sin(\omega_c t + \Delta f \sin \omega_m t)$$
  where $$\Delta f = K_p E_m$$
- $$BW_{PM} = 2(\Delta f)$$
- The continuous variation of the amplitude, width, or position of the pulses in a pulse train to represent an information signal is defined as **analog pulse modulation.**
- In **Pulse Amplitude Modulation (PAM),** relative amplitudes of successive pulses serve as a measure of signal amplitudes at the corresponding sampling instants.
- The durations or widths of the pulses are varied (from the nominal value) according to the samples.
- In **Pulse Position Modulation (PPM),** the amplitude and width of the pulses are kept constant, while the position of each pulse, with reference to the position of a reference pulse, is changed according to the instantaneous sampled value of modulating signal.
- The source signals are generally referred to as ***baseband*** signals.
- **Signal element :** A **signal element** carries data element and it is the shortest unit (time wise) of a digital signal.

- **Data element** is the smallest entity that can represent a piece of information known a **bit.**
- The **data rate** is defined the number of data element (or bits) sent in one second. Its unit is bps (Bits per second).
- The **signal rate** is the number of signal element sent in one second. Its unit is baud.
- **Amplitude shift keying (ASK)** in the context of digital communications is a modulation process that imparts to a sinusoid two or more discrete amplitude levels.
- $s(t) = A\ m(t) \cos 2\pi fct,\ 0 = t = T$
- The amplitude of a carrier is switched or keyed by the binary signal $m(t)$. This is sometimes called **on-off keying (OOK)**.
- For ASK we use one bit either 0 or 1 to represent one symbol, so the rate of symbol transmission (baud rate) N_b will be the same as bit rate.

$$\text{Baud rate } (N_b) = \text{Bit rate}$$

- $BW_{max} = f_b$ Hz
- FSK modulation is achieved using just two frequencies to represent binary data.
- FSK has the advantage of being very simple to generate, simple to demodulate and due to the constant amplitude can utilize a non-linear PA.
- Baud = $\dfrac{f_b}{N}$
- $B = |(f_s - f_m)| + 2f_b$
- $B = 2(\Delta f + f_b)$
- In **Phase Shift Key Modulation** method the phase of the carrier changes between different phases determined by the logic states of the input bit stream.
- The bandwidth of QPSK signal will be

$$\text{BW} = 2 \times \frac{1}{2T_b} = f_b$$

- The error probability QPSK system is expressed by

$$P_e = \frac{1}{2} efr_c \sqrt{\frac{E_b}{N_0}}$$

- Quadrature Amplitude Modulation or QAM is a form of modulation which is widely used for modulating data signals onto a carrier used for radio communications.
- Quadrature Amplitude Modulation, QAM is a signal in which two carriers shifted in phase by 90 degrees are modulated and the resultant output consists of both amplitude and phase variations.
- 8-QAM is an M-ary encoding technique where M is equal to 8.In 8-QAM, the output signal from an 8QAM modulator is not a constant-amplitude signal as in 8 PSK.
- QAM is in many radio communications and data delivery applications. However some specific variants of QAM are used in some specific applications and standards.
- The advantage of using QAM is that it is a higher order form of modulation and as a result it is able to carry more bits of information per symbol.
- **Bandwidth Efficiency** is the ratio of the transmission bit rate to the minimum bandwidth required for a particular modulation scheme.

$$B\eta = \text{Transmission bit rate (Bps)/ minimum bandwidth (Hz)}$$

- **Carrier recovery** is the process of extracting a phase coherent reference carrier from a receiver signal and sometimes it is referred as *phase referencing*.
- Costas loop carrier recovery circuit.

➡ **Differential phase shift keying (DPSK)** is an alternative form of digital modulation where the binary input information is contained in the difference between two successive signaling elements rather than the absolute phase.

➡ Bandwidth of DPSK Signal is given as

$$BW = \frac{2}{T} = \frac{1}{T_b} = f_b$$

➡ **Clock recovery** in the process of regenerating the timing or synchronization clock, at the receiver, so that transmitter and receiver are in synchronous in the digital radio system.

➡ The simplest form of phase modulation is binary (two level) phase modulation.

➡ **Probability of error** : Probability of error is a theoretical expectation of the bit error rate for a given system.

➡ BER is an empirical record of a system's actual bit error performance.

➡ In telecommunication, **trellis modulation** (also known as **trellis coded modulation**, or simply **TCM**) is a modulation scheme which allows highly efficient transmission of information over band-limited channels such as telephone lines.

➡ The name *trellis* was coined because a state diagram of the technique, when drawn on paper closely resembles the trellis lattice used in rose gardens.

Review Questions

1. What do you understand by term modulation? Describe the need of the modulation in communication.
2. How we can classify the modulation? Explain analog modulation in brief.
3. What is meant by term amplitude modulation?
4. Derive an expression for amplitude modulated wave.
5. Draw the sideband frequency spectrum for a single tone amplitude modulated wave.
6. Derive the power relations for amplitude modulated wave.
7. What are frequency components in an AM wave? Explain the bandwidth of AM wave.
8. What do you understand by vestigial sideband amplitude modulation?
9. What is meant by term frequency modulation?
10. Derive an expression for frequency modulated wave.
11. What is meant by term phase modulation?
12. Derive an expression for phase modulated wave.
13. Compares the frequency and phase modulation.
14. Write the difference between AM and FM.
15. What do you understand by pulse modulation? Write various types of pulse modulation.
16. Describe pulse amplitude modulation with the waveform representation.
17. List the steps that take an analog signal to digital PCM signal.
18. What is the difference between encoding and modulating?
19. What is digital-to-digital encoding?
20. What is digital-to-analog and analog-to-digital conversion?
21. What is analog-to-analog conversion?
22. Why is frequency modulation superior to amplitude modulation?
23. What is the advantage of QAM over ASK or PSK?
24. What is the purpose of a carrier signal in modulation?

25. How is QAM related to ASK and PSK?
26. What are the major factors that make PSK superior to ASK?
27. How does
    (*a*) AM differ from ASK? (*b*) FM differs from FSK?
28. What basically PAM mean? Explain briefly.
29. Explain ASK and FSK with an example. Draw their frequency spectrum.
30. What is PSK? Explain 2-PSK and 4-PSK. Can 8-PSK be generated? If yes, then explain how?
31. Explain the basic concept of Amplitude Modulation and Frequency Modulation.
32. What would be the data stream if the above figure is pseudoternary encoding of the data stream?
33. What is the sampling rate for PCM if the frequency ranges from 1000 to 4000 Hz?
34. Using Nyquist theorem, calculate the sampling rate for an analog signal with frequencies from 2000 Hz to 6000 Hz and for a signal with a horizontal line in the time domain representation and for a signal with a vertical line in the time domain representation?
35. A signal is sampled. Each sample represents each of the four levels. How many bits are needed to represent each sample? If the sampling rate is 8000 samples per second, what is the bit rate?
36. Calculate the bandwidth for each of the following AM stations. Disregard the FCC rules:
    (*a*) Modulating signal with a bandwidth of 8 kHz.
    (*b*) Modulating signal with frequencies of 2000 to 3000 Hz.
37. Calculate the bandwidth required for each of the following FM stations. Disregard
38. FCC rules:
    (*a*) Modulating signal with a bandwidth of 12 kHz.
    (*b*) Modulating signal with frequencies of 2000 to 3000 Hz.
39. How the "Information" is transmitted using AM?
40. Define modulation index of an AM wave.
41. What is the maximum value of "m" for distortion free transmission?
42. What is the difference between percent modulation and modulation index?
43. Write down the expression for an AM wave
44. What all frequency components are present in AM wave spectrum?
45. Which of these frequency components contain the message information?
46. What is the minimum bandwidth of an AM wave?
47. How to vary amplitude of sidebands?
48. When does over modulation take place and what is its effect?
49. Explain how modulation index can be measured on CRO.
50. Write down the expression for carrier power.
51. Why do we consider the rms carrier voltage to calculate the carrier power?
52. Differentiate between QAM and QPSK in detail. What are the advantages of QAM over PSK? **[MDU Dec. 2^08]**
53. Explain frequency modulation in detail ? Compare FM bandwidth with AM bandwidth in terms of modulating signal. **[MDU Dec. 2008]**
54. Explain in detail about the following :
    (*i*) Clock recovery
    (*ii*) Trellis encoding
    (*iii*) Carrier recovery. **[MDU May 2009]**
55. Differentiate between QAM and QPSK in detail ? What are the advantages of QAM over QPSK. **[MDU May 2009]**
56. Describe briefly about the QAM. **[MDU May 2009]**
57. Explain BPSK and how it is implemented. **[MDU Dec. 2009]**
58. Define the term clock recovery and carrier recovery. **[MDU Dec. 2009]**

59. Explain the following in detail:
 (*a*) Frequency shift keying
 (*b*) Phase shift keying. **[MDU Dec. 2009]**

60. Differentiate between ASK and PSK. What are the advantages of QAM over ASK and PSK. **[MDU May 2010]**

61. Draw and explain the DPSK transmitter and receiver. **[MDU Dec. 2010]**

62. Derive the relationship between minimum bandwidth, bit rate, mark and space frequency for an FSK system. Determine bandwidth and baud with a mark frequency of 32 kHz, space frequency of 24 kHz and bit rate of 4 Kbps. **[MDU Dec. 2010]**

63. Write short note on Trellis encoding. **[MDU Dec. 2010]**

64. Why carrier recovery is needed in QAM and QPSK systems ? Draw and explain in detail all the methods of carrier recovery. **[MDU May 2011]**

65. Which of the four digital to analog techniques is most susceptible to noise ? Defend your answer. Which is the best technique and why ? **[MDU May 2011]**

66. For an 8 PSK system operating at a bit rate of 24 Kbps determine baud, minimum bandwidth and bandwidth efficiency. **[MDU May 2011]**

67. Write the expression for sideband power in AM.

68. What is the maximum value of sideband power in AM?

69. What is the maximum transmitted power in AM?

70. What is the effect of "m" on carrier power in AM?

71. Define transmission efficiency. What is its maximum value for AM?

72. Describe pulse width modulation with the waveform representation.

73. Describe pulse position modulation with the waveform representation.

74. Compares the PAM, PWM and PPM.

75. Explain the steps involves in pulse code modulation.

76. What do you understand by digital modulation? Write the various types of digital modulation.

77. Explain amplitude shift keying with the waveforms.

78. Explain frequency shift keying with the waveforms.

79. Explain phase shift keying with the waveforms.

80. Compares the ASK, FSK and PSK.

81. Differentiate between ASK and PSK in detail. What are the advantages of QAM over ASK and PSK? **[MDU May 2009]**

82. Explain BPSK and how it is implemented? **[MDU Dec. 2009]**

83. Differentiate between QAM and QPSK in detail. What are the advantages of QAM over QPSK? **[MDU Dec. 2009]**

84. Explain in detail the following: **[MDU Dec. 2009]**
 (*i*) Frequency shift keying
 (*ii*) Amplitude shift keying
 (*iii*) Clock recovery
 (*iv*) Trellis encoding
 (*v*) Carrier recovery

85. Write shorts on following:
 (*i*) Modulation Index in AM
 (*ii*) Frequency deviation
 (*iii*) Carrier swing
 (*iv*) Pulse modulation
 (*v*) PAM
 (*vi*) PDM
 (*vii*) Sampling
 (*viii*) Encoding
 (*ix*) Quantization
 (*x*) MODEM

❑❑❑

3

CHAPTER

Data Communication

3.1. INTRODUCTION

INSIDE THIS CHAPTER

What is Data Communications?

The advancement in technology around the world has occurred at very high rate, transforming the telecommunication industry into an extremely dynamic and highly sophisticated field. Due the advent of VLSI chips and the accompanying low cost microprocessors, microcontrollers, computers have dramatically increased need for the exchange of digital information.

Data Communications is the transfer of data or information between a source and a receiver. The source transmits the data and the receiver receives it. The actual generation of the information is not part of Data Communications nor is the resulting action of the information at the receiver.

Data Communication is interested in the transfer of data, the method of transfer and the preservation of the data during the transfer process.

In the data communication world, **data** means the information stored in digital form. A single unit of data is **datum** and the data is plural. *Data Communications* is the process of transferring the binary (digital) information between two or more points.

The purpose of Data Communications is to provide the rules and regulations that allow computers with different disk operating systems, languages, cabling and locations to share resources.

The rules and regulations are called protocols and standards in Data Communications.

3.2. HISTORY OF DATA COMMUNICATION

Though data communication may seem a relatively new innovation, the history of data connections dates to the early 19th century. Through an array of technological developments that enjoyed commercial, government and military contributions, data communication evolved from simple wired connections to a robust global exchange of information.

Definition : Before studying the history of data communication, it is first important to understand exactly what constitutes this form of information exchange. Because the history of data communication extends well beyond modern computer connections and wireless devices, some might find the communication channel's earliest roots surprising. According to Dr. Carl Rebman, professor of Data Communications and Computer Networks at the University of San Diego, lexicographers have defined data communication as "Any process that permits the passage from a sender to one or more receivers of information of any nature, delivered in any easy to use form by any electromagnetic system." This definition helps historians trace the roots of data communication considerably further than modern digital equipment.

Early History

According to the History of Computing organization, data communication has its earliest roots in Samuel Morse's 1837 exhibition of a telegraph system. An account of data communication history posted by telecommunications experts at General Telecom, LLC also points to a telegraph patent that inventor Charles Wheatstone filed that same year. By 1843, telegraph service had become adopted by the Great Western Railway, an endorsement that allowed the service to expand across the nation.

Telecom Developments

Improving on the telegraph, according to the History of Computing, Alexander Graham Bell introduced the telephone in 1876.

Though standard telephone lines did not carry data traffic until nearly a hundred years later, the development of early telecommunications—coupled with an 1895 invention by Guglielmo Marconi, the radio—laid the groundwork for numerous subsequent developments in communication technology. In 1947, Bell Labs introduced the transistor, a device that found integration in myriad subsequent electronic products. The U.S. government expanded on these technologies in 1958 with its launch of a communications-oriented satellite, and the first facsimile transmission over standard telephone lines occurred four years later.

Computer Connections

After the first fax transmission in 1962, the modulation of data into sound for transmission across telephone lines spread in popularity for several years.

Though modulation/demodulation, or modem, technology continued to carry slower data traffic for the remainder of the 20th century, according to the History of Computing, the 1969 development of Internet Protocol (IP) marked a significant milestone in data communication history.

Within the following decades, early packet communication technologies like Asynchronous Transfer Mode (ATM), Frame Relay and Integrated Services Digital Network (ISDN) emerged as a viable solution for commercial and high-end residential data needs. By 1991, more than 1 million servers had come online using Internet Protocol technology, and the World Wide Web emerged as the primary component of the Internet by the mid-1990s.

Wireless

As wired data communication expanded, a separate form of data exchange that required no wires experienced a concurrent development. According to the wireless communication reference website WirelessCommunication.nl, wireless technologies developed rapidly during World War I. By 1921, the first commercial radios voice broadcast, a communication channel that transmitted data in the form of modulated radio waves, had taken place and police dispatchers had adopted the technology. In 1974, the Federal Communication Commission (FCC) began allocating wireless spectrums for wireless communication, and wireless companies began integrating radio packet data as early as 1984. This technology evolved into the various digital wireless packet data protocols used into the 21st century.

Three main developments that led to the growth of data communications systems:

- Large-scale integration of circuits reduced the cost and size of terminals and communication equipments.
- Developments of software systems made establishment of communication networks easy and
- Competition among providers of transmission facilities reduced the cost of data circuits

Summary of History of Data Communication

- 1838: Samuel Morse & Alfred Veil Invent Morse Code Telegraph System
- 1876: Alexander Graham Bell invented Telephone
- 1910:Howard Krum developed Start/Stop Synchronisation
- 1930: Development of ASCII Transmission Code
- 1945: Allied Governments develop the First Large Computer
- 1950: IBM releases its first computer IBM 710
- 1960: IBM releases the First Commercial Computer IBM 360

Main Contributors of Data Communication

Transmission Technology

- Packet Switching Technology
- Internet
  - 1967: ARPANET by Advanced Research

Project Agency (ARPA) of U.S.

  - 1975: TCP/IP protocol
- LAN Technology
  - DIX-Ethernet & IEEE 802 Networks
- **WAN :** 1976: ISO releases HDLC & CCITT releases X.25 (PSPDN)
- In 70's & 80's main thrust in Wide Area Networking (WAN) was to put *Data on Voice Circuits* using Modem & on ISDN lines
- In 90's the trend is reverse. Major Efforts were on putting *Voice OverData* using:
  - Voice Over Frame Relay
  - Voice Over Internet
  - Voice Over ATM etc

3.3. STANDARD ORGANIZATION FOR DATA COMMUNICATION

The importance of standards in the field of communication cannot be overstressed. Standards enable equipment from different manufacturer and with different operating characteristics to become components of the same network. Standards also enable different networks in different geographical locations (*e.g.,* different countries and continents) to be interconnected.

Standards are developed by national and international organizations established for this exact purpose. We will discuss a number of important standards developed by various organizations as following:

1. The **International Standards Organization** (ISO) is a voluntary organization with representations from national standards organizations of member countries, major vendors and end-users. ISO is active in many areas of science and technology, including information technology. ISO standards are published as *ISO serial-no* (*e.g.*, ISO 8632).
2. The **International Telecommunication Union-Telecommunication Sector (ITU-T)** formerly the Consultative Committee for International Telegraph and Telephone (CCITT), is one of four permanent part of the International Telecommunication Union based in Geneva, Switzerland. It is a standards organization devoted to data and telecommunication, with representations from governments, major vendors, telecommunication carriers, and the scientific community. The ITU-T has developed three sets of specifications as following :
   (*i*) **The V series :** It is for modem interfacing and data transmission over telephone lines.
   (*ii*) **The X series :** It is for data transmission over public digital networks, E-mail and directory services.
   (*iii*) **The I & Q series :** For Integrated services Digital Network (ISDN) and its extension Broadband ISDN.

   The ITU-T is separated into 14 study groups that prepare the recommendations for the defined topics. These are revised and republished every four years. ITU-T standards are very influential in the field of telecommunications and are adhered to by most manufacturers and carriers.
3. The **Institute of Electrical and Electronic Engineers** (IEEE) is a US standards organization with members throughout the world. IEEE is active in many electric, electronics, computer, and communication related areas. The IEEE standards for local area networks are widely adopted. IEEE standards are published as *IEEE serial-no* (*e.g.*, IEEE 908). The IEEE works closely with ANSI (American National Standards Institute) to develop communications and information processing standards with the ultimate object of advancing theory, creativity and product quality in any field related to electrical engineering.
4. The **American National Standards Institute (ANSI)** is a completely non profit private organization comprised of manufactures and users of data processing equipment and services. It is the official standards agency for the United States and is the US voting representative for the ISO.

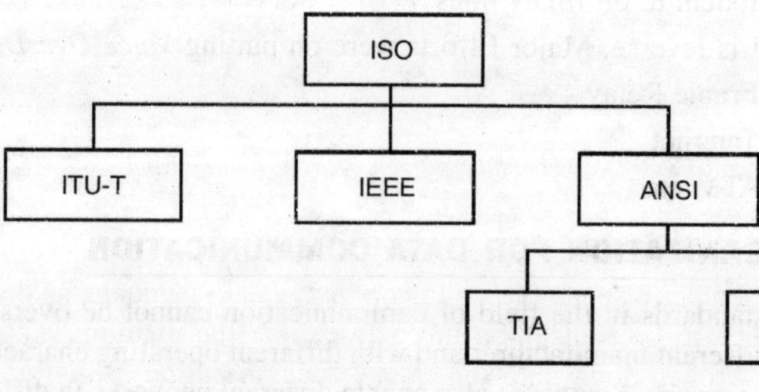

Fig. 3.1. *Standard Organization for data communication*

5. The **Electronic Industries Association** (EIA) is a US trade association best known for its EIA-232 standard. EIA is responsible for developing the RS (recommended standards) series of standards for data and telecommunications. EIA activities include standards development, increasing public awareness and lobbying.
6. The **Telecommunication Industry Association (TIA)** is the major trade association in the information technology and communications industry. The TIA represents manufactures of communications and information technology products and service providers for the global marketplace through its core competencies.
7. The **European Computer Manufacturers Association** (ECMA) is a standards organization involved in the area of computer engineering and related technologies. ECMA directly cooperates with ISO and CCITT.

In addition to these organizations, and because of their global market influence, large vendors occasionally succeed in establishing their products as *de facto* standards.

3.4. DATA COMMUNICATION CIRCUITS

The basic purpose of a **data communications circuit** is to transfer digital information from one place to another using electronic circuit.

It provides a transmission path between source and destination. Thus data communications can be summarized as the transmission, reception and processing of digital information.

A **station** is simply an endpoint where subscribers get access to the circuit; sometimes it may be called a **node**, which is the location of computers, computer terminals and other digital computing equipment.

We have as many types of data communication circuits as there are types of data communication equipments. These data communication circuits employing electronic communications equipment and facilities to interconnect digital computer equipment.

Communication facilities are the path of transmission from one station to another station either physically (using transmission cables) or wireless (using microwave, IR etc) transmission media.

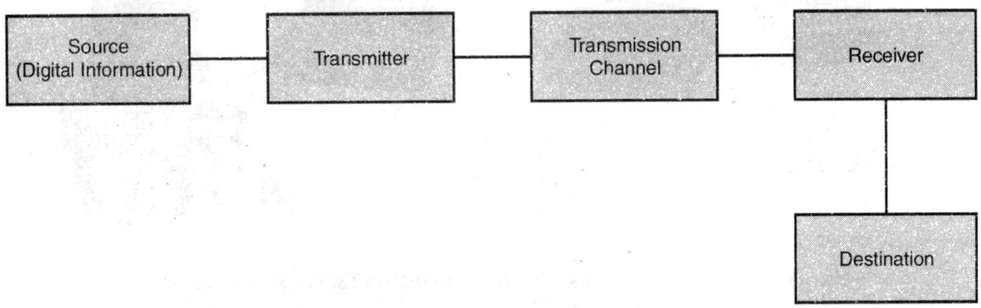

Fig. 3.2. *Data communication circuit block diagram.*

Data Communication Model

Often the source is not able to speak the same communication language as the medium and the data communications needs to be translated into a form that the medium can understand. Data communication is further broken down into the ***Data Communication Model*** which consists of 5 parts as shown in figure :

1. DTE

The Data Terminal Equipment (DTE) is the source and receiver's equipment which is generating the data stream.

Examples of DTE devices are: PCs, terminals, end user devices.

2. DCE

The interface between the source and the medium, and the medium and the destination or receiver is called the Data Communication Equipment (DCE) and is a physical piece of equipment. Its job is to provide the interface from the DTE device to the medium. Examples of DCE devices are modems, channel banks, and routers.

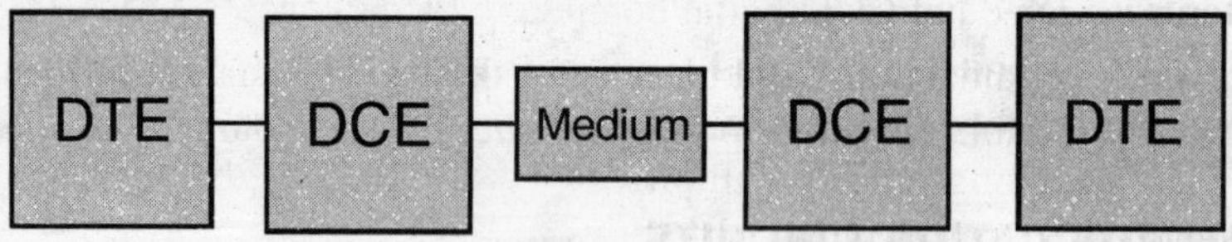

Fig. 3.3. *Data Communication Model*

An example of this would be your PC dialing into Internet Service Provider (ISP):

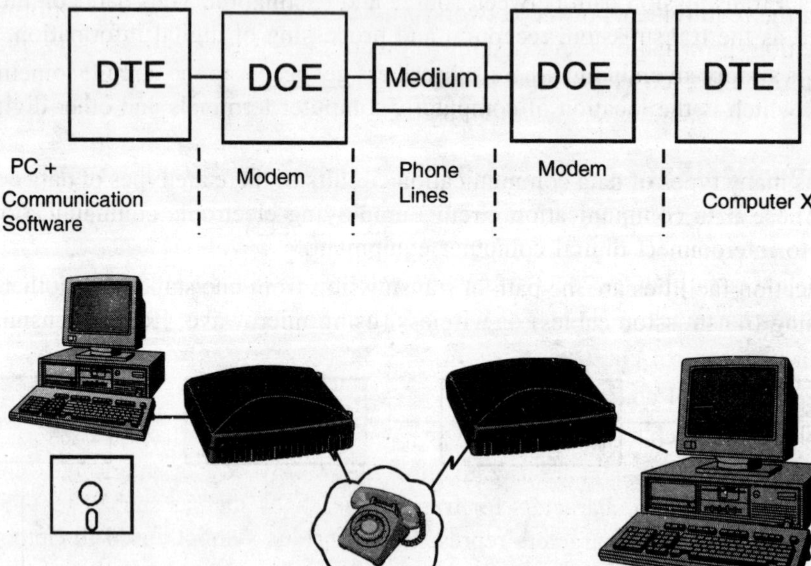

Fig. 3.4. *An Example of PC Dialing into internet service provider (ISP).*

The output of the DTE device (the PC) is an RS232 signal which is a digital communication consisting of logic ones and zeros.

The medium (the phone line) is made for carrying analog communications consisting of voice transmission.

The output of the PC and the requirements of the phone line are mutually incompatible. The phone line doesn't understand digital signals and the PC doesn't understand analog transmission.

The DCE device (the modem) provides the conversion (the interface) so that digital data can communicate across the analog medium.

3.5. DATA COMMUNICATION CODES

We have already discussed about the term Data *i.e.*, defined as information that is stored in digital format, usually in the form of binary digits. Data may be numeric, alphabetic or symbolic in nature. Information can be encoded in several formats, such as binary coded digits(BCD), binary-weighted numbers, alphanumeric characters, digitized analog signals (voice or video) and many more. As we pressed the key on a standard keyboard, the symbol (letter, digit etc) that the key represents is converted to a unique binary code. In simple term keyboard is an encoder and each key has a different codes. There are various types of keyboards such as ASCII keyboard produces ASCII codes, hex keyboards produces hex codes.

Data are communicated between digital computers as sequences of bits. To provide meaning to a sequence of bits, the bits are grouped to form a data character and an encoding scheme, or translation table, is provided to allow a computer system to translate each group of bits into a character. The ideal encoding scheme will provide a unique code for every possible character to be communicated and stored in the computer, but this requires that each group have a sufficient number of bits for each data character.

Information sources generate signals which are not in a form suitable for transmission over a data communication networks. Therefore, it is must that information be formatted (encoded) to comply with the requirements of the network prior to transmission and it is unformatted (decoded) at the destination. Information converted into a binary code is generally known as data transmission code. Data transmission codes facilitate people to communicate people and machines to communicate with machines over data communication networks.

Data Communication Character Codes

Data communication codes are normally used to represent characters and symbols, such as digits, letters and punctuation marks. So these types of data communications codes are called symbol codes, character codes, character sets or character languages. Mainly, there are three types of characters used with character codes:

1. Data-link control characters
2. Graphic control characters
3. Alphanumeric characters

Data-link control characters are used to allow the orderly flow of data from source to destination it includes special function characters for framing messages such as start of text (STX) and end of text (ETX). Alphanumeric characters represent the various symbols used for letters, numbers and punctuation and graphic control characters involves the syntax or presentation of the data at receiver includes special characters for displaying text on printers and monitors.

Most commonly used types of character sets are as follows:

1. Morse Code
2. Baudot Code
3. ASCII Code
4. Extended Binary Decimal Interchange Code

1. Morse Code

Morse code is a character encoding for transmitting telegraphic information, using standardized sequences of short and long elements to represent the letters, numerals, punctuation and special characters of a given message. The short and long elements can be formed by sounds, marks or pulses, in on off keying and are commonly known as "dots" and "dashes" or "dits" and "dahs". The Morse code is an example of a variable-length source code, where the code length is varying from character to character.

Originally created for Samuel F. B. Morse's electric telegraph in the early 1840s, Morse code was first used in 1844 when Morse established the words first digital communication link between Washington, D.C. and Baltimore, Maryland, using the telegraph system. Morse code is designed to be easily read by humans without a computer, making it appropriate for sending automated digital data in voice channels, as well as making it ideal for emergency signaling, such as by way of improvised energy sources that can be easily "keyed" by supplying and removing electric power,for example, by flipping a switch or turning a flashlight on and off.

Fig. 3.5.

Morse code was extensively used for early radio communication beginning in the 1890s. For the first half of the twentieth century, the majority of high-speed international communication was conducted in Morse code, using telegraph lines, undersea cables, and radio circuits. However, the variable length of the Morse characters made it hard to adapt to automated circuits, so for most electronic communication it has been replaced by more machinable formats, such as Baudot code and ASCII. It actually requires the reasoning ability of a human brain to decode Morse code. Morse code also has an inadequate selection of graphic and data link control characters to facilitate the transmission and presentation of the data used normally in computer applications.

Fig. 3.6.

A typical "straight key" shown in Fig. 3.6, this U.S. model, known as the J-38, was manufactured in huge quantities during World War II, and remains in widespread use today. In a straight key, the signal is "on" when the knob is pressed, and "off" when it is released. Length and timing of the *dits* and *dahs* are entirely controlled by the operator.

The Morse code was developed so that operators could translate the indentations marked on the paper tape into text messages. In his earliest code, Morse had planned to only transmit numerals, and use a dictionary to look up each word according to the number which had been sent. However, the code was soon expanded to include letters and special characters, so it could be used more generally. The shorter marks were called "dots", and the longer ones "dashes", and the letters most commonly used in the English language were assigned the shortest sequences.

Morse code was an integral part of international aviation. Commercial and military pilots were required to be familiar with it, both for use with early communications systems and identification of navigational beacons which transmitted continuous three letter ID's in Morse code. As late as the

1990s, aeronautical charts listed the three letter ID of each airport in Morse and sectionals still show the Morse signals for Vortac and NDB used for in flight navigation.

Morse code has been in use for more than 160 years — longer than any other electronic encoding system. What is called Morse code today is actually somewhat different from what was originally developed by Vail and Morse. The Modern International Morse code, or *continental code*, was created by Friedrich Clemens Gerke in 1848 and initially used for telegraphy between Hamburg and Cuxhaven in Germany. After some minor changes, in 1865 it was standardized at the International Telegraphy congress in Paris (1865), and later made the norm by the International Telecommunication Union (ITU) as International Morse code. Morse's original code specification, largely limited to use in the United States, became known as **American Morse code** or "railroad code." American Morse is now very rarely used except in historical re-enactments.

Table 3.1. Morse Code

Character	Code	Character	Code	Character	Code	Character	Code	Character	Code	Character	Code
A (info)	·—	J (info)	·— — —	S (info)	···	1 (info)	·— — — —	Period [.]	·— ·— ·—	Colon [:]	— — —···
B (info)	—···	K (info)	—·—	T (info)	—	2 (info)	··— — —	Comma [,]	— — ··— —	Semi colon [;]	—·— ·—·
C (info)	—·—·	L (info)	·—··	U (info)	··—	3 (info)	···— —	Question mark [?]	··— —··	Double dash [=]	—··· —
D (info)	—··	M (info)	— —	V (info)	···—	4 (info)	····—	Apostrophe [']	·— — — — — ·	Plus [+]	·— ·—·
E (info)	·	N (info)	—·	W (info)	·— —	5 (info)	·····	Exclamation mark [!]	—·— ·— —	Hyphen, Minus [-]	—·· ··—
F (info)	··—·	O (info)	— — —	X (info)	—··—	6 (info)	—····	Slash [/], Fraction bar	—·· —·	Underscore [_]	··— —· —
G (info)	— —·	P (info)	·— —·	Y (info)	—·— —	7 (info)	— —···	Parenthesis open [(]	—· — —·	Quotation mark ["]	·—· ·—·
H (info)	····	Q (info)	— — ·—	Z (info)	— — ··	8 (info)	— — —··	Parenthesis closed [)]	—·— —·—	Dollar sign [$]	···— ··—
I (info)	··	R (info)	·—·	0 (info)	— — — — —	9 (info)	— — — — ·	Ampersand [&], Wait	·— ···	At sign [@]·	— — ·—·

1. Baudot Code

A code used early in the data communications industry is the Baudot (Pronounced "baw-doh") code or sometimes called the Telex code. The baudot code was used extensively in telegraph systems. It is a five bit code invented by the frenchman Emile Baudot in 1870. It has fixed length of character code and developed for machine rather than for people. Baudot code was used primarily for low speed teletype equipment, such as Telex system and radio Teletype.

Baudot uses five bits per character, thus allowing up to 2^5, or 32 distinct characters only, which is insufficient to represent the 26 letters of the alphabet, 10 digits and many punctuation and control characters. As a technique used to extend this limitation, the code uses up-shift and down-shift modes as is used on a typewriter. In the Baudot code, each five bits transmitted must be interpreted according to whether they are up-shifted (figures) or down-shifted (letters). Each character is preceded by a start bit, and followed by a stop bit. It is an asynchronous code.

For example, the bit pattern 11111 represents up-shift and the bit pattern 11011 represents down-shift characters. All characters transmitted after the sequence 11111 but before the shifted sequence 11011 are treated as up-shift characters. All characters transmitted after the sequence 11011 are treated as down-shift characters until the pattern 11111 is recognized. The complete BAUDOT code (modified for this problem) is shown in the table at the end of this problem.

Input

The input consists of two parts. The first part is the Baudot character set: line one contains the 32 down-shift characters and line two contains the 32 up-shift characters. Spaces are inserted for the shift characters. The remainder of the input file consists of one or more messages encoded using the Baudot code. Each message will be on a line in the input file. Each line will consist of 1's and 0's, with no characters between the bits. There can be up to 80 bits per message.

The input file will be terminated by end-of-file. The initial state of each message should be assumed to be in the down-shift state.

Output

The output should consist of one line of text for each message. The output should contain the character representation, as translated using BAUDOT, of each of the messages.

Table 3.2. The BAUDOT code

Bit Pattern	Down-Shift Character	Up-Shift Character	Decimal
00000	<	>	0
00001	T	5	1
00010	*	@	2
00011	O	9	3
00100	Space	Space	4
00101	H	%	5
00110	N	,	6
00111	M	.	7
01000	=	+	8
01001	L	)	9
01010	R	4	10

Bit Pattern	Down-Shift Character	Up-Shift Character	Decimal
01011	G	&	11
01100	I	8	12
01101	P	0	13
01110	C	:	14
01111	V	;	15
10000	E	3	16
10001	Z	"	17
10010	D	$	18
10011	B	?	19
10100	S	#	20
10101	Y	6	21
10110	F	!	22
10111	X	/	23
11000	A	-	24
11001	W	2	25
11010	J	'	26
11011	Shift Down	Shift Down	27
11100	U	7	28
11101	Q	1	29
11110	K	(	30
11111	Shift Up	Shift Up	31

2. ASCII Code

American Standard Code for Information Interchange (**ASCII**), pronounced **æski** is a character encoding based on the English alphabet. ASCII codes represent text in computers, communications equipment, and other devices that work with text.

ASCII includes definitions for 128 characters: 33 are non-printing, mostly obsolete control characters that affect how text is processed; 94 are printable characters (excluding the space. The American National Standards Institute (then called the American Standards Association or ASA, and later the United States of America Standards Institute or USASI developed ASCII based on earlier teleprinter encoding systems. The standards committee decided that ASCII required at least a seven-bit code.

ASCII itself first entered commercial use in 1963 as a seven-bit teleprinter code for American Telephone & Telegraph's TWX (Teletype Wide-area eXchange) network. TWX originally used the earlier five-bit Baudot code, which was also used by the competing Telex teleprinter system.

ASCII has been incorporated into the Unicode character set as the first 128 symbols, so the ASCII characters have the same numeric codes in both sets. This allows UTF-8 to be backward compatible with ASCII, a significant advantage.

Binary	Oct	Dec	Hex	Abbr	PR	CS	CEC	Description
000 0000	000	0	00	NUL	?	^@	\0	Null character
000 0001	001	1	01	SOH	?	^A		Start of Header
000 0010	002	2	02	STX	?	^B		Start of Text
000 0011	003	3	03	ETX	?	^C		End of Text
000 0100	004	4	04	EOT	?	^D		End of Transmission
000 0101	005	5	05	ENQ	?	^E		Enquiry
000 0110	006	6	06	ACK	?	^F		Acknowledgment
000 0111	007	7	07	BEL	?	^G	\a	Bell
000 1000	010	8	08	BS	?	^H	\b	Backspace
000 1001	011	9	09	HT	?	^I	\t	Horizontal Tab
000 1010	012	10	0A	LF	?	^J	\n	Line feed
000 1011	013	11	0B	VT	?	^K	\v	Vertical Tab
000 1100	014	12	0C	FF	?	^L	\f	Form feed
000 1101	015	13	0D	CR	?	^M	\r	Carriage return
000 1110	016	14	0E	SO	?	^N		Shift Out
000 1111	017	15	0F	SI	?	^O		Shift In
001 0000	020	16	10	DLE	?	^P		Data Link Escape
001 0001	021	17	11	DC1	?	^Q		Device Control 1 (oft. XON)
001 0010	022	18	12	DC2	?	^R		Device Control 2
001 0011	023	19	13	DC3	?	^S		Device Control 3 (oft. XOFF)
001 0100	024	20	14	DC4	?	^T		Device Control 4
001 0101	025	21	15	NAK	?	^U		Negative Acknowledgement
001 0110	026	22	16	SYN	?	^V		Synchronous Idle
001 0111	027	23	17	ETB	?	^W		End of Trans. Block
001 1000	030	24	18	CAN	?	^X		Cancel
001 1001	031	25	19	EM	?	^Y		End of Medium
001 1010	032	26	1A	SUB	?	^Z		Substitute
001 1011	033	27	1B	ESC	?	^[	\e	Escape
001 1100	034	28	1C	FS	?	^\		File Separator
001 1101	035	29	1D	GS	?	^]		Group Separator
001 1110	036	30	1E	RS	?	^^		Record Separator
001 1111	037	31	1F	US	?	^_		Unit Separator
111 1111	177	127	7F	DEL	?	^?		Delete

Some Features of ASCII

ASCII has some interesting features which are as follows:

1. ASCII has 7 bit code that designed to provide code for 128 symbols.
2. The first code (0) is the null character that means lack of any character.

3. The space character is at position 32 and it is a printable character.
4. The uppercase letters start from 65 (A) and the lowercase letters start from 97.
5. The uppercase and lowercase letters differs by only one bit in the 7-bit code.
6. The uppercase letters are not immediately followed by lowercase letters. We have some punctuation characters in between.

3. Extended Binary Coded Decimal Interchange Code (EBCDIC)

Extended Binary Coded Decimal Interchange Code (**EBCDIC**) is an 8-bit character encoding used on International Business Machines Corporations (IBM) mainframe operating systems such as z/OS, OS/390, VM and VSE, as well as IBM minicomputer operating systems such as OS/400 and i5/OS. With the 8 bits, 256 codes are possible, although only the 139 of the 256 codes are assigned character.

EBCDIC was devised in 1963 and 1964 by IBM and was announced with the release of the IBM System/360 line of mainframe computers. It was created to extend the Binary-Coded Decimal encoding that existed at the time. It is an 8-bit character encoding, in contrast to, and developed separately from, the 7-bit ASCII encoding scheme.

One of the main advantages of this code is that it provides error detection in addition to the parity technique and also provides some control functions. It is to be noted that systems using these different codes (ASCII and EBCDIC) will not be able to interact directly, as the codes have different meanings for the last bit of the code. However, with the help of an ASCII / EBCDIC translator or converter, the conversion between these codes is done inherently by the system and is considered to be part of communication protocols.

3.6. TRANSMISSION ERRORS

No physical medium can guarantee error-free transmission of data. Errors occur because of interference from the environment surrounding the transmission medium or because of artifacts introduced by the equipment or the medium itself. In case of copper wires, for example, thermal noise, electrical noise, and impulse noise are some of the most common causes. The networks must be able to transfer data from one device to another device with the complete accuracy. Many factors can alter or wipe out one or more bits of given data.

Transmission errors are caused by:

1. Thermal noise (Shannon)
2. Impulse noise (*e.g.,* arcing relays)
3. Signal distortion during transmission (attenuation)
4. Crosstalk
5. Voice amplitude signal compression (companding)
6. Quantization noise (PCM)
7. Jitter (variations in signal timings)
8. Receiver and transmitter out of synch.

Reliable systems must have a mechanism for detecting and correcting such errors. *The most important role of the data link layer is to provide error-free transmission.* To do this, it uses an error checking scheme to detect errors, and overcomes them by retransmitting corrupted frames. Many different error detection and correction methods have been devised.

DO YOU KNOW

Errors occur because of interference from the environment surrounding the transmission medium.

3.6.1. Types of Errors

There are two types of errors occur in data communication, namely:

1. Single-Bit Error
2. Burst Error

In case of **single-bit error** only one bit of the data is corrupted and rest all the bits are received correctly. That means only one bit from the data unit is changed either from 1 to 0 or 0 to 1.

For example, suppose the data frame 1111 is sent over the transmission line and at the receiver end the data frame is received as 1**0**11. The second bit is changed from 1 to 0.

Data sent 1111

Data received 1**0**11

↑

Bit 1 is changed to 0

Single-bit error most likely to occurs when we send the data using parallel transmission mode. Suppose, if seven wires are connected to send all the 7 bits of a data unit at the same time and one of the transmission wire is noisy, one bit can be corrupted in each data unit. It is rare that we have single bit error in serial transmission mode. Imagine a sender sends data at 1Mbps, that means each bit lasts only 1/1000000s *i.e.,* 1ms. Now for a single bit error to occur, the noise duration must be 1ms, but practically the noise duration is much longer than this.

In case of **burst error**, multiples bits are corrupted and changed. For example, suppose data sent is 1111, received 0001. Three bits are changed from 1 to 0.

Example 1:

Data sent 1111

Data received **000**1

↑

3 bits are changed from 1 to 0

Let us discuss one more example of burst error, in this case data is sent 0100010001, but received as 0101110101. Three bits are changed from 0 to 1. From the example, a burst length also can be seen, *i.e.,* 5 bits in this case.

Example 2:

The **burst length** is measured from the first corrupted bit to the last corrupted bit. Some bits in between may be received correctly.

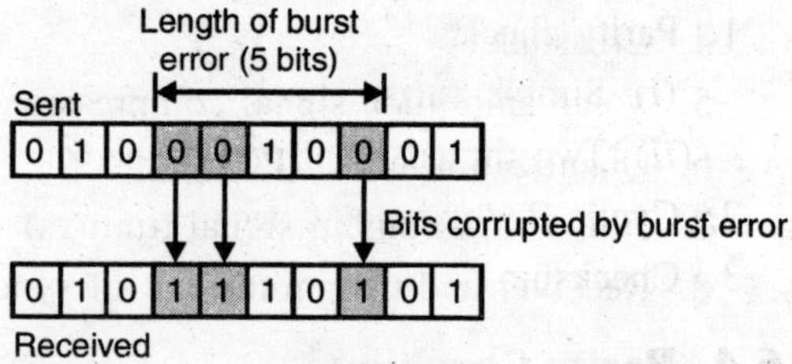

Fig. 3.7.

As we know that noise duration is more than the duration of one bit that means when noise affects data, it affects a set of bits. Serial transmission is most likely to be affected by the noise. It is obvious that the number of bits affected depends on the duration of noise and the data rate. Suppose we have data rate 1 kbps and a noise duration is 1/100s, than 10 bits are affected.

3.6.2. Error Detection

Error detection is the first stage of error correction. Error detection is simpler and easy than error correction. There are some mechanism used to detect the error in data communication based a concept of **redundancy**. In this concept a shorter group of bits may be appended to the end of each data unit before the transmission and they are discarded as soon as accuracy of the transmission has been confirmed.

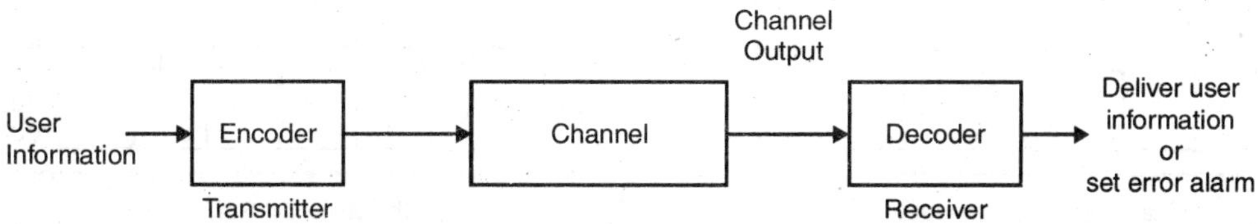

Fig. 3.8. *General error detection system.*

Figure 3.8 shows a general error detection system. In this figure, it is shown that user information is fed to encoder which will add some extra bits to the information and at the receiving end decoder, check the pattern.

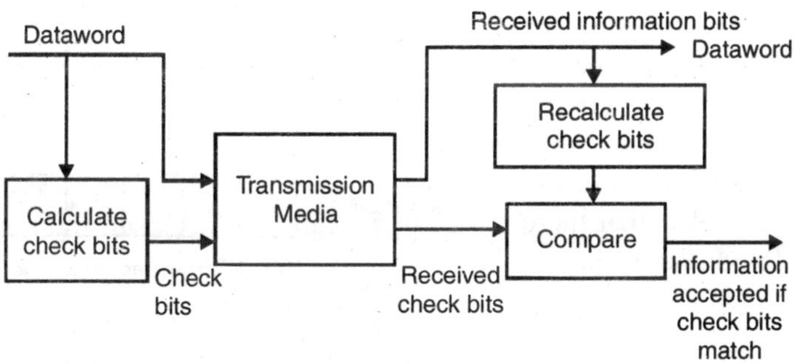

Fig. 3.9. *Detection system using check bits.*

Figure 3.9 shows the detection system using check bits. In this system check bits are calculated and appended to the dataword before transmitting to the receiver. At the receiver side, check bits are recalculated and with the received check bits. If check bits are same information is accepted otherwise it is discarded.

3.6.3. Types of Error Detection Techniques

Types of error detection techniques using Redundancy concept are as follows:

1. Parity check
   (*i*) Simple parity check
   (*ii*) Two dimensional parity check
2. Cyclic Redundancy Check
3. Checksum

3.6.4. Parity Checking

Parity checking is a simple error detection method used with character oriented protocols. There are two types of parity checking techniques, such as **Even parity** and **Odd parity**. One bit in every character bit sequence is reserved for parity.

This bit is set by the transmitter to 0 or 1 so that the total number of 1 bits in the character is always even (in case of even parity checking) or always odd (in case of odd parity checking). The receiver checks the parity bit of each character to see if it is as expected. Figure 3.10 illustrates the method for an even and odd parity checking schemes.

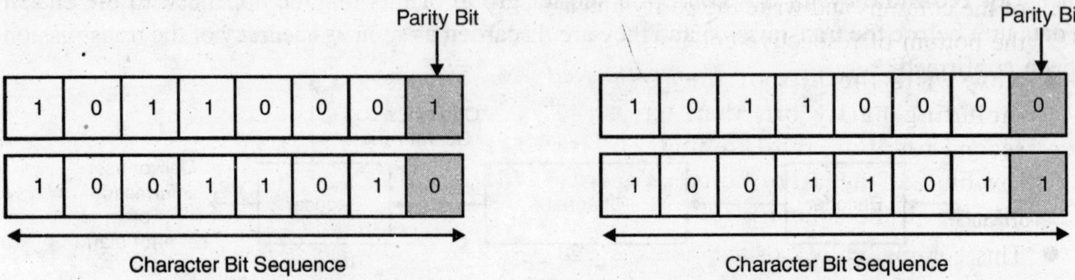

(a) Even parity error checking *(b) Odd parity error checking*

Fig. 3.10.

Simple parity checking technique mainly used to detect the single error only. However, it can also detect the multiples errors, when the number of errors is odd. In case when the number of errors is even, this technique fails, refer Fig. 3.11, where two bits are changed. The receiver takes that there is no errors because odd parity is still maintained.

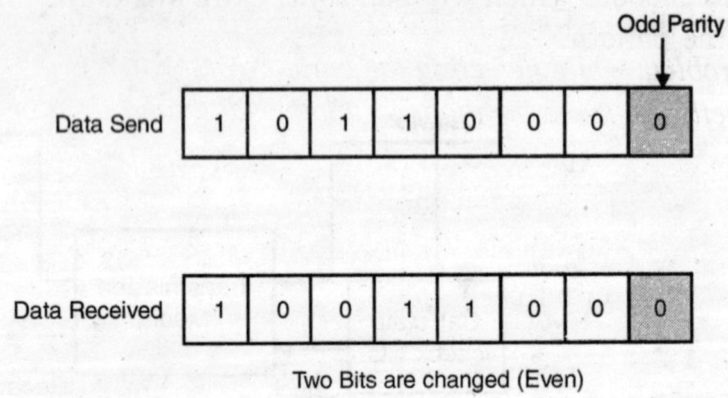

Fig. 3.11. *Case of multiple errors such as 2 errors.*

DO YOU KNOW

There are two types of parity checking techniques, such as **Even parity** and **Odd parity**.

Two Dimensional Parity Check

As we have seen that simple parity check technique is not enough when numbers of errors are even. Another approach is the **Two-Dimensional Parity Check**; in this we have many data units and they are organized in rows and columns. In this technique we calculate the parity bit for each data unit. For example, we have four data units shown in four rows and five (suppose 5 bits in a data unit) columns, in Fig. 3.12. We then calculate the even parity bit for each column and write a new row of five bits, they are the parity bits for the whole block. It is to be noted that first parity bit in the fifth row is determined based on all first bits and second parity bit is based on all second bits and so on. Now we append the parity bits to the original data and then send to the receiver through channel. After appending the parity bits the data word is referred as codeword.

Example: Suppose we have 4 data units; 10010, 01000, 10010, 11011.

Steps:

- Arrange the data units into row and columns as shown below.
- Calculate the even parity bit for first row and write at the extreme right.
- Now calculate the even parity bits for rest of the rows.

- Calculate the even parity bit for first column and write at the bottom of that column.
- Likewise calculate the even parity for rest of the columns and write the parity bits at the bottom of respective columns.
- Now write the bits of the first row including parity bit, than write the second row bits, third row bits, fourth row bits and the last row bits (*i.e.*, parity bits for all the columns).
- This complete block of data units is sent to the receiver.

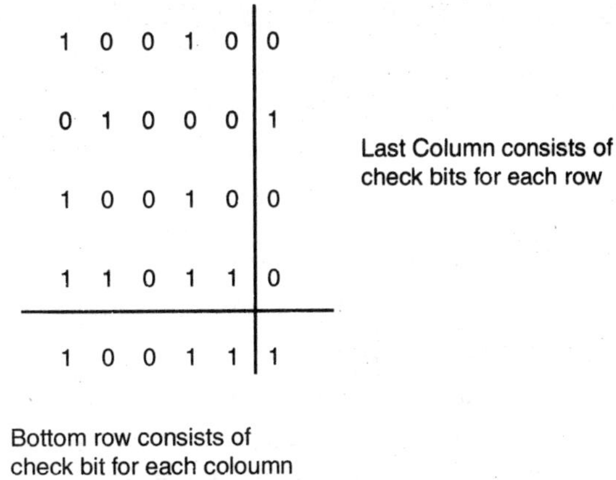

Fig. 3.12. *Two dimensional parity check.*

Data and Parity bits to sent —10010 **0** 01000 **1** 10010 **0** 11011 **0** **100111**

Two dimensional parity check increases the possibility of detecting burst errors. Still *there is one problem while detecting the burst errors, when two bits in one data unit are corrupted and two bits in exactly the same positions in another data unit are damaged, the checker will not detect an error.*

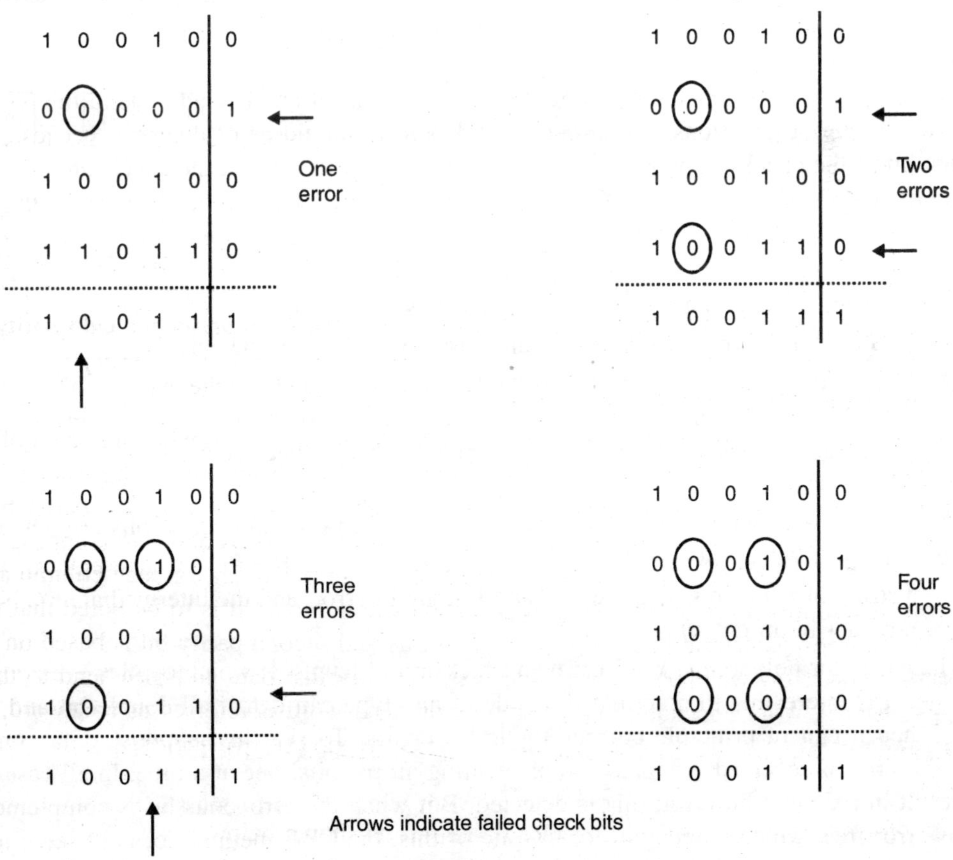

Fig. 3.13.

3.6.5. Cyclic Redundancy Check (CRC)

The CRC is a very powerful but easily implemented technique to obtain data reliability. The **CRC** technique is used to protect blocks of data called Frames. Using this technique, the transmitter appends an extra n- bit sequence to every frame called **Frame Check Sequence (FCS)**. The FCS holds redundant information about the frame that helps the transmitter detect errors in the frame. The CRC is one of the most used techniques for error detection in data communications. The technique gained its popularity because it combines three advantages:

- Extreme error detection capabilities.
- Little overhead.
- Ease of implementation.

Cyclic Redundancy Check (CRC) is a more sophisticated method used with bit-oriented protocols. It is also called the **polynomial code** — so named because it treats the single bits in a bit sequence as the coefficients of an imaginary polynomial.

For example, the bit sequence 101001 represents the polynomial:

$$\mathbf{1}x^5 + \mathbf{0}x^4 + \mathbf{1}x^3 + \mathbf{0}x^2 + \mathbf{0}x^1 + \mathbf{1}x^0 = x^5 + x^3 + x^0$$

This polynomial is said to be of *order* 5, because its highest-order term is x^5.

CRC relies on the use of a **generator polynomial**, the choice of which is fairly arbitrary, except that the high and low order bits of its bit sequence must be 1 (*e.g.*, 10111 is acceptable, but 0101 and 10110 are not) and that the transmitter and the receiver must use exactly the same generator polynomial.

Choosing a "good" generator polynomial is something of an art, and beyond the scope of this text. Two simple observations: For an *r*-bit checksum, *G* should be of degree *r*, because otherwise the first bit of the checksum would always be 0, which wastes a bit of the checksum. Similarly, the last coefficient should be 1 (that is, *G* should not be divisible by *x*), because otherwise the last bit of the checksum would always be 0 (because $Mx^r = QG + R$ if *G* is divisible by *x*, then *R* must be also).

The following facts about generator polynomials are proved :

- If *G* contains two or more terms, all single-bit errors are detected.
- If *G* is not divisible by *x* (that is, if the last term is 1), and *e* is the least positive integer such that *G* evenly divides $x^e + 1$,then all double errors that are within a frame of *e* bits are detected. A particularly good polynomial in this respect is $x^{15} + x^{14} + 1$ for which $e =$ 32767.
- If $x + 1$ is a factor of *G*; all errors consisting of an odd number of bits are detected.
- An *r*-bit CRC checksum detects all burst errors of length = $\leq r$. (A burst error of length *r* is a string of *r* bits in which the first and last are in error, and the intermediate *r*-2 bits may or may not be in error.)

The generator polynomial $x + 1$ create a checksum of length 1, which applies even parity to the message. It is interesting to note that if a code of any type can detect all double-bit and single-bit errors, then it can in principle correct single-bit errors. To see this, suppose data containing a single-bit error is received. Imagine complementing all the bits, one at a time. In all cases but one, this result in a double-bit error, this is detected. But when the erroneous bit is complemented, the data is error-free, which is recognized. In spite of this, the CRC method does not seem to be used for single-bit error correction. Instead, the sender is requested to repeat the whole transmission if any error is detected.

Generator Polynomials used by Some Common CRC Standards

Table 3.3 shows the generator polynomials used by some common CRC standards.

The "Hex" column shows the hexadecimal representation of the generator polynomial; the most significant bit is omitted, as it is always 1.

Table 3.3. Generator polynomials used by some common CRC standards.

Common Name	r	Generator Polynomial	Generator Hex
CRC-12	12	$x^{12}+x^{11}+x^{3}+x^{2}+x+1$	80F
CRC-16	16	$x^{16}+x^{15}+x^{2}+1$	8005
CRC-CCITT	16	$x^{16}+x^{12}+x^{5}+1$	1021
CRC-32	32	$x^{32}+x^{26}+x^{23}+x^{22}+x^{16}+x^{12}+x^{11}+x^{10}$ $+x^{8}+x^{7}+x^{5}+x^{4}+x^{2}+x+1$	04C11D87

Operation of CRC

Given a message (bit sequence) m represented by the polynomial $m(x)$, and a generator polynomial $g(x)$ of order k, m is appended with exactly k zero bits. The result is equivalent to the polynomial $p(x) = xkm(x)$. $p(x)$ is divided by $g(x)$ and the remainder is added to $p(x)$ to produce the final polynomial $q(x)$:

$$q(x) = p(x) + (p(x) \textbf{ rem } g(x))$$

It follows that $q(x)$ is divisible by $g(x)$. The transmitter calculates q using m and g and then transmits q instead of m. If any of the bits in q are changed due to transmission errors, $q(x)$ has a very high probability of no longer being divisible by $g(x)$. This is checked by the receiver to detect transmission errors. If there are no errors, the receiver extracts m from q (by discarding the k low-order bits) and this will be identical to the original message.

The polynomial calculation procedures described above serve only as a conceptual model. In practice, equivalent binary operations are used instead. The binary operations are surprisingly simple, to the point that the whole process is usually implemented in hardware. Calculating q involves the following steps:

1. Given a bit sequence m and a generator polynomial bit sequence g of order k, append m with exactly k zero bits to produce bit sequence p. For example:

$$m = 100101110101$$
$$g = 10111 \ (k = 4)$$
$$p = 1001011101010000$$

2. Divide p by g, using modulo-2 arithmetic, to produce a remainder r. Modulo-2 arithmetic operations are similar to binary operations, except that there are no borrows or carries. Addition and subtraction in modulo 2 are identical to the binary *exclusive or* operation. Continuing with our example:

$$r = 1110$$

3. Add r to p to produce q:

$$q = 1001011101010000 + 1110$$
$$= 1001011101011110$$

The effectiveness of CRC depends to a large extent on the choice of the generator polynomial. In practice, polynomials of order 16 are widely used. Hence 16 bits (two octets3) are needed for storing the remainder, which appears as the checksum field in the frame. Such polynomials are more than 99% effective in catching a variety of errors.

In brief CRC can detect following types of errors:

- CRC can detect all burst error that affects an odd numbers of bits.
- CRC can detect, all burst error of length less than or equal to the degree of the polynomial.
- CRC can detect all single bit error and double bits errors.

Important Steps:

1. Always divisor will be given in the examination problem.
2. A string of zeros to be added to the data before staring the division. The number of zeros must be one less than divisor. Refer example-1, where number of bits in the divisor is five and number of zeros added in the data are 4 (5-1=4).
3. While dividing, divisor is subtracted from corresponding bits of dividend bits.
4. Left most bit is always to be discarded after the subtraction from the divisor.
5. If next to the left most bit is zero than instead of subtracting divisor subtract a string of zeros and the value of quotient is zero.
6. Divisor should always begin with the left most bit 1.
7. If the left most bit is zero in the remainder is zero in intermediate steps of division, it is discarded. But if the again next bit is also zero then it is retained and subtract from the string of zeros by pulling a higher bit from dividend to remainder.
8. The remainder, which we get is the CRC parity bits and to appended at the end of data. For the valid remainder CRC, the number of bits in CRC should be one less than divisor.

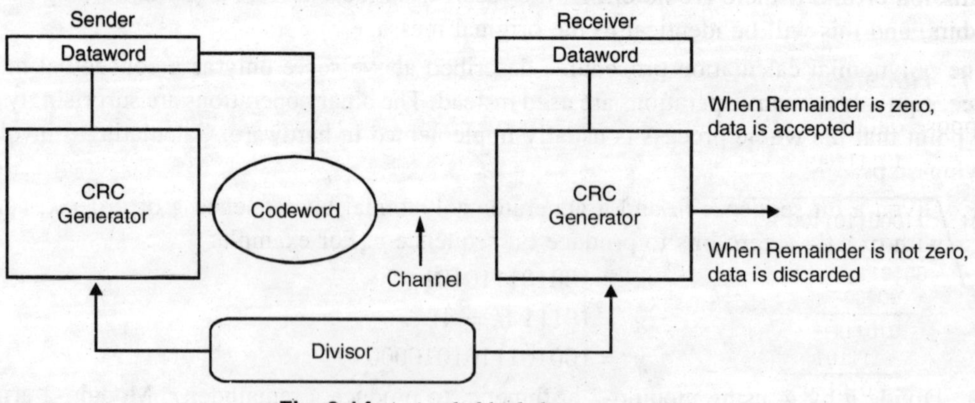

Fig. 3.14. *A simple block diagram of CRC*

9. The data is to be transmitted is the data strings followed by the CRC parity bits (remainder).
10. Now at the receiver end, data string followed by CRC parity is to be divided by the same divisor. If the remainder comes all zeros than data is correct and accepted as shown in example-1. In case remainder is not zeros, that means data received is not correct and discarded.
11. Figure 3.14, shows a simple block diagram of CRC error detection method. In this at the sender end, CRC generator will do the modulo-2 division and remainder is appended to the dataword that is then known as codeword. Codeword is send through the channel to the receiver. At the

receiver end, CRC checker will perform the operation of division. Codeword is divided by the same divisor and if the remainder is zero, data is accepted and if the remainder is not zero than data is incorrect and discarded.

Example 1: Let a data - 11000110 is to send and the divisor- 11101, find the CRC (redundancy bits).

```
                          Quotient

                          10111100
                        ____________
        Divisor - 11101 ) 110001 0000  <------ Nos. of zeros added in data
                          11101                is one less than divisor
                          -----
                          001011
                            ------------------------ When left most bit is
                        /   00000                    zero, divisor will be 00000
Discarded left most bit     ------
                             010111
                              11101
                             ------
                              010100
                               11101
                              ------
                               010010
                                11101
                                011110
                                 11101
                                 000110
                                  00000
                                 ------
                                  001100
                                   00000
                                  ------
                                    1100 | Remainder
```

Now, CRC parity bits, *i.e.,* remainder is 1100. Before sending these bits are to be appended to the data. The codeword (data along with the redundancy bits) will become 11000110**1100** and to be transmitted.

At the receiver end, the received data is to be divided by the same divisor and verify the remainder.

Case-1: Received correct data

Suppose data received is correct then remainder will be all zeros as shown below.

```
          10111100
       ________________
11101 / 110001101100
        11101
        -----
        001011
          00000
        ------
          010111
            11101
          ------
            010100
              11101
            ------
              010011
                11101
              ------
                011101
                  11101
                ------
                  00000 0
                    00000
                  ------
                    00000 0
                      00000
                    ------
                        0000        Remainder
                                    (All bits are zeros, i.e. means data received is correct)
```

Case-2: Received incorrect data

Suppose data received is incorrect then remainder will not be all zeros as shown below.

```
          10111101
11101 / 110001111100
        11101
        001011
          00000
          010111
            11101
            010101
              11101
              010001
                11101
                011001
                  11101
                  00100 0
                    00000
                    01000 0
                      11101
                      01101     Remainder
                                (All bits are not zeros, i.e. means data received is incorrect)
```

Example 2: Let a data 1 0 1 0 0 1 1 0 1 0 0 0 0 0 is to send and the divisor 1 0 0 1 1 1, find the CRC (redundancy bits).

```
                    1 0 1 1 1 1 1 0          Discard quotient
1 0 0 1 1 1 | 1 0 1 0 0 1 1 0 1 0 0 0 0 0
              1 0 0 1 1 1
                1 1 1 0 1 0
                1 0 0 1 1 1
                  1 1 1 0 1 1
                  1 0 0 1 1 1
                    1 1 1 0 0 0
                    1 0 0 1 1 1
                      1 1 1 1 1 0
                      1 0 0 1 1 1
                        1 1 0 0 1 0
                        1 0 0 1 1 1
                          1 0 1 0 1 0
                          1 0 0 1 1 1
                              1 1 0 1 0   <== Remainder
```

The codeword that is to be transmitted will became 1 0 1 0 0 1 1 0 1 **1 1 0 1 0.**

At the receiver end, the codeword is again divided by the divisor 1 0 0 1 1 1.

```
                  101111110   <== Discard qoutient
        ___________________
100111 | 10100110111010
         100111
         ------
           111010
           100111
           ------
            111011
            100111
            ------
             111001
             100111
             ------
              111101
              100111
              ------
               110100
               100111
               ------
                100111
                100111
                ------
                     0        <== Remainder equals zero (no errors)
```

3.6.6. Checksum Codes

One popular method for error detection is the use of checksums. **Checksums** have the advantage of operating on a long sequence of data. Typically the data is 8 bit character data (or ASCII data) in a file or character frame. Each character byte is summed into a checksum variable. As the data is summed into the checksum variable, the carries are ignored. The checksum variable is usually appended to the end of the data field and is called the **checksum field**. The checksum field can be one or more bytes. The ability to detect errors in a multi byte sequence is affected by the number of bytes in the sequence and the size of the checksum field.

In ones complement arithmetic, a negative integer $-A$ is represented as the complement of A; that is each bit of A is inverted. When adding numbers in ones complement arithmetic, a carryout from the MSB needs to be added to the result.

Checksum Generator

At the transmitter end, the checksum generator divides the data unit in to equal segments of bits (usually 16). These segments are added using ones complement arithmetic in such a way so that the

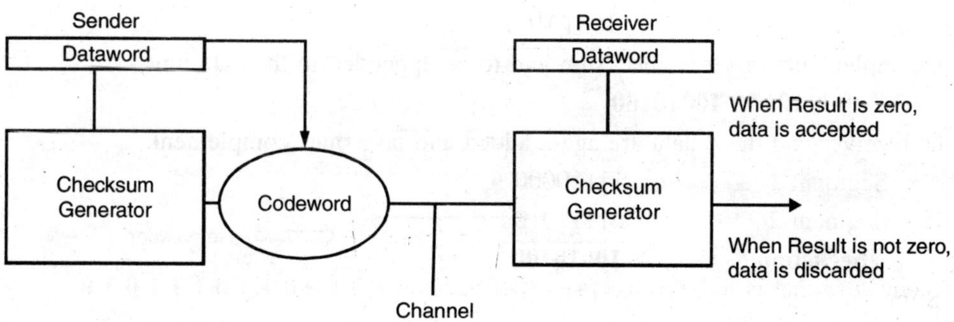

Fig. 3.15. *A simple block diagram of checksum error detection method*

total will be having the same number of bits. The total sum is complemented and appended to the end of the data unit as redundancy bits, called the **checksum field**. The codeword is now transmitted across the network. (**Refer Fig. 3.15**)

Checksum Checker

At the receiver end, the codeword is again subdivided in to equal segments and added all segments, as it was done at the transmitter end. Than the sum is complemented and gets the result. If the data unit received at the receiver is correct than the checksum field should be zero and if the result is not zero, the data unit contains an error and receiver discard it.

Steps

At the transmitter and the receiver ends following steps are taken:

1. The data unit is divided into *m* sections. The number of bits in each segment is *n*.
2. All the segments are added using ones complement to get the sum.
3. The total is complemented and becomes the checksum. The checksum is send with the data.
4. The data unit is divided into *m* sections. The number of bits in each segment is *n*.
5. All the segments are added using ones complement to get the sum.
6. The total is complemented and if the result is zero, the data is accepted otherwise they are rejected.

There is one limitation of this technique is as follows:

If one or more bits of a section are corrupted and the corresponding bit or bits of opposite value in a second section is also damaged. In this case the sum of those columns will remained same and error is not detected by the receiver.

Example 1: Let us take the data unit to be transmitted is

1111000001111010.

Data unit has 16 bits, divides in two segments of 8 bits and add them.

Segment-1	11110000
Segment-2	01111010
	01101010
	1
Sum	01101011

Now complement the sum,

	01101011
Ones complement	10010100

Ones complements result is checksum and to be appended to the data unit,

11110000 01111010 **10010100**

At the receiver end these data are again added and take ones complement.

Segment-1	11110000
Segment-2	01111010
checksum	**10010100**
Sum	11111110
	1
	11111111

Ones complement 00000000 Result is zero that means there is no error.

Let us consider that one of the bits of the data unit is changed.

```
Bit changed from 1 to 0
           ↓
          11010000
         01111010
         10010100
         --------
         11011110
                1
         --------
         11011111
         --------
```

Ones complement 00100000; Result is not zero that means there is some error.

3.6.7. Error Correction

There are two approaches for the error corrections, namely:

1. Feedback Error Correction
2. Forward Error Correction

Feedback Error Correction

In **feedback error correction**, only error detection technique is used and whenever errors occur, sender will resend that frame or packet to the receiver.

Error detection is adding enough "extra" bits to deduce that there is an error but not enough bits to correct the error.

If only error detection is employed in a network transmission than retransmission is necessary to recover the frame (data link layer) or the packet (network layer). At the data link layer, this is referred to as **ARQ (Automatic Repeat reQuest).**

> **DO YOU KNOW**
>
> In **feedback error correction**, only error detection technique is used and whenever errors occur, sender will resend that frame or packet to the receiver.

ARQ (Automatic Repeat reQuest)

In the **ARQ system** of error control, when an error is detected, a request is made for the retransmission of the data. Therefore a feedback channel is required for sending the request for retransmission.

ARQ system is slightly different from the forward error correction systems in following aspects:

(*i*) In ARQ system less number of check bits is required to be sent.

(*ii*) In order to implement repeat transmission of codeword's, a return transmission path and additional hardware will be needed.

(*iii*) The bit rate of forward transmission must make consideration for the backward repeat transmission.

A simple block diagram of the basic ARQ system is shown in Fig. 3.16.

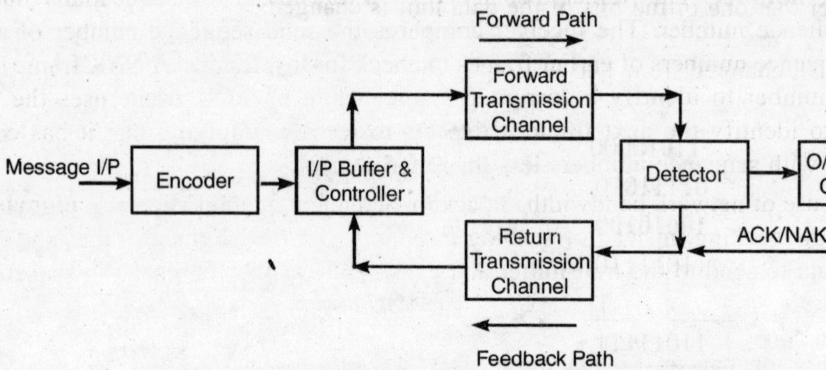

Fig. 3.16. *A simple block diagram of the basic ARQ system*

The encoder generated the codewords for each message signal at its input. Each codewords at the encoder output is stored temporily and transmitted over the forward transmission path. At the receiving end a decoder will decode the codewords and check for errors. If there is no error then decoder will send a positive acknowledgement (ACK) and if in case of error, it will send negative acknowledgement (NAK). On receiving a negative acknowledgement signal through return transmission path, the controller will send that message signal again from the words stored in the input buffer. The output controller and buffer on the receiver side assemble the output bit stream from the codewords accepted by the decoder.

There are various types of ARQ systems/ Flow Control Protocols

3.7. VARIOUS TYPES OF FLOW CONTROL PROTOCOLS

1. Unrestricted simplex protocol,
2. Simplex stop- and -wait protocol,
3. Sliding window protocol;
4. Go-Back N ARQ
5. Selective repeats

Before proceeding to the different types of flow control protocol, let us discuss some important terms related to the flow-control mechanism.

3.7.1. Important Terms Related to the Flow Control

(a) Acknowledgments

Data link layer does not provide end-to-end accountability; this is a responsibility of higher layers. Instead, each station is responsible for ensuring that the data received by it is correctly passed to the next station. This is facilitated by the use of acknowledgments, of which there are two types:

- An **acknowledgment** (ACK) is sent by a receiver to a transmitter to indicate that a frame has been received seemingly intact.
- A **negative acknowledgment** (NAK) is sent by a receiver to a transmitter to indicate that a corrupt frame has been received, or that a frame is missing (NAK is used in Selective Repeats protocol).

In character-based protocols, acknowledgments are implemented as control characters. In bit-oriented protocols, they are implemented as **acknowledgment frames**.

The control field of a frame can contain two sequence numbers: a **send sequence number** and a **receive sequence number**. The frames dispatched by a transmitter are sequentially numbered using the send sequence number. The receiver compares the send sequence number of a frame against the send sequence numbers of earlier frames to check for lost frames. A NAK frame uses the receive sequence number to identify a corrupted frame, while an ACK frame uses the receive sequence number to identify the next frame it expects to receive, implying that it has correctly received all frames with sequence numbers less than that number.

To improve the use of network bandwidth, an acknowledgment method known as **piggybacking** is often used. In piggybacking, instead of sending a separate acknowledgment frame, the receiver waits until it has data to send to the transmitter and embeds the acknowledgment in that frame.

(b) Timers

Timers provide a mechanism for placing a time limit on certain operations to be completed. A timer is set by a process and has a predefined expiry period. When the timer expires, the corresponding process is notified to take appropriate action. Link protocols make use of a number of different timers, the exact number and function of which is protocol-dependent. However, two important timers (called T1 and T2) are used by most protocols.

T1 is set by a transmitter upon transmitting a frame. The transmitter expects to receive an acknowledgment frame before T1 expires. Otherwise, the transmitter resets T1 and retransmits the frame. The protocol allows a certain number of retries before the problem is propagated to a higher layer.

T2 is set by a receiver to ensure that an acknowledgment frame is sent to the transmitter before its T1 timer expires. This is intended to avoid the transmitter having to retransmit frames because the receiver has been slow in acknowledging their receipt.

(c) Retransmission

Data link layer's approach to dealing with corrupted or lost frames is to retransmit them. Two different methods exist for this purpose. In the first method, called **Selective Reject**, upon encountering a faulty frame, the receiver requests the retransmission of that specific frame. Since additional frames may have followed the faulty frame, the receiver needs to be able to temporarily store these frames until it has received a corrected version of the faulty frame, so that frame order is maintained.

Now we discuss the various types of flow control protocols in brief.

3.8. UNRESTRICTED SIMPLEX PROTOCOL

Unrestricted simplex protocol is the simplest form of data communication. Unrestricted simplex protocol is totally unrealistic protocol. In this mechanism data are transmitted in one direction only. Both the sender and receiver network layers are always ready. We have infinite buffer space and the processing time is ignored. There is no loss or damage of frame in the communication channel or we can say that communication channel is error free. There is no errors or flow control restrictions in this protocol.

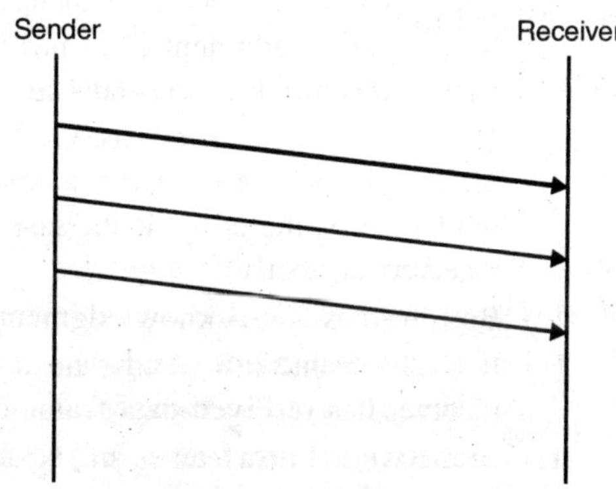

Fig. 3.17. *Unrestricted simplex protocol*

We have two distinct procedures, a sender and a receiver. There is no sequence numbers and acknowledgements are used in this protocol. The sender is just sits in an infinite loop pumping out the data onto the line as fast as it can. The body of loop performed three actions:

(*i*) Go fetch a packet from the network layer,

(*ii*) Construct an outbound frame using the variable S, and

(*iii*) Send the frame.

The receiver is very simple because it has to just wait for the arrival of an undamaged frame. Figure 3.17 shows the mechanism of Unrestricted simplex protocol.

3.9. SIMPLEX STOP- AND -WAIT PROTOCOL

The stop-and-wait protocol also provides unidirectional flow of data from sender to receiver. The communication channel is also assumed to be error free as it was in unrestricted simplex protocol. However, in this, the receiver has only a finite buffer capacity and a finite processing speed, so that protocol prevent the sender from overflowing the receiver with data faster than it can handled.

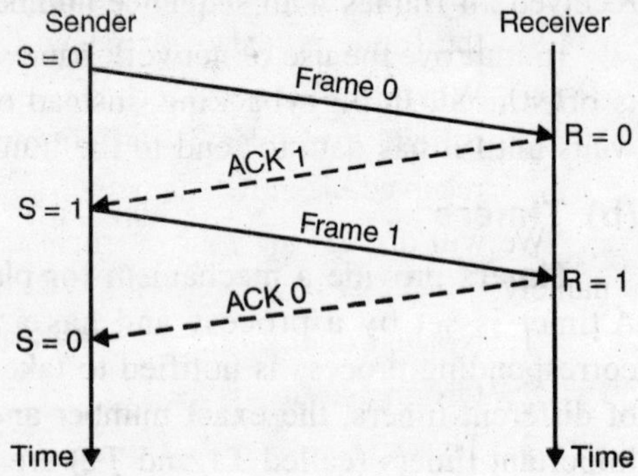

Fig. 3.18. *Simplex Stop-and-Wait*

Protocols in which the sender sends one frame and then waits for an acknowledgement before sending next frame are called **stop-and-wait**. The sending data link layer needs not to inspect the incoming frame; there is only one possibility. The incoming frame is always an acknowledgement. Figure 3.18 shows an example of a simplex stop-and-wait protocol.

DO YOU KNOW

Protocols in which the sender sends one frame and then waits for an acknowledgement before sending next frame are called **stop-and-wait**.

3.10. STOP-AND-WAIT ARQ

Now let us consider the practical situation of a communication channel with the noise that makes errors. Frames may be lost completely or damaged. The simplest ARQ (Automatic Repeat Request) scheme is the stop-and-wait protocol.

(*i*) In **stop-and-wait ARQ**, after transmitting one frame, the sender waits for an acknowledgement before transmitting the next frame.

(*ii*) If the acknowledgment does not arrive after a certain period of time, the sender time out and retransmits the original frame.

(*iii*) The sending device keeps a copy of the last frame transmitted until it gets an acknowledgment for that frame. In case the acknowledgment does not arrive after a certain period, the sender keeps the copy of the last transmitted frame so that it can retransmit until they are received correctly.

(*iv*) Both frames and acknowledgments are numbered alternatively 0 and 1.

(*v*) If a data frame 0 is send than it will be acknowledgment by ACK 1, indicating that the receiver has received data frame 0 and is now expecting next data frame 1.

(*vi*) Numbering of data frames and acknowledgment frames allows for identification of data frames in case of duplicate transmission. It is useful in the case of lost or delayed acknowledgment.

(*vii*) The sender has control variable, S, that hold the number of recently frame sent (0 or 1). The receiver has a control variable, R, which holds the number of the next frame expected (0 or 1).

(*viii*) If the receiver detects an error in the received data frame, it discards the frame and it will not send acknowledgment frame to the sender.

(*ix*) If the receiver receives a frame which is out of order (1 instead of 0 or 0 instead of 1) that means lost of data frame, it simply discards the out of order received data frame.

(*x*) Timer at sender side start counting the time as the sender sends a data frame and if the acknowledgment is not received within a specified time period, the sender considers that the frame is lost or damaged. Sender resends the particular data frame.

(*xi*) Whenever the receiver receives the data frame safe and sound, a positive acknowledgement is send to sender. If data frame 1 is received, acknowledgment 0 is sent and if frame 0 is received, acknowledgment 1 is received.

We will discuss the four practical scenarios present in data communication over a noisy channel, namely:

1. Normal operation, when ACK is received before the timer expires.
2. Original frame is lost.
3. The ACK is lost.
4. The ACK is delayed

(a) Case-1: Normal operation

Figure 3.19 illustrate the normal operation in stop-and-wait ARQ.

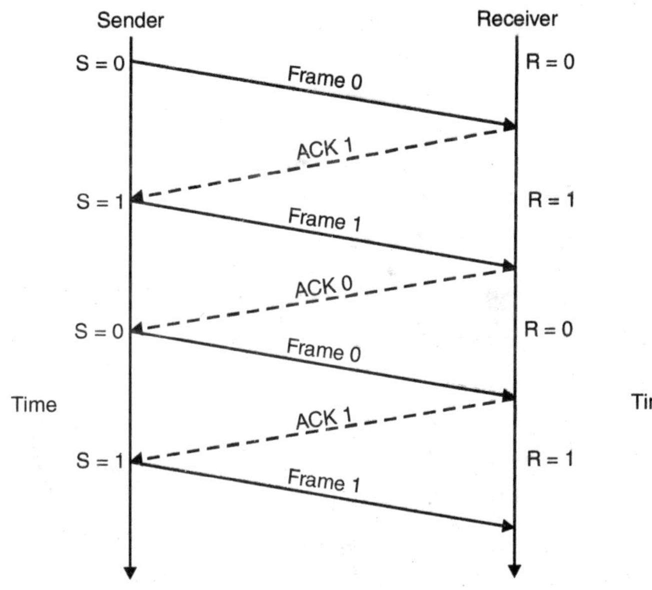

Fig. 3.19. *Illustrate the normal operation in stop-and-wait ARQ.*

Working operation

(*i*) Sender sends the data frame-0 to the receiver, having control variable $S = 0$.

(*ii*) Receiver control variable $R = 0$, that means expected data frame 0.

(*iii*) Receiver receives the data frame-0, safe and sound. Receiver sends the ACK-1 to the sender within the allotted time period.

(*iv*) After receiving the ACK-1, the sender control variable changes to $S = 1$, that means next data frame to be sent is number 1. Sender sends the data frame-1.

(*v*) Receiver control variable is changed to $R = 1$, that means receiver is expecting frame-1. Receiver received the data frame-1 and sends the ACK-0 to the sender.

(*vi*) Likewise, the transmission of data will continue.

It is to be noted that normal operation of stop-and-wait ARQ is similar to the simplex stop-and-wait protocol.

(b) Case-2: Original frame is lost

Figure 3.4 illustrate the condition when original frame is lost in stop-and-wait ARQ.

Working Operation

(*i*) Sender sends the data frame - 0 to the receiver, having control variable $S = 0$.

(*ii*) Receiver control variable R = 0, that means expected data frame 0.

(*iii*) Receiver receives the data frame-0, safe and sound. Receiver sends the ACK-1 to the sender within the allotted time period.

(*iv*) After receiving the ACK-1, the sender control variable changes to S = 1, that means next data frame to be sent is number 1. Sender sends the data frame-1.

(*v*) Data frame-1 is lost in the transmission media. Now sender will wait for ACK-0 until the timed-out at the sender and send the frame-1 again.

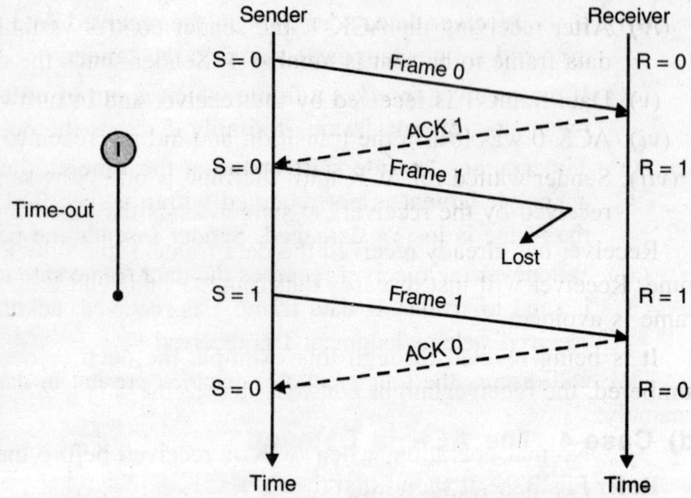

Fig. 3.20. *Illustrate the condition when original frame is lost in stop-and-wait ARQ.*

(*vi*) Receiver is expecting frame and it receives the frame-1. Receiver sends the ACK-0 and it is being received by the sender within the allotted time period.

(c) Case 3: The ACK is lost

Figure 3.21 illustrate the condition when ACK is lost in stop-and-wait ARQ

Working Operation

(*i*) Sender sends the data frame-0 to the receiver, having control variable $S = 0$.

(*ii*) Receiver control variable $R = 0$, that means expected data frame 0.

(*iii*) Receiver receives the data frame-0, safe and sound. Receiver sends the ACK-1 to the sender within the allotted time period.

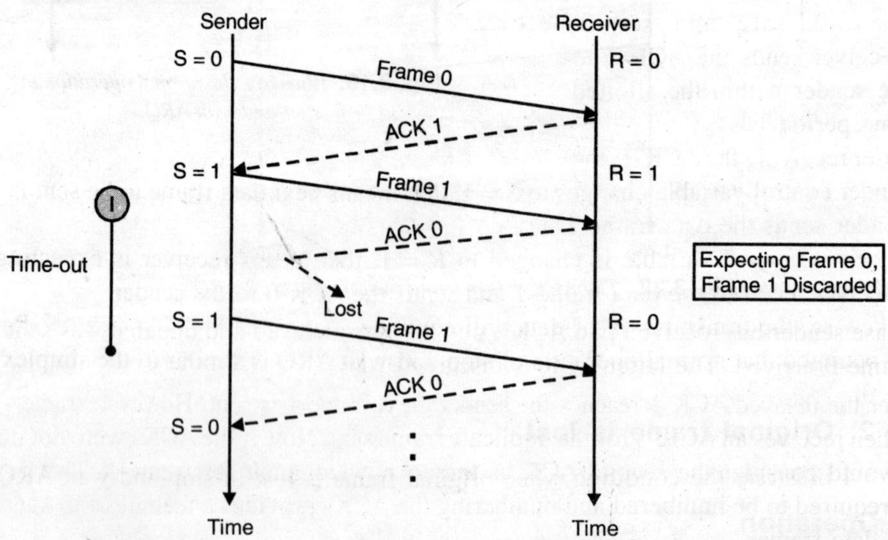

Fig. 3.21. *Illustrate the condition when ACK is lost in stop-and-wait ARQ.*

(*iv*) After receiving the ACK-1, the sender control variable changes to $S = 1$, that means next data frame to be sent is number 1. Sender sends the data frame-1.

(*v*) Data frame-1 is received by the receiver and in turn sends the ACK-0.

(*vi*) ACK-0 was lost in the transition and did not reach to the sender.

(*vii*) Sender waited for ACK until the time is out. Now sender assumes that data frame-1 is not received by the receiver, so sender sends the same data frame-1 again.

Receiver has already received the data frame-1 and now expecting data frame-0 ($R = 0$) should come. Receiver will just discards that frame-1 and send the ACK-0 again. Like this duplicating of frame is avoided.

It is being reveled through this example the need to number frames. If the frames were not numbered, the receiver might consider that frame-1 is a new frame (not duplicate), keeps it.

(d) Case-4: The ACK is Delayed

Refer Fig. 3.22, it shows that the ACK-1 is delayed that means ACK-1 is not received within the allotted time period, sender assumed that the frame-0 is lost and retransmit the same frame-0. But the receiver has already received that frame-0 and expecting next frame-1, so receiver will discards the duplicate frame-0. Receiver will sends the ACK-1 again but the ACK-1 for the same frame-0 is already being received by the sender, so now receiver will discard the duplicate ACK-1.

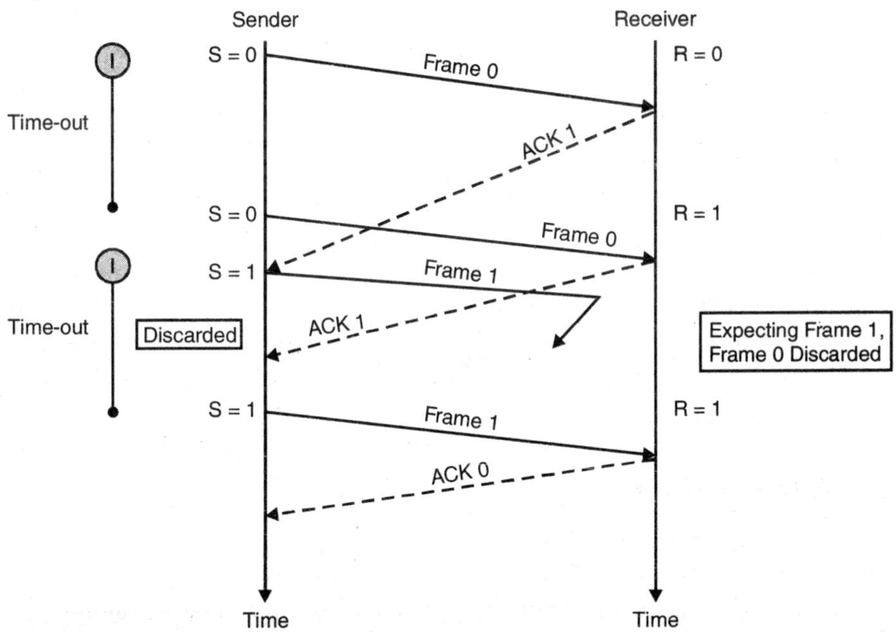

Fig. 3.22. *The ACK-1 is delayed in stop-and-wait ARQ*

In this case sender has received two ACKs, one that was delayed and one that was sent after the duplicate frame-0 arrived. The latter ACK is discarded.

Now after the delayed ACK-1 reaches the sender, next frame-1 is sent. However, frame -1 is lost. The sender then receives an ACK-1 for the duplicate frame sent. Now if the ACKs were not numbered, the sender would consider the second ACK as the acknowledgment for frame-1. That is why the ACKs were required to be numbered and numbering the ACKs provides a technique to keep track of the received data frames.

The primary **advantage of stop-and-wait flow control** is its simplicity.

The main **disadvantage** is speed, as time lapse between each frame is wasted time. Each frame takes essentially twice as long to transmit as necessary because both the message and the acknowledgment cover the entire length of the data link before the next frame can be sent.

3.10.1. Piggybacking

However, stop-and-wait mechanism is unidirectional but we may also have bidirectional transmission if sender and receiver both have two separate channels for full duplex transmission or share the channel for half duplex transmission. The method used for bidirectional transmission is known as **Piggybacking.** In this case each party needs both sender (S) and receiver (R) control variables to track frames sent and expected. Figure 3.23, shows the example of piggybacking method in stop-and-wait protocol.

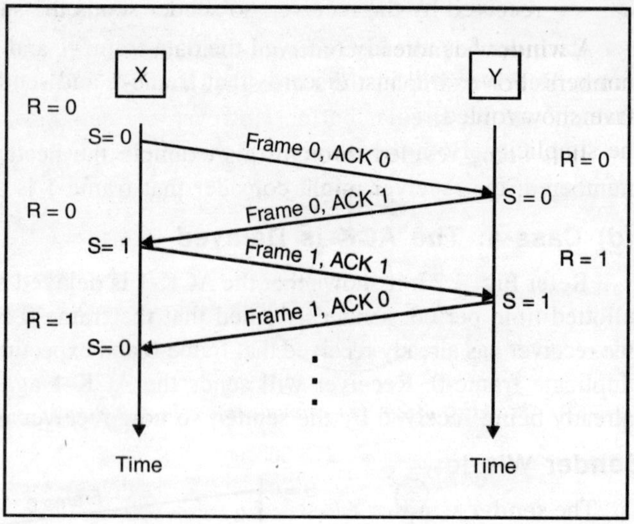

Fig. 3.23. *The example of piggybacking method in stop-and-wait protocol*

In this method a data frame with an ACK is combined. As shown in Fig. 3.23, we have two parties X and Y both have data to send. Instead of sending data frames or ACKs alone, we combine both of them. Station-X sends a data frame that includes an acknowledgment and in station -Y also behaves in a similar manner.

We can save bandwidth in piggybacking method, because the overhead from a data frame and acknowledgment frame can be combined into one frame.

> **DO YOU KNOW**
>
> The method used for bidirectional transmission is known as **Piggybacking.**

3.11. SLIDING WINDOW PROTOCOL

A Sliding window protocol is used to accomplish the flow-control in TCP. Sliding window is a general protocol used with bit-oriented protocols. In this method both sender and receiver use a window for each connection. In this protocol, the sender (transmitter) maintains a variable, *S*, which denotes the sequence number of the next frame to be transmitted. Similarly, the receiver maintains a variable, *R*, which denotes the sequence number of the next frame it expects to receive. Both variables are restricted to a limited range of values (*e.g.*, 0 through 7) by using modulo arithmetic (*e.g.*, modulo 8).

The window is referred as a **sliding window** because it can slide over the buffer as data and acknowledgements are sent and received.

Let us discuss the case in which there is no sliding window protocol. In this situation, the sender can go ahead and send all the bytes in its buffer, without considering the condition of the receiver. The size of the receiver is limited; it could completely fill up because the receiving process is not consuming data fast enough. The excess data discarded by the receiver will require retransmission. The sender must adjust itself to the number of locations available at the receiver side.

DO YOU KNOW

Sliding window protocol is used to accomplish the flow-control in TCP.

A **window** denotes a continuous sub range within the permitted range of values for the sequence numbers. For example, the ranges 0-3 and 6-1 both represent windows of size 3 (Fig. 3.24). We have shown the circular buffer. However, we can show a flat buffer instead of circular buffer for the simplicity. Both the sender and the receiver have their own window.

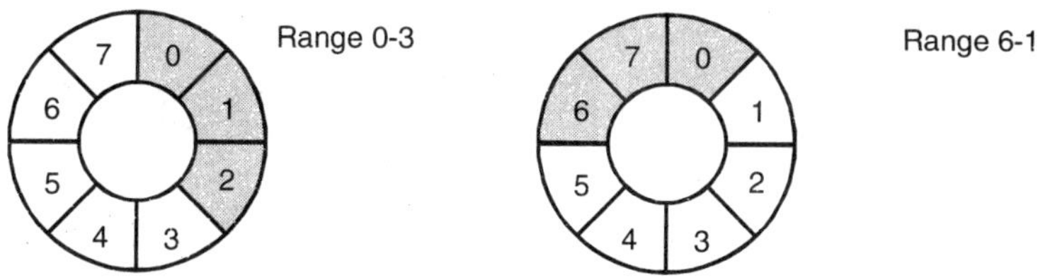

Fig. 3.24. *For example, the ranges 0-3 and 6-1 both represent windows of size 3.*

Sender Window

The **sender window** denotes the frames that have been transmitted but remain unacknowledged. This window can vary in size, from empty to the entire range. The sender must have enough buffer space to store the maximum possible number of unacknowledged frames. The size of the sender window should be equal or less than the size of the receiver window. This window includes the bytes sent and not acknowledgement and those that can be sent. Figure 3.25 shows the sender buffer with the sender window.

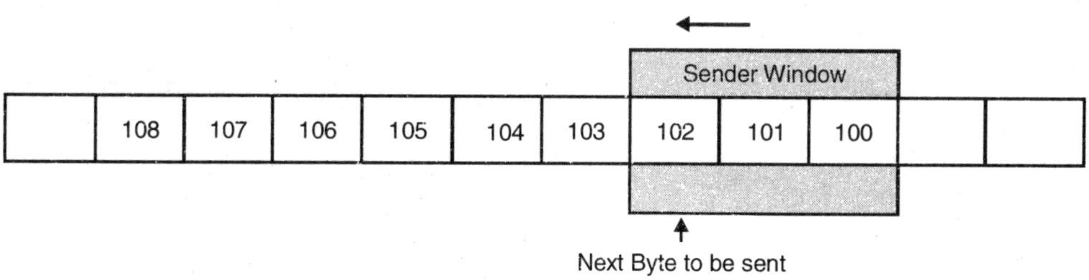

Fig. 3.25. *The sender buffer with the sender window.*

The size of the sender window is equal to the size of the receiver window (3 in our case). However, this does not mean that the sender can send 3 more bytes; it can send only one more bytes because it has already sent two bytes. It is also to be noted that although bytes 103 to 108 are in sending buffer but they also cannot be sent until more acknowledgements arrives from the receiver.

Receiver Window

The **receiver window** denotes frames that are expected to be received. Figure 3.26 shows the receiver buffer and that shows that the next byte to be consumed by the process is byte 100. The receiver expects to receive byte 103 from the sender. A receiver window size of 1 means that frames must be received in transmission order. Larger window sizes allow the receiver to receive as many frames out of order. The receiver must have enough buffer space to store the maximum possible number of frames that can be received out of order. How many more bytes can the receiver store? Now if suppose the total size of the receiving buffer is N and the locations are occupied, M, then only N-M more bytes can be received and this value is called the receiver window. For example, if $N = 10$ and $M = 7$, this means that the value of the receiver window is 3.

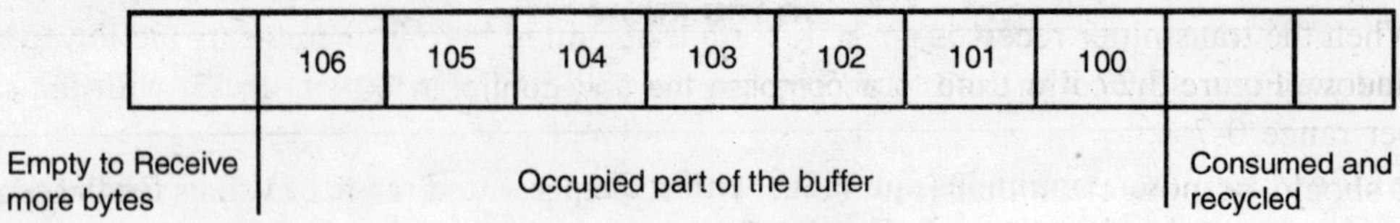

Fig. 3.26. *Receiver buffer*

Sliding the Window

The sliding window protocol operates as follows.

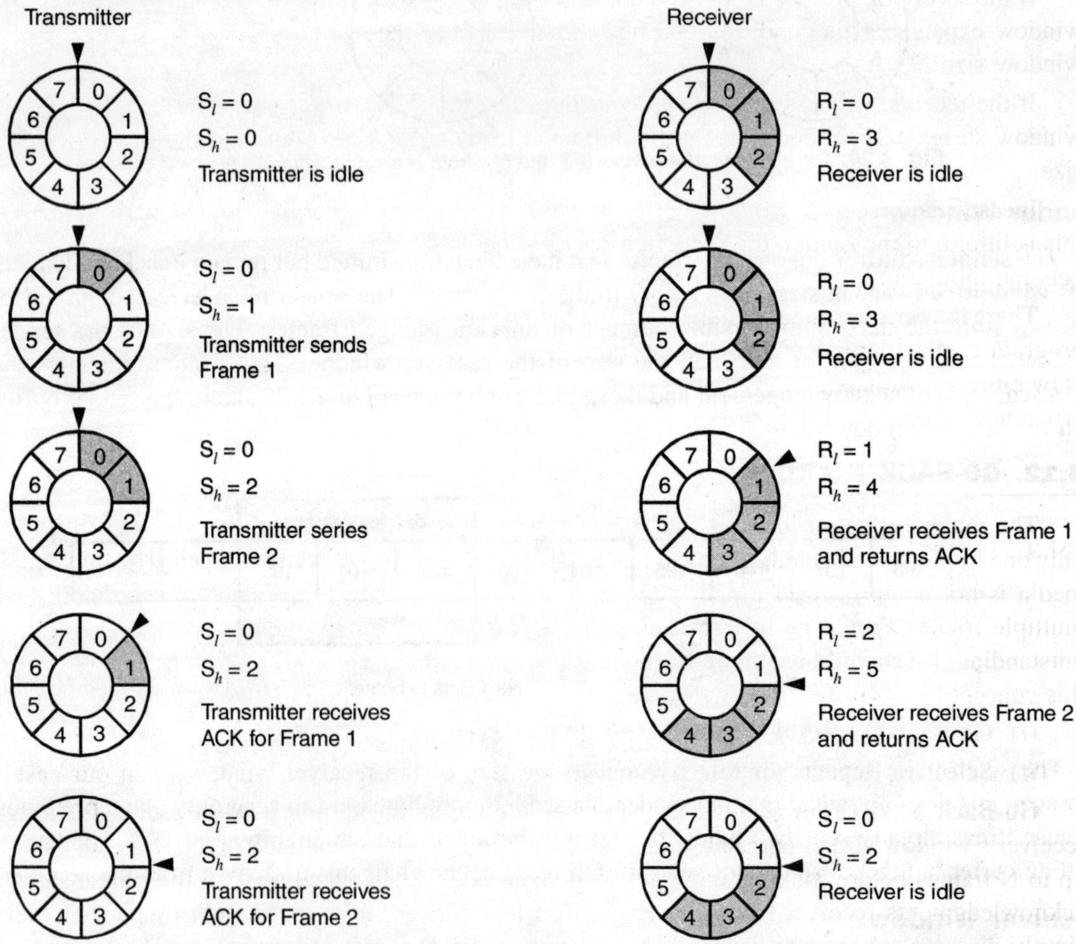

Fig. 3.27. *Illustrates the protocol for a sliding window of size 3*

When the transmitter sends a byte, it increment S (and the upper bound of its window). When the receiver receives a byte whose sequence number falls within its window, it increments R and sends an ACK to the transmitter. If the byte's sequence number matches any position other than the lower bound of the window, it notes the fact that the corresponding byte has now been received. If the byte's sequence number matches the lower bound of the window, the window is rotated clockwise by one position (or more positions if succeeding bytes within the window have already been received).

When the transmitter receives an ACK for a transmitted byte, it increments the lower bound of its window. Figure 3.27 illustrates the protocol for a sliding window of size 3, with the sequence number range 0-7.

It should be now clear that flow control is straightforward in the sliding window protocol. If the receiver withholds ACK frames, the transmitter soon reaches its maximum window size and has to stop transmitting. Once it receives further ACK frames, it can reduce its window size and transmit more frames.

Expanding and Shrinking the Sender Window

If the receiving process is fast and consumes more data than it receives, the size of the receiver window expands. This condition can be intimated to the sender, resulting in the increase of the window size.

If the receiver process is slow and consumes less data than it receives, the size of the receiver window shrinks. This condition can be intimated to the sender, resulting in to reduce the window size.

Incase when receiver window buffer is totally filled, the receiver window value is zero. When this is inform to the sender, the sender closes its window. The sender cannot send anything before it gets the intimation of the nonzero window value.

There is a serious problem in the sliding window protocol that when either the sending application program creates data slowly or the receiving application program consumes data slowly, or both. Now either of these situations results in the sending of data in very small segment that reduces the efficiency of the operation.

3.12. GO-BACK N ARQ

The major limitation of stop-and-wait ARQ is that at any point in time for a sender, there is only one frame, the outstanding frame, that is sent and waiting to be acknowledged. The transmission media is not being utilized properly. To have the optimum utilization of the transmission media, multiple frames should be in transition while waiting for acknowledgment. Instead of one frame outstanding, we should have more than one frame outstanding. There are two types of protocols use this concept :

(*i*) Go-Back N ARQ

(*ii*) Selective Repeats (or selective reject)

Go-Back N ARQ corresponds to the example given in the sliding window protocol where the receiver's window size is one and sender window size is greater than one. In this case we can send up to N frames before looking about acknowledgments. We keep a copy of all the frames until the acknowledgments receive. In this case, if an error occurs in receiving the nth frame the receiver simply discards subsequent frames and does not send the acknowledgement corresponding to the nth frame. This compels the sender to resend all frames starting with the nth and thereafter. This deteriorates the data rate.

In this method some additional features are required to be incorporated in the stop-and-wait protocol. In this mechanism frames are numbered sequentially at the sender side. However, it is required to include the sequence number of each frame in the header, so we need to set a limit. If suppose the header of the frame allows n bits for the sequence number then the sequence numbers range from 0 to $2^n - 1$. For example, if $n = 4$, the only sequence numbers are 0 through 15 inclusive. However, we can repeat the sequence.

0, 1, 2, 3, 4, 5, 6, 7, 8, 9, 10, 11, 12, 13, 14, 15, 0, 1, 2,………

Sender Sliding Window

We have to store the outstanding frames in a buffer until they are acknowledged at the sender side; a concept of window is used. The frames to the left of the window are those which are already sent and received the acknowledgements for the same, these frames can be removed from the buffer. All those frames which are right of the window cannot be sent till the window slides over them. The window slides over the unsent frames when the correct acknowledgments are received. Like this it is seen that window at sender side is sliding and referred as sliding window.

The size of the sender window is fixed that is at most $2^n - 1$(why?). For example, refer Fig. 3.28(*a*), the size of the window is 7. The frames 0 to 6 have been sent. Now refer Fig. 3.28(*b*), where the window slides over next two frames (7 and 0) as soon as it received the acknowledgements for the frames 0 and 1.

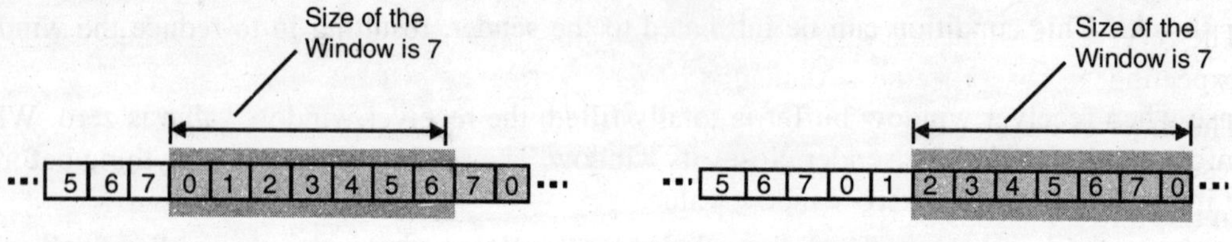

(*a*) *Sender sliding window (before)* (*b*) *Sender sliding window (after)*

Fig. 3.28.

Receiver Sliding Window

In this protocol the size of the receiver window is always 1and the receiver is looking for a specified frame. If the frame is arriving out of order, that frame is being discarded and needs to send again. The receiver window is also sliding as shown in Fig. 3.29. In Fig. 3.29(*a*), the receiver is waiting for frame-3 and when it is received, the windows slides over to frame-4.

(*a*) Before Sliding (*b*) After Sliding

Fig. 3.29. *(a) (before) and (b) (after) Receiver sliding Window*

Control Variables

We have three variables at sender site, namely S, S_F and S_L as shown in Fig. 3.30.

S = The sequence number of the recently sent frame,

S_F = The sequence number of the first frame in the window.

S_L = The sequence number of the last frame in the window.

The size of the window is N, where $N = S_L - S_F + 1$.

We have only one variable, *R*, at the receiver end and hold the sequence number of the frame which is expected to receive. Receiver accept the frame if the value of received frame is same as the value of *R*; if not, it is discarded.

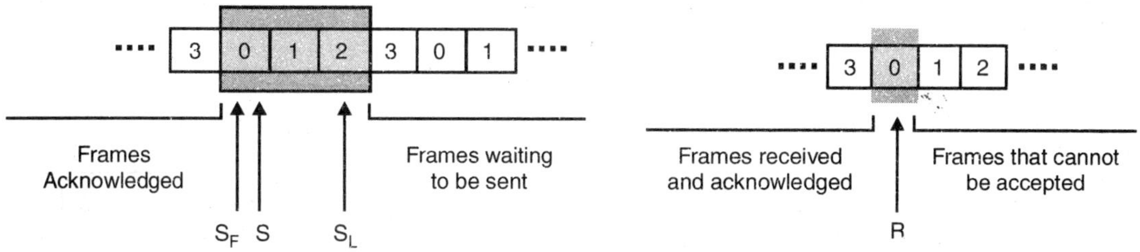

(*a*) Sender window control variables (*b*) Receiver window control variable

Fig. 3.30.

Acknowledgment and Retransmission of Frame

(*i*) A positive acknowledgment is send by the receiver if a frame has been received safe and sound in order.

(*ii*) The receiver is silent and discards all subsequent frames until it receives the one which is expecting, in case a frame is damaged or received out of order.

(*iii*) The timer of the unacknowledged frame is expired. Now it causes the sender to go back and resend all frames, beginning with the one with the expired timer.

(*iv*) In this ARQ, the receiver does not have to acknowledge each frame received but it can send one cumulative acknowledgment for many frames.

(*v*) In case, when a frame is damaged, the sender goes back and retransmits a set of frames starting from the damaged one up to the last frame sent.

(*vi*) Go-Back-N ARQ also can be bidirectional as it was in Stop-N-Wait ARQ.

(*vii*) Piggybacking can also be used to improve the efficiency of the transmission.

Pipelining

When a task is begun before the previous task has finished, this is referred as **pipelining** in the networking. There was no pipelining in Stop-N-wait ARQ protocol because we need to wait for the acknowledgment before the next frame can be sent. But in Go-Back-N ARQ and selective repeat ARQ, several frames can be sent before we receive the ACKs for the previous frames. It improves the efficiency of the transmission.

Let us take one example, suppose the sender has already sent frame 7, but the timer is expired for the frame 4. That means frame 4 has not been acknowledged, so the sender has to go back and resends the frames 4, 5, 6 and 7. That is why this protocol is known as Go-Back-N ARQ.

Working Operation

Let us discuss the working operation of various conditions in Go-Back-N ARQ protocol.

(*i*) Normal operation

(*ii*) Damaged and Lost frame

(*iii*) Damaged and Lost Acknowledgment

(*iv*) Delayed Acknowledgment

(a) Case-1: Normal Operation

Refer Fig. 3.31, that shows a normal operation of Go-Back-N ARQ protocol.

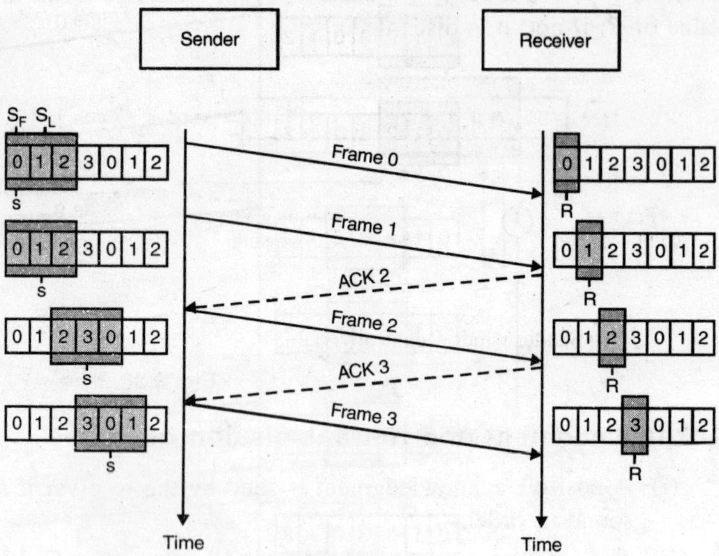

Fig. 3.31. *A normal operation of Go-Back-N ARQ protocol*

1. The sender site window size is 3. Frame-0 is the first frame of the sequence (S_F) and frame-2 is last frame of the sequence (S_L).
2. Sender sends the first frame-0 to the receiver. Receiver control variable (R) is set to 0 that means receiver is expecting frame-0, which is the same one and it is accepted.
3. Sender sends the next frame-1, without waiting for the ACK for previous frame. Receiver control variable is changed to $R = 1$, so the frame-1 is also accepted.
4. Receiver sends the cumulative ACK-2; it means previous two frames are received safe and sound.
5. Sender is continuing to send the next frame-2 and likewise receiver will send the cumulative acknowledgment accordingly.

The sender keeps track of the outstanding frames and updates variables and windows as the acknowledgments arrive.

(b) Case-2: Damaged and Lost frame

Figure 3.32 illustrates the working of operation, when a frame is lost or damaged in Go-Back-N ARQ. Let us discuss this case point by point.

1. The sender site window size is 3. Frame-0 is the first frame of the sequence (S_F) and frame-2 is last frame of the sequence (S_L).
2. Sender sends the first frame-0 to the receiver. Receiver control variable (R) is set to 0 that means receiver is expecting frame-0, which is the same one and it is accepted.
3. Sender sends the next frame-1, without waiting for the ACK for previous frame. Receiver control variable is changed to $R = 1$, so the frame-1 is also accepted.
4. Receiver sends the cumulative ACK-2; it means previous two frames are received safe and sound.
5. Till now there is no confusion, it working as normal operation. Let us suppose the frame-2 is lost in the transition.
6. Next Frame-3 is also send by the sender but the receiver control variable is not changed, it remains $R = 2$; that means frame-3 is not accepted by the receiver and it is discarded.
7. Frame-2 timer is expired. Sender goes back and resends the frame-2 and 3.

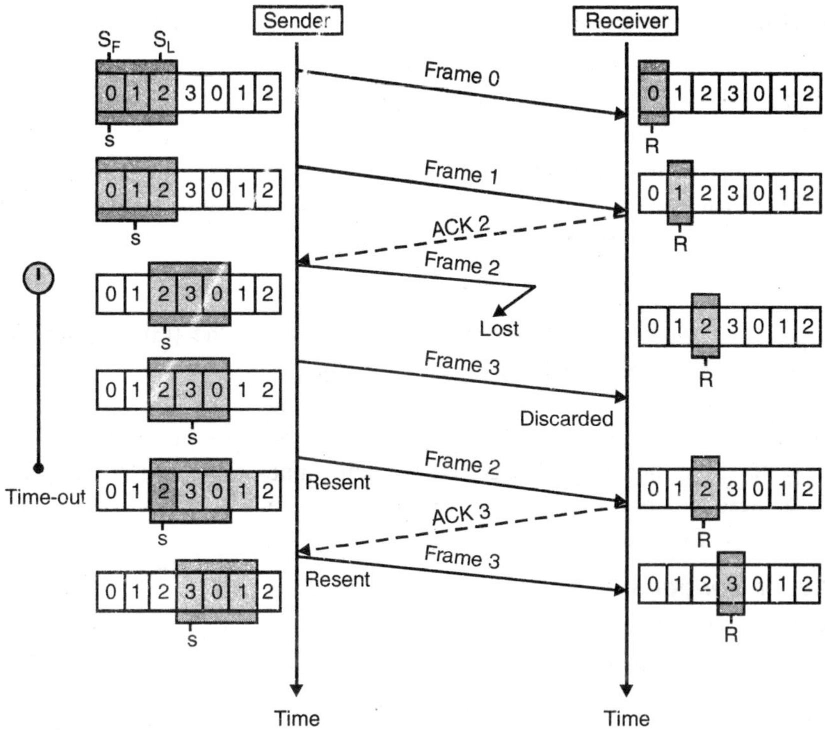

Fig. 3.32. *The working of operation, when a frame is lost or damaged in Go-Back-N ARQ*

(c) Case-3: Damaged and Lost Acknowledgments

Figure 3.33 illustrates the working of operation, when the ACK is lost or damaged in Go-Back-N ARQ. Let us discuss this case point by point.

1. The sender site window size is 3. Frame-0 is the first frame of the sequence (S_F) and frame-2 is last frame of the sequence (S_L).
2. Sender sends the first frame-0 to the receiver. Receiver control variable (R) is set to 0 that means receiver is expecting frame-0, which is the same one and it is accepted.
3. Sender sends the next frame-1, without waiting for the ACK for previous frame. Receiver control variable is changed to R =1, so the frame-1 is also accepted.

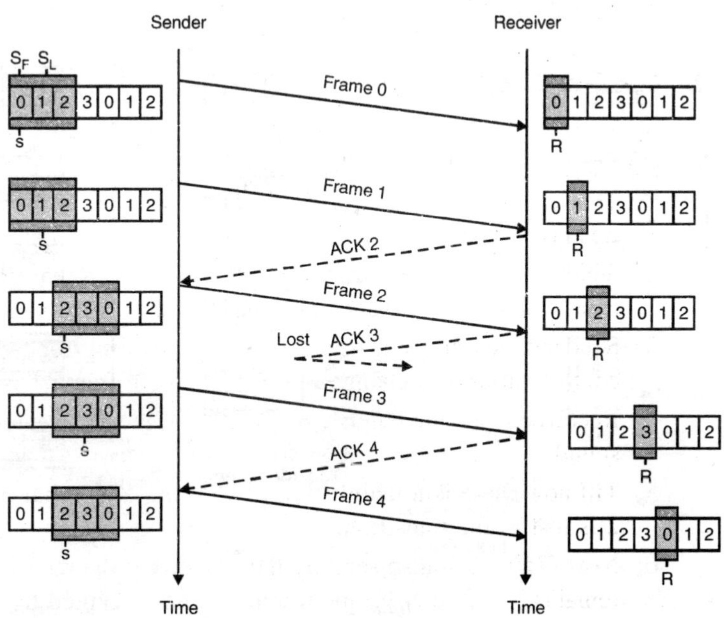

Fig. 3.33. *ACK is lost or damaged in Go-Back-N ARQ*

4. Receiver sends the cumulative ACK -2; it means previous two frames are received safe and sound.
5. Sender sends the frame-2 and frame-3. Receiver sends the ACK-3 that is lost in the transition but before the timer expired for the frame-2, receiver sends the ACK-4 which is received by the sender. In this situation there is no need of retransmission of frames because acknowledgments are cumulative in this protocol. ACK-4 means ACK-1 to ACK-4. So if ACK-2 is lost, ACK-4 covers that.

(d) Case-4: Delayed Acknowledgment

If the ACK is received after the time-out, the frame and all the frames after that are resent. A delayed acknowledgment results the resending of frames.

Go-Back-N ARQ protocols works well if errors are rare, but if the line is poor, it wastes a lot of bandwidth on retransmitted frames. Because this protocol has multiple outstanding frames, it logically needs multiples timers, one per outstanding frame. Each frames times out independently of all the other ones. All of these timers can easily be simulated in software, using a single hardware clock that causes interrupts periodically.

Why the size of sender's window is less than $2^n - 1$?

The size of the sender window is fixed that is at most $2^n - 1$. Let us see this with an example, where we select $n = 2$, that means the size of window can be $2^n - 1 = 2^2 - 1 = 3$.

And if we select $n = 2$ and take the size of window $2^n = 2^2 = 4$. Let us compares a window size of 3 and 4, refer to Fig. 3.34 (*a*) and (*b*).

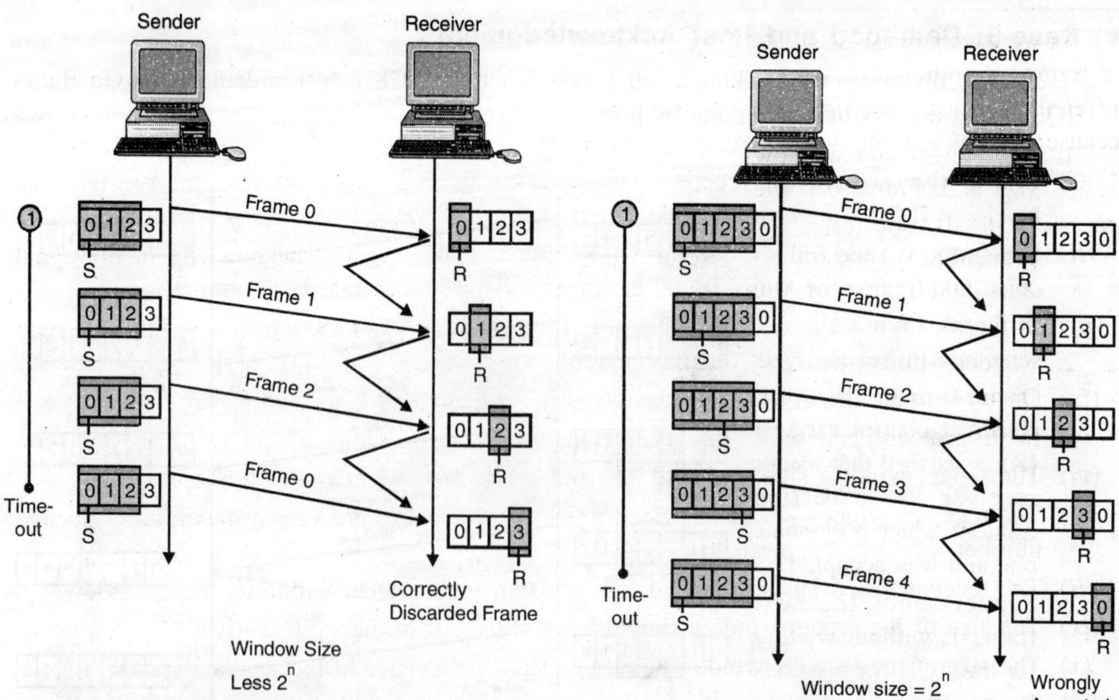

(*a*) *The size of the window is less than; $2^n - 1 = 3$*

(*b*) *the size of the window is equal to $2^n = 2^2 = 4$.*

Fig. 3.34.

Case-1: If the size of window is $2^n - 1$

(*i*) In the Fig. 3.34(*a*) the size of the window is less than; $2^n - 1 = 2^2 - 1 = 3$.

(*ii*) In case if all the three acknowledgments are lost and the timer expires for frame-0.

(*iii*) All the three frames (frame-0, 1 and 2) are resent.

(*iv*) Receiver window is expecting frame-3, not frame-0, so that duplicate frame is discarded correctly.

Case-2: If the size of window is 2^n (4)

(*i*) In the Fig. 3.34(*b*) the size of the window is equal to $2^n = 2^2 = 4$.

(*ii*) In this case if all the three acknowledgments are lost and the timer expires for frame-0.

(*iii*) All the three frames (frame-0, 1 and 2) are resent.

(*iv*) However, this time receiver window expect frame-0 and receiver will accept the duplicate frame-0 as the first frame in the next cycle.

(*v*) This is an error.

With the above mentioned examples, it is justified that the size of the sender window must be less than 2^n.

DO YOU KNOW

The size of the sender window is fixed that is at most $2^n - 1$.

3.13. SELECTIVE REPEAT

As we have already seen that if the line is poor, it wastes a lot of bandwidth just to retransmits the frames. An alternatively mechanism for handling errors is to allow the receiver to accept and buffer the frames following a damaged or lost one. This protocol does not discard frames just because an earlier frame was damaged or lost.

(*i*) In **selective repeat**, the receiver's window size is more than one, and the receiver can receive a fixed number of out of order frames equal to the window size.

(*ii*) If a frame is received with an error it simply sends a 'negative acknowledgement' signal corresponding to the faulty frame prompting the sender to resend the frame.

(*iii*) In the meantime, the recipient can store the subsequent frames, which have already been sent, in a buffer and wait for the remaining frame to be received.

(*iv*) Once received, the frames are transferred in sequence to the network layer.

(*v*) In this protocol, both sender and receiver maintain a window of acceptable sequence numbers.

(*vi*) The sender window size starts out at 0 and grows to some predefined maximum.

(*vii*) The receiver window is always fixed in size and equal to predefined maximum sequence number.

(*viii*) The receiver has a buffer reserved for each sequence number within its fixed window.

(*ix*) The size of the sender window should be at most half of the value 2^n (why?).

(*x*) The size of the receiver window should be same as the size of the sender window.

(*xi*) The size of the receiver window specifies the range of the accepted received frame.

(*xii*) The configuration of control variables at sender is same as in Go-Back-N ARQ.

(*xiii*) The receiver has two control variables, R_F and $R_{L,}$

(*xiv*) R_F is the first frame of the window and is last frame of the window. Receiver will look for the range of frames not a specified frame.

(*xv*) Selective-N-Repeat ARQ also can be bidirectional as it was in Stop-N-Wait ARQ and Go-Back-N ARQ.

(*xvi*) Piggybacking can also be used to improve the efficiency of the transmission.

Figure 3.35 (*a*) and (*b*) shows the sender and receiver windows along with the control variables.

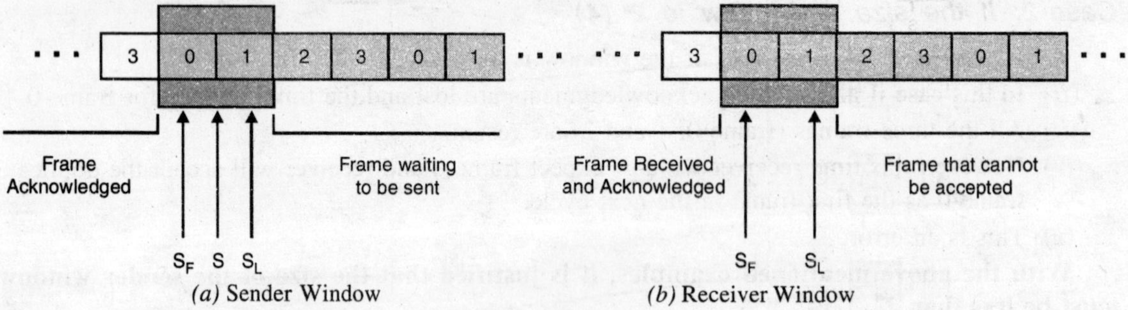

Fig. 3.35. *The sender and receiver windows along with the control variables*

Working Operation

1. Normal Operation : Let us now discuss the normal working operation of Selective and Repeat method with an example, as shown in Fig. 3.36.

(*i*) Frame-0 and 1are sent and being accepted by the receiver because they are within the range of the receiver window.

(*ii*) Receiver sends the ACK-2.

Now further frame-2 and 3 are send and being acknowledgements by receiver.

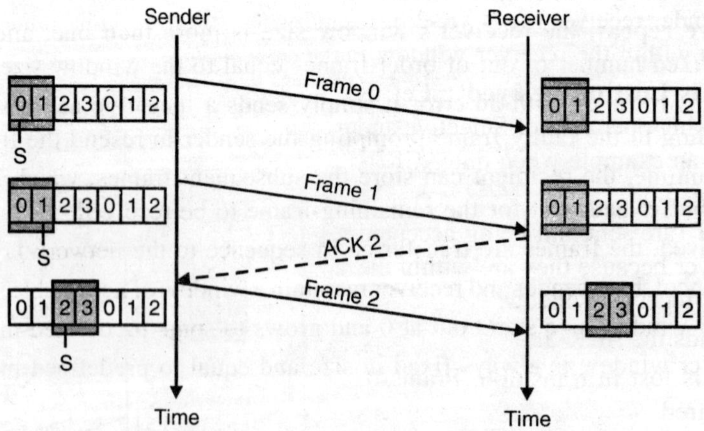

Fig. 3.36. *The normal working operation of Selective Repeat*

2. When Frame is Lost : Let us now discuss the working operation of Selective and Repeat method with an example when one frame is lost, as shown in Fig. 3.37.

(*i*) Frame-0 and 1are sent and being accepted by the receiver because they are within the range of the receiver window.

(*ii*) Receiver sends the ACK-2.

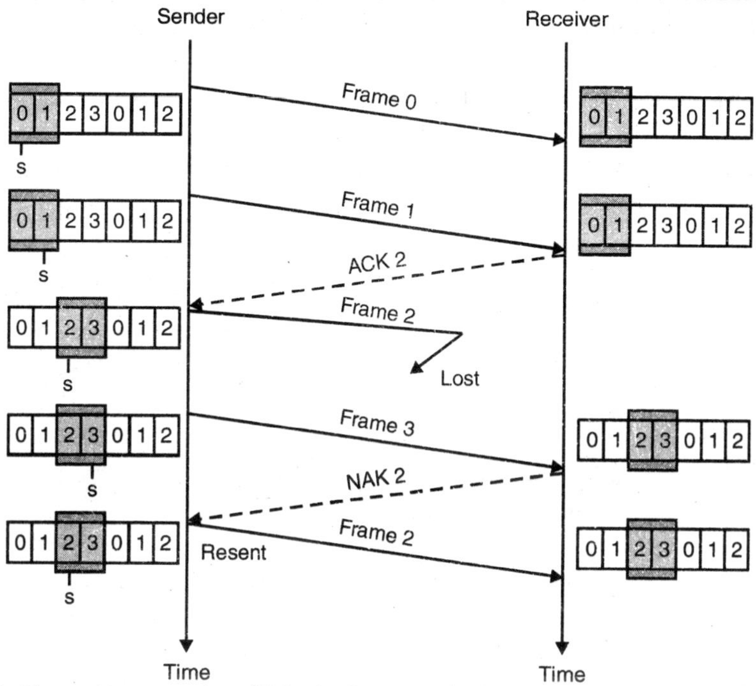

Fig. 3.37. *The working operation of Selective Repeat method with an example when one frame is lost*

(*iii*) Frame-2 is send but it is lost in transition.

(*iv*) Frame-3 is send and being accepted by receiver because it is in the same range of receiver window.

(*v*) However, the receiver sends a negative acknowledgment (NAK-2) for frame-2, which was lost.

(*vi*) When the sender receives the NAK-2, it resends only frame-2. This frame is then accepted because it is within the receiver window range.

3. When ACK is Lost or Delayed : Let us now discuss the working operation of Selective and Repeat method with an example when one ACK is lost, as shown in Fig. 3.38.

(*i*) Frame-0 and 1are sent and being accepted by the receiver because they are within the range of the receiver window.

(*ii*) Receiver sends the ACK-2.

(*iii*) But ACK-2 is lost in transition. Frame-0, timer is expired.

(*iv*) Frame-0 is sent again but now receiver is expecting frame-2.

(*v*) So the copy of frame-0 is discarded.

Figure 3.38 the working operation of Selective and Repeat method with an example when one ACK is lost.

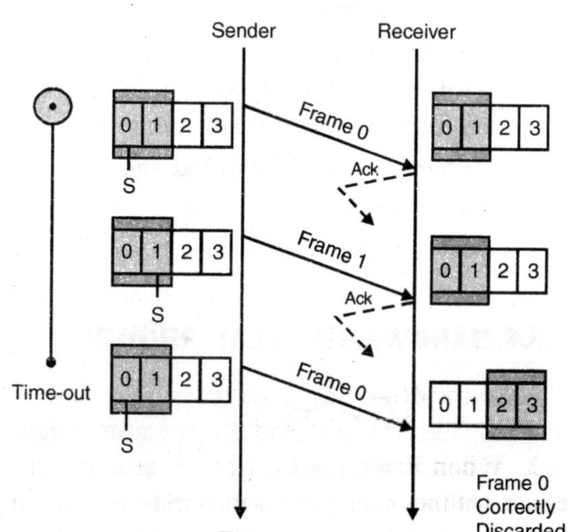

Fig. 3.38. *The working operation of selective repeat method with an example when one ack is lost*

Why the size of windows must be at most one-half of 2^n ?

For example, if we select $n = 2$ that means the size of the window should be one-half of 2^n or $4/2 = 2$. Refer Fig. 3.39 (*a* and *b*), which compares a window size of 3 with a window size of 2.

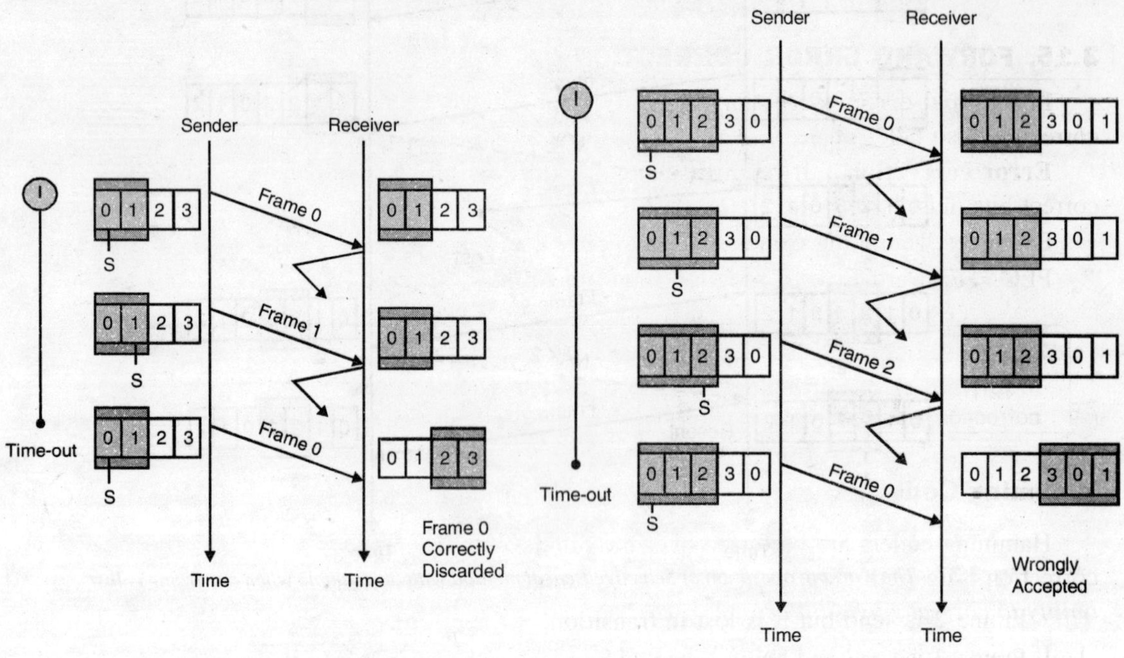

(*a*) In case when window size of 2 (*b*) In case when window size of 3

Fig. 3.39.

In case when window size of 2 :

(*i*) Suppose frames (0 and 1) are sent and all the acknowledgments are lost.

(*ii*) Frame-0 timer is expired and it is resend.

(*iii*) However receiver is waiting for frame-2, so it is discarded and avoids receiving the duplicate frame-0.

In case when window size of 3 :

(*i*) Suppose frames (0 and 1) are sent and all the acknowledgments are lost.

(*ii*) Frame-0 timer is expired and it is being resend.

(*iii*) However this time window of the receiver expect frame-0 (0 is the part of window), so it accept the frame-0 not as a duplicate but as first frame in next cycle. This frame-0 is accepted wrongly.

3.14. BANDWIDTH-DELAY PRODUCT

Bandwidth-delay-product is a measure of the efficiency of an ARQ system and it is a product of bandwidth (in bps) and the round-trip delay (in seconds). If the bandwidth is sufficient in the link, and then the sender will exhaust its window fast and will wait for the acknowledgments to come. On the other hand if the delay is more, the sender can also exhaust its window while waiting. In other words, it is a measure of the number of bits we can send out of our system while waiting for acknowledgment from the receiver.

$$\text{BW delay product} = \text{BW} \times \text{Round trip delay}$$

> **DO YOU KNOW**
>
> **Bandwidth-delay-product** is a measure of the efficiency of an ARQ system and it is a product of bandwidth (in bps) and the round-trip delay (in seconds).

3.15. FORWARD ERROR CORRECTION

In **forward error correction**, not only error detection technique is used but automatic error correction also takes place.

Error correction : It requires enough additional bits (redundant bits) to deduce what the correct bits must have been.

Examples, Hamming Codes

FEC = Forward Error Correction found in MPEG-4.

> **DO YOU KNOW**
>
> In **forward error correction**, not only error detection technique is used but automatic error correction also takes place.

Hamming Code

Hamming coders are preferred over the Single-Parity Check coders when detection as well as correction is desired. **Hamming codes** *are capable of detecting up to two bits in error and have the ability to correct single bit errors.*

If more error-correcting bits are included with a message, and if those bits can be arranged such that different incorrect bits produce different error results, then bad bits could be identified. In a 7-bit message, there are seven possible single bit errors, so three error control bits could potentially specify not only that an error occurred but also which bit caused the error.

A mathematician named Richard W. Hamming developed the Hamming code at working at Bell Telephone Laboratories in 1950. It is an error-correcting code used for correcting transmission errors in synchronous data streams. However, the hamming code is used to detect only single bit errors but there is process that can be used to detect burst errors.

Now, to understand how errors can be handled, it is necessary to look closely at what an error really is. The Hamming code requires the addition of overhead to message, thus increasing the length of a transmission. The concept of underlying error correction can be easily understood by examine the simplest that is single bit errors. Single bit errors can be detected by adding a redundant or parity bit.

> **DO YOU KNOW**
>
> **Hamming codes** are capable of detecting up to two bits in error and have the ability to correct single bit errors.

A single redundant bit can detect single bit errors in any sequence bits because it must distinguish between only two states either error or no error. The two state of bit, *i.e.,* 0 and 1, are sufficient for this level of detection. These two states are enough only to identify that either error is there or not but not sufficient to correct it. The secret of error correction therefore is to locate that invalid bit or bits. An error occurs when the receiver reads a '0' bit as '1' or '1' bit as '0' bit. To correct the error, the receiver simply reverses the value of the altered bits.

Hamming bits are inserted into a character at random locations. Altogether, the combination of the data bits and hamming bit is known as **hamming code.**

To calculate the number of redundent Hamming bits needed for a given character length, relationship between the character bits and the Hamming bits must be established. Normally a frame consists of '*m*' data (message) bits and 'r' redundant or Hamming or check bits. Let the total lengths be '*n*' *i.e.,* $n = m + r$. A n-bit unit containing data and check bits is often referred to as n-bit code word.

Let us suppose, we have two code word say, 10001001 and 10110001, it is possible to determine how many corresponding bits differ. In this case, we determine, how many bits differ, just EX-OR the two code word and count the number of 1 bits in the result.

From the above example,

$$\begin{array}{r} 10001001 \\ 10110001 \\ \hline 00111000 \\ \hline \end{array}$$

The number of bit positions in which two codewords differ is called **Hamming distance**. The significance of this is that if two code words are Hamming distance '*d*' a part, it will require '*d*' single bit errors to convert are in to the other. In this case Hamming distance in 3.

Figure 3.40 shows a data contain '*m*' character bits and '*r*' Hamming bits. Therefore the total number of bits in one data unit, is $m + r$, since the Hamming bits should be able to identity which bit is in error, '*r*' Hamming bits must be able to indicate at least $m + r + 1$ different codes.

One code indicates that no errors have accumed and the remaining $m + r$ codes indicates the bit position where an error has occured. Therefore, $m + r$ bit position must be identified with '*r*' bits. Since '*r*' bits can produce $2r$ different codes, $2r$ must be equal to greater than $m + r + 1$.

$$2r \geq m + r + 1$$

where m = Number of bits in each data character.

r = Number of redundancy or Hamming bits.

For example, to correct a single bit error in an ASCII character, the error correction code must be able to determine which of the '*r*' bits has changed. In this case, we have to distinguish between eight different states; No error, error in position '1' error in position '2' and so on, up to error in position 7.

To do so, we requires enough redundency code to show eight different states. It seems that a 3-bit redundency code should be enough because 3-bits can show eight different states *i.e.,* 000 to 111 and can therefore indicate the location of eight different posibilities.

Seven bits of data (ASCII code) and 3-bits of redundancy will become 10-bits. Three bits (redundancy bits) however can identity only 8-possibility. Additional bits are require to cover all posible error locations.

If the total number of bits in a transmittable unit is $m + r$, than '*r*' must be able to indicate at least $m + r + 1$ different states.

So $m + r + 1$ states must be discoverable by r bits; and r bits can indicate $2z$ different states. If the value of $m = 7$, than the smallest r value that can saitisfy is 4.

$$2^r = 7 + 4 + 1$$

The following Table 3.4, shows some possible m values and the corresponding r values.

Table 3.4.

Number of data bits (m)	Number of redundancy bits (r)	Total number of bits ($m + r$)
1	2	3
2	3	5
3	3	6
4	3	7
5	4	9
6	4	10
7	4	11

The bits of the code word are numbered consecutively starting with bit 1 at the left end, bit 2 to its immediate right, and so on. The bits which are power of 2 (such as 1, 2, 4, 8, 16, 32 etc.) are Hamming bits and the rest (such as 3, 5, 7, 9, etc.) are filled up with the m data bits, as shown in Fig. 3.40.

11	10	9	8	7	6	5	4	3	2	1
d_1	d_2	d_3	r_8	d_4	d_5	d_6	r_4	d_7	r_2	r_1
			2^3				2^2		2^1	2^0

Fig. 3.40

d = Data bits,

r = Redundancy bits (2^0, 2^1, 2^2, 2^3...)

In the Hamming code, each r bit is the parity bits for one combination of data bits and that can be determined by the Table 3.5.

We can easily get the position of data bits to be considered for calculating the value of redundancy bit (r). For example, let us get the positions of data combination for determining the value r_1 bit. In table 3.5, check the 1's at LSB (r_1), wherever the value of LSB bit is 1, note down the corresponding decimal number, such as 1, 3, 5, 7, 9 and 11. These decimal numbers are denoting the data combination positions to be considerable to determine the value of r_1 bit. Similarly, get the data combination positions for r_2, r_4 and r_8.

Following bits positions to be considered to determine the parity bits (even) values.

$r_1 \rightarrow 1, 3, 5, 7, 9, 11$

$r_2 \rightarrow 2, 3, 6, 7, 10, 11$

$r_4 \rightarrow 4, 5, 6, 7$

$r_8 \rightarrow 8, 9, 10, 11$

Table 3.5.

MSB			LSB	Decimal Number
0	0	0	0	0
0	0	0	1	1
0	0	1	0	2
0	0	1	1	3
0	0	0	0	4
:	:	:	:	:
1	1	1	1	15

Steps Involved for Calculating the Values of Redundancies Bits

(*i*) The first step is to place each bit of the original character in its appropriate position. For example, we have ASCII code for upper case letter '*A*';

1000001

11	10	9	8	7	6	5	4	3	2	1
1	0	0	r_8	0	0	0	r_4	1	r_2	r_1

Hamming bits r_1, r_2, r_4 and r_8.

(*ii*) We calculate the even parity bits for the various bit combinations. Determine the value of r_1 by considering the following bit positions

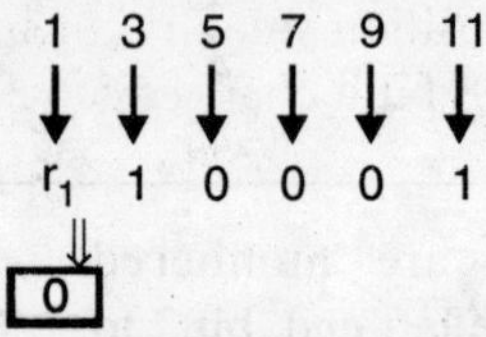

The value of r_1 will be '0' if we calculate the even parity bit for the above combination.

$$r_1 = 0$$

(*iii*) Replace the r_1 with the parity bit '0'.

(*iv*) Same way, determine the value of r_2 by considerity the following bit positions:

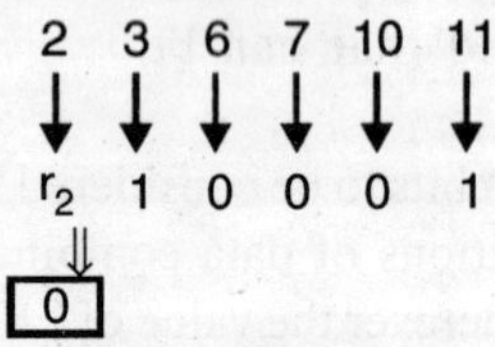

The value of r_2 will be '0' if we calculate the value of even parity bit for the above combination.

$$r_2 = 0$$

(*v*) Replace the r_2 with the parity bits '0'.

(*vi*) The value of r_4 will be calculated by considering the following bit positions

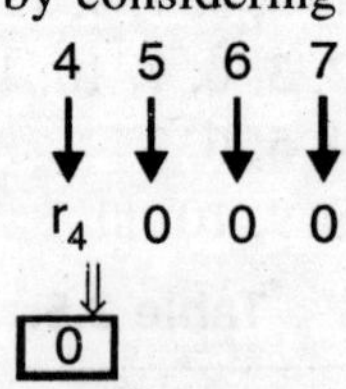

$$r_4 = 0$$

(*vii*) Replace the value of r_4 with the '0' bit.

(*viii*) Again calculate the value of r_8 by considering the following bit positions.

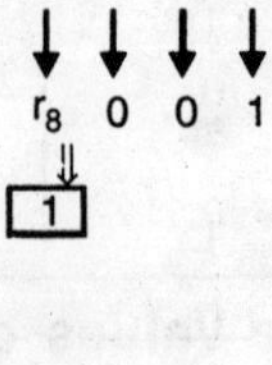

The value of r_8 will be '1' bit.

$$r_8 = 1$$

(*ix*) Replace the r_8 value by '1'.

(*x*) Finally, we get the following redundancy bits, $r = 0$, $r_2 = 0$, $r_4 = 0$, $r_8 = 1$

11	10	9	8	7	6	5	4	3	2	1
1	0	0	1	0	0	0	0	1	0	0

Data → 1000001

Codeword → 10010000100 (To be transmitted)

Steps Involved for Error Detection and Corrections

Let us consider that the 9th bit position value is corrupted when it reached to the receiver. The value of 9th position of the bit was '0' bit it changed to '1' during the transmission. The code word received is

bit values changed from '0' to '1'

1 0 [1] 1 0 0 0 0 1 0 0

(*i*) Recalculate the values of 4 parity bits, using the same sets of bits used by the sender plus the relavent parity *r* bit for each set.

For r_1

1	3	5	7	9	11
0	1	0	0	1	1

Since the value of 9th position is changed '1' that's why the parity value is also changed from '0' to 1, so here

$$r_1 = 1$$

For r_2

2	3	6	7	10	11
0	1	0	0	0	1

The value of r_2 in same as earlier *i.e.*, '0'.

$$r_2 = 0$$

For r_4

4	5	6	7
0	0	0	0

$$r_4 = 0$$

For r_8

8	9	10	11
1	1	0	1

Again, find parity bit value is coming wrong because 9th position bit is changed from 0 to 1.

So, $r_8 = 1$

(*ii*) Then assembles the new parity values in to a binary number in order of *r* position *i.e.*, r_8, r_4 r_2, r_1. In case there in no error then redundancy bits will be (0000). The method is used only for single bit errors.

In our examples,

r_8	r_4	r_2	r_1
1	0	0	1

↓

Decimal value is '9'

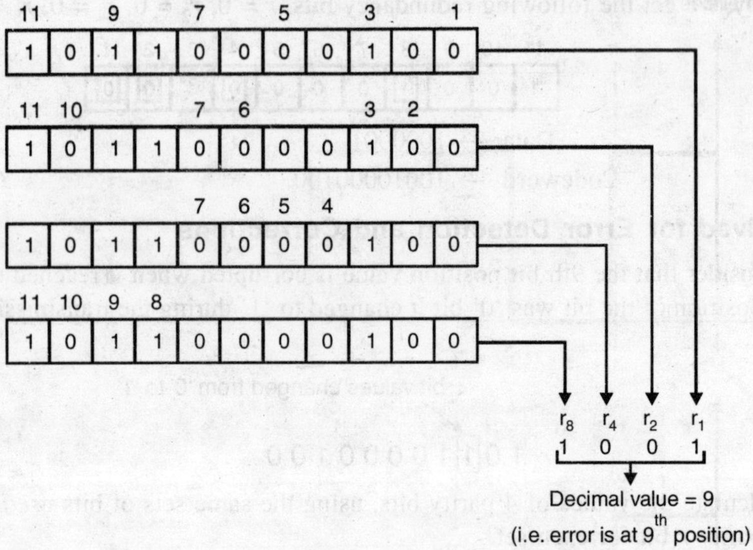

In the example, this step gives us the binary number of 1001 (9 in decimal), which is the precise location of the bit in error.

(*iii*) Once the bit is identified, it is very simple to correct it just reversing its value. So, by reversing the 9th bit position from '1' to '0'. 9th positions bit value changed from 1 to 0.

Finally,

9th position bit value changed from '1' to '0'

1 0 0 1 0 0 0 0 1 0 0

3.16. SYNCHRONIZATIONS

Serialized data is not generally sent at a uniform rate through a channel. Instead, there is usually a burst of regularly spaced binary data bits followed by a pause, after which the data flow resumes. Packets of binary data are sent in this manner, possibly with variable-length pauses between packets, until the message has been fully transmitted. In order for the receiving end to know the proper moment to read individual binary bits from the channel, it must know exactly when a packet begins and how much time elapses between bits. When this timing information is known, the receiver is said to be **synchronized** with the transmitter, and accurate data transfer becomes possible. Failure to remain synchronized throughout a transmission will cause data to be corrupted or lost.

In other words when two devices are about to communicate, the transmitter should somehow notify the receiver as to when to expect to receive data. This allows the receiver to prepare it for receiving the data. Furthermore, such notifications should occur frequently enough so that both devices maintain an agreement about the exact distribution of data over time. This process is called **synchronization**.

3.16.1. Asynchronous vs. Synchronous Transmission

There are two basic methods of synchronization: synchronous transmission and asynchronous transmission. In **synchronous transmission**, a clock signal is used as a common source of reference by both the transmitter and the receiver. By tying the data signal to the clock signal, either device can look at the clock signal to know where data bits may begin or end. The clock signal may be

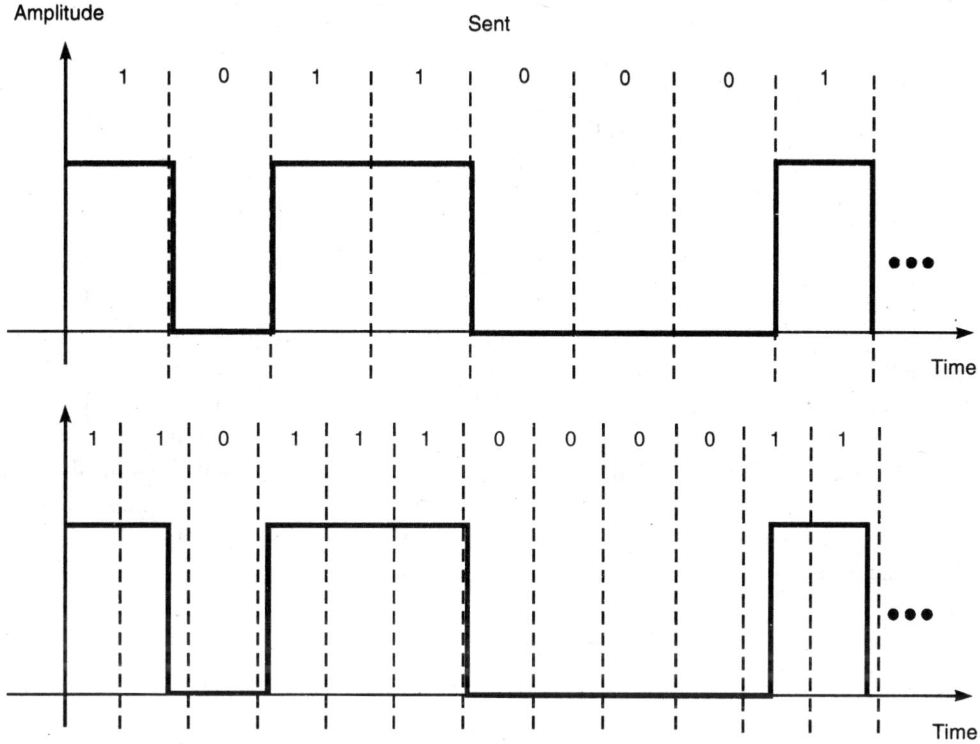

Fig. 3.41. *Lack of synchronisation.*

provided on a separate line, or be embedded in the data signal itself (refer Fig. 3.42). Because having a separate clock line increases the costs, it is only used for covering very short distances (*e.g.,* for connecting personal computers).

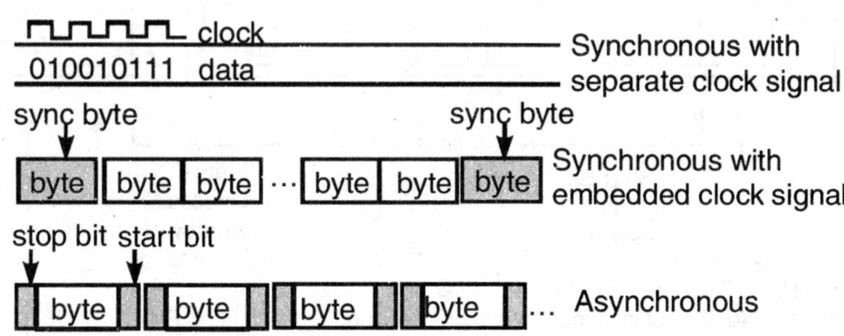

Fig. 3.42. *Synchronous and Asynchronous pattern.*

In synchronous systems, separate channels are used to transmit data and timing information. The timing channel transmits clock pulses to the receiver. Upon receipt of a clock pulse, the receiver reads the data channel and latches the bit value found on the channel at that moment. The data channel is not read again until the next clock pulse arrives. Because the transmitter originates both the data and the timing pulses, the receiver will read the data channel only when told to do so by the transmitter (via the clock pulse), and synchronization is guaranteed.

Techniques exist to merge the timing signal with the data so that only a single channel is required. This is especially useful when synchronous transmissions are to be sent through a modem.

Two methods in which a data signal is self-timed are nonreturn-to-zero and biphase Manchester coding. These both refer to methods for encoding a data stream into an electrical waveform for transmission.

In **asynchronous transmission**, the beginning and end of each byte of data is marked by start and stop bits. This enables the receiver to work out the byte boundaries (see Figure 3.43). Because of its simplicity, asynchronous transmission is cheaper to implement and is therefore more widely used.

In **asynchronous systems**, a separate timing channel is not used. The transmitter and receiver must be preset in advance to an agreed-upon baud rate. A very accurate local oscillator within the receiver will then generate an internal clock signal that is equal to the transmitters within a fraction of a percent. For the most common serial protocol, data is sent in small packets of 10 or 11 bits, eight of which constitute message information. When the channel is idle, the signal voltage corresponds to a continuous logic '1'. A data packet always begins with a logic '0' (the start bit) to signal the receiver that a transmission is starting. The start bit triggers an internal timer in the receiver that generates the needed clock pulses. Following the start bit, eight bits of message data are sent bit by bit at the agreed upon baud rate. The packet is concluded with a parity bit and stop bit. One complete packet is illustrated below:

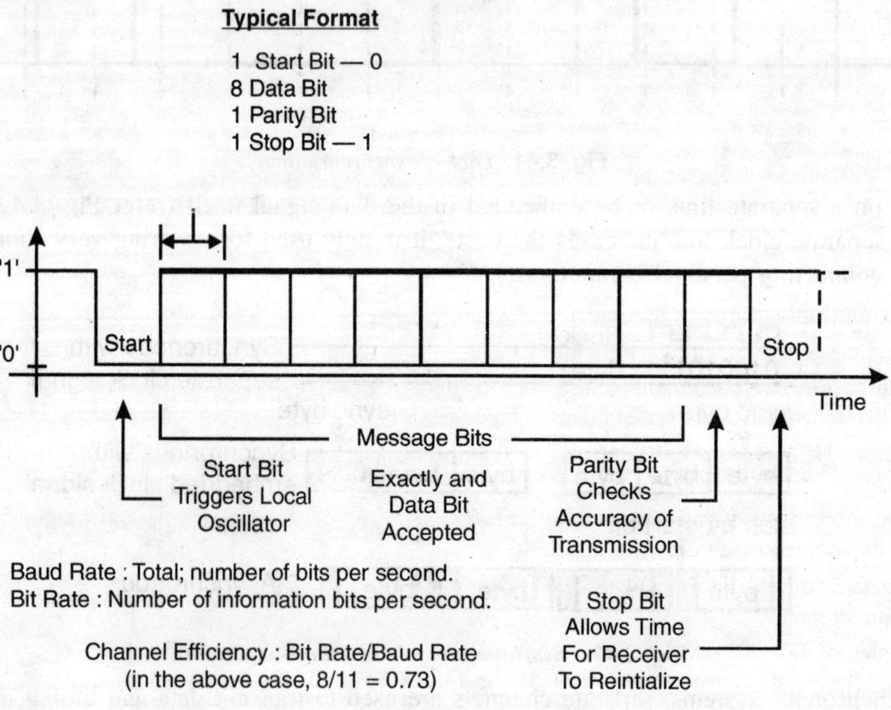

Fig. 3.43.

The packet length is short in asynchronous systems to minimize the risk that the local oscillators in the receiver and transmitter will drift apart. When high-quality crystal oscillators are used, synchronization can be guaranteed over an 11-bit period. Every time a new packet is sent, the start bit resets the synchronization, so the pause between packets can be arbitrarily long. Note that the EIA232 standard defines electrical, timing, and mechanical characteristics of a serial interface.

3.16.2. Isochronous and Asynchronous Transmissions

At first glance, the terms *isochronous* and *synchronous*, as applied to network transmissions, describe essentially the same process. Upon further investigation, however, small but significant differences emerge.

An *isochronous* channel has a constant time interval (or integer multiples thereof) between similar "significant instants." This basic definition means that each isochronous packet in a stream begins a constant time interval after the previous packet in that stream. In IEEE 1394, isochronous packets begin 125 μs apart (at least on average). This period is controlled by a clock packet, transmitted by the IEEE 1394 bus master, which signals devices on the network that it is time to send the next isochronous packet. Another example of an isochronous channel is the 270 Mbits/s serial digital video interface, where a line of video begins every 64 μs (625/50) or 63.56 μs (525/60).

The opposite of isochronous is *anisochronous*. An anisochronous channel does not maintain constant time intervals between significant instants. A good example of this is Ethernet, a technology that does not synchronize data transmission from nodes on the network, but instead relies on signaling when conflicts occur.

The term *synchronous* refers to a timing relationship between two different channels. If two "significant instants" (*e.g.*, packet starts) on two different channels occur at the same time every time, the channels are synchronous. A good example of synchronous channels would be a 4:2:24:4 SDI signal carried on two coaxial cables, where lines of video are completely time aligned on the two cables.

The opposite of synchronous is *asynchronous*. If "significant instants" on two channels do not consistently line up, then the channels are asynchronous.

Summary

- **Data Communications** is the transfer of data or information between a source and a receiver.
- In the data communication world, **data** means the information stored in digital form.
- The purpose of Data Communications is to provide the rules and regulations that allow computers with different disk operating systems, languages, cabling and locations to share resources.
- Standards are developed by national and international organizations established for this exact purpose.
- The basic purpose of a **data communications circuit** is to transfer digital information from one place to another using electronic circuit.
- The Data Terminal Equipment (DTE) is the source and receiver's equipment which is generating the data stream.
- Examples of DTE devices are: PCs, terminals, end user devices.
- The interface between the source and the medium, and the medium and the destination or receiver is called the Data Communication Equipment (DCE) and is a physical piece of equipment.
- Data communication codes are normally used to represent characters and symbols, such as digits, letters and punctuation marks. So these types of data communications codes are called symbol codes, character codes, character sets or character languages.
- No physical medium can guarantee error-free transmission of data.
- *The most important role of the data link layer is to provide error-free transmission.*
- **Errors** occur because of interference from the environment surrounding the transmission medium.

- In case of **single-bit error** only one bit of the data is corrupted and rest all the bits are received correctly.
- In case of **burst error**, multiples bits are corrupted and changed.
- The **burst length** is measured from the first corrupted bit to the last corrupted bit. Some bits in between may be received correctly.
- **Error detection** is the first stage of error correction.
- There are some mechanism used to detect the error in data communication based a concept of **redundancy**.
- **Parity checking** is a simple error detection method used with character oriented protocols. There are two types of parity checking techniques, such as **Even parity** and **Odd parity**.
- There are two types of parity checking techniques, such as **Even parity** and **Odd parity**.
- The CRC is a very powerful but easily implemented technique to obtain data reliability. The **CRC** technique is used to protect blocks of data called Frames.
- One popular method for error detection is the use of checksums. **Checksums** have the advantage of operating on a long sequence of data.
- In **feedback error correction**, only error detection technique is used and whenever errors occur, sender will resend that frame or packet to the receiver.
- In the **ARQ system** of error control, when an error is detected, a request is made for the retransmission of the data.
- In **forward error correction**, not only error detection technique is used but automatic error correction also takes place.
- In **forward error correction**, not only error detection technique is used but automatic error correction also takes place.
- Hamming coders are preferred over the Single-Parity Check coders when detection as well as correction is desired. *Hamming codes are capable of detecting up to two bits in error and have the ability to correct single bit errors.*
- Altogether, the combination of the data bits and hamming bit is known as **hamming code**.
- The number of bit positions in which two codewords differ is called **Hamming distance**.
- When this timing information is known, the receiver is said to be **synchronized** with the transmitter, and accurate data transfer becomes possible.
- There are two basic methods of synchronization: synchronous transmission and asynchronous transmission. In **synchronous transmission**, a clock signal is used as a common source of reference by both the transmitter and the receiver.
- In **asynchronous transmission**, the beginning and end of each byte of data is marked by start and stop bits.
- In **asynchronous systems**, a separate timing channel is not used.
- An *isochronous* channel has a constant time interval (or integer multiples thereof) between similar "significant instants."
- The opposite of isochronous is *anisochronous*. An anisochronous channel does not maintain constant time intervals between significant instants.

Review Questions

1. Given a remainder of "111", a data unit of "10110011", and a divisor of 1001, is there an error is data unit. **[MDU Dec. 2008]**
2. Given the data word "10110011" and one divisor 1001.

   (*a*) Generate the codeword at the sender site.

   (*b*) Generate the dataword from the codeword (assume no error and use binary division).

   [MDU May 2009]

3. Generate the hamming code for the codeword "1000". **[MDU Dec. 2009]**
4. Describe in detail about the data communication circuits. **[MDU Dec. 2009]**
5. Construct the hamming code for one bit sequence 10011101. **[MDU May 2010]**
6. Given the data word 1010011010 and the divisor 10111. Show the conversion of the codeword at the sender side using CRC and check the codeword at the receiver side. **[MDU May 2011]**
7. What do you understand by data communication? Explain various communication modes.
8. Discuss the various standards in data communication.
9. Describe in detail about data communication circuits. **[MDU Dec. 2009]**
10. What are the three popular ARQ mechanisms?
11. Explain the following protocols with the suitable examples in flow control:
    (*i*) Stop-and-wait (*ii*) Sliding window
    (*iii*) Go-back-N-ARQ (*iv*) Selective repeats
12. Compare the Go-back-N-ARQ and selective repeats flow control protocols.
13. Explain stop-and-wait protocol and write its advantages and disadvantages.
14. What do you understand by transmission error and explain the causes of that?
15. What are the types of errors and explain burst error?
16. What do you understand by error correction? Explain the two main approaches of error correction.
17. What are the various types of error detection techniques? Explain parity check with the help of examples. Write the limitations of this method.
18. Explain two-dimensional parity check method and discuss the case when this technique fails.
19. Explain CRC method with the help of example.
20. Explain checksum method with the help of example.
21. What do you understand by forward error correction and explain the method used in this approach.
22. Explain Hamming code method with the help of example.
23. How can a parity bit detect the damaged data unit?
24. Discuss the parity check and the types of errors it can and cannot detect.
25. Explain what are transmission errors? How they can be reduced? Suggest measures.
26. List various transmission errors. Discuss feedback error control approach.
27. Explain error detection and parity check.
28. Write technical notes on hamming codes and their importance.
29. How the error can be detected? Explain any one method for the same.
30. Write the algorithm for computing checksum in CRC and illustrate with example.
31. How does a single bit error differ from a burst error?
32. Discuss the concept of redundancy in error detection.
33. How can a parity bit detect a damaged data unit?
34. What is the difference between even parity and odd parity?
35. Discuss VRC and the types of errors it can and cannot detect.
36. What does the CRC generator append to the data unit?
37. How is VRC related to LRC?
38. Discuss LRC and the types of errors it can and cannot detect?
39. What is the relationship between the size of the CRC remainder and the divisor?
40. How does the CRC checker know that the received data unit is undamaged?
41. What are the conditions for the polynomial used by CRC generator?
42. How is CRC superior to LRC?
43. What is the error detection method used by upper layer protocols?
44. What kind of arithmetic is used to add segments in the checksum generator and checksum checker?
45. List the steps used in creating a checksum.
46. How does the checksum checker know that the received data unit is undamaged?
47. What kind of error is undetectable by the checksum?
48. What is the formula to calculate the number of redundancy bits required to correct a bit error in a given number of data bits?

49. What is the purpose of hamming code?

50. Assuming even parity, find the parity bit for each of the following data units:

1001011

1110111

51. A receiver receives the bit pattern 01101011. If the system is using even parity VRC, is the pattern in error?

52. Find the LRC for the following block of data:

10011001 01101111

53. Given a 10-bit sequence 1010011110 and a divisor of 1011, find CRC. Check your answer.

54. Given a remainder of 111, a data unit of 10110011, and a divisor of 1001, is there an error in the data unit?

55. Find the checksum for the following bit sequence. Assume a 16-bit segment size:

1001001110010011

1001100001001101

56. Find the complement of 1110010001110011.

57. Add 11100011 and 00011100 in one's complement. Interpret the result.

58. For each data unit of the following sizes, find the minimum number of redundancy bits needed to correct one single-bit error:

(*i*) 12

(*ii*) 16

(*iii*) 24

(*iv*) 64

59. Construct the hamming code for the bit sequence 10011101.

60. Calculate the VRC and LRC for the following bit pattern using even parity:

0011101 1100111 1111111 0000000

61. A sender sends 01110001; the receiver receives 01000001. If only VRC is used, can the receiver detect the error?

62. If a divisor is 101101, how many bits long are the CRC?

63. Find the binary equivalent of $x^8 + x^3 + x + 1$.

64. Find the polynomial equivalent of 100001110001.

65. A receiver receives the code 11001100111. When it uses the hamming encoding algorithm, the result is 0101. Which bit is in error? What is the correct code?

66. The code 11110101101 was received. Using hamming code what is the original code sent?

67. In a single-bit error correction, a code of three bits can be in one of the four states: no error, first bit in error, second bit in error, third bit in error. How many of these bits should be redundant to correct this code? How many bits can be the actual data?

68. Write short notes on following :

(*i*) Error correction and detection

(*ii*) Parity check

(*iii*) Frame check sequence

(*iv*) ARQ

(*v*) Checksum

(*vi*) Hamming codes

(*vii*) Cyclic redundancy check **[MDU May 2005, 2008]**

69. Write short notes on following :

(*i*) Acknowledgement and negative Acknowledgement

(*ii*) Timer

(*iii*) ARQ

(*iv*) Sliding window

(*v*) Size of sliding window in selective repeats.

❑❑❑

4

CHAPTER

Data Communication Hardwares Devices

4.1. INTRODUCTION: NETWORK HARDWARE DEVICES

INSIDE THIS CHAPTER

In this topic we focus on one particular set of network hardware devices—network interfaces, repeater/hubs, switches, bridges, routers, and firewalls. These devices interconnect individual computers and ensure that they communicate efficiently.

In order to understand the role of these devices, it is helpful to consider analogy between networks and information highways. In this analogy, the network manager is like a city planner. The manager defines the resources that will be available to the network town—for example, student laboratories, teachers' computers, accounting and so forth. The manager also decides where to locate those resources and the capacity of the roads leading to them. For example, you network will contain small streets (less costly to implement and maintain) for areas with low traffic, large streets with traffic lights for areas of moderate traffic, and highways with limited access for areas of high traffic density.

Finally, the manager considers how and when to restrict access to some resources that should remain private.

Network interfaces, hubs, bridges, switches, routers, and firewalls work together in a number of ways to create these different kinds of network roadways.

4.2. REPEATERS

Repeaters are physical hardware devices that have a primary function to regenerate the electrical signal by:

- Reshaping the waveform
- Amplifying the waveform
- Retiming the signal

As signals travel along a network cable (or any other medium of transmission), they degrade and become distorted in a process that is called attenuation. If a cable is long enough, the attenuation will finally make a signal unrecognizable by the receiver.

A Repeater enables signals to travel longer distances over a network. Repeaters work at the OSI's Physical layer. A repeater regenerates the received signals and then retransmits the regenerated (or conditioned) signals on other segments.

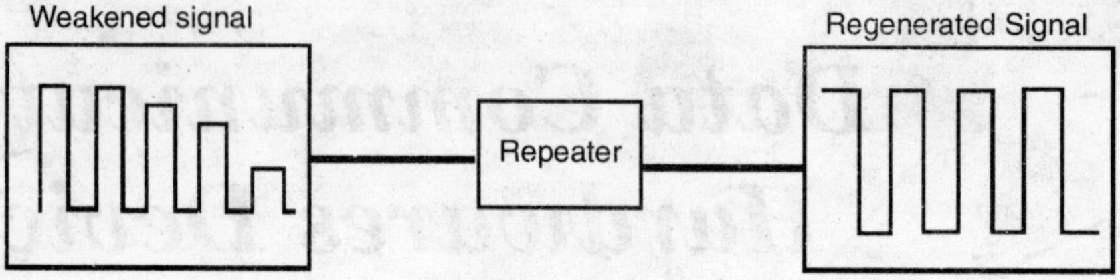

Fig. 4.1. *Repeater.*

4.3. HUBS

In the networking technology, the special type of network device called hub can be seen around homes and business networks. The introduction to hubs in networking can be explained as a small rectangular box that is usually made of plastic. This hub which joins multiple computers and networks receives its power from an ordinary wall outlet and forms a single network segment. An important contribution that emerges out of the hub is the facility that it provides for a direct communication which all computers on the same network can access to.

A networking hub that provides direct accessibility to all computers on the system comprises of a series of ports. Each of such ports accepts a network cable and hence provide for the accessibility of all the linked computers. The introduction of hubs is not complete till one gets an insight into its basic characteristics. A reasonably advanced IT infrastructure, literacy, multi language capabilities, professionals with communication skills and creativity, credibility and rapport with the user community and series of ports are important components that form part of hubs.

Hubs are also called Multiport Repeaters or Concentrators. They are physical hardware devices. In the physical form, a hub is a small rectangular box usually made of plastic.

A hub that joins multiple computers and networks receives its power from an ordinary wall outlet and forms a single network segment. Hub as a facility provides for a direct communication channel which all computers on the same network can access to and use the system accordingly.

Some Hubs are basic hubs with minimum intelligence - no microprocessors. Intelligent Hubs can perform basic diagnostics and test the nodes to see if they are operating correctly. If they are not, the Smart Hubs or Intelligent Hubs will remove the node from the network. Some Smart Hubs can be polled and managed remotely.

Purpose of Hubs

Hubs are used to provide a Physical Star Topology. The Logical Topology is dependent on the Medium Access Control Protocol. At the center of the star is the Hub with the network nodes on the tips of the star.

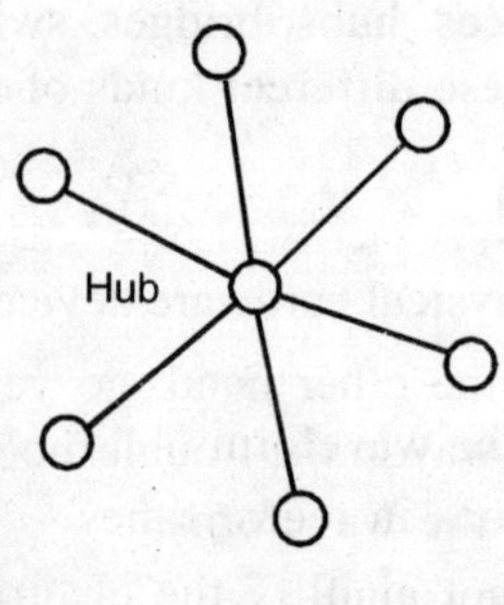

Fig. 4.2. *Star topology hub.*

The Hub is installed in a central wiring closet with all the cables extending to the network nodes. The advantage of having a central wiring location is that it is easier to maintain and troubleshoot large networks. All of the network cables come to the central hub, it is especially easy to detect and fix cable problems. You can easily move a workstation in a star topology by changing the connection to the hub at the central wiring closet.

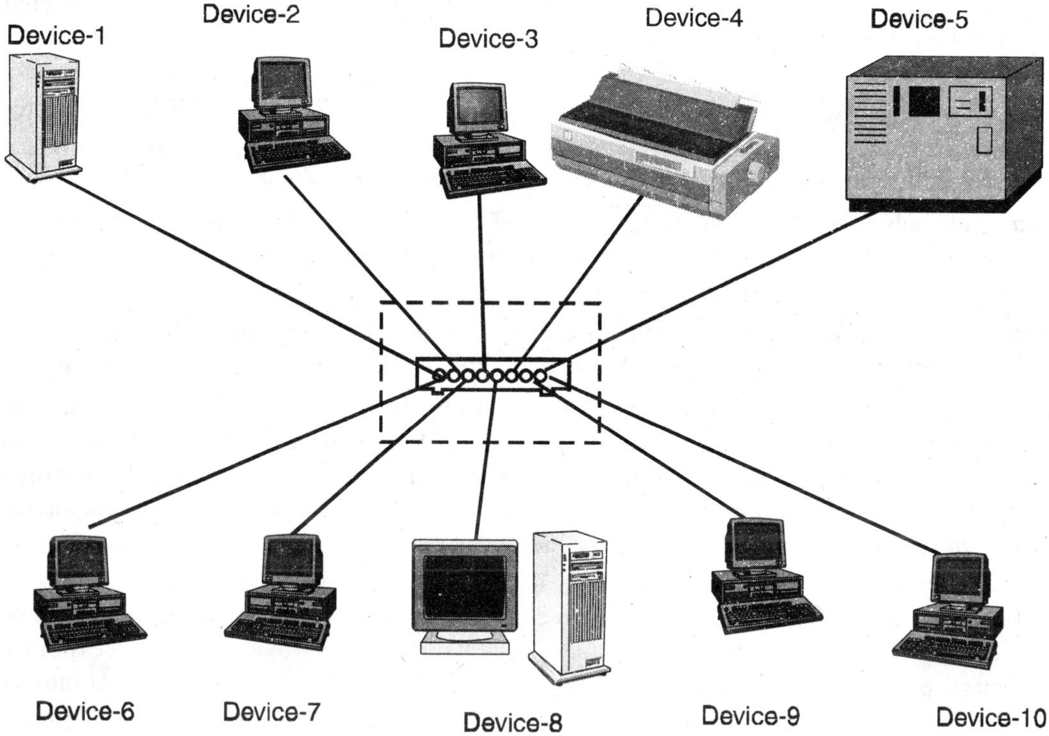

Fig. 4.3. *Various devices connected through Hub.*

Types of Hub

Hubs are the layer 1 devices while switches and routers are layer 2 and layer 3 devices respectively. All kinds of hubs have some common features that are determined primarily by the types of cabling attached to the system. Usually, it could be regarded as a network device that works within the standard parameters of the specific network that it actually is working within.

Hubs are three types:

(1) Passive hubs

(2) Active Hubs

(3) Smart Hubs.

(1) Passive Hubs

A passive hub on the other hand, do very little to enhance the performance of the network. Neither, it helps in any way in the troubleshooting operations which have become an integral part of the networking operations in recent times

Passive hubs do not amplify the electrical signal of incoming packets before broadcasting them out to the network.

These hubs are nothing more than point contacts for the wires that make up the physical network. An example of this would be a punchdown block that is a simple plastic, unpowered box used to plug network cables into it.

From the cost perspective, most of the passive hubs are easily obtainable at a lesser cost. A passive hub when upgraded from 10base-2 then the performance becomes quite impressive and can match with the best.

(2) Active Hubs

Active hubs, on the other hand, do perform this amplification, as does a different type of dedicated network device called a repeater. Some people use the terms **concentrator** when referring to a passive hub and **multiport repeater** when referring to an active hub.

An active hub possesses all the usual features of a passive hub besides having some more. An active hub takes a larger role in Ethernet communications with the help of technology called store and forward.

Here the hub actually goes through the data before they are transmitted. In cases wherever needed, an active hub repair the data and reschedule the distribution of databases accordingly.

In a situation of data received being weak but readable, the active hub restores the signal before rebroadcasting the same. It enables a certain number of devices which are not operating at their optimal capacity still to be used in the network. Another aspect of an active hub is that it provides information on devices on the network which are not yet fully functional. An active hub also provides certain performance advantages in addition to, certain diagnostic capabilities.

These hubs usually come in configurations of 4, 8, 16 and 24 ports, providing link & activity led's (lights) to show which devices are currently connected, powered on and transmitting or receiving data.

All Ethernet hubs are active hubs.

(3) Smart Hubs

Smart Hubs or **Intelligent hubs** add extra features to an active hub that are of particular importance to businesses.

Intelligent hub is another form of hub that is increasingly being used. An advanced version that comprises the best of both active and passive hubs, it provides with the ability to manage the network from one central location. With the help of an intelligent hub, one can easily identify, diagnose problems and even come up with remedial solutions. This troubleshooting of a large enterprise scale network is possible with the help of an intelligent hub. In addition to, an intelligent hub can offer flexible transmission rates to various devices. With standard transmission rates of 10, 16 and 100Mbps to desktop systems using popular technologies like Ethernet, FDDI or Token Ring, an intelligent hub easily incorporates the better of the other two hubs in terms of features and benefits. No wonder, hubs have become an integral part of the current networking systems.

An intelligent hub typically is stackable (built in such a way that multiple units can be placed one on top of the other to conserve space). It also typically includes remote management capabilities via SNMP and virtual LAN (VLAN) support.

4.4. BRIDGES

Bridges are both hardware and software devices. They can be standalone devices - separate boxes specifically designed for bridging applications, or they can be dedicated PCs with 2 NICs and bridging software. Most servers' software will automatically act as a bridge when a second NIC card is installed.

The purposes of a Bridge are:

- Isolates networks by MAC addresses
- Manages network traffic by filtering packets
- Translate from one protocol to another

Like a repeater, a bridge can join segments or workgroup LANs. However, a bridge can also divide a network to isolate traffic or problems. For example, if the volume of traffic from one or two computers or a single department is flooding the network with data and slowing down entire operation, a bridge can isolate those computers or that department.

In the following Fig. 4.4, a bridge is used to connect two segments; segment 1 and segment 2.

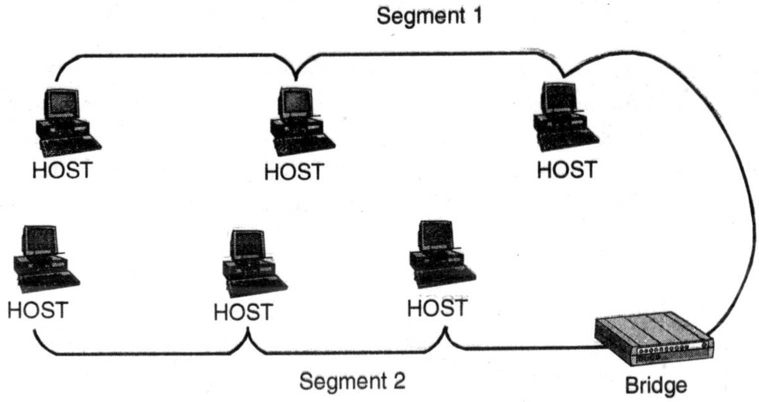

Fig. 4.4. *Bridges.*

Bridges can be used to:

- Expand the distance of a segment.
- Provide for an increased number of computers on the network.
- Reduce traffic bottlenecks resulting from an excessive number of attached computers.

Bridges work at the Data Link Layer of the OSI model. Because they work at this layer, all information contained in the higher levels of the OSI model is unavailable to them. Therefore, they do not distinguish between one protocol and another.

When a packet is received by the bridge, the bridge determines the destination and source segments. If the segments are the same, the packet is dropped ("filtered"); if the segments are different, then the packet is "forwarded" to the correct segment. Additionally, bridges do not forward bad or misaligned packets.

Bridges are also called "store-and-forward" devices because they look at the whole Ethernet packet before making filtering or forwarding decisions. Filtering packets and regenerating forwarded packets enable bridging technology to split a network into separate collision domains. This allows for greater distances and more repeaters to be used in the total network design.

Bridges simply pass all protocols along the network. Because all protocols pass across the bridges, it is up to the individual computers to determine which protocols they can recognize.

A bridge works on the principle that each network node has its own address. A bridge forwards the packets based on the address of the particular destination node.

As traffic passes through the bridge, information about the computer addresses is then stored in the bridge's RAM. The bridge will then use this RAM to build a routing table based on source addresses.

4.5. ROUTERS

In an environment consisting of several network segments with different protocols and architecture, a bridge may not be adequate for ensuring fast communication among all of the segments. A complex network needs a device, which not only knows the address of each segment, but also can determine the best path for sending data and filtering broadcast traffic to the local segment. Such device is called a **Router.**

Routers are hardware and software devices. They can be cards that plug into a collapsed backbone, stand-alone devices (rack mount or desktop) or software that would run on a file server with 2 NICs.

Routers work at the Network layer of the OSI model meaning that the Routers can switch and route packets across multiple networks. They do this by exchanging protocol-specific information between separate networks. Routers have access to more information in packets than bridges, and use this information to improve packet deliveries. Routers are usually used in a complex network situation because they provide better traffic management than bridges and do not pass broadcast traffic.

Routers can share status and routing information with one another and use this information to bypass slow or malfunctioning connections.

Routers do not look at the destination node address; they only look at the network address. Routers will only pass the information if the network address is known. This ability to control the data passing through the router reduces the amount of traffic between networks and allows routers to use these links more efficiently than bridges Routers filter out network traffic by specific protocol rather than by packet address. Routers also divide networks logically instead of physically. An IP router can divide a network into various subnets so that only traffic destined for particular IP addresses can pass between segments. Network speed often decreases due to this type of intelligent forwarding. Such filtering takes more time than that exercised in a switch or bridge, which only looks at the Ethernet address. However, in more complex networks, overall efficiency is improved by using routers.

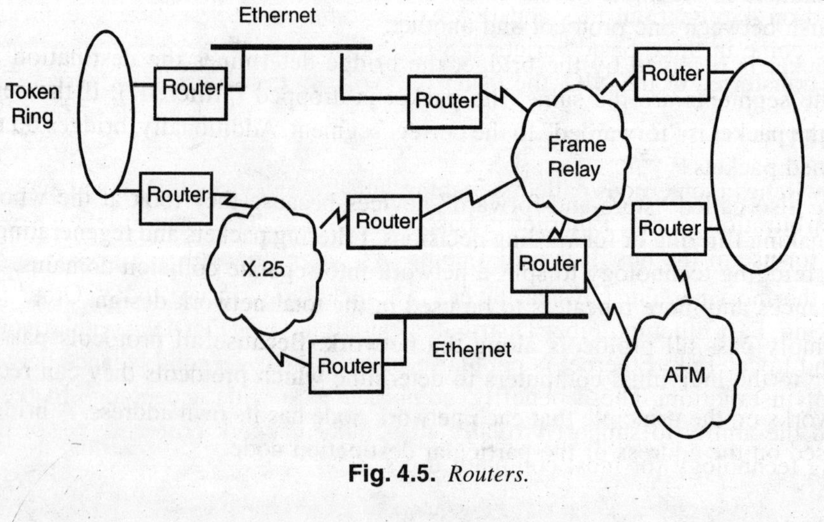

Fig. 4.5. *Routers.*

Purpose of Routers

The purpose of a router is to connect nodes across an internet work regardless of the Physical Layer and Data Link Layer protocol used. Routers are hardware and topology independent. Routers are not aware of the type of medium or frame used (Ethernet, Token Ring, FDDI, X.25, etc...). Routers are aware of the Network Layer protocol used: Novell's IPX, UNIX's IP, XNS, Apples DDP, etc.

4.6. BROUTERS (BRIDGE/ROUTERS)

Brouters are protocol dependant devices. When a brouter receives a frame to be forwarded to the remote segment, it checks to see if it recognizes the Network layer protocol. If the Brouter does, it acts like a router and finds the shortest path. If it doesn't recognize the Network layer protocol, it acts like a bridge and forwards the frame to the next segment.

The key advantage to Brouters is the ability to act as both a bridge and a router. It can replace separate bridges and routers, saving money. This is, of course, provided that the Brouter can accomplish both functions.

4.7. GATEWAYS

Gateways make communication possible between different architectures and environments. They repackage and convert data going from one environment to another so that each environment can understand the other's environment data. One definition of a Gateway is the Hardware/Software device that is used to interconnect LANs & WANs with mainframe computers such as DECnet and IBM's SNA.

Often the router that is used to connect a LAN to the Internet will be called a gateway. It will have added capability to direct and filter higher layer protocols (layer 4 and up) to specific devices such as web servers, ftp servers and e-mail servers.

4.8. NIC (NETWORK INTERFACE CARD)

A NIC or **Network Interface Card** is a circuit board or chip, which allows the computer to communicate to other computers on a Network. This board when connected to a cable or other method of transferring data such as infrared can share resources, information and computer hardware.

The NIC provides a physical connection between the networking cable and the computer's internal bus. Different computers have different bus architectures; PCI bus master slots are most commonly found on 486/Pentium PCs and ISA expansion slots are commonly found on 386 and older PCs. NICs come in three basic varieties: 8-bit, 16-bit, and 32-bit. The larger the number of bits that can be transferred to the NIC, the faster the NIC can transfer data to the network cable.

4.9. ETHERNET

Ethernet Switches (sometimes called switching hubs) are devices that have changed the way Ethernet's bus arbitration works. Ethernet is based on CSMA/CD, a shared bus where all nodes listen and fight for use of the bus. It is a first come first serve protocol.

Ethernet is the most popular physical layer LAN technology in use today. Other LAN types include Token Ring, Fast Ethernet, Fiber Distributed Data Interface (FDDI), Asynchronous Transfer Mode (ATM) and Local Talk. Ethernet is popular because it strikes a good balance between speed, cost and ease of installation. These benefits, combined with wide acceptance in the computer marketplace and the ability to support virtually all popular network protocols, make Ethernet an ideal networking technology for most computer users today.

4.10. CONNECTORS

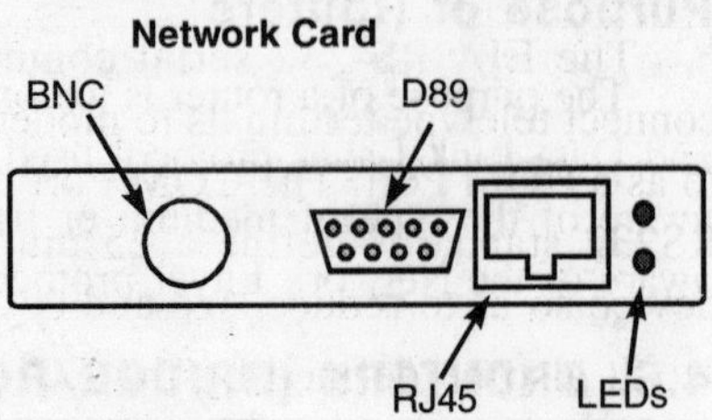

Fig. 4.6. *Connectors.*

Network cards have three main types of connectors. Below is an example of what a network card may look like.

BNC connector : As illustrated in the above picture the BNC connector is a round connector, which is used for thin net or 10Base-2 Local Area Network.

DB9 (RJ45 JACK): The DB9 connector not to be confused with the Serial Port or sometimes referred to as the RJ45 JACK not to be confused with the RJ45 connection is used with Token Ring networks.

DB15 Connector: The DB15 connector is used for a Thick net or 10Base-5 Local area network.

RJ45 connector: Today one of the most popular types of connections used with computer networks. RJ45 looks similar to a phone connector or RJ11 connector however is slightly larger.

LED : The LED's as shown in the above illustration indicates if it detects a network generally by a green light which may flash as it communicates and then a red light which indicates collisions which will generally flash or not flash at all.

4.11. PHYSICAL LAYER INTERFACE STANDARDS

Founded in 1924, the EIA (Electronic Industries Alliance) is a U.S. organization of electronics manufacturers. The EIA has published a number of standards related to telecommunication and computer communication, and works closely with other associates such as ANSI and the ITU.

The primary EIA standards for telecommunication define the serial interface between modems and computers. The most popular are the RS-232-C, RS-449, RS-422, and RS-423 serial interfaces. The physical layer specifications define 37-pin (DB-37), 25-pin (DB-25), and 9-pin (DB-9) connectors and associated cable, as well as electrical characteristics such as the type of signal used on each pin and the timing of those signals.

Note that the EIA RS-232 standard is also the CCITT standard V.24. CCITT is part of the ITU. The CCITT V series protocols are more commonly used than the EIA standards, partly because they are used in Europe where government standards dictate the type of protocol to be used. For the most part, EIA standards have CCITT equivalents.

It is important to note that the various standards introduced define not only the form of the electrical signals to be used, but also a complete range of additional signals, which control the timing and order of data transfer across the appropriate interfaces. At large, these signals are said to form the physical layer interface standard. We have many types of physical layer interfaces such as, RS-232, X.21, RS-449, RS-550, V-35, 1.430, 1.431, T1, E1, etc. We shall discuss some of the important physical layer interfaces.

4.11.1. RS-232 Interface

RS232 interface is used to connect Data terminal equipment (DTE) and Data circuit terminating equipment (DCE). It is used for serial interface. In United States, a standard interconnection between business machine and modem is given by the RS-232 interface. In 1962 EIA has standardized some set of standard specifications to ensure compatibly between data sets (DCE) and terminal equipment (DTE). The interface uses a 25 pin connector, where each of the 25 pins has been assigned a specific function by Electronic Industries Association (EIA).

The RS-232 interface specifications limits the interconnecting cable to length of 15 meter with data rate of 20kbps.

The EIA RS-232 serial communication standard is a universal standard, originally used to connect teletype terminals to modem devices. In a modern computer the RS-232 interface is referred to as a COM Port. The COM Port uses a 9 pin D-Type connector to attach to the RS-232 cable. The RS232 standards define a 25 pin D-Type connector but IBM reduced this connector to a 9 pin device so as to reduce size and cost.

It also be defined by the EIA that the voltage levels with which data and control signals are exchanged between DCE and DTE. Each pin in the connector carries either binary 1 or 0 to indicate activation or deactivation of control functions or data values. The RS-232 interface can accommodate many different type of circuit operation, using different combination of circuit lines. For example, point to point dedicated system will require a minimum number of control lines in the interface.

Signaling Levels

As stated earlier the physical layer converts the bit stream 0's and 1's into disturbances. Within the RS 232 standards the disturbances are represented as voltages along the transmission medium, usually copper wire between the sending and receiving DTE/DCE. The strength of the voltage travelling along the copper wire determines whether it recognized as a 0 or 1 by the receiving physical layer. This is graphically represented as shown in Fig. 4.7.

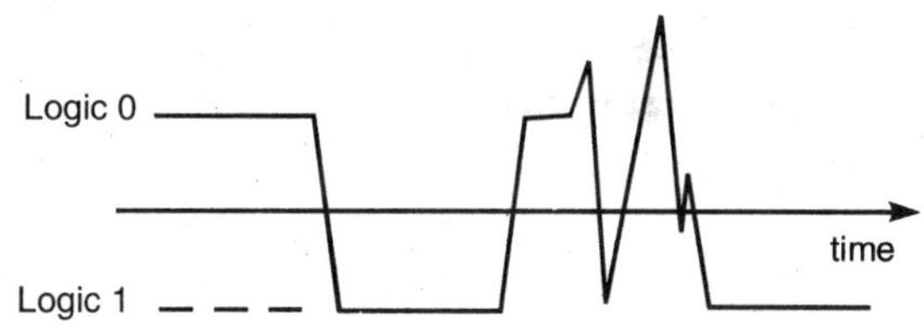

Fig. 4.7. *Representations of 0 and 1 in RS 232.*

The diagram shows that a 1 is represented by the electrical signal which is a negative voltage and a 0 is a positive voltage. The receiving physical layer is able to recognize these positive and negative voltages and convert them into a recognizable bit stream of which the data link layer can interpret. The passing of disturbances by the physical layer is graphically represented as follows.

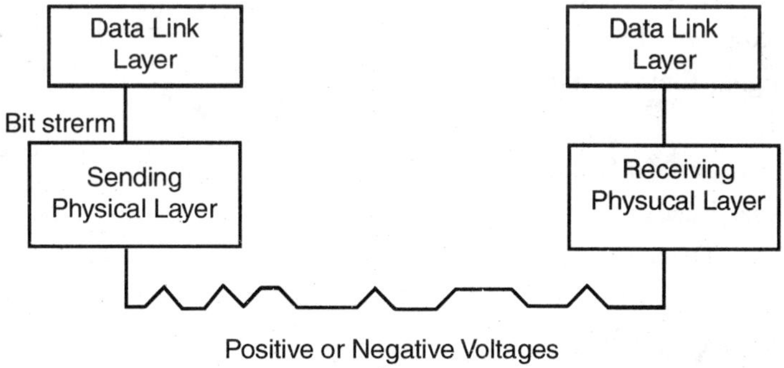

Fig. 4.8. *The passing of disturbances by the physical layer*

As can be seen in this illustration, the sending physical layer converts the bit stream into voltages, transmits them along the transmission medium, the receiving physical layer recognizes the voltages as either positive or negative and converts them back into 0's and 1's so they are recognizable by the data link layer.

The differences in the voltage become important when cable maximum lengths are being reached between DTE and DCE. The larger range of voltage at the receiver means it may be able to pick up a weaker signal which may have been caused by a long cable.

The specific voltage levels for the RS 232 standard are :

Table 4.1. Voltage Levels for the RS 232 Standard

State	Voltage Level TX	Voltage Level RX
Logic 0	+5v to +15v	+3v to +25v
Logic 1	–5v to –15v	–3v to –25v

RS-232 Signals

The number preceding each signal name correspond to the pin number defined in the standard

1. Protective Ground
2. Transmitted Data
3. Received Data
4. Request to Send
5. Clear To Send
6. Data Set Ready
7. Signal Ground
8. Received Line Signal Detect (Carrier Detect)
9. +P (for testing only)
10. -P (for testing only)
11. (unassigned)
12. Secondary Received Line Signal Detect
13. Secondary Clear To Send
14. Secondary Transmitted Data
15. Transmission signal element Timing
16. Secondary Received Data
17. Receiver Signal Element Timing
18. (unassigned)
19. Secondary Request To Send
20. Data Terminal Ready
21. Signal Quality Detector
22. Ring Indicator
23. Data Signal Rate Selector
24. Transmitter Signal Element Timing (unassigned)

RS-232 Signals Functional Description

The first letter of the EIA signal name categorizes the signal into one of five groups, each representing a different "circuit":

1. A - Ground
2. B - Data
3. C - Control
4. D - Timing
5. S - Secondary channel

Pin-1 : Protective Ground

This pin is usually connected to the frame of one of the devices, either the DCE or the DTE, which is properly grounded. The sole purpose of this connection is to protect against accidental electric shock and usually this pin should not be tied to Signal Ground.

This pin should connect the chassis (shields) of the two devices, but this connection is made only when connection of chassis grounds is safe and it is considered optional.

Pin-2 : Transmit Data

Serial data (primary) is sent on this line from the DTE to the DCE. The DTE holds this line at logic 1 when no data are being transmitted. A "On" (logic 0) condition must be present on the following signals, where implemented, before data can be transmitted on this line: CA, CB,CC and CD (Request To Send, Clear To Send, Data Set Ready, Data Terminal Ready).

Pin-3 : Receive Data

Serial data (primary) is sent on this line from the DCE to the DTE. This pin is held at logic 1 (Mark) when no data are being transmitted, and is held "Off" for a brief interval after an "On" to "Off" transition on the Request To Send line, in order to allow the transmission to complete.

Pin-4 : Request To Send

This pin enables transmission circuits. The DTE uses this signal when it wants to transmit to the DCE. This signal, in combination with the Clear To Send signal, coordinates data transmission between the DTE and the DCE.

A logic 0 on this line keeps the DCE in transmit mode. The DCE will receive data from the DTE and transmit it on to the communication link.

The Request To Send and Clear To Send signals relate to a half- duplex telephone line. A half duplex line is capable of carrying signals on both directions but only one at a time. When the DTE has data to send, it raises Request To Send, and then waits until the DCE changes from receive to transmit mode. This "On" to "Off" transition instructs the DCE to move to "transmit" mode, and when a transmission is possible, the DCE sets Clear To Send and transmission can begin.

On a full duplex line, like a hard-wired connection, where transmission and reception can occur simultaneously, the Clear To Send and Request To Send signals are held to a constant "On" level.

An "On" to "Off" transition on this line instructs the DCE to complete the transmission of data that is in progress, and to move to a "receive" (or "no transmission") mode.

Pin-5 : Clear To Send

Clear To Send signal is an answer signal to the DTE. When this signal is active, it tells the DTE that it can now start transmitting (on Transmitted Data line). When this signal is "On" and the Request To Send, Data Set Ready, and Data Terminal Ready are all "On", the DTE is assured that its data will be sent to the communications link. When "Off", it is an indication to the DTE that the DCE is not ready, and therefore data should not be sent.

When the Data Set Ready and Data Terminal Ready signals are not implemented, in a local connection which does not involve the telephone network, the Clear To Send and Request To Send signals are sufficient to control data transmission.

Pin-6 : Data Set Ready

On this line the DCE tells the DTE that the communication channel is available (*i.e.,* in an automatic calling system, the DCE (modem) is not in the dial, test or talk modes and therefore is available for transmission and reception). It reflects the status of the local data set, and does not indicate that an actual link has been established with any remote data equipment.

Pin-7 : Signal Ground

This pin is the reference ground for all the other signals, data and control.

Pin-8 : Receive Line Signal Detect or Data Carrier Detect

The DCE uses this line to signal the DTE that a good signal is being received (a "good signal" means a good analog carrier that can ensure demodulation of received data).

Pin-9 : +P

This pin is held at +12 volts DC for test purposes.

Pin-10 : –P

This pin is held at -12 volts DC for test purposes.

Pin-12 : Secondary Receive Line Signal Detect

This signal is active when the secondary communication channel is receiving a good analog carrier (same function as the Receive Line Signal Detect signal).

Pin-13 : Secondary Clear To Send

Secondary Clear To Send signal is an answer signal to the DTE. When this signal is active, it tells the DTE that it can now start transmitting on the secondary channel (on the Secondary Transmitted Data line).

Pin-14 : Secondary Transmitted Data

Serial data (secondary channel) is sent on this line from the DTE to the DCE. This signal is equivalent to the Transmitted Data line except that it is used to transmit data on the secondary channel.

Pin-15 : Transmission Signal Element Timing

The DCE sends the DTE a clock signal on this line. This enables the DTE to clock its output circuitry which transmits serial data on the Transmitted Data line.

The clock signal frequency is the same as the bit rate of the Transmitted Data line. A "On" to "Off" transition should mark the center of each signal element (bit) on the Transmitted Data line.

Pin-16 : Secondary Receive Data

Serial data (secondary channel) is received on this line from the DCE to the DTE. When the secondary channel is being used only for diagnostic purposes or to interrupt the flow of data in the primary channel, this signal is normally not provided.

Pin-17 : Receiver Signal Element Timing

The DCE sends the DTE a clock signal on this line. This clocks the reception circuitry of the DTE which receives serial data on the Received Data line.

The clock signal frequency is the same as the bit rate of the Received Data line (BB). The "On" to "Off" transition should indicate the center of each signal eleme (bit) on the Received Data line.

Pin-19 : Secondary Request To Send

The DTE uses this signal to request transmission from the DCE on the secondary channel. It is equivalent to the Request To Send signal.

When the secondary channel is only used for diagnostic purposes or to interrupt the flow of data in the primary channel, this signal should turn "On" the secondary channel unmodulated carrier.

Pin-20 : Data Terminal Ready

When on, tells the DCE that the DTE is available for receiving. This signal must be "On" before the DCE can turn Data Set Ready "On", thereby indicating that it is connected to the communications link.

The Data Terminal Ready and Data Set Ready signals deal with the readiness of the equipment, as opposed to the Clear To Send and Request To Send signals that deal with the readiness of the communication channel.

When this signal is "Off", it causes the DCE to finish any transmission in progress and to be removed from the communication channel.

Pin-21 : Signal Quality Detector

This line is used by the DCE to indicate whether or not there is a high probability of an error in the received data. When there is a high probability of an error, it is set to "Off", and is "On" at all other times.

Pin-22 : Ring Indicator

On this line the DCE signals the DTE that there is an incoming call. This signal is maintained "Off" at all times except when the DCE receives a ringing signal.

Pin-23 : Data Signal Rate Selector

The DTE uses this line to select the transmission bit rate of the DCE. The selection is between two rates in the case of a dual rate synchronous connection, or between two ranges of data rates in the case of an asynchronous connection.

Typically, when this signal is "On", it tells the DCE (modem) that the receive speed is greater than 600 baud.

Pin-24 : Transmitter Signal Element Timing

The DTE sends the DCE a transmit clock on this line. This is only when the master clock is in the DTE.

An "On" to "Off" transition should indicate the center of each signal element (bit) on the Transmitted Data line.

It is to be noted that the pin names are the same for the DCE and DTE. The Transmit Data (pin number 2) is a transmit line on the DTE and a receive line on the DCE, Data Set Ready (pin number 6) is a receive line on the DTE and a transmit line on the DCE, and so forth.

Electrical Signal Characteristics

Voltage levels defined in the standard

Data signals	"0", "Space"	"1", "Mark"
Driver (Required)	5 to 15	–5 to –15 Volts
Terminator (expected)	3 to 25	–3 to –25 Volts
Control signals	"Off"	"On"
Driver (Required)	–5 to –15	5 to 15 Volts
Terminator (expected)	–3 to –25	3 to 25 Volts

Physical Connectivity

The RS 232 standard defines the requirements of the physical connectors which interconnect the DTE to the DCE and to the transmission medium. Having the correct connectors allows the equipment to be interconnected but does not necessarily make it work.

Common connectors in use are as follows:

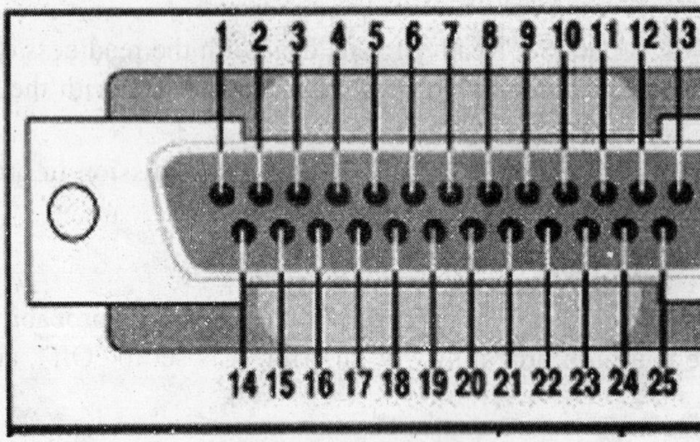

Fig. 4.9. *Pin 25 DB-25 connector*

Pin description		EIA Circuit	From DCE	To DCE
1	Frame Ground	AA		
2	Transmitted Data	BA		D (Data)
3	Received Data	BB	D	
4	Request to Send	CA		C (Control)
5	Clear to Send	CB	C	
6	Data Set Ready	CC	C	
7	Signal Gnd/Common Return	AB		
8	Received Line Signal Detector	CF	C	
11	Undefined			
12	Secondary Received Line Signal Detector	SCF	C	
13	Secondary Clear to Send	SCB	C	
14	Secondry Transmitted Data	SBA	C	
15	Transmitter Signal Element Timing	DB	T (Timing)	
16	Secondary Received Data	SBB	D	
17	Receiver Signal Element TIming	DD	T	
18	Undefined			
19	Secondary Request to Send	SCA		C
20	Data Terminal Ready	CD		C
21	Signal Quality Detector	CG		C
22	Ring Indicator	CE	C	
23	Data Signal Rate Selector (DCE)	CI	C	
24	Transmitter Signal Element Timing	DA		T
25	Undefined			

Fig. 4.10. *The 25 pin RS 232 connector defined in the standard*

The 25 pin RS 232 connector is the one defined in the standard shown in Fig. 4.11.

This connector is nine pin RS 232 connector referred as DB-9 and developed by IBM. On a PC, this is commonly referred to as the serial port.

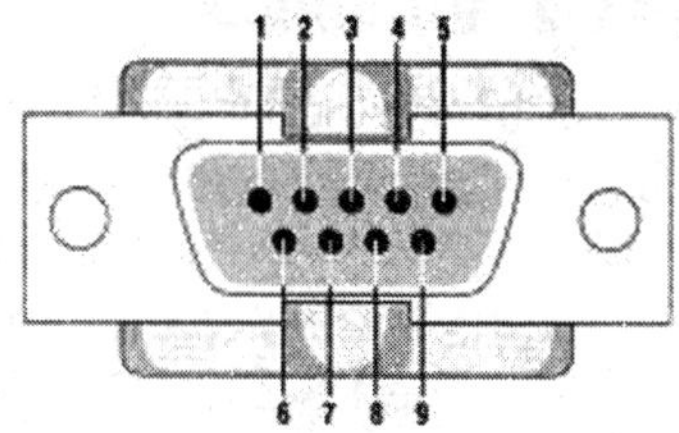

Fig. 4.11. *Pin 9 DB-9 Connector*

Table 4.2. DB 9 Pin Assignments

DB-9 Pin	Corresponding DB-25	Signal	Function
2	3	RxD	Received Data
3	2	TxD	Transmited Data
8	5	CTS	Clear to send
7	4	RTS	Request to Send
6	6	DSR	Data set ready
5	7	SG	Signal Ground
4	20	DTR	Data Terminal Ready
1	8	DCD	Data Carrier Detect

The connectors allow the physical components to interconnect; however, how the cabling is wired to the various pins determines whether the sending and receiving physical layers will be able to communicate.

A 9 pin connector will be used to explain how the physical layers communicate.

DTE

- At the physical layer it can be a terminal, computer, printer, fax etc
- Sends data on transmitted data (TXD) line
- Receives on received data (RXD) line
- Asserts request to send (RTS) when ready to receive
- Waits for clear to send (CTS) to be asserted before sending
- Asserts data terminal ready (DTR) when ready and waiting (to receive)
- Expects data set ready (DSR) to be asserted before performing (sending)
- Expects Data Carrier Detect (DCD) to be asserted before receiving

DCE

- At the physical layer the DCE takes the bit stream converts them into a signal (voltages) and then introduces the signal onto the telecommunications medium. Normally the DCE is known as a modem.
- The wiring is the inverse as that of the DTE.
- The most important lines are RxD, TxD, and SG; the others are used with modems, printers and plotters to indicate internal states.

DO YOU KNOW

RS232 interface is used to connect Data terminal equipment (DTE) and Data circuit terminating equipment (DCE).

4.11.2. EIA-232-D

An industry trade group, the Electronic Industries Association (EIA), defined it originally for teletypewriter devices. In 1987, the EIA released a new version of the standard and changed the name to EIA-232-D. Many people, however, still refer to the standard as RS-232C, or just RS-232.

The RS-232D Serial Interface Standard added the mechanical characteristics to the RS-232C Standard. The RS-232D Standard defines:

- The mechanical characteristics of the interface
- The electrical characteristics of the interface
- The function of each signal
- Subsets of the signals for certain applications

The European version of RS-232D is defined in:

- V.24 - mechanical standard
- V.28 - electrical standard

Mechanical Characteristics of the RS-232D

(*i*) The connector is a DB25 connector. DB9 is not universally accepted.

(*ii*) The connector gender is male at the DTE and female at the DCE.

(*iii*) The assignments of signals to pins.

(*iv*) The maximum cable length is 50 ft.

(*v*) The maximum cable capacitance = 2500 pF. Typical cable has 50 pF/foot capacitance.

Electrical Characteristics of the RS-232D

The transmitter side generates a voltage between +5 and +25 volts for a space (digital 0 or low) and generates a voltage between –5 and –25 volts for a mark (digital 1 or high). (Fig. 4.12)

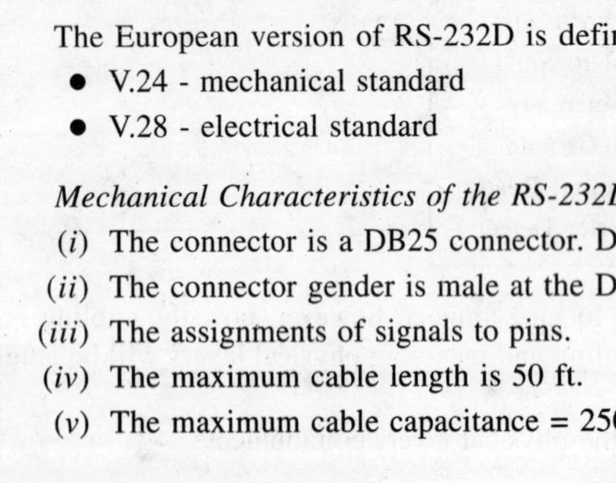

Fig. 4.12

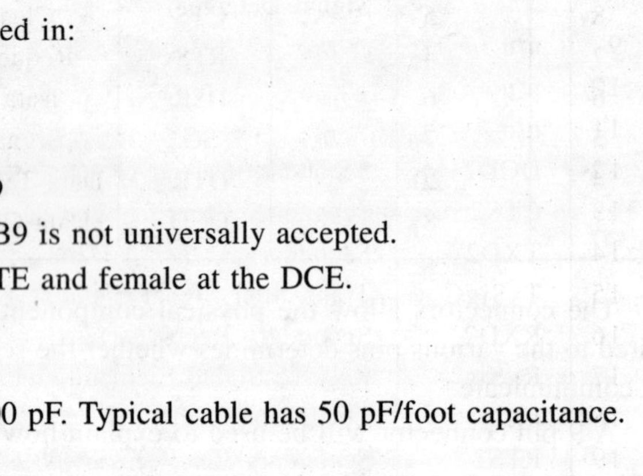

Fig. 4.13

The receiving side recognizes a space as any voltage between +3 and +25V and a mark as any voltage between –3 and –25 V. The standard allows for a voltage loss through the cable and noise immunity by reducing the receive requirements to +/-3 volts. (Fig. 4.13)

Function of Each Signal

Pin	Name	Description	EIA Circuit
1	*GND*	Chassis Ground	AA
2	*TXD*	Transmit Data (TXD)	BA
3	*RXD*	Receive Data (RXD)	BB
4	*RTS*	Ready to Send	CA
5	*CTS*	Clear to Send	CB
6	*DSR*	Data Set Ready (DCE Ready)	CC
7	*SGND*	Signal ground	AB
8	*DCD*	Carrier Detect (CD or RLSD)(RLSD - Received Line Signal Detector)	CF
9	n/u		
10	n/u		
11	n/u		
12	DCD2	Secondary Carrier Detect (SRLSD)	SCF
13	CTS2	Secondary Clear to Send	SCB
14	TXD2	Secondary Transmit Data	SBA
15	TxSigC	Transmitter Signal Element Timing - DCE	DB
16	RXD2	Secondary Receive Data	SBB
17	RxSig	Receive Signal Element Timing - DCE	DD
18	LL	Local Loopback	
19	RTS2	Secondary Ready to Send	SCA
20	*DTR*	Data Terminal Ready (DTE Ready)	CD
21	SQ/RL	Signal Quality/Remote Loopback	CG
22	*RI*	Ring Indicator	CE
23	DSRS	Data Signal Rate Selector	CH/CI
24	TxSigT	Transmitter Signal Element Timing - DTE	DA
25	TM	Test Mode	

The signals in bold/italic are required for a basic asynchronous modem connection.

Subsets of Signals for Certain Applications

Data Signals

2	*TXD*	Transmit Data Data generated by DTE	BA
3	*RXD*	Receive Data Data generated by DCE	BB

Control Signals

4	*RTS*	Ready to Send	DTE wishes to transmit
5	*CTS*	Clear to Send	DCE ready to receive
6	*DSR*	Data Set Ready (DCE Ready)	DCE powered on & ready to go
20	*DTR*	Data Terminal Ready (DTE Ready)	DTE powered on & ready to go
22	*RI*	Ring Indicator	Phones ringing

Test Modes

18	LL	Local Loopback	Initiate Local Loopback Self-Test
21	SQ/RL	Signal Quality/Remote Loopback	Initiate Remote Loopback Self-Test
25	TM	Test Mode	Initiate Test Mode

Synchronous Control Signals

21	SQ/RL	Signal Quality/Remote Loopback	Error in received data!
23	DSRS	Data Signal Rate Selector	DTE can dynamically select 1 of 2 data rates

Synchronous Timing Signals

15	TxSigC	Transmitter Signal Element Timing - DCE	DCE generated
17	RxSig	Receive Signal Element Timing - DCE	DCE generated
24	TxSigT	Transmitter Signal Element Timing - DTE	DTE generated

Ground/Shield

1	*GND*	Chassis ground	Shield DTE side only for noise protection. Do NOT connect to signal ground!
7	*SGND*	Signal ground	Signal return path

Limitations of RS-232

RS-232 has some serious shortcomings as an electrical interface.

Firstly, the interface presupposes a common ground between the DTE and DCE. This is a reasonable assumption where a short cable connects a DTE and DCE in the same room, but with longer lines and connections between devices that may be on different electrical busses, this may not be true. We have seen some spectacular electrical events causes by "uncommon grounds".

Secondly, a signal on a single line is impossible to screen effectively for noise. By screening the entire cable one can reduce the influence of outside noise, but internally generated noise remains a problem. As the baud rate and line length increase, the effect of capacitance between the cables introduces serious crosstalk until a point is reached where the data itself is unreadable.

Using low capacitance cable can reduce crosstalk. Also, as it is the higher frequencies that are the problem, control of slew rate in the signal (*i.e.,* making the signal more rounded, rather than square) also decreases the crosstalk. The original specifications for RS-232 had no specification for maximum slew rate.

Voltage levels with respect to ground represent the RS 232 signals. There is a wire for each signal, together with the ground signal (reference for voltage levels). This interface is useful for point-to-point communication at slow speeds. For example, port COM1 in a PC can be used for a mouse, port COM2 for a modem, etc. This is an example of point-to-point communication: one port, one device. Due to the way the signals are connected, a common ground is required.

This implies limited cable length-about 30 to 60 meters maximum. (Main problems are interference and resistance of the cable.) Shortly, RS 232 was designed for communication of local devices, and supports one transmitter and one receiver.

4.11.3. EIA-449/ RS 449

The RS449 or RS-449 interface is a further enhancement of RS232 and RS423. It is aimed at catering for very fast serial data communications at speeds up to 2 Mbps. In order to achieve this RS449 makes some changes when compared to RS232 to the way in which the signals are referenced, while still being able to retain some compatibility with RS232.

The **RS-449** specification, also known as **EIA-449** or **TIA-449**, defines the functional and mechanical characteristics of the interface between data terminal equipment and data communications equipment.

RS 449 is General purpose 37-Position Interface for Data terminal Equipment and Data Circuit-Terminating Equipment employing serial binary data interchange.

A serial mechanical interface standard for transmission of balanced and unbalanced signals between a variety of higher-end computer, media, and multimedia peripherals.

The standard specifies DC-37 and DE-9 for the primary and secondary data circuits.

EIA-449 allows a maximum data rate of 10 Mbps and uses a 37- or 9-pin connector. EIA-449 cabling uses EIA-422 differential pairs to send and receive data, and EIA-423 single-ended lines to send and receive control signals.

Interface Characteristics

RS-449 is a differential communications interface used for synchronous communications with some single-ended link management signals, typically limited to a maximum throughput of 10Mbps. Communications over distances exceeding 1000m is possible at low bit rates, the actual performance being mostly dependent on cable specification. Separate clock lines are used for receiving and transmitting data.

Interface Applications

RS-449 interfaces are commonly found on communications equipment in some parts of the world where high throughput and/or long distances are required. The interface also offers good noise immunity enabling reliable communications in environments where there are high levels of EMI (electromagnetic interference).

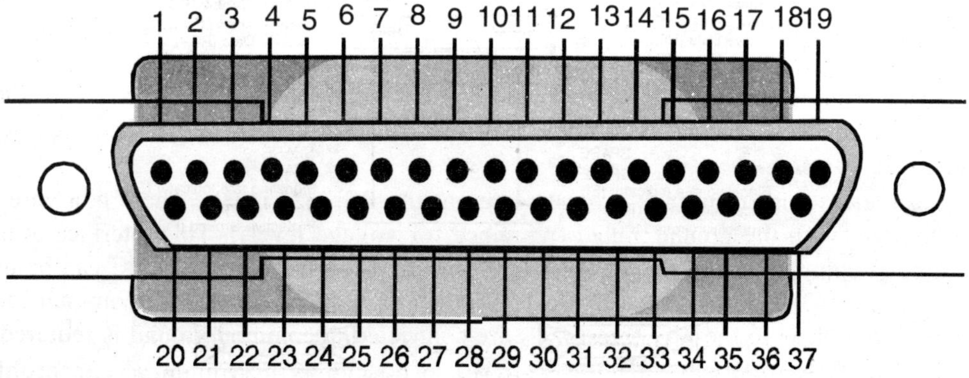

Fig. 4.14.

Pin description		EIA Circuit	From DCE	To DCE
1	Frame Ground	AA		
2	Transmitted Data	BA		D (Data)
3	Received Data	BB	D	
4	Request to Send	CA		C (Control)
5	Clear to Send	CB	C	
6	Data Set Ready	CC	C	
7	Signal Gnd/Common Return	AB		
8	Rcvd. Line Signal Detector	CF	C	
11	Undefined			
12	Secondary Rcvd. Line Sig. Detector	SCF	C	
13	Secondary Clear to Send	SCB	C	
14	Secondry Transmitted Data	SBA	C	
15	Transmitter Sig. Element Timing	DB	T (Timing)	
16	Secondary Received Data	SBB	D	
17	Receiver Sig. Element TIming	DD	T	
18	Undefined			
19	Secondary Request to Send	SCA		C
20	Data Terminal Ready	CD		C
21	Sig. Quality Detector	CG		C
22	Ring Indicator	CE	C	
23	Data Sig. Rate Selector (DCE)	CI	C	
24	Transmitter Sig. Element Timing	DA		T
25	Undefined			

Fig. 4.15. *The 25 pin RS 232 connector defined in the standard*

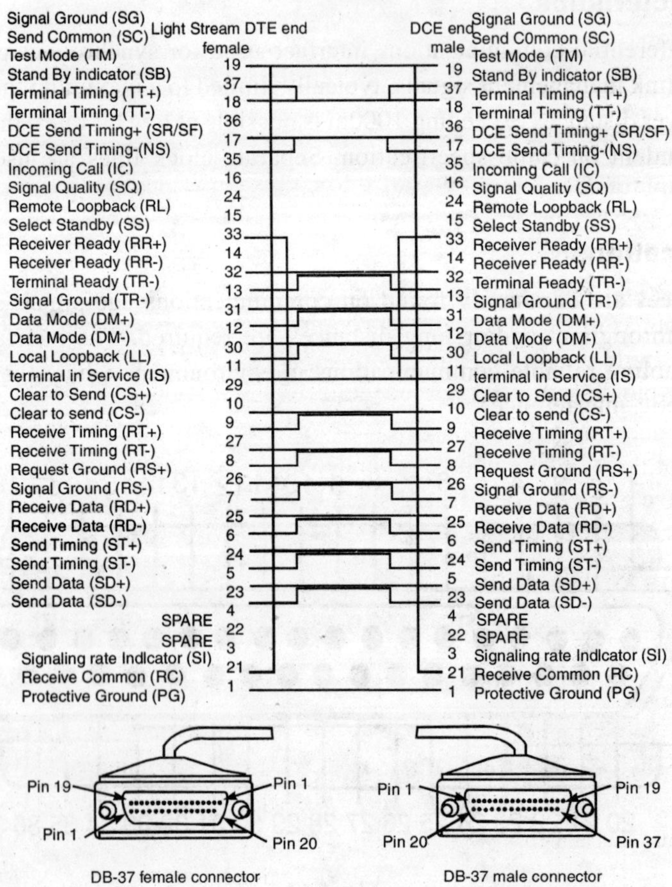

Fig. 4.16. *Interfacing of Rs 449 DB-37 Connector*

Table 4.3. DB 9 Pin Assignments

DB-9 Pin	Corresponding DB-25	Signal	Function
2	3	RxD	Received Data
3	2	TxD	Transmited Data
8	5	CTS	Clear to send
7	4	RTS	Request to Send
6	6	DSR	Data set ready
5	7	SG	Signal Ground
4	20	DTR	Data Terminal Ready
1	8	DCD	Data Carrier Detect

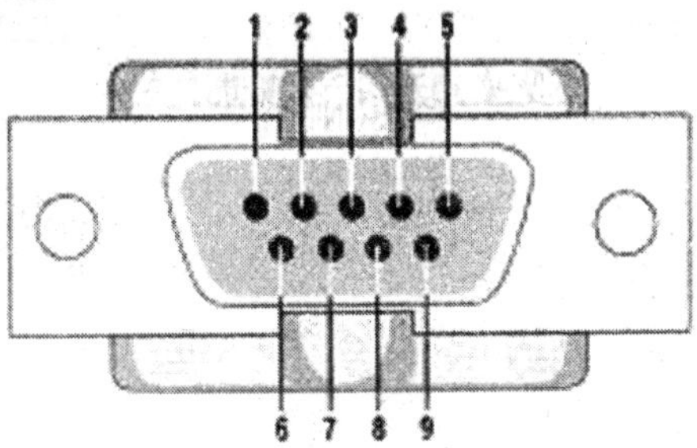

Fig. 4.17. *Pin 9 DB-9 Connector*

RS449 Cable Length Recommendations : The cable length will determine maximum bus speed, or the maximum transfer rate is determined by the cable length (examples): 2Mbps out to 50 feet, 1Mbps to 100 feet, 512kbps to 200 feet, 256kbps to 400 feet, 128kbps to 800 feet, 56kbps to 1600 feet, 1.2kbps to 3000 feet.

Normally EIA422 and 423 systems may not be connected together. EIA-449 cabling of 422 sends and receives data as differential pairs and control signal as single-ended, but for 423 cabling it sends and receives single-ended data and control signals. Receiving the single ended signal in 423 is accomplished by grounding the 'B' side of the differential receiver at the connector.

So if the system does not follow the EIA-449 (cabling) specification than one (B) side of the differential receiver of the 423 side will be grounded at the connector forcing the differential driver on the 422 side to drive ground. The EIA422 side also uses a termination resistor between the ends of its differential receiver, providing a serious 120 ohm short to ground for the 423 driver.

Pin Configurations of RS 449

Table 4.4. RS-449- 9 Pin Connector Pin Out

Pin	Signal Name	Signal Function	Pin	Signal Name	Signal Function
1	Shield	Ground	—	—	—
2	Sec Receive Ready	Control from DCE	**6**	Receive Common	Ground
3	Sec Send Data	Data to DCE	**7**	Sec Request to Send	Control from DCE
4	Sec Receive Data	Data from DCE	**8**	Sec Clear to Send	Control from DCE
5	Signal Ground	Ground	**9**	Send Common	Ground

EIA-449-1 was rescinded in January, 1986. Superseded by EIA/TIA-530-A, the final version EIA-449-1 was withdrawn in September, 2002.

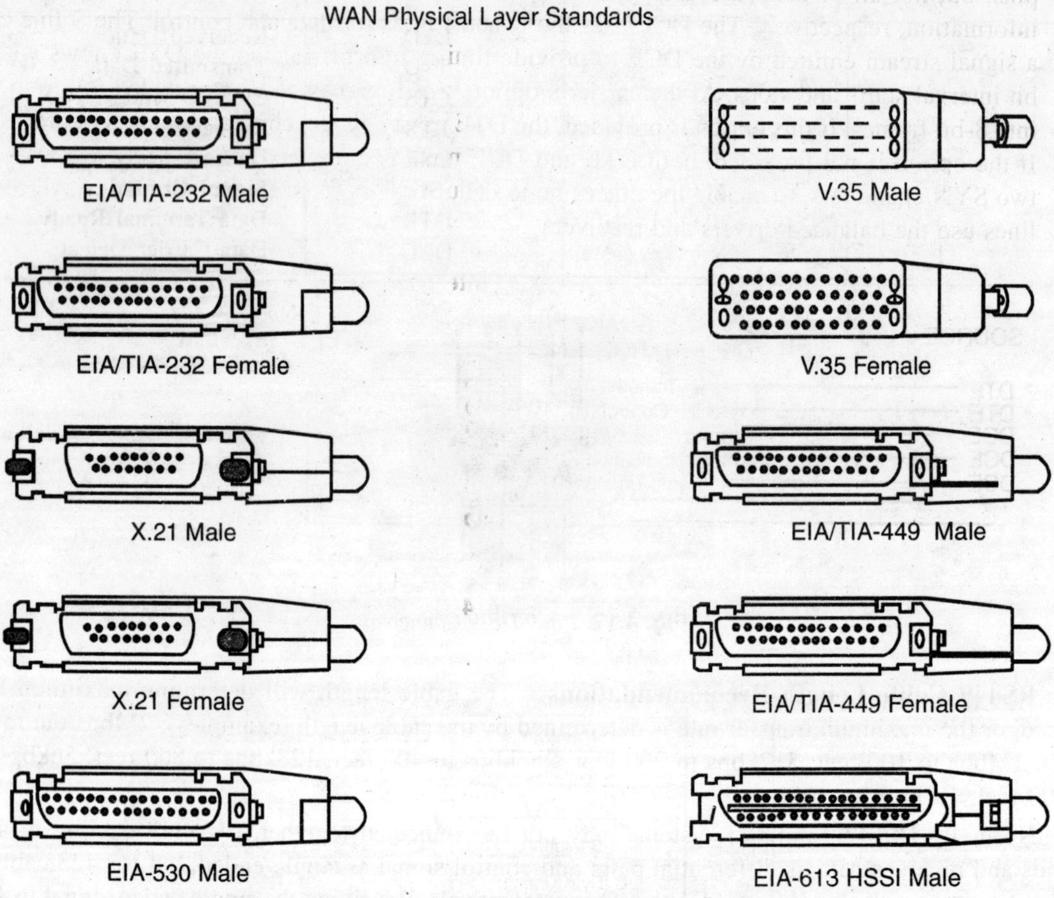

Fig. 4.18. *Various Physical Layer Standards*

4.11.4. Digital Interface: X.21

X.21 Physical Layer Interface

The X.21 interface was recommended by the CCITT in 1976. It is defined as a digital signalling interface between customers (DTE) equipment and carrier's equipment (DCE). And thus primarally used for telecom equipment.

All signals are balanced. Meaning there is always a pair (+/-) for each signal, like used in RS422. The recommendation specifies how the DTE, sets up and clears calls by exchanging signals with the carrier's equipment, the DCE.

Voltages: +/- 0.3Vdc

Speeds: Max. 100Kbps (X.26)
Max. 10Mbps (X.27)

Figure 4.50 shows the connections between the DTE and DCE.

X.21 interface is also used as the terminations interface for digital leased circuits of n x 64kbps. The connector, which is known as DB-15 is shown in the Fig. 4.19. The names and functions of the eight wires defined by X.21 are given in the following Fig. 4.20. The physical connector has 15 pins, but not all of them are used. The DTE uses the T and C lines to transmit data and control information, respectively. The DCE uses the R and I lines for data and control. The S line contains a signal stream emitted by the DCE to provide timing information, so the DTE knows when each bit interval starts and stops. At the carrier's option, a B line may also be provided to group the bits into 8-bit frames. If this option is provided, the DTE must begin each character on a frame boundary. If the option is not provided, both DTE and DCE must begin every control sequence with at least two SYN characters, to enable the other one to deduce the implied frame boundaries. All the signal lines use the balanced drivers and receivers.

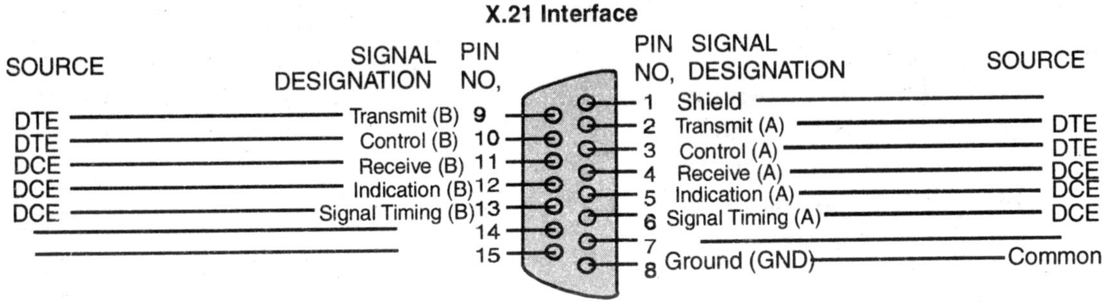

Fig. 4.19.

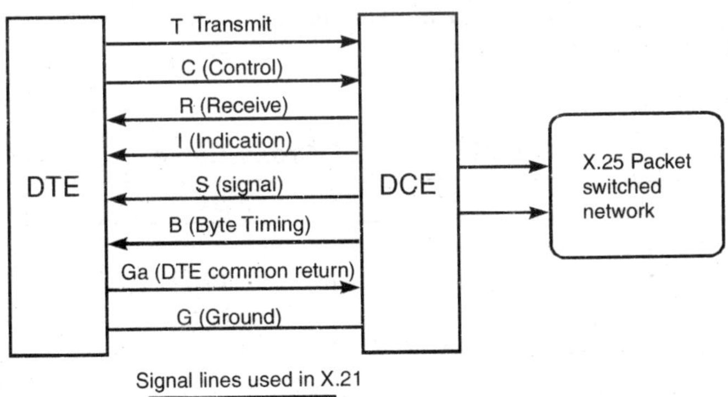

Fig. 4.20. *X.21 Interface Interconnection.*

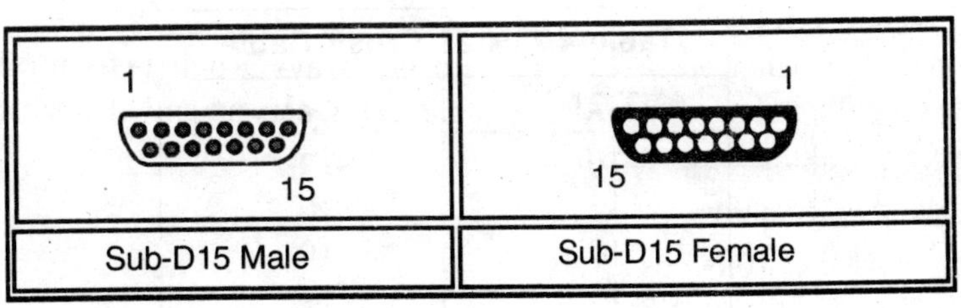

Fig. 4.21. *D-15 X.21.*

Table 4.5. Pinning according to ISO 4903.

Pin	Signal	abbr.	DTE	DCE
1	Shield		—	—
2	Transmit (A)		Out	In
3	Control (A)		Out	In
4	Receive (A)		In	Out
5	Indication (A)		In	Out
6	Signal Timing (A)		In	Out
7	Unassigned			
8	Ground		—	—
9	Transmit (B)		Out	In
10	Control (B)		Out	In
11	Receive (B)		In	Out
12	Indication (B)		In	Out
13	Signal Timing (B)		In	Out
14	Unassigned			
15	Unassigned			

Functional Description

As can be seen from the pinning specifications, the Signal Element Timing (clock) is provided by the DCE. This means that your provider (local telco office) is responsible for the correct clocking and that X.21 is a synchronous interface. Hardware handshaking is done by the Control and Indication lines. The Control is used by the DTE and the Indication is the DCE one.

Cross-Cable pinning

Table 4.6. X.21 Cross Cable

X.21	X.21
1	1
2	4
3	5
4	2
5	3
6	7
7	6
8	8
9	11

Table 4.7. X.21 Cross Cable

X.21	X.21
10	12
11	9
12	10
13	14
14	13
15	

A variation of X.21 known as X.21 bis is a CCIR recommendation that defines the analog interface to allow access to the digital circuit switched networks using an analog circuit. X.21 bis provides procedures for sending and receiving addressing information which enable a DTE to establish switched circuits with other DTEs.

DO YOU KNOW

The CCIR recommendation X.21 defines the physical layer interfaces between a node (DTE) and the network connection (DCE).

4.11.5. EIA/TIA-530

EIA/TIA-530; High Speed 25-Position Interface for Data Terminal Equipment and Data Circuit-Terminating Equipment, Including Alternative 26-Position Connector EIA530 Defines a cabling interface between a DTE and DCE device, with pin outs. A replacement for EIA-449 that uses a DB-25 (EIA-232-D) connector instead of a 37-pin connector, while keeping the most important electrical signals intact. A 26 pin connector may also be used. EIA-530 is to be used in conjunction with either EIA-422-A or EIA-423, which defines the electrical drivers and receivers. Nominal maximum signaling rate is 2.1Mbps for both synchronous and non synchronous serial binary data. The maximum cable distance is 60 meters.

Two types are circuits are used over EIA-530:

Category 1, which have the balanced electrical characteristics of EIA-422-A, and Category 2, which have the unbalanced electrical characteristics of EIA-423.

Both types of circuits [depending on the signal] go in both directions. Signal ground is passed between the two devices, but the cable shield is connected to ground on the DTE but not connected on the DCE side.

Category I	Category II
Transmitted Data	DCE Ready
Received Data	DTE Ready
Transmit Signal Timing, DTE Source	Ring Indicator
Transmit Signal Timing, DCE Source	Local Loopback
Receive Signal Timing, DCE Source	Remote Loopback
Request to Send	Test Mode
Clear to Send	——
Received Line Single Detector	——
Ready for Receiver	——

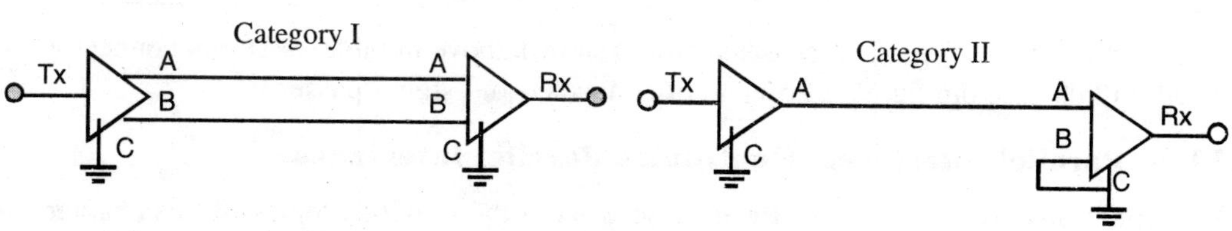

Fig. 4.22.

Pin Configurations of EIA-530 25 Pin Connector

Pin	Circuit	Category	Connect	Direction	Description
1	Shield	—	—	From DTE	Cable Shield, no connect on DCE
2	BA	I	A - A'	To DCE	Tx, Transmit Data
3	BB	I	A - A'	From DCE	Rx, Receive Data
4	CA/CJ	I	A - A'	To DCE	Request to Send / Ready For Receiving
5	CB	I	A - A'	From DCE	Clear to Send
6	CC	II	A - A'	From DCE	DCE Ready
7	AB	—	C - C'	—	Ground
8	CF	I	A - A'	From DCE	Received Line Signal Detector
9	DD	I	B - B'	From DCE	Receive Signal Timing
10	CF	I	B - B'	From DCE	Received Line Signal Detector
11	DA	I	B - B'	To DCE	Transmit Signal Timing
12	DB	I	B - B'	From DCE	Transmit Signal Timing
13	CB	I	B - B'	From DCE	CTS, Clear to Send
14	BA	I	B - B'	To DCE	TXD, Transmit Data
15	DB	I	A - A'	From DCE	Transmit Signal Timing
16	BB	I	B - B'	From DCE	RXD, Receive Data
17	DD	I	A - A'	From DCE	Receive Signal Timing
18	LL	II	A - A'	To DCE	Local Loopback
19	CA/CJ	I	B - B'	To DCE	Request to Send / Ready For Receiving
20	CD	II	A - A'	To DCE	DTE Ready
21	RL	II	A - A'	To DCE	Remote Local
22	CE	II	A - A'	From DCE	Ring Indicator
23	AC	—	C - C'	—	Ground
24	DA	I	A - A'	To DCE	Transmit Signal Timing
25	TM	II	A - A'	From DCE	Test Mode

The pin out for a 25 pin D type connector is shown above in table. A 26 pin connector may also be used, in this case pin number 26 is a not connected [no signal passed].

4.11.6. Parallel Interfaces: Centronics Parallel Interfaces

Parallel transmission mode is already discussed in the previous topics in this chapter. Now we will be discussing about the Parallel Interfaces (centronics). Dr. An Wang, Robert Howard and Prentice Robinson developed the Centronics parallel interface at Wang Laboratories.

Centronics parallel is generally compliant with IEEE 1284 compatibility mode. The original Centronics implementation called for the busy lead to toggle with each received line of data (busy by line), whereas IEEE 1284 calls for busy to toggle with each received character (busy by character). Some host systems or print servers may use a strobe signal with a relatively low voltage output or a fast toggle. Any of these issues might cause no or intermittent printing, missing or repeated characters or garbage printing. Some printer models may have a switch or setting to set busy by character; others may require a handshake adapter.

HP adopted Centronics parallel on their printer models and introduced a bidirectional version known as *Bitronics* on the LaserJet 4 in 1992. The Bitronics and Centronics interfaces were superseded by the IEEE 1284 standard in 1994.

Fig. 4.23. *A DB-25 parallel printer port*

Fig. 4.24. *Micro ribbon 36 pin female, such as on printers*

A **parallel port** is a type of interface found on computers for connecting various peripherals. It is also known as a **printer port** or **Centronics port**. The IEEE 1284 standard defines the bi-directional version of the port.

Data products introduced a very different implementation of the parallel interface for their printers. It used a DC-37 connector on the host side and a 50 pin connector on the printer side—either a DD-50 (sometimes incorrectly referred to as a "DB50") or the block shaped M-50 connector; the M-50 was also referred to as Winchester. Data products parallel was available in a short-line for connections up to 50 feet (15 m) and a long-line version for connections from 50 feet (15 m) to 500 feet (150 m). The Data products interface was found on many mainframe systems up through the 1990s, and many printer manufacturers offered the Data products interface as an option.

IBM released the IBM Personal Computer in 1981 and included a variant of the Centronics interface— only IBM logo printers (rebranded from Epson) could be used with the IBM PC. IBM standardized the parallel cable with a DB25F connector on the PC side and the Centronics connector on the printer side. Vendors soon released printers compatible with both standard Centronics and the IBM implementation.

At the consumer level, the USB interface—and in some cases Ethernet—has effectively replaced the parallel printer port. Many manufacturers of personal computers and laptops consider parallel to be a legacy port and no longer include the parallel interface. USB to parallel adapters are available to use parallel-only printers with USB-only systems.

Port addresses

Traditionally IBM PC systems have allocated their first three parallel ports according to the configuration in the table below.

PORT NAME	Interrupt	Starting I/O	Ending I/O
LPT1	IRQ 7	0x378	0x37f
LPT2	IRQ 7	0x278	0x27f
LPT3	IRQ 5	0x3bc	0x3bf

If there is an unused LPTx slot, the port addresses of the others are moved up. (For example, if a port at 0x3bc does not exist, the port at 0x378 will then become LPT1.) The IRQ lines, however, remain fixed (therefore, 0x378 at LPT1 would use IRQ 7). Unfortunately the default IRQ used by the first two addresses is the same, and it's difficult to get correct interrupt behaviour if both of these addresses are in use. The port addresses assigned to each LPTx slot can be determined by reading the BIOS Data Area (BDA) at 0000:0408.

Bit to Pin Mapping for the Standard Parallel Port (SPP):

Address		MSB							LSB
	Bit:	7	6	5	4	3	2	1	0
Base	Pin:	9	8	7	6	5	4	3	2
Base+1	Pin:	~11	10	12	13	15			
Base+2	Pin:					~17	16	~14	~1

~ indicates a hardware inversion of the bit.

Pinouts

Pinouts for a DB25 connector are:

Pin No (DB25)	Signal name	Direction	Register - bit	Inverted
1	nStrobe	Out	Control-0	Yes
2	Data0	In/Out	Data-0	No
3	Data1	In/Out	Data-1	No
4	Data2	In/Out	Data-2	No
5	Data3	In/Out	Data-3	No
6	Data4	In/Out	Data-4	No
7	Data5	In/Out	Data-5	No
8	Data6	In/Out	Data-6	No
9	Data7	In/Out	Data-7	No
10	nAck	In	Status-6	No
11	Busy	In	Status-7	Yes
12	Paper-Out	In	Status-5	No
13	Select	In	Status-4	No
14	Linefeed	Out	Control-1	Yes
15	nError	In	Status-3	No
16	nInitialize	Out	Control-2	No
17	nSelect-Printer	Out	Control-3	Yes
18-25	Ground	—	—	—

In early parallel ports the data lines were mono directional (data out only) so it was not easily possible to feed data in to the computer. However, a workaround was possible by using 4 of the 5 status lines. A circuit could be constructed to split each 8-bit byte into two 4-bit nibbles which were fed in sequentially through the status lines. Each pair of nibbles was then re-combined into an 8-bit byte.

Summary

- Network interfaces, hubs, bridges, switches, routers, and firewalls work together in a number of ways to create these different kinds of network roadways.
- Repeaters are physical hardware devices that have a primary function to regenerate the electrical signal by:
  (*i*) Reshaping the waveform
  (*ii*) Amplifying the waveform
  (*iii*) Retiming the signal
- A networking hub that provides direct accessibility to all computers on the system comprises of a series of ports.
- A networking hub that provides direct accessibility to all computers on the system comprises of a series of ports.
- Hubs are also called Multiport Repeaters or Concentrators.
- Hubs are used to provide a Physical Star Topology.
- Hubs are three types:
  (1) Passive Bubs.
  (2) Active Hubs.
  (3) Smart Hubs.
- **Passive hubs** do not amplify the electrical signal of incoming packets before broadcasting them out to the network.
- **Active hubs**, on the other hand, do perform this amplification, as does a different type of dedicated network device called a repeater.
- Smart Hubs or **Intelligent hubs** add extra features to an active hub that are of particular importance to businesses.
- **Bridges** are both hardware and software devices.
- The purposes of a Bridge are:
  - Isolates networks by MAC addresses
  - Manages network traffic by filtering packets
  - Translate from one protocol to another
- Routers are hardware and software devices.
- The purpose of a router is to connect nodes across an internet work regardless of the Physical Layer and Data Link Layer protocol used.
- **Brouters** are protocol dependant devices.
- The key advantage to Brouters is the ability to act as both a bridge and a router.
- **Gateways** make communication possible between different architectures and environments.
- A NIC or **Network Interface Card** is a circuit board or chip, which allows the computer to communicate to other computers on a Network.
- **Ethernet Switches** (sometimes called switching hubs) are devices that have changed the way Ethernet's bus arbitration works. Ethernet is based on CSMA/CD, a shared bus where all nodes listen and fight for use of the bus.
- Network cards have three main types of connectors.

- The media used to transmit and receive the information from one point to another point, is called the **Transmission Media**.
- A **transmission medium** may be broadly defined as anything can carry information from source to destination.
- Each type of transmission media has special characteristics that make it suitable for a specified type of service.
- Transmission media are classified in two categories such as :
  1. Guided transmission media
  2. Unguided transmission media
- **Twisted pair** cable consists of two insulated copper wires, typically 1mm thick, twisted together.
- The number of twists per unit of length effects on the quality of the cable.
- Cables without a shield are called **Unshielded Twisted Pair** or UTP.
- Cables with a shield are called **Shielded Twisted Pair** and commonly abbreviated STP.
- Coaxial cabie gets its name because two conductors share a common axis; Coax is shorthand for Co-axial.
- Coaxial cables are categorized by their Radio Government (RG) rating.
- **Fiber optic transmission** is the transmission of information using light waves as a carrier signal that travel through an optical waveguide (optical fiber).
- The EIA has published a number of standards related to telecommunication and computer communication, and works closely with other associates such as ANSI and the ITU.
- RS232 interface is used to connect Data terminal equipment (DTE) and Data circuit terminating equipment (DCE). It is used for serial interface.
- The EIA has published a number of standards related to telecommunication and computer communication, and works closely with other associates such as ANSI and the ITU.
- RS232 interface is used to connect Data terminal equipment (DTE) and Data circuit terminating equipment (DCE). It is used for serial interface.
- A **registered jack** (**RJ**) is a standardized physical network interface — both jack construction and wiring pattern — for connecting telecommunications or data equipment to a service provided by a local exchange carrier or long distance carrier.
- Ethernet cables are terminated with RJ45 connectors.
- RJ45 is a keyed connector, meaning the connector can be inserted in only one way.
- The **BNC connector (Bayonet Neill-Concelman connector)** is a common type of RF connector used for the coaxial cable which connects to various types of radio, television, and other radio-frequency electronic equipment.
- The **RS-449** specification, also known as **EIA-449** or **TIA-449**, defines the functional and mechanical characteristics of the interface between data terminal equipment and data communications equipment.
- The X.21 interface was recommended by the CCITT in 1976. It is defined as a digital signalling interface between customers (DTE) equipment and carrier's equipment (DCE). And thus primarally used for telecom equipment.
- **EIA/TIA-530**; High Speed 25-Position Interface for Data Terminal Equipment and Data Circuit-Terminating Equipment, Including Alternative 26-Position Connector EIA530 Defines a cabling interface between a DTE and DCE device, with pin outs.
- Category 1, which have the balanced electrical characteristics of EIA-422-A, and Category 2, which have the unbalanced electrical characteristics of EIA-423.

- Dr. An Wang, Robert Howard and Prentice Robinson developed the Centronics parallel interface at Wang Laboratories. Centronics parallel is generally compliant with IEEE 1284 compatibility mode.
- HP adopted Centronics parallel on their printer models and introduced a bidirectional version known as *Bitronics* on the LaserJet 4 in 1992. The Bitronics and Centronics interfaces were superseded by the IEEE 1284 standard in 1994.
- A **parallel port** is a type of interface found on computers for connecting various peripherals. It is also known as a **printer port** or **Centronics port**.

Review Questions

1. What do you understand by term transmission media? Name various types of transmission media used in communication.
2. Define guided transmission media and explain advantages of twisted pair cable over open wire cable.
3. What do you understand by term transmission impairment in transmission media?
4. Explain CCITT X.21 in detail. **[MDU Dec. 2008]**
5. Explain DDD networks in detail. **[MDU Dec. 2008]**
6. Explain RS-530 serial interface in detail. **[MDU Dec. 2008]**
7. Explain AT and T's FDM in detail. **[MDU Dec. 2008]**
8. Differentiate between serial and parallel interface ? Explain RS-449 serial interface in detail. **[MDU May 2009]**
9. Differentiate between synchronous and asynchronous modems. Explain synchronous modem in detail. **[MDU Dec. 2009]**
10. Explain in RS-232 serial interface in detail. **[MDU May 2010]**
11. Differentiate between synchronous and asyncrhonous modems. Explain Modem synchronisation. **[MDU May 2010]**
12. Explain DDD networks in detail. **[MDU May 2010]**
13. What is a null modem connection ? Discuss RS 449 serial interface explaining its unbalanced and balanced modes and category I and category II pins. **[MDU Dec. 2010]**
14. Write short note on Data Modems. **[MDU Dec. 2010]**
15. Draw and explain the electrical, mechanical and functional specification of the Centronics parallel interface. **[MDU May 2011]**
16. Explain various types of transmission impairments.
17. What is delay distortion? How it affects the data rate of channel? **[MDU Dec. 2008]**
18. Define and explain Shannon limit theorem. **[MDU Dec. 2008]**
19. What is the maximum data rate of channel? Explain the theorems related to it.
20. What do you understand by physical layer interface and explain the functions of it.
21. Describe physical layer interface device RS-232.
22. What is the effect of limited bandwidth on digital signal? **[MDU May 2007]**
23. What are the functions of a DTE? What are the functions of a DCE? Give an example of each.
24. What standard organizations are involved in DTE-DCE interface standards? Name some popular DTE-DCE standards.

25. What implementations of EIA-232 are available? How are they different?
26. What is the purpose of null modem? Describe the data pins of null modem.
27. How is X.21 able to eliminate most of the control circuits of EIA standards?
28. What factors affect the data rate of a link?
29. Define the bandwidth of a line. What is the bandwidth of a traditional telephone line?
30. What is an intelligent modem?
31. Explain the asymmetry of 56K modems.
32. What is the difference between a balanced circuit and an unbalanced circuit?
33. What is the relationship between the data rate and the distance that the data can reliably travel on an EIA interface?
34. What does the mechanical specifications, electrical specifications and functional specifications of EIA-232 describe?
35. Write short notes on following:
    (*i*) Transmission impairments
    (*ii*) Physical layer interface
    (*iii*) RS-232
    (*iv*) X.21
    (*v*) Data rate of channel
    (*vi*) Repeater
    (*vii*) Hub
    (*viii*) Ethernet
    (*ix*) Bridges
    (*x*) Router

□□□

5

CHAPTER

The Telephone Networks

5.1. INTRODUCTION

INSIDE THIS CHAPTER

The largest communication network on this earth is the global Public Switched Telephone Network (PSTN). The telephone network had its beginnings in the late 1876s.

The entire network, which is referred to as plain old telephone set (POTS) was originally designed to carry analog voice signals, the advent of computers and data communication in the 1980s, has lead to it being increasingly used to also carry computer data traffic.

Modern telecommunication networks are designed to carry information in a variety of forms (*e.g.*, voice, data, video and fax). Telecommunication and data communication are fast converging.

5.2. PUBLIC TELEPHONE NETWORK

The Public Telephone Network includes mainly two types of subscribers:

1. Private
2. Public

1. Private : All those customers who lease transmission media, equipments and services from the telephone companies on permanent basis have known as **private subscribers** or private line circuit or dedicated circuits.

The leased circuits are specifically designed and configured for their use, for example, banks do not like to share their communication network with other users. Therefore, they lease facilities and equipments from public telephone companies and operate a private telephone or network within the public telephone network.

Public telephone companies also referred as service providers. They provide services and lease equipments to other private companies, government agencies and organizations.

2. Public : Subscribers to the public sector of the telephone network share facilities and equipments that are available to all the public subscribers to the network.

Anyone with the telephone number is a subscriber to the public sector of the public telephone network.

All the subscribers to public network are interconnected temporarily through the switches. So they are also known as **public switched telephone network (PSTN)** or dial-up network. PSTN refers to the combination of transmission lines and switches (may be central offices) which forms a system of electrical routes through the network.

5.2.1. Basic Concepts

The most familiar component of a telephone network is the telephone set provided for each user. It is a relatively simple device which can exchange control signals with the network to help establish and release calls, and to send and receive a user's speech signal.

Users are referred to as **subscribers**, because they subscribe to a service provided by the telephone company.

Typically, many subscribers are attached to the same network, and each subscriber can contact every other subscriber.

The system employs **switches** to facilitate connections between subscribers.

The arrangement of switches and subscribers and the interconnections between them determines the structure of the network.

5.2.2. Main Components

Figure 5.1 shows the major components of the telephone network. They are as follows:

1. Local Loops
2. Trunks
3. Switching offices (end offices, tandem offices and regional offices)

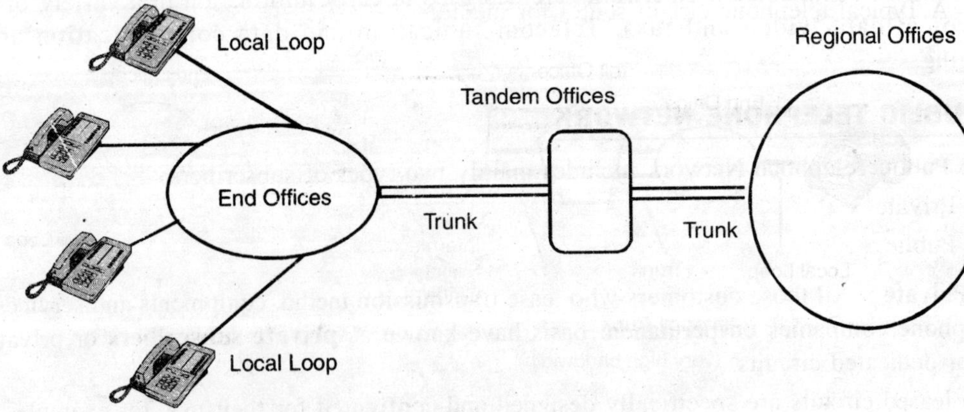

Fig. 5.1. *The major components of the telephone network.*

5.2.3. Local loop

The circuit which connects each subscriber (via a pair of wires) to the switch is called a **subscriber loop** or **local loop**.

A twisted pair cable which connects the subscriber telephone set to the nearest end office (may be referred as local central office).

The twisted pair has its limited bandwidth but enough for voice transmission which needs only 4 kHz.

Normally the range of local loop is from 1 to 10 km.

5.3. TRUNK LINES

A trunk circuit is similar to a local loop except trunk circuits are used to interconnect two telephone offices.

The two exchanges are connected using a set of interexchange lines called **trunk lines.**

The main difference between a local loop and a trunk is that a local loop is permanently associated with a particular station, whereas a trunk is a common usage.

Trunk lines are transmission media used to connects the interoffice.

The transmission media used for trunk lines may be twisted copper wire or optical fibers.

A trunk circuit could be a wireless communication channel (satellite links). Because it has to handles large numbers of connections using multiplexing, it requires higher bandwidth.

5.4. SWITCHING OFFICES OR EXCHANGE

Switching offices or exchange is a central location where subscribers are interconnected, permanently or temporarily basis.

Switching offices are connected with several local loops or trunk lines and provide a connection between different subscribers.

Switching machine are located in exchange, switching machines are programmable matrices that provide temporary signal paths between two subscribers.

This is done to avoid the having permanent physical connection between any two subscribers. Data modems are connected through local loops to switching machines located in exchange.

A Typical telephone circuit route (for medium distance call)

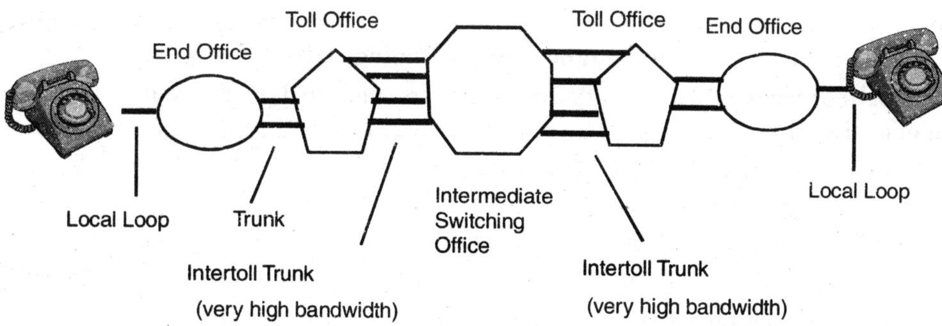

Fig. 5.2.

5.5. PRIVATE TELEPHONE NETWORKS

Private telephone networks coexist with the public telephone network to address the specific telecommunication needs of independent organizations.

Private telephone networks come in different configurations and sizes, with varying degrees of connectivity or integration with public telephone networks and data networks. Most private networks revolve around a private exchange, which we will look at first.

There are numbers of private telephone networks applications. Few of them are as discussed below:

1. Private Branch Exchange
2. Corporate Networks
3. Intelligent Networks

5.5.1. Corporate Networks

Many large corporations operate their own distributed private telephone network, which consists of a number of PBX systems at different sites, interconnected through leased trunk lines.

Each user is provided with a network-wide number which can be used to contact them, independent of their geographic location.

The network also provides an interface to the public network for contacting people outside the corporation.

By concentrating traffic on leased trunk lines, significant reductions in running costs can be achieved.

5.5.2. Intelligent Networks

CCITT's Signaling System Number 7 (SS7) opens the scope for user-defined procedures for the way calls are handled. This leads to the concept of **Intelligent Networks (IN)**.

Network access is facilitated through a set of **Service Switching Points** (SSPs). The signaling link between the switches is controlled through a set of **Signaling Transfer Points** (STPs).

A **Service Control Point** (SCP) is in charge of dictating how calls should be handled and routed. For its operation, the SCP uses a service database which contains service profile definitions provided by the customer.

5.6. THE DISTANCE DIRECT DIALING (DDD) NETWORK

The North American Telephone Numbering Plan (NANP) was established to provide a telephone numbering system for the United States, Mexico, and Canada that would allow any subscriber in North America to direct dial virtually any other subscriber without the assistant of an operator. The network is often referred to as the **Distance Direct Dialing (DDD) Network.**

The numbering plan is used to identify the subscribers connected in a telecommunication network. The main objective of numbering plan by any nation is to standardize the number length wherever practical according to CCITT recommendations.

Other objectives includes

(*a*) to meet the challenges of the changing telecom environment

(*b*) to meet subscriber needs for a meaningful and user friendly scheme

(*c*) to reserve numbering capacity to meet the undefined future needs.

A **telephone numbering plan** is a type of numbering scheme used in telecommunications to allocate and route telephone numbers in a telephone network. A closed numbering plan, such as found in North America, features fixed length area codes and local numbers. An open numbering plan features variance in length of area code or local number, or both.

A **dial plan** establishes the expected number and pattern of digits for a telephone number. This includes country codes, access codes, area codes and all combinations of digits dialed. For instance, the North American public switched telephone network (PSTN) uses a 10-digit dial plan that includes a 3-digit area code and a 7-digit telephone number. Most PBXs support variable-length dial plans that use 3 to 11 digits. Dial plans must comply with the telephone networks to which they connect.

In early telephone systems, connections were made in the central office by telephone operators using patch cords to connect one party to another. If a person wanted to make a phone call, he or she would pick up a phone and wind a crank on the side. The crank was a small generator that would light a lamp at the central office. An operator would see the light and insert their patch cord into a socket and assist the customer with the call connection. The operator would use patch cords to connect the caller to the person being called. If the party being called was in another exchange, the operator would use a patch cord to connect to another exchange where an operator elsewhere would finish the connection. As technology advanced, electro-mechanical switches were introduced and calls were made using "rotary dials".

Initial use of area codes in the United States began in the 1950s with large cities. By 1966, the system was nationwide.

Area codes were assigned based on the length of time a rotary dial phone took to dial the area code. Densely populated areas like New York City, Chicago, Los Angeles, and Detroit had huge incoming call volume and were assigned numbers (212, 312, 213, 313) that could be quickly dialed from a rotary dial phone. On a rotary dial phone low digits (1, 2, 3, 4) could dial quickly as the time the rotary dial took to return to the home position was minimal. High digit numbers (7, 8, 9, 0) on rotary dial phones took much longer to return to the home position and were usually used in less densely populated areas like Vermont (802), rural Texas (915), Tennessee (901), and the Canadian Maritimes (902). This system became unnecessary when touch-tone phones arrived, as the tone allowed instant entry of digits.

Standards

Although the International Telecommunication Union (ITU) has attempted to promote common standards among nation states, numbering plans take different formats in different parts of the world. For example, the ITU recommends that member states adopt **00** as their international access code. However, as these recommendations are not binding on member states, some have not, such as the United States, Canada, and other countries and territories participating in the North American Numbering Plan.

5.7. MODEMS

*A modem is a **mo**dulator/**dem**odulator.* The analog signals encode the digital information at the time of modulation and decode it back during demodulation to transmit the data.

Modems allow digital transmissions over analog telephone lines. They allow people to connect their computers with other computers, corporate office LANs, and the Internet. Modems are used to establish connections of the telephone company's *local loop,* which is the analog copper cable that runs between homes and businesses to the telephone company central office (CO). The phone company then switches the call like any other voice call to create a point-to-point link to the destination.

It converts digital signals from a computer to analog signals so they can be sent over a telephone line. The receiving computer will normally have a modem which will convert the analog signals back to digital signals. A modem may be an internal modem which is a card inside your computer or it may be external with a connection to a serial RS-232 line on your computer.

The telephone jack that will plug into the modem is called an RJ-11 jack. Speed of the modem is measured in bits per second (bps). There are a set of V-series standards developed by the International Telecommunications Union which indicate the speed of the modem.

DO YOU KNOW

A device which carries out the two important tasks of modulating and demodulating the analog signals is called a **modem.**

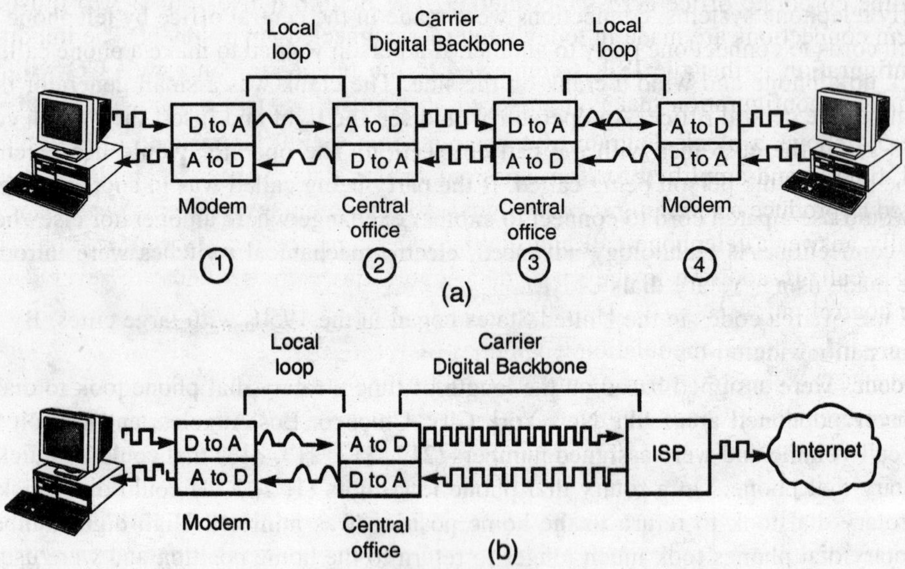

Fig. 5.3. *Role of MODEM*

Before the Internet, the usual connection was for one user to dial another user, or for a user to dial into a modem at a bulletin board, place of business, or a school (rather than connect through the Internet, as discussed next). In these cases, the connection is modem to modem across the PSTN, as shown at the top in Fig. 5.3(*a*). Note that in the figure, the computer's outgoing digital signal is converted to an analog signal for transmission across the copper local loop to the CO. At the CO, the analog signal is converted back to digital for transmission across its digital backbones. At the destination CO, the signal is once again converted from digital to analog and transmitted to the end user's modem, where it is converted back to digital. Remember all those conversions.

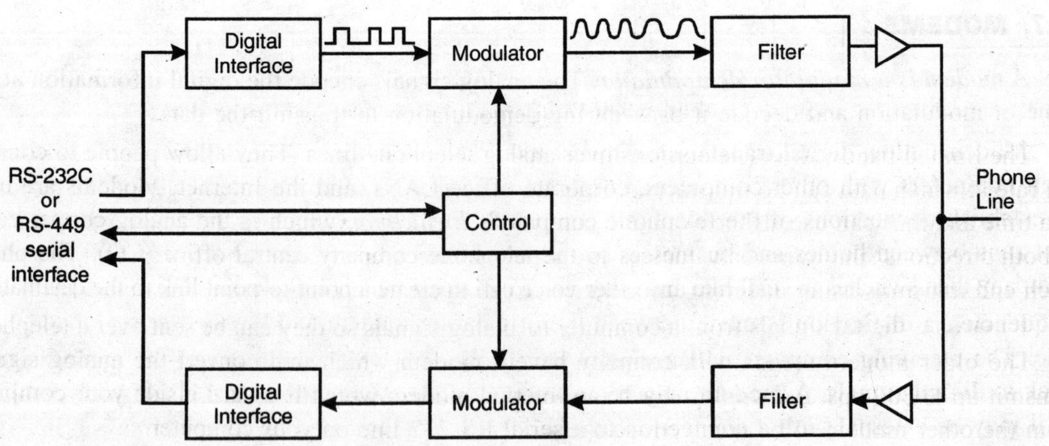

Fig. 5.4. *Modem block diagram.*

Today, most users dial into the Internet, where they connect with other users via e-mail, collaborative applications, instant messaging, and so on. Users also connect with their home networks over the Internet by using secure VPN technology. This allows users to dial into a local number and connect to the corporate office across the Internet. The bottom half of Fig. 5.3(*b*) illustrates how most modem connections are made in today's Internet-connected environment. The important point of this configuration is that the ISP is connected to the telephone company's CO with a digital connection. This configuration makes high-speed 56-Kbit/sec modem technology possible.

Figure 5.4 shows a block diagram of a basic modem. Serial digital data to be transmitted is modulated, filtered and amplified for transmission. Analog data received is amplified, filtered and demodulated to produce a serial digital data signal. Modem often includes ring detector so they can automatically answer a telephone call and dialing tone (dual-tone multifrequency, or DTMF) for originating a call. In addition to the serial data inputs and outputs, there are several handshake signals for control purposes.

There are many digital modulation techniques such as FSK, QPSK, DPSK, QAM, etc are used in the Modem. The simplest form of modulation is frequency–shift keying (FSK), in which one frequency represents a '0' and another '1'.

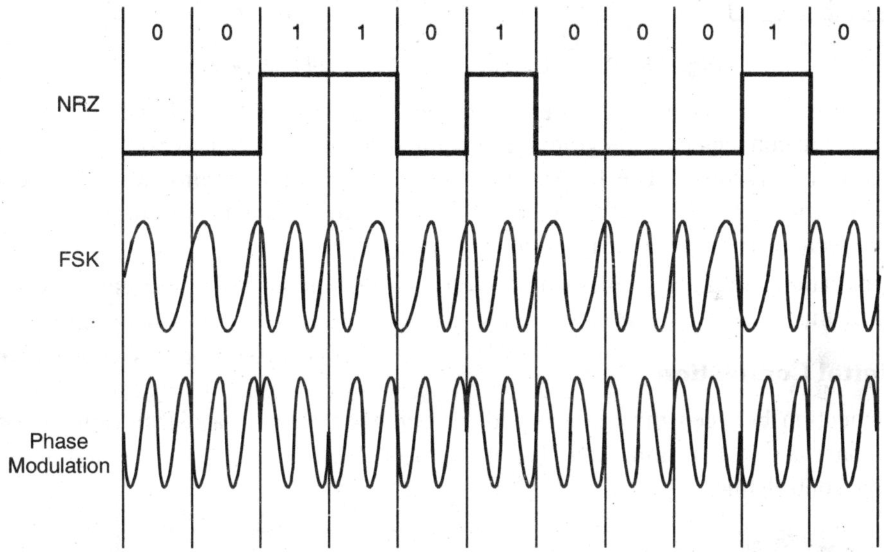

Fig. 5.5. *FSK and phase modulation.*

The limited bandwidth of telephone lines (300 Hz to 3 kHz for a standard dial-up line) restricts the rate at which data can be transmitted. Half-duplex modems can transmit in only in one direction at a time and thus can use all the available bandwidth. Full duplex modems, however, must transmit in both directions simultaneously. Unless two separate lines are used, the bandwidth must be shared. Each end transmits using different frequencies. To identify which end should transmit using which frequencies, a distinction is made between the originating modem and the answering modem.

The older modem standard in common use is the Bell 103, which uses FSK modulation to transmit up to 300 bps, full duplex. Each modem contains receiver filters that allow only the signal from the other modem to be received, so the modem is not confused by its own transmissions.

It uses the following frequency assignments:

Originate Modem: 0 = 1070 Hz
1 = 1270 Hz

Answer Modem: 0 = 2025 Hz
1 = 2225 Hz

Another common standard is Bell 212A, which provides 1200 bps full duplex operation. It uses the dibit phase modulation. The transmit carrier frequency is 1200 Hz for the originate modem and 2400 Hz for the answering modem. The term dibit refers to the fact that the two bits are encoded in each phase shift. Figure 5.6 shows the four phase shift angles used in 212 A type modem and an example waveform with a 2400 Hz carrier. The baud rate of the phase encoded signal is the rate of the rate of phase changes. Since the two data bits are encoded into a single phase change, the 1200 bps data stream produces a 600 baud analog signal.

Dibit	Phase Shift
00	90°
01	0°
10	180°
11	270°

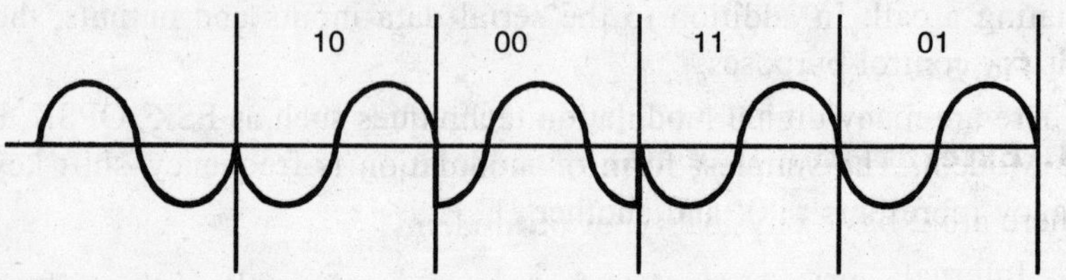

Fig. 5.6. *Bell 212A Phase Modulation Dibit waveforms.*

Today's modems are used for different functions. They act as textual and voice mail systems, facsimiles, and are connected or integrated into cellular phones and in notebook computers enabling sending data from anywhere. The future might lead to new applications. Modem speeds are not expected to be increased much over today's 28.8 kbps. Further dramatic speed increases will require digital phone technology such as ISDN and fiber optic lines.

New applications might be implemented such as simultaneous voice and data. Videophones are an example of this.

5.7.1. Digital Connection

The connection between the modem and terminal/computer is a digital connection. A basic connection consists of a Transmit Data (TXD) line, a Receive Data (RXD) line and many hardware hand-shaking control lines.

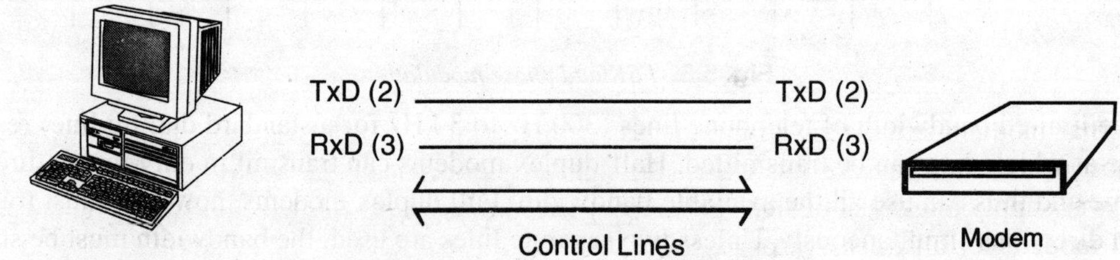

Fig. 5.7. *Digital connection between the modem and terminal*

The control lines determine whose turn it is to talk (modem or terminal), if the terminal/computer is turned on, if the modem is turned on, if there is a connection to another modem, etc.

5.7.2. Analog Connection

The connection between the modem and outside world (phone line) is an analog connection. The Voice Channel has a bandwidth of 0-4 kHz but only 300 - 3400 Hz is usable for data communications.

The modem converts the digital information into tones (frequencies) for transmitting through the phone lines. All of the tones are restricted to the narrow 300-3400 Hz Voice Band.

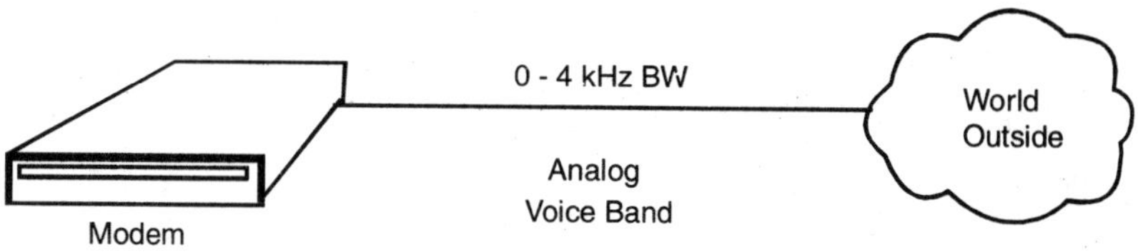

Fig. 5.8. *Analog connection. The connection between the modem and outside world (phone line).*

5.7.3. External and Internal Modems

There are 2 basic physical types of modems:

1. Internal
2. External modems.

External modems sit next to the computer and connect to the serial port using a straight through serial cable.

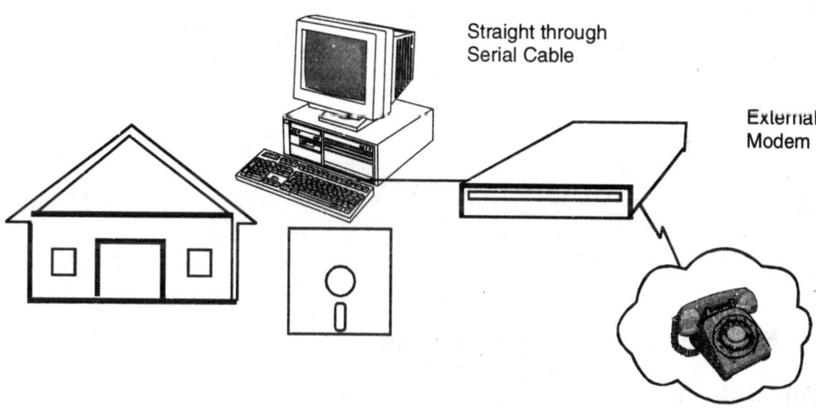

Fig. 5.9. *Using External Modem.*

Internal modems are a plug-in circuit board that sits inside the computer. It incorporates the serial port on-board. They are less expensive than external modems because they do not require a case, power supply and serial cable. They appear to the communication programs as if they were an external modem for all intensive purposes.

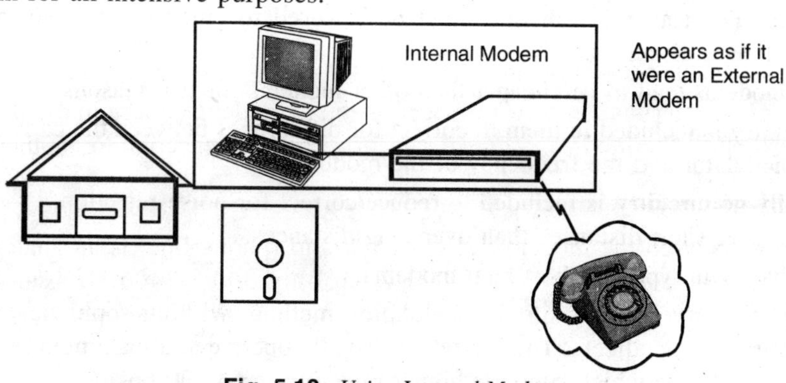

Fig. 5.10. *Using Internal Modem.*

5.7.4. Classification of Modems

The modems can be classified according to their characteristics:

1. Range
   (i) Short Haul
   (ii) Voice Grade (VG)
   (iii) Wideband
2. Line Type
   (i) Dial-up
   (ii) Leased
   (iii) Private
3. Operation Mode
   (i) Half Duplex
   (ii) Full Duplex
   (iii) Simplex
4. Synchronization
   (i) Asyr hronous
   (ii) Sync onous
5. Modulatio
   (i) AM
   (ii) FM/FS
   (iii) PM
   (iv) TCM
6. Data Rate

1. Classifying Modems according to: Range

(i) Short Haul

Short haul modems are cheap solutions to system of short ranges (up to 15 km), which use private lines and are not part of a public system. Short haul modems can also be used, even if the end-to-end length of the direct connection is longer than 15 km, when both ends of the line are served by the same central office in the telephone system. These lines are called "local loops". Short haul modems are distance-sensitive, because signal attenuation occurs as the signal travels through the line. The transmission rate must be lowered to ensure consistent and error-free transmission on longer distances.

Short haul modems tend to be cheaper than other modems for two reasons:

(a) No circuitry is included in them to correct for differences between the carrier frequency of the demodulator and the frequency of the modulator.

(b) Generally no circuitry is included to reduce/correct for noise rejection, n is less of a problem over short distances than over long distances.

There are two main types of short haul modems:

1. Analog modems, using a simple modulation method, without sophisticated devices for error control or equalizers. These modems usually operate at a maximum rate of 9600 bps, but there are some which supports higher rates (up to 64,000 bps).

2. Line drivers increase the digital signal, which transmit to the communication channel and do not transmit the carrier signal, as conventional modems. Line drivers are very cheap and tiny and connect to the RS232 connector of the terminal (since they lack a power supply, they use the signal voltage of the DTE-DCE interface for DC power supply).

(ii) Voice Grade (VG)

Voice-grade modems are used for unlimited destination, using a moderate to high data rate. These modems are expensive and their maintenance and tuning are sophisticated. Communication channels are leased lines and dial-up.

Voice-band telephone network is used for data transmission. A user-to-user connection may be either dedicated or dialed. The links in the connection are the same in the two cases, and the only difference for the user is that for some impairment (particularly attenuation and delay distortion), a dedicated (*private* or leased) line is guaranteed to meet certain specifications, whereas a dialed connection can only be described statistically.

(iii) Wideband

Wideband modems are used in large-volume telephone-line multiplexing, dedicated computer-to-computer links. These modems exceed high data rates.

2. Classifying Modems according to: Line Type

(i) Leased, Private

Leased, private or dedicated lines (usually 4-wire) are for the exclusive use of "leased-line" modems - either pair (in a simple point-to-point connection) or several (on a multidrop network for polling or a contention system). If the medium is the telephone network, their transmission characteristics are usually guaranteed to meet certain specifications, but if the link includes any radio transmission, the quality of it may be as variable as that of a switched (*i.e.,* nondedicated) line.

(ii) Dial-Up

Dial-up modems can establish point-to-point connections on the PSTN by any combination of manual or automatic dialing or answering. The quality of the circuit is not guaranteed, but all phone companies establish objectives. The links established are almost always 2-wire because 4-wire dialing is tedious and expensive.

(iii) Two and Four-Wires Lines

A four-wire (4W) line is a pair of two-wire (2W) lines, one for transmitting and one for receiving, in which the signals in the two directions are to be kept totally separate. Perfect separation can be maintained only if the four-wire configuration is sustained from transmitter to receiver. The lines may be combined in a 4W/2W network (often called a *hybrid* or a *hybrid transformer*) at any point in the signal path. In this case impedance mismatches will cause reflections and interference between the two signals.

3. Classifying Modems According to : Operation Mode

(i) Half Duplex

Half duplex means that signals can be passed in either direction, but not in both simultaneously. A telephone channel often includes an echo-suppressor, allowing transmission in only one direction;

this renders the channel half-duplex. Echo suppressors are slowly being replaced by echo cancellers, which are theoretically full-duplex devices.

When a modem is connected to a two-wire line, its output impedance cannot be matched exactly to the input impedance of the line, and some part of its transmitted signal (usually badly distorted) will always be reflected back. For this reason half- duplex receivers are disabled (received data is clamped) when their local transmitter is operative.

Half-duplex modems can work in full-duplex mode.

(ii) Full Duplex

Full duplex means that signals can be passed in either direction, simultaneously. Full duplex operation on a two-wire line requires the ability to separate a receive signal from the reflection of the transmitted signal. This is accomplished by either FDM (frequency division multiplexing) in which the signals in the two directions occupy different frequency bands and are separated by filtering, or by Echo Canceling (EC).

The implication of the term *full-duplex* is usually that the modem can transmit and receive simultaneously at *full* speed. Modems that provide a low-speed reverse channel are sometimes called *split-speed* or *asymmetric* modems.

Full duplex modems will not work on half-duplex channels.

(iii) Simplex

Simplex means that signals can be passed in one direction only. A remote modem for a telemetering system might be simplex and a 2-wire line with a common unidirectional amplifier is simplex.

(iv) Echo Suppressor and Echo Canceller

At the junction between the local loop, which is usually a 2-wire circuit, and the trunk, which is a 4-wire circuit, echoes can occur. The effect of the echo is that a person speaking on the telephone hears his own words after a short delay. Psychological studies have shown that this is annoying to many people, often making them stutter or become confused. To eliminate the problem of echoes, echo suppressors are installed on lines longer than 2000 km. (On short lines the echoes come back so fast that people cannot detect them). An *echo suppressor* is a device that detects human speech coming from one end of the connection and suppresses all signals going the other way. The device compares the levels at its two input ports, and if it decides, for example that the other end is talking, it inserts an attenuator in the return (echo) path, and vice versa.

Echo suppressors have several properties that are undesirable for data communication. First, they prevent full-duplex data transmission, which would otherwise be possible, even over the 2-wire local loop (by allocating part of the bandwidth to the forward channel and part to the reserve channel). Even if half-duplex transmission is adequate, they are a nuisance because the time required to switch directions can be substantial. Double-talking totally confuses them, and the attenuation may be switched in and out repeatedly. Furthermore, they are designed to reverse upon detecting human speech, not digital data.

To reduce these problems, when echo suppressors detect a specific tone they shut down, and remain shut down as long as the carrier is present (this is an example of *inband signaling*, where control signals that activate and deactivate internal control functions lie within the band accessible to the user). This disabling is usually done during initial handshaking by one modem transmitting an *answer* tone in either 2100 Hz (CCITT standard) or 2225 Hz (modems following the old Bell 103 standard).

Echo suppressor are slowly being replaced by ECs, which allow a certain amount of double-talking and do not require "capture" time for any one talker to assume control of the connection.

4. Classifying Modems according to : Synchronization

(i) Asynchronous Modems

The common modem used today is asynchronous modem. In **asynchronous** transmission, data is coded into a series of pulses, including a start bit and a stop bit. A start bit is sent by the sending modem to inform the receiving modem that a character is to be sent. The character is then sent, followed by a stop bit designating that the transfer of that bit is complete. Each byte is placed between a stop and a start bit. Each modem must operate with the same start and stop bit sequence Each modem must operate at the same baud rate. Each modem must have the same parity settings for the data checking in order to communicate correctly. In asynchronous transmission parity checking is defined.

Most of the modems that operate in slow and moderate rates, up to 1800 bps, are asynchronous (using asynchronous data). Asynchronous modems operate in FSK modulation and use two frequencies for transmission and another two for receiving. Asynchronous modems can be connected in different options to the communication media:

- Using 2-wire or 4-wire interface.
- Using switched lines or leased lines.
- Using interface to call unit/automatic answer, when dialing-up.

In a 2-wire line, full duplex operation can be achieved by splitting the channel into two sub-channels.

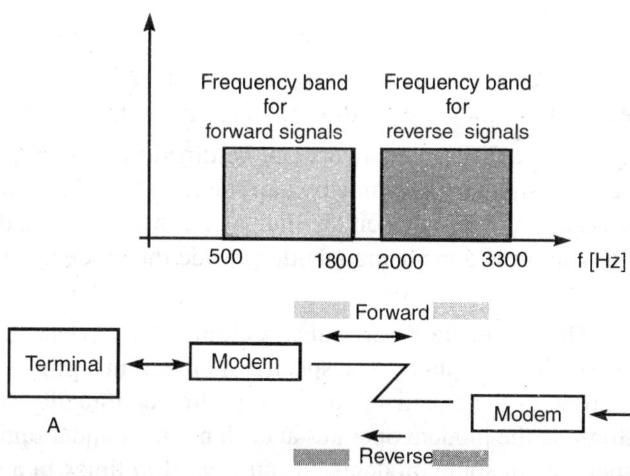

Fig. 5.11. *Modem-1*

Asynchronous Data

Asynchronous data is not accompanied by any clock, and the transmitting and receiving modems know only the nominal data rate. To prevent slipping of the data relative to the modems' clocks, this data is always grouped in very short blocks (characters) with framing bits (start and stop bits). The most common code used for this is the seven-bit ASCII code with even parity.

(i) Synchronous Modems

Synchronous modems can be faster than asynchronous. They depend on timing to communicate. In **synchronous** data transmission, data is sent via a bit-stream, which sends a group of characters in a single stream. Data is transmitted in frames with synchronization bits which are used to be sure the timing of the transmission and reception of data is accurate. In order to do this, modems gather groups of characters into a buffer, where they are prepared to be sent as such a stream. In order for the stream to be sent, synchronous modems must be in perfect synchronization with each other. They accomplish this by sending special characters, called synchronization, or syn, characters. When the clocks of each modem are in synchronization, the data stream is sent. Synchronous modems are normally used on dedicated leased lines.

Synchronous modems are one of binary synchronous communications protocol (bisync), high level data link control (HDLC), or synchronous data link control (SLDC). Three methods can be used to control synchronization:

(*a*) Additional clock signal

(*b*) Guaranteed state change - Clocking is part of the data signal.

(*c*) Over sampling - The receiver samples the signal much faster than the data is sent. The extra samples can be used to be sure the clock is synchronized

Synchronous modems operate in the audio domain, at rates up to 28800 bps in audio lines, used in telephones systems (using synchronous data). The usual modulation methods are the phase modulation and integrated phase and amplitude (at higher rates than 4800 bps).

In synchronous modems, equalizers are used, in order to offset the misfit of the telephone lines. These equalizers are inserted in addition to the equalizers that sometimes already exist in the telephone lines.

These equalizers can be classified into three main groups:

(*a*) ***Fixed/statistical equalizer*** : These equalizers offset the signal according to the average of the known attenuation in each frequency. Tuning the equalizer is sometimes done in the factory and stays fixed, usually they are used to operate at low rates in a dial up line.

(*b*) ***Manually adjusted equalizer*** : These equalizers can be tuned to optimal performance to a given line. These equalizers should be re-tuned when the line is replaced and periodically. Specially, it should be tuned frequently when the line is of a low quality and it's parameters are changed frequently. Tuning is done using a button inside the modem (or on the external board).

(*c*) ***Automatic equalizer*** : These equalizers are tuned automatically when the connection is established. Depending on the line quality in a specific moment, in a process of about 15ms to 25ms, after the first tuning, the equalizer samples the line continually and adjusts itself to the changed conditions, so the modem operates at each moment under optimal conditions. The fitness process operates, in some modems, at rates of 2400 times in a second.

Synchronous modems operate in the same manner asynchronous modems. However, synchronous modems operate at higher rates and since the requirements to transmit at these rates are increasing, most of the innovations are implemented for synchronous modems.

In synchronous modems the channel can be split for several consumers at various speeds. Modems that have this ability are called **SSM - Split System Modem**. These modems can use a simple split or a split using multipoint connection.

Synchronous data is accompanied by a clock signal. Synchronous data is almost always grouped in blocks, and it is the responsibility of the data source to assemble those blocks with framing codes and any extra bits needed for error detecting and/or correcting according to one of many different protocols (BISYNC, SDLC, HDLC, etc.). The data source and destination expect the modem to be transparent to this type of data, conversely, the modem can ignore the blocking of the data.

5. Classifying Modems according to : MODULATION

Communication channels like telephone lines are usually analog media. Analog media is a bandwidth limited channel. In the case of telephone lines the usable bandwidth frequencies is in the range of 300 Hz to 3300 Hz.

Data communication means moving digital information from one place to another through communication channels. These digital information signals have the shape of square waves and the meaning of "0" and "1".

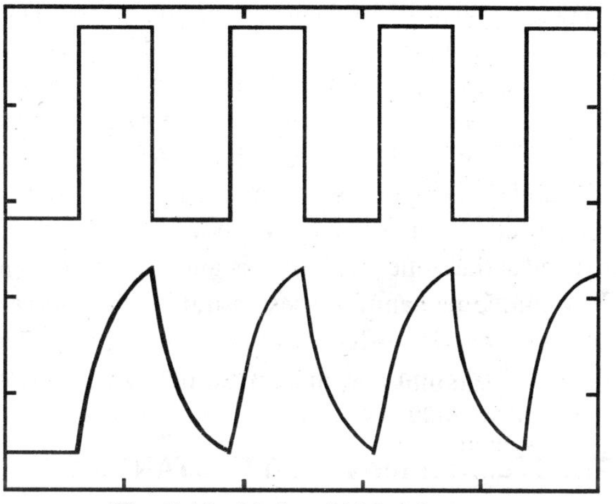

Fig. 5.12. *Modem-2* Modulation*

If such digital signals were transmitted on analog media the square waves of the digital signals would be distorted by the analog media as shown in figure Modem-2. The receiver which receives these distorted signals will be unable to interpret accurately the incoming signals. These digital signals must be converted into analog signals so that the communication channels can carry the information from one place to another. The technique which enables this conversion is called ***modulation.***

Modulation is a technique of modifying some basic analog signal in a known way in order to encode information in that basic signal. Any measurable property of an analog signal can be used to transmit information by changing this property in some known manner and then detecting those changes at the receiver end. The signal that is modulated is called the carrier signal, because it carries the digital information from one end of the communication channel to the other end.

The device that changes the signal at the transmitting end of the communication channel is called the ***MODULATOR***. The device at the receiving end of the channel, which detects the digital information from the modulated signal, is called the ***DEMODULATOR.***

Modulation can use any of these three measurable and changeable properties of the sine wave for encoding purposes.

There are three modulation techniques; each of them changes one of the properties of the basic analog signal.

5.7.5. STANDARDS and PROTOCOLS of Modem

Communication between two devices might work only when the interface is defined and agreed. For modems, the standards define techniques used for modulation, for error correction for data compression, and other attributes. There are some standard organizations for the development of interface standards. The ITU - International Telecommunications Union an agency of the United Nations (Geneva, Switzerland), ISO - International Standards Organization, and CCITT - International Telegraph and Telephone Consultative Committee a group of ITU.

The modem standards were developed during the years and published as V series of recommendations. In the United States the primary standards body is the ANSI - American National Standards Institute. Its committees, concerned with information processing and data communication, are designated X3 and X3S3, respectively. The organizations deal with De Jure standards. There are also De Facto standards which were developed by a specific manufacturer, using new features in his products, while they were not defined yet by the standard committees. When the definitions were adapted by other manufacturers for compatibility they became a De Facto standard. Some examples of De Facto standards follow. The Bell-100 standards established by the Bell System with their 100 and 200 series of modems; Bell Dataphone 103 of 300 bps, that was introduced at 1958 was the first modem used to transfer data over telephone lines. The modem command language, "AT" commands, invented by Hayes but used by all modem vendors ("AT" stands for "Attention" and each command begun with an "AT" command). This command language enabled the control of the modem operation set from a simple one, such as dialing a phone number and from a complicated one, such as answering the phone only after 15 rings. Since only a few modem manufactures existed at that time, "Hayse" became a leading company and the "Hayse" commands became standard. New modem manufactures imitated the "Hayse " commands. Most of the communication SW companies today offer "Hayse" compatibility. The MNP protocols that define error correction and data compression schemes, were developed by Microcom Inc. but are widely accepted throughout the modem industry.

The ITU (formerly CCITT) STANDARDS

Standard	Date (Ratified)	Speed (Bps)	Hdx/Fdx	PSTN/ Private	Modulation
V.21	1964	200	FDX(FDM)	PSTN	FSK
V.22	1980	1200	FDX(FDM)	PSTN	PSK
V.22 bis	1984	2400	FDX(FDM)	PSTN	QAM
V.23	1964	1200	HDX	PSTN	FSK
V.26	1968	2400	HDX	Private	PSK
V.26 bis	1972	2400	HDX	PSTN	PSK
V.26 ter	1984	2400	FDX(EC)	PSTN	PSK
V.27	1972	4800	HDX	Private	PSK
V.27 bis	1976	4800	HDX	Private	PSK
V.27 ter	1976	4800	HDX	PSTN	PSK
V.29	1976	9600	HDX	Private	QAM
V.32	1984	9600	FDX(EC)	PSTN	QAM
V.32 bis	1991	14400			TCM
V.32 Ter	19200				TCM
V.34 (V.fast)	1994	28800			TCM

FDM means Frequency Division Multiplexing

EC means Echo Canceller

There are other standards that deal with the modem's related functions. Some of these standards and protocols are listed here.

Data compression involves different methods such as *Huffman coding* and *run length coding*. The first method deals with the frequent characters being transmitted in fewer bits than the other characters. The second method transmits the bit value and the chain length, instead of a chain of identical succeeding bits. The main characteristic of data compression protocols is that they buffer the desired data to be transferred, compress it and only then transfer it to the second modem. The second modem must do the opposite work. The data compressing algorithms are similar to the ARC, ZIP or ARJ programs. They are located in the modem ROM and compress the data in real time. The compression depends on the characteristics of the data. For example, PC '.EXE' or '.COM' files may be compressed up to 40-50%. Text files may be compressed up to 100%.

Protocols used by modems to transfer files :

The widespread protocols for transferring files are:

1. Xmodem,
2. Ymodem,
3. Zmodem

The following is a brief description of the protocols.

1. X-Modem Protocol

The **X-modem protocol** divides the data into blocks. Each block contains a sequence number of the block, 128 Bytes of the data and 4 Bytes of checksum. The checksum is computed for the 128 Data Bytes.

The protocol on the other side is synchronized by checking the sequence number of currently transferred block and then calculating the checksum of 128 bytes of data and comparing it to the transferred checksum. In the case of error, it requests to send the same block again.

2. Y-modem protocol

In the Y-modem protocol, each block contains 1024 Bytes. The checksum size is 4 Bytes. It is faster than the X-modem protocol. It also may transfer a batch of files and information about each file to be transferred and its size. This helps the user on the other side to see the time left of the transfer.

3. Z-modem

Z-modem is a public domain program written by Chuck Forsberg at Omen Technology. There are several main advantages of this protocol. The block size varies from 16 to 1024 Bytes. The protocol dynamically finds the optimal block size for the file transfer on the current phone line. It starts with a data block size of 1 Kbyte. It reduces it automatically in case the phone line is noisy or enlarges the block size when the line disturbance disappears. The transfer rate is fast with big block sizes but remember that in case of an error the whole block is retransmitted and time is wasted. Since this protocol adjusts the block size to the line quality it combines the best characteristics on each side. It resumes the transfer after line disconnection. The checksum size is 8 Bytes (CRC/32), therefore, it raises the protocol error correction authenticity.

Establishing Connection

Establishing a connection between two modems involves a handshaking process of sending and receiving coded signals to coordinate the connection. The FallBack method is used to find a common way of communication. The calling modem first tries to connect at its highest speed (or best error-correction or data compression scheme). If the called modem doesn't signal back that is can handle that protocol, the calling modem falls back to a slower speed or less effective scheme and tries to connect again. This cycle continues until a common ground is found or they run out of options.

5.7.6. Today's Status and Future Trends

Today in addition external modems, there are internal modems which are included as an additional board within the computer. There are advantages to each type.

Actual transfer rates are limited due to type of phone lines. Using slower phone trunks, international circuits where half the normal bandwidth is used, and the slow cellular connections where it might run at only 14.4 kbps (without compression) although the modem itself enable 28.8kbps.

More and more users are accessing the Internet and on-line services such as Compuserve, so, use of modems has increased dramatically. The more powerful processors such as Pentium and PowerPC in workstations and PC's, enable the modem h/w to be less complicated. Part of the functions done in the DSP or microcontroller might be performed by the host. So, modems might drop in price.

The advent of semiconductor modems will enable a wide range of applications to be implemented: Vending machines will call up when they need more goods to vend, or elevators will call when they require service, and so on.

The last approved standard of V.34 with 28.8 kbits/s speed will enable the Digital Simultaneous Voice and Data (DSDV) applications. DSVD is a modem specification that lets voice and data to be shared over a single dial-up connection. The data is multiplexed into packets, much like an ATM stream.

V.34 is approaching the theoretical speed limit of an analog line - estimated to be in the low 30kbps range. This may be the last of the new modem protocols which doubled the previous speeds.

5.7.7. Features of Modems

1. Speed

The speed at which the modem can send data in bps (bits per second). Typically modem speeds are: 300, 600, 1200, 2400, 4800, 9600, 14.4k, 19.2k, 28.8k, 33.6k and 56k bps. (Where little k = kilo = 1000 and big K = 1024.)

2. Auto Dial/Redial

Smart Modems can dial the phone number and & auto redial if a busy signal is received.

3. Auto Answer

Most modems can automatically answer the phone when an incoming call comes in. They have Ring Detect capability.

4. Self-Testing

New modems have self-testing features. They can test the digital connection to the terminal / computer and the analog connection to a remote modem. They can also check the modem's internal electronics.

5. Voice Over Data

Voice over Data modems allow a voice conversation to take place while data is being transmitted. This requires both the source and destination modems to have this feature.

6. Synchronous or Asynchronous Transmission

Newer modems allow a choice of synchronous or asynchronous transmission of data. Normally, modem transmission is asynchronous. We send individual characters with just start and stop bits. Synchronous transmission or packet transmission is used in specific applications.

Table 5.1. MODEM Speeds

Baud Rate	Modulation Method
300 - 1200	FSK - Frequency Shift Keying
2400	QAM - Quadrature Amplitude Modulation
9.6 - 14.4	DPSK - Dual Phase Shift Keying
28.8 - 56.6	DPSK - Dual Phase Shift Keying (Dial-up, AOL)
128K	ISDN - Integrated Services Digital Network
380K	DSL - Digital Subscriber Line (SBC)
1.544M	T1
6M	Cable Modem - DOCSIS (Comcast)
16M	Backplane bus - VME Bus
44.7M	T3
155M	Sonet - Synchronous Optical Network
622M	ATM - Asnychronous Transfer Mode

Summary

- The largest communication network on this earth is the global Public Switched Telephone Network (PSTN).
- The Public Telephone Network includes mainly two types of subscribers:
  1. Private
  2. Public
- **Private :** All those customers who lease transmission media, equipments and services from the telephone companies on permanent basis have known as **private subscribers** or private line circuit or dedicated circuits.
- **Public :** Subscribers to the public sector of the telephone network share facilities and equipments that are available to all the public subscribers to the network.
- All the subscribers to public network are interconnected temporarily through the switches. So they are also known as **public switched telephone network (PSTN)** or dial-up network. PSTN refers to the combination of transmission lines and switches (may be central offices) which forms a system of electrical routes through the network.
- Users are referred to as **subscribers**, because they subscribe to a service provided by the telephone company.
- The system employs **switches** to facilitate connections between subscribers.
- The circuit which connects each subscriber (via a pair of wires) to the switch is called a **subscriber loop** or **local loop**.
- A twisted pair cable which connects the subscriber telephone set to the nearest end office (may be referred as local central office).

- A trunk circuit is similar to a local loop except trunk circuits are used to interconnect two telephone offices.
- **Switching offices** or exchange is a central location where subscribers are interconnected, permanently or temporarily basis.
- Private telephone networks come in different configurations and sizes, with varying degrees of connectivity or integration with public telephone networks and data networks. Most private networks revolve around a private exchange, which we will look at first.
- CCITT's Signaling System Number 7 (SS7) opens the scope for user-defined procedures for the way calls are handled. This leads to the concept of **Intelligent Networks (IN)**.
- The North American Telephone Numbering Plan (NANP) was established to provide a telephone numbering system for the United States, Mexico, and Canada that would allow any subscriber in North America to direct dial virtually any other subscriber without the assistant of an operator. The network is often referred to as the **Distance Direct Dialing (DDD) Network.**
- A **telephone numbering plan** is a type of numbering scheme used in telecommunications to allocate and route telephone numbers in a telephone network.
- A **dial plan** establishes the expected number and pattern of digits for a telephone number. This includes country codes, access codes, area codes and all combinations of digits dialed.
- *A modem is a* ***modulator/demodulator***. The analog signals encode the digital information at the time of modulation and decode it back during demodulation to transmit the data.
- A device which carries out the two important tasks of modulating and demodulating the analog signals is called a **modem**.
- There are many digital modulation techniques such as FSK, QPSK, DPSK, QAM, etc are used in the Modem.
- Half-duplex modems can transmit in only in one direction at a time and thus can use all the available bandwidth. Full duplex modems, however, must transmit in both directions simultaneously.
- The older modem standard in common use is the Bell 103, which uses FSK modulation to transmit up to 300 bps, full duplex.
- There are 2 basic physical types of modems:
  1. Internal
  2. External modems.
- Short haul modems are cheap solutions to systems of short ranges (up to 15 km), which use private lines and are not part of a public system.
- Voice-grade modems are used for unlimited destination, using a moderate to high data rate.
- Leased, private or dedicated lines (usually 4-wire) are for the exclusive use of "leased-line" modems - either pair (in a simple point-to-point connection) or several (on a multidrop network for polling or a contention system).
- Dial-up modems can establish point-to-point connections on the PSTN by any combination of manual or automatic dialing or answering.
- An *echo suppressor* is a device that detects human speech coming from one end of the connection and suppresses all signals going the other way.
- Echo suppressor are slowly being replaced by ECs, which allow a certain amount of double-talking and do not require "capture" time for any one talker to assume control of the connection.
- In **asynchronous** transmission, data is coded into a series of pulses, including a start bit and a stop bit.
- Synchronous modems can be faster than asynchronous. They depend on timing to communicate. In **synchronous** data transmission, data is sent via a bit-stream, which sends a group of characters in a single stream.

- Three methods can be used to control synchronization:
  - (*a*) Additional clock signal
  - (*b*) Guaranteed state change - Clocking is part of the data signal.
  - (*c*) Over sampling - The receiver samples the signal much faster than the data is sent. The extra samples can be used to be sure the clock is synchronized
- Communication between two devices might work only when the interface is defined and agreed. For modems, the standards define techniques used for modulation, for error correction for data compression, and other attributes.
- The **X-modem protocol** divides the data into blocks. Each block contains a sequence number of the block, 128 Bytes of the data and 4 Bytes of checksum.
- In the Y-modem protocol, each block contains 1024 Bytes. The checksum size is 4 Bytes.
- **Z-modem** is a public domain program written by Chuck Forsberg at Omen Technology.

Review Questions

1. Explain the public telephone networks in brief.
2. List the names of all the synchronous and asynchronous protocols. What is the major drawback of byte oriented protocols that is overcome by bit oriented protocol ? Draw and explain the frame format of BSC protocol. **[MDU May 2011]**
3. Explain the public switched telephone networks with the block diagram in brief.
4. Explain the need of MODEM for the data communication.
5. Explain the functions of MODEM with the block diagram.
6. What type of modulation is used for the MODEM?
7. Differienate between the synchronous and asynchronous MODEMs. Explain synchronous MODEM in details. **[MDU Dec. 2009]**
8. Explain the Distance Direct Dialing (DDD) Network.
9. Explain any two types of Telephone systems.
10. Write short notes on following:
    - (*i*) Components of PSTN
    - (*ii*) Local loop and Trunk line
    - (*iii*) PBX
    - (*iv*) Private telephone network
    - (*v*) MODEM
    - (*vi*) Modern MODEM
    - (*vii*) Distance Direct Dialing (DDD) Network

❑❑❑

6

CHAPTER

Data Communications Protocols

6.1. INTRODUCTION

INSIDE THIS CHAPTER

A network is any collection of independent computers that communicate with one another over a shared network medium. A **computer network** is the infrastructure that allows two or more computers (called **hosts**) to communicate with each other. When these computers are joined in a network, people can share files and peripherals such as modems, printers, or CD-ROM drives. When networks at multiple locations are connected using services available from phone companies, people can send e-mail, share links to the global Internet, or conduct videoconferences in real time with other remote users. As companies rely on applications like electronic mail and database management for core business operations, computer networking becomes increasingly more important.

The network achieves this by providing a set of rules for communication, called **protocols**, which should be observed by all participating hosts. The need for a protocol should be obvious: it allows different computers from different manufactures and with different operating characteristics to 'speak the same language'.

This chapter introduces the fundamental concepts of a reference model for network protocol architectures which we will expand upon throughout the rest of this book.

In this chapter a reference model referred as Open System Interconnection (OSI) model is discussed in brief. Then various data transmission protocols such as synchronous and asynchronous protocols are described.

6.2. OPEN SYSTEMS INTERCONNECTION (OSI) REFERENCE MODEL

Established in 1947, the International Standards Organization (ISO) is a multinational body dedicated to worldwide agreement on international standards.

An ISO standard that covers all aspects of network communications is the Open Systems Interconnection (OSI) model.

It was first introduced in the late 1970s.

DO YOU KNOW

ISO is the Organization.
OSI is the model.

Definition

A set of protocols is open if:

- protocol details are publicly available
- changes are managed by an organization whose membership and transactions are open to the public

A system that implements open protocols is called an open system.

International Organization for Standards (ISO) prescribes a standard to connect open systems.

The **OSI model** defines internetworking in terms of a vertical stack of seven layers. The upper layers of the OSI model represent software that implements network services like encryption and connection management. The lower layers of the OSI model implement more primitive, hardware-oriented functions like routing, addressing, and flow control.

The OSI reference model was introduced in 1984 by International Organization for Standardization (ISO). Although it was designed to be an abstract model, the OSI model remains a practical framework for today's key network technologies like Ethernet and protocols like IP.

An open system is a set of protocols that allows any two different systems to communicate regardless of their underlying architecture. The purpose of this model is to show how to facilitate communication between different systems without requiring changes to the logic of the underlying software and hardware.

The *Open System Interconnection (OSI) reference model* describes how information from a software application in one computer moves through a network medium to a software application in another computer. The OSI reference model is an obsolete conceptual model composed of seven layers, each specifying particular network functions.

The OSI model should be used as a guide for how data is transmitted over the network. It is an abstract representation of the data pathway and should be treated as such.

OSI model contains seven layers that build on one another. Each layer provides specific services and makes the results available to the next layer. Each layer should be independent of all others.

The OSI model was specifically made for connecting open systems. These systems are designed to be open for communication with almost any other system. The model was made to break down each functional layer so that overall design complexity could be lessened. The model was constructed with seven layers for the flow of information.

This is a top-down explanation of the OSI Model, starting with the user's PC and what happens to the user's file as it passes though the different OSI Model layers. The top-down approach was selected specifically (as opposed to starting at the Physical Layer and working

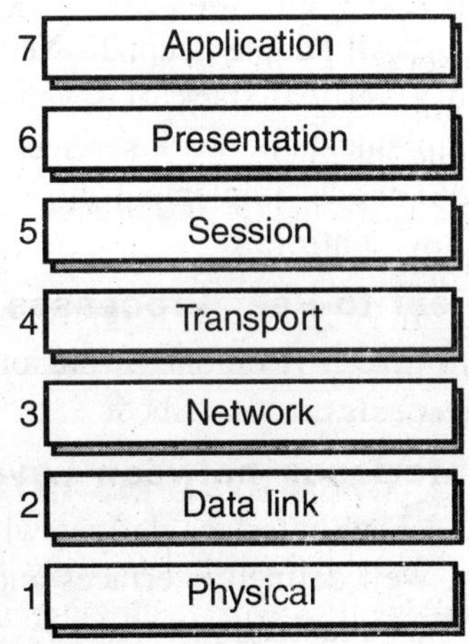

Fig. 6.1.

up to the Application Layer) for ease of understanding of how the user's files are transformed through the layers into a bit stream for transmission on the network.

The following list details the seven layers of the Open System Interconnection (OSI) reference model in Fig. 6.1.

7	Application	Message	Mutually-agreeable meaning of application data (common semantics).
6	Presentation	Message	Mutually-agreeable binary representation of application data (common syntax).
5	Session	Message	Negotiation of the establishment and termination of connections (sessions).
4	Transport	Message	Efficient and cost-effective transportation of data across the network.
3	Network	Packet	Routing of data packets within the network and across multiple networks.
2	Data Link	Frame	Provision of a reliable communication line to the network layer.
1	Physical	Bit	Transmission of raw data bits over communication lines.

- Layer 7—Application (Top Layer)
- Layer 6—Presentation
- Layer 5—Session
- Layer 4—Transport
- Layer 3—Network
- Layer 2—Data link
- Layer 1—Physical (Bottom Layer)

Control is passed from one layer to the next, starting at the application layer in one station, and proceeding to the bottom layer, over the channel to the next station and back up the hierarchy.

Remember

A convenient aid for remembering the OSI layer names is to use the first letter of each word in the phrase:

"All People Seem To Need Data Processing." The **OSI model** defines internetworking in terms of a vertical stack of seven layers. *The upper layers* of the OSI model represent software that implements network services like encryption and connection management. *The lower layers* of the OSI model implement more primitive, hardware-oriented functions like routing, addressing, and flow control.

Peer-to-Peer Processes

Layer X on one machine communicates with layer x on another machine - called *Peer-to-Peer Processes.*

Interfaces between Layers

Each interface defines what information and services a layer must provide for the layer above it. Well defined interfaces and layer functions provide modularity to a network.

Organizations of the layers.

Network support layers : Layers 1, 2, 3

User support layer : Layer 5, 6, 7

It allows interoperability among unrelated software systems.

Transport layer (Layer 4) : links the two subgroups.

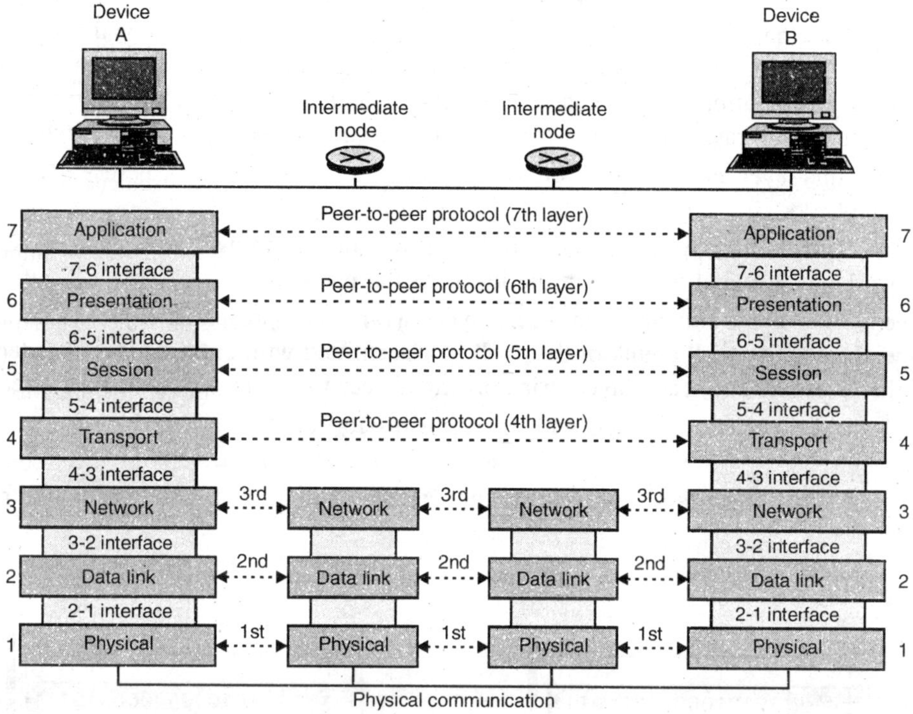

Fig. 6.2. *The interaction between layers in the OSI model.*

The data portion of a packet at level N – 1 carries the whole packet from level N. The concept is called *encapsulation.*

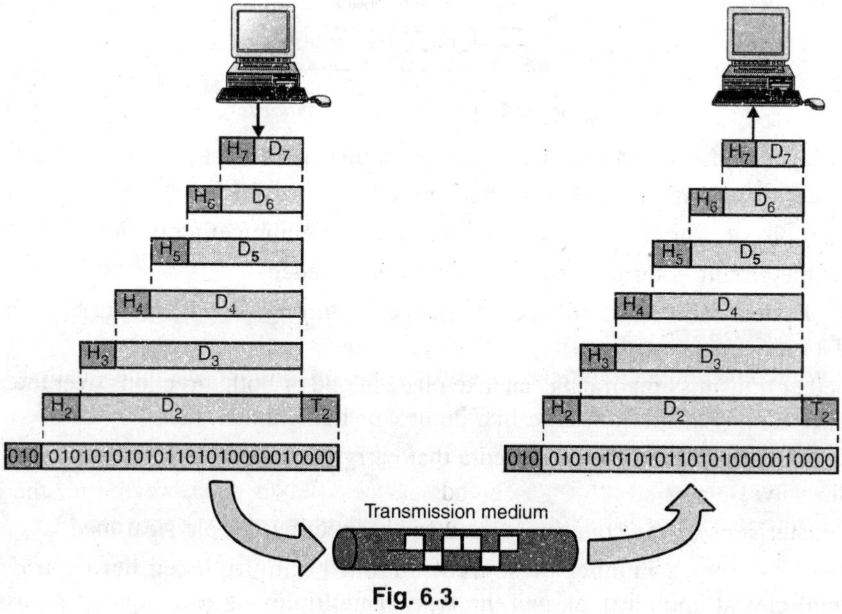

Fig. 6.3.

6.3. PHYSICAL LAYER

Physical layer handles the bit-level electrical/light communication across the network channel. The physical layer coordinates the function required to carry a bit stream over a physical medium. The physical layer defines the electrical, mechanical, procedural, and functional specifications for activating, maintaining, and deactivating the physical link between communicating network systems. Physical layer specifications define characteristics such as media, voltage levels, timing of voltage changes, physical data rates, maximum transmission distances, and physical connectors.

Basically, this layer ensures that a bit sent on one side of the Network is received correctly on the other side.

Data travels from the application layer of the sender, down through the levels, across the nodes of the network service, and up through the levels of the receiver.

To keep track of the transmission, each layer "wraps" the preceding layer's data and header with its own header. A small chunk of data will be transmitted with multiple layer headers attached to it. On the receiving end, each layer strips off the header that corresponds to its respective level.

The physical layer is concerned with the transmission of raw data bits over communication lines.

- The physical layer is responsible for movements of individual bits from one hop (node) to the next.

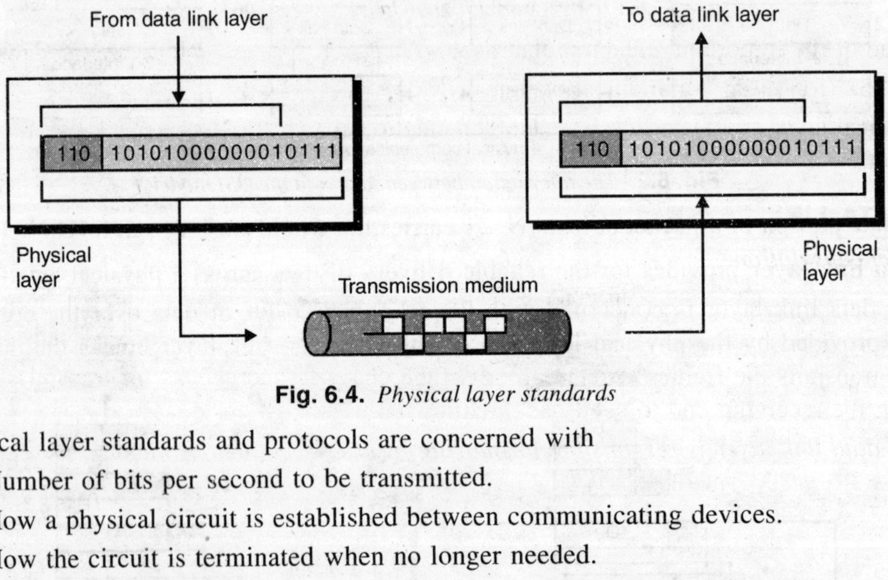

Fig. 6.4. *Physical layer standards*

Physical layer standards and protocols are concerned with

1. Number of bits per second to be transmitted.
2. How a physical circuit is established between communicating devices.
3. How the circuit is terminated when no longer needed.
4. The physical form (*e.g.*, voltages, frequencies, timing) in which data bits (binary values 0 and 1) are represented.
5. Whether transmission of data can take place in one or both directions over the same physical connection. Transmission type like duplex or half-duplex etc.
6. Characteristics of the physical media that carry the signals (*e.g.*, copper wire, optical fiber, radio waves).
7. Characteristics of the connectors used for connecting the physical media.
8. How data from a number of sources should be multiplexed before transmission and demultiplexed upon arrival, and the type of multiplexing technique to be used.

9. The type of modulation to be used for transmitting digital data over analog transmission lines.

The physical layer accounts for much of the tangible components of a network, including cables, satellites, earth stations, repeaters, multiplexers, concentrators, and modems. Physical layer protocols and standards are of mechanical, electrical, functional, and procedural nature.

This layer tells the driver software for the MAU (media attachment unit, ex. network interface cards (NICs, modems, etc.)) what needs to be sent across the medium.

Frequently used Physical layer protocols :

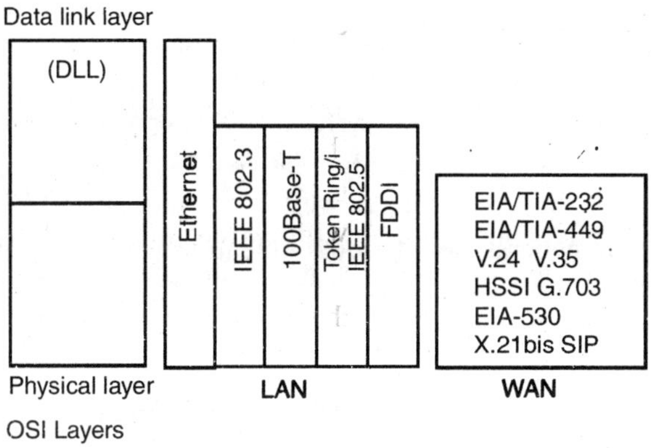

Fig. 6.5. *Physical layer protocols.*

Some of the important standards that deal with physical layer specifications are:

RS-232 (for serial communication lines), X.21, EIA 232, and G730.

Physical layer and Data link layer implementations can be categorized as either LAN or WAN specifications.

6.4. DATA LINK LAYER

Data link layer provides for the reliable delivery of data across a physical network.

The data link layer is concerned with the reliable transfer of data over the communication channel provided by the physical layer. To do this, the data link layer breaks the data into **data frames**, transmits the frames sequentially over the channel, and checks for transmission errors by requiring the receiving end to send back **acknowledgment frames**.

The data link layer is responsible for moving frames from one hop (node) to the next.

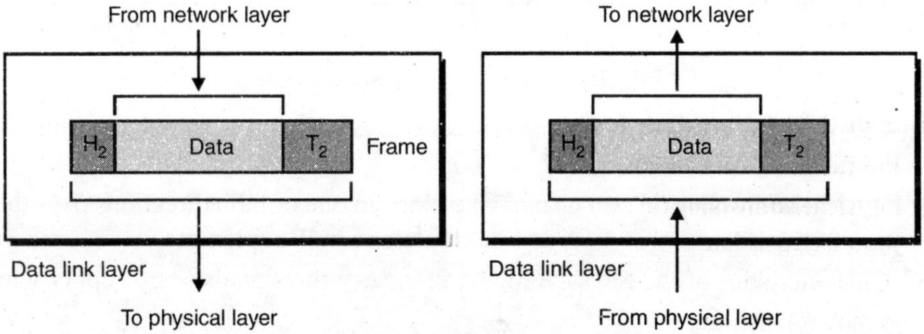

Fig. 6.6. *Data link layer.*

This layer deals with issues such as flow regulation, error detection and control, and frames. This layer has the important task of creating and managing what frames are sent out on the network.

The network data frame, or packet, is made up of checksum, source address, destination address, and the data itself. The largest packet size that can be sent defines the maximum transmission Unit (MTU).

Data link protocols are concerned with the following issues:

1. How to divide the data into frames.
2. How to delimit frames by adding special bit patterns to the beginning and end of each frame. This allows the receiving end to detect where each frame begins and where it ends.
3. **Error detection :** Some form of error check is included in the frame header. This is constructed by the transmitting end based on the contents of the frame, and checked for integrity by the receiving end. A change in the frame bits can be detected in this way.
4. **Error correction :** When a frame arrives corrupted or is for any reason lost in the network, it is retransmitted. Lost acknowledgment frames may result in duplicate frames, which need to be detected and corrected as well.
5. **Flow control :** In general, not all communication devices in a network operate at the same speed. Flow control provides a means of avoiding a slow receiver from being swamped by data from a fast transmitter.

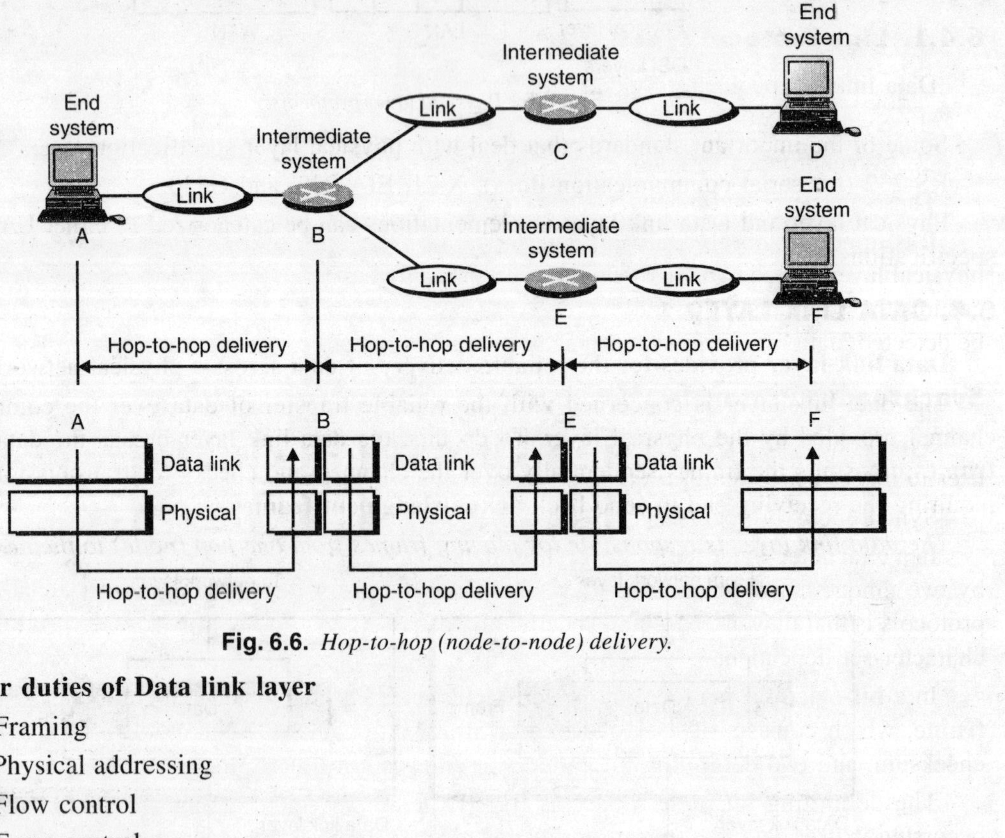

Fig. 6.6. *Hop-to-hop (node-to-node) delivery.*

Major duties of Data link layer

(*a*) Framing

(*b*) Physical addressing

(*c*) Flow control

(*d*) Error control

(*e*) Access control

This layer consists of two sub layers:

Logical Link Control (LLC) defines how data is transferred over the cable and provides data link service to the higher layers.

Medium Access Control (MAC) defines who can use the network when multiple computers are trying to access it simultaneously (*i.e.,* Token passing, Ethernet [CSMA/CD]).

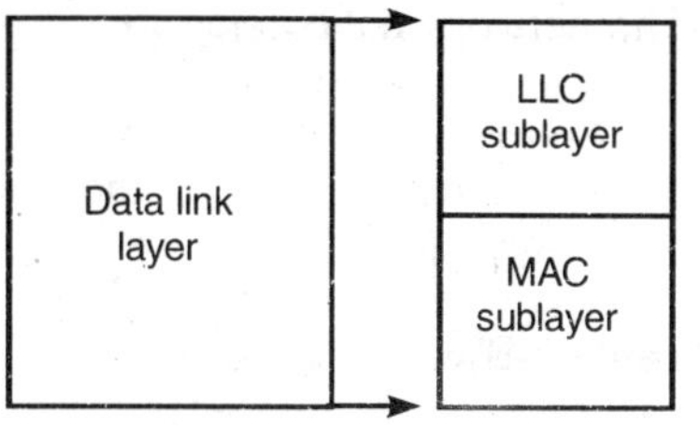

Fig. 6.7. *Data link layer.*

The data link layer provides reliable transit of data across a physical network link. Different data link layer specifications define different network and protocol characteristics, including physical addressing, network topology, error notification, sequencing of frames, and flow control. Physical addressing (as opposed to network addressing) defines how devices are addressed at the data link layer. Network topology consists of the data link layer specifications that often define how devices are to be physically connected, such as in a bus or a ring topology. Error notification alerts upper-layer protocols that a transmission error has occurred, and the sequencing of data frames reorders frames that are transmitted out of sequence. Finally, flow control moderates the transmission of data so that the receiving device is not overwhelmed with more traffic than it can handle at one time.

The protocols used in Data link layer are SLIP, PPP, MTU, and CSLP.

6.4.1. Link Protocol Types

Data link protocols are divided into two basic categories:

1. Synchronous Protocol
2. Asynchronous Protocol

These are described below.

It is important not to confuse synchronization at the data link layer with synchronization at the physical layer. These two are distinct and essential in their own right: physical layer synchronization ensures that the transmitter and the receiver share a common clock signal so that bit boundaries can be detected; data link layer synchronization ensures that user data is not confused with control data.

Synchronous Protocols

Synchronous protocols operate by delimiting user data with unique bit patterns which the receiver uses to detect where the user data begins and where it ends.

Synchronous protocols may be character-oriented or bit-oriented.

In a **character-oriented** protocol, user data consists of a sequence of characters and is delimited by two unique control characters (SYN and EOT). The biggest disadvantage of character-oriented protocols is that they are based on specific character sets (*e.g.*, ASCII or EBSDIC) and are therefore character-set dependent.

In a **bit-oriented protocol**, no specific character set is assumed. The unit of transmission is a **frame**, which consists of user data (an arbitrary bit sequence), control data, address data, error checksum, and two delimiting bit patterns at either end of the frame.

The delimiting bit pattern used is 01111110 and is called a **flag**. To avoid this bit pattern occurring in user data, the transmitter inserts a 0 bit after every five consecutive 1 bits it finds. This is called **bit stuffing**; its effect is canceled by the receiver, which removes every 0 bit that occurs after every five consecutive 1 bits.

Bit-oriented protocols are by comparison more recent than other protocols and have dominated the market.

Asynchronous Protocols

Asynchronous protocols are character-oriented and operate by having the transmitter surround each character with a start and a stop bit. These two bits indicate to the receiver where the character starts and where it ends. The receiver extracts the user data by removing the start and stop bits.

Master-Slave Protocols

Master-slave protocols are used in situations where a primary station (such as a host) controls one or more secondary stations (such as peripheral devices).

Communication between the master and its slaves is governed by a general technique called **polling**. Polling is typically used in a multidrop line configuration where, for example, a set of terminals are connected to a host via a shared communication line.

In its simplest form, polling works as follows. The master can send data in form of a message to any of the slaves. The message contains an address which uniquely identifies the slave for which it is intended. To receive data from the slaves, the master 'polls' each slave in turn by sending a message to it and asking if it has any data to send. The slave responds by either sending data or by sending a rejection message to indicate it has nothing to send.

6.5. NETWORK LAYER

The network layer is concerned with the routing of data across the network from one end to another. To do this, the network layer converts the data into **packets** and ensures that the packets are delivered to their final destination, where they can be converted back into the original data.

Network Layer is responsible for the routing of data (packets) through the network; handles the addressing and delivery of data. This layer provides for congestion control, accounting information for the network, routing, addressing, and several other functions. IP (Internet Protocol) is a good example of a network layer protocol. Network layer does not deal with lost messages.

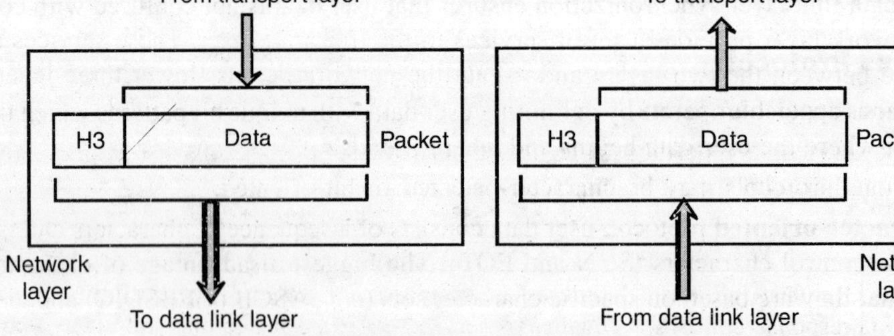

Fig. 6.8. *Network layer.*

The network layer is responsible for the delivery of individual packets from the source host to the destination host.

6.5.1. Important Features of Network Layer Protocols

1. Concerned with the transmission of packets.
2. Choose the best path to send a packet (routing).
3. The routing may be complex in a large network (*e.g.*, Internet).
4. Routing packets through a network may be accomplished by using simple static routes or by using complex dynamic routing algorithms.

5. The interface between two hosts across the network.
6. Routing of packets across the network, including the allocation of a route and handling of congestion.
7. Correct ordering of packets to reflect the original order of data.
8. Collection of statistical information (*e.g.,* number of transmitted packets) for performance measurement and accounting purposes.
9. Internetworking: communication between two or more networks.

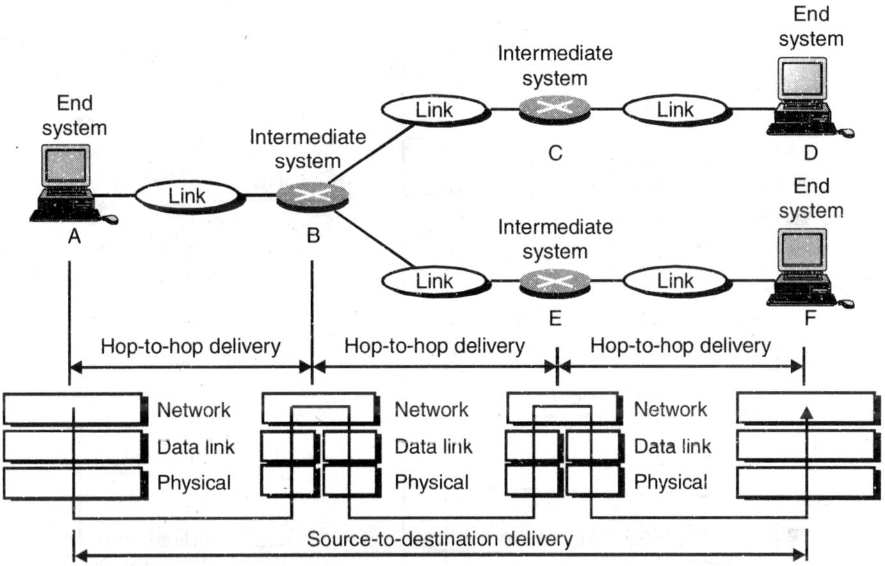

Fig. 6.9.

6.5.2. Network Services

The network layer provides a set of services to the transport layer. These services characterize the interface between the two layers and isolate the network details (lower three layers) from the network users (upper four layers).

Network services are defined in terms of network service primitives. Figure —illustrate the primitives together with their possible types and parameters.

Primitive	Types	Parameters	Purpose
N-Connect	Request indicate response confirm	(addresses, options, quality of service, user data)	Used for initiating a connection (connections are always initiated by service users).
N-Data	Request indicate	(user data, confirm option)	Used for normal data transfer (with or without confirmation)
N-Experi-dated-Data	Request indicate	(user data)	Used for high priority data transfer.
N-Ack-Data	Request indicate	()	Used for confirmation of data transfer.
N-Reset	Request indicate response confirm	()	Used by the service user or provider for resetting the network service.
N-Disconnect	Request indicate	(addresses, reason, user data)	Used by the service user or provider for disconnecting a connection.

Fig. 6.10. *Network services.*

Addresses refer to Network Service Access Points (NSAP); these denote entities at the network layer that act as the interface to service users. *Quality Of Service* (QOS) denotes a set of parameters (such as error rate, delays, cost, failure likelihood, and throughput) which collectively describe the quality of the network service.

User data refers to actual user data provided by service users for transfer by the service provider. *Options* denote a variety of options that may be activated with some commands (*e.g.*, whether data transfers can be confirmed).

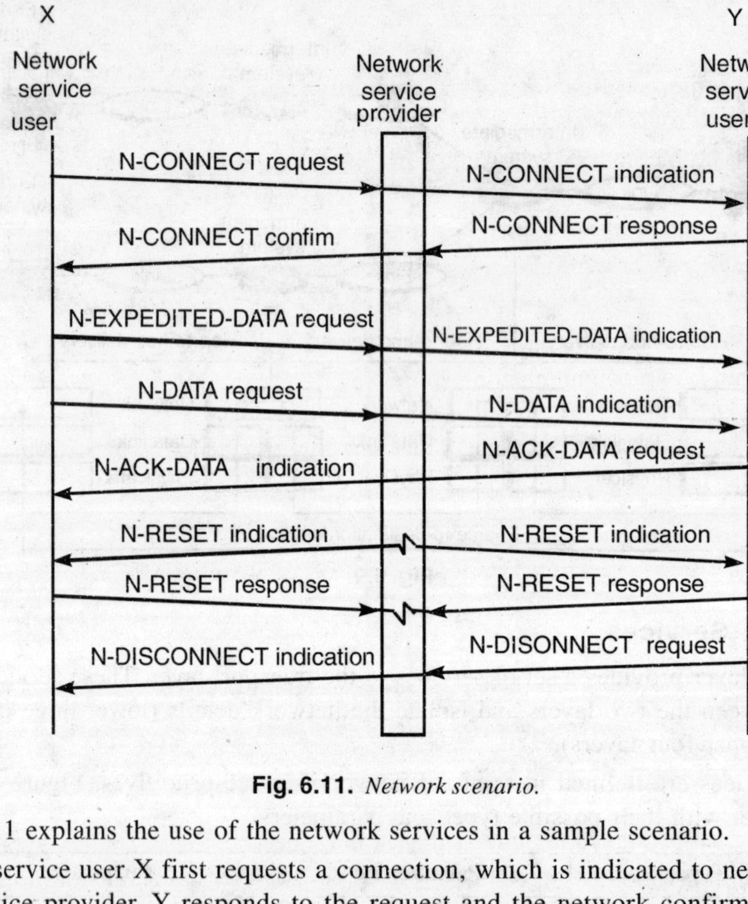

Fig. 6.11. *Network scenario.*

Figure 6.11 explains the use of the network services in a sample scenario.

Network service user X first requests a connection, which is indicated to network service user Y by the service provider. Y responds to the request and the network confirms with X. Then X expedites some data which is indicated to Y by the network. A normal data transfer from X to Y follows which includes confirmation. Then the network encounters an error and simultaneously sends reset indications to X and Y, which both respond to. Finally, Y requests a disconnection, which is indicated to X, and terminates the connection.

6.5.3. Switching Methods

Switching is the generic method for establishing a path for point-to-point communication in a network. It involves the nodes in the network utilizing their direct communication lines to other nodes so that a path is established in a piecewise pattern (see Figure 6.12). Each node has the capability to 'switch' to a neighboring node (*i.e.*, a node to which it is directly connected) to further stretch the path until it is completed.

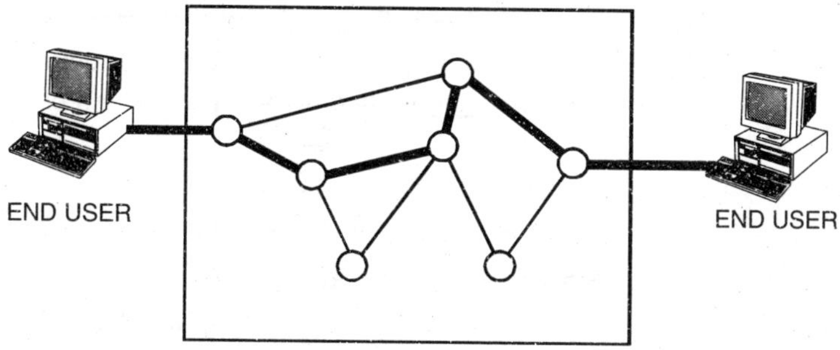

Fig. 6.12.

One of the most important functions of the network layer is to employ the switching capability of the nodes in order to route messages across the network.

There are two basic methods of switching:

1. Circuit switching
2. Packet switching.

These are separately discussed below.

Circuit Switching

In circuit switching, two communicating stations are connected by a *dedicated* communication path which consists of intermediate nodes in the network and the links that connect these nodes. What is significant about circuit switching is that the communication path remains intact for the duration of the connection, engaging the nodes and the links involved in the path for that period.

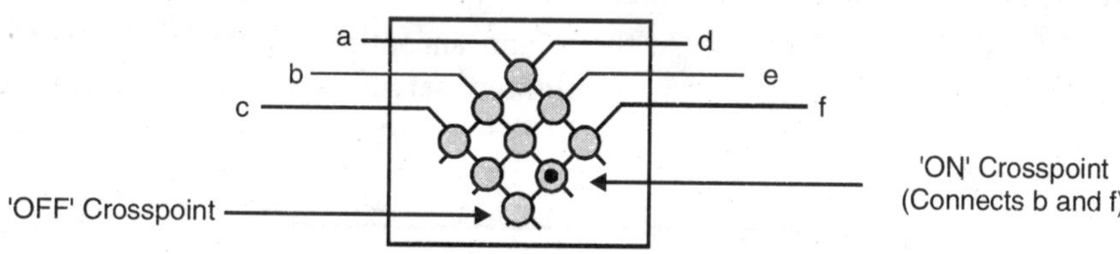

Fig. 6.13. *A simple circuit switch.*

Packet Switching

Packet switching was designed to address the shortcomings of circuit switching in dealing with data communication. Unlike circuit switching where communication is continuous along a dedicated circuit, in packet switching, communication is discrete in form of packets. Each packet is of a limited size and can hold up to a certain number of octets of user data. Larger messages are broken into smaller chunks so that they can be fitted into packets. In addition to user data, each packet carries additional information (in form of a header) to enable the network to route it to its final destination.

A packet is handed over from node to node across the network. Each receiving node temporarily stores the packet, until the next node is ready to receive it, and then passes it onto the next node. This technique is called **store-and-forward** and overcomes one of the limitations of circuit switching. A packet-switched network has a much higher capacity for accepting further connections. Additional connections are usually not blocked but simply slow down existing connections, because they increase the overall number of packets in the network and hence increase the delivery time of each packet.

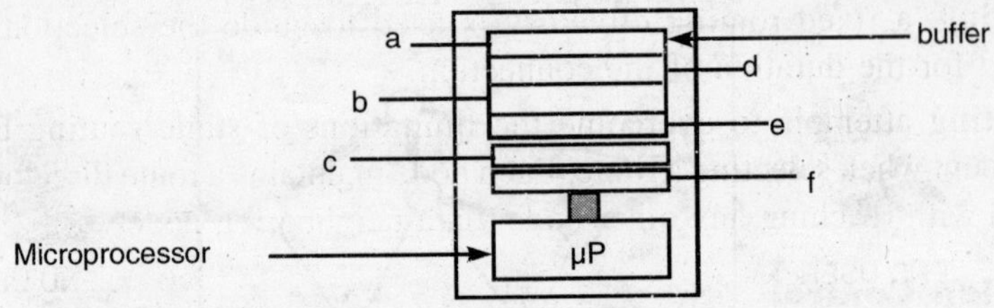

Fig. 6.14. *A simple packet switch.*

Two variations of packet switching exist: virtual circuit and datagram.

The **virtual circuit** method (also known as **connection-oriented**) is closer to circuit switching. Here a complete route is worked out prior to sending data packets.

The **datagram** method (also known as **connectionless**) does not rely on a pre established route; instead each packet is treated independently. Therefore, it is possible for different packets to travel along different routes in the network to reach the same final destination. As a result, packets may arrive out of order, or even never arrive (due to node failure).

6.5.4. Routing

Routing is the task of selecting a path for the transport of packets across the network, and is one of the most important functions of the network layer. Routing is generally viewed as an optimization problem with the objective of choosing an optimal path according to certain criteria:

- **Transmission cost :** measured in terms of tied up network resources.
- **Transmission delay :** measured as the delay involved in delivering each packet.
- **Throughput :** measured as the total number of packets delivered per unit of time.

The overall cost depends on all these three, and an optimal route is one that minimizes the overall cost. This can be represented by a weighted network, where an abstract cost figure is associated with each link, as illustrated in Fig. 6.15. The cost of a route is the sum of the cost of its links.

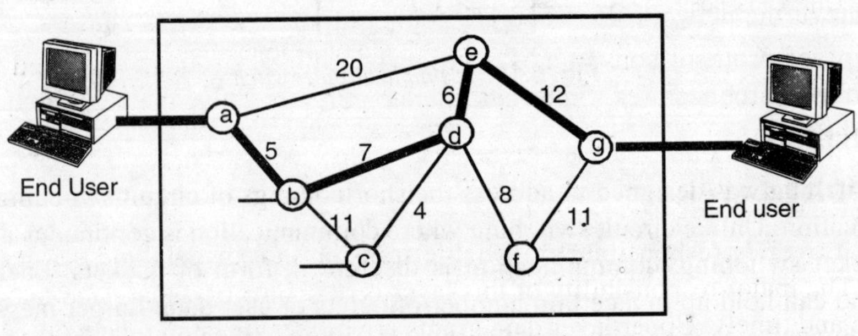

Fig. 6.15. *A weighted network.*

There are three classes of routing algorithms:

1. Flooding
2. Static Routing
3. Dynamic Routing

In **flooding**, every possible path between the source and the destination station are exercised. Each node, upon receiving a packet, forwards copies of it to all its neighboring nodes (except the one from which it received the packet).

In **static routing**, a fixed routing directory is used to guide the selection of a route which remains unchanged for the duration of the connection.

Dynamic routing attempts to overcome the limitations of static routing by taking network variations into account when selecting a route. Each node maintains a route directory which describes the cost associated with reaching any other node via a neighboring node.

6.5.5. Congestion Control

A network has a certain carrying capacity, denoted by the maximum number of packets that it can hold at any point in time. When this limit is approached, considerable delays are experienced in packet delivery, and the network is said to be **congested**. Congestion can occur in all types of networks. Uncontrolled congestion can lead to outright network failure.

6.5.6. Error Handling

The extent to which the network layer attempts to deal with errors is largely dependent on the type of service being offered. The datagram service offers little in this respect. If packets do not arrive due to being corrupted or lost, or if they arrive out of order, it is the responsibility of the network user to deal with these problems. The virtual circuit service, however, handles these problems transparently.

The usual approach to dealing with corrupt or lost packets is to request retransmission.

6.5.7. Internetworking

It is an inevitable aspect of communication networks that because of business, organizational, political, and historical reasons, there exist many different types of networks throughout the world. More and more, it is becoming desirable, if not necessary, to interconnect these networks. The problem of interconnecting a set of independent networks is called **internetworking**. Each of the participating networks is referred to as a **subnetwork** (or **subnet**). The role of the **Interworking Units** (IWU) is to carry out protocol conversion between the subnets. IWUs are used for interconnecting networks that use the same architecture but employ different protocols. Because of this, they operate at the network layer and are commonly referred to as **routers**.

Another type of protocol converter is a **gateway**. Unlike a router, it is used to interconnect networks of different architectures. Gateways operate at layers above the network layer and usually encompass all the layers.

6.6. TRANSPORT LAYER

Transport Layer is responsible for reliable transmission of data and service specifications between hosts. The major responsibility of this layer is data integrity; that data transmitted between hosts is reliable and timely. Upper layer data grams are broken down into network-sized data grams if needed and then implemented using appropriate transmission control. The transport layer creates one or more than one network connection, depending on conditions. This layer also handles what type of connection will be created.

The transport layer is responsible for the delivery of a message from one process to another.

6.6.1. Important Features of Transport Layer

1. Transport layer ensures reliable service.
2. Establishment and termination of host-to-host connections.

3. Efficient and cost-effective delivery of data across the network from one host to another.
4. Multiplexing of data, if necessary, to improve use of network bandwidth, and demultiplexing at the other end.
5. Splitting of data across multiple network connections, if necessary, to improve throughput, and recombining at the other end.
6. Flow control between hosts.
7. Addressing of messages to their corresponding connections. The address information appears as a part of the message header.
8. Type of service to be provided to the session layer (*e.g.*, error-free versus error prone connections, whether messages should be delivered in the order received or not).
9. Breaks the message (from session's layer) into smaller packets, assigns sequence number and sends them.
10. Reliable transport connections are built on top of X.25 or IP.
11. In case IP, lost packets arriving out of order must be reordered.

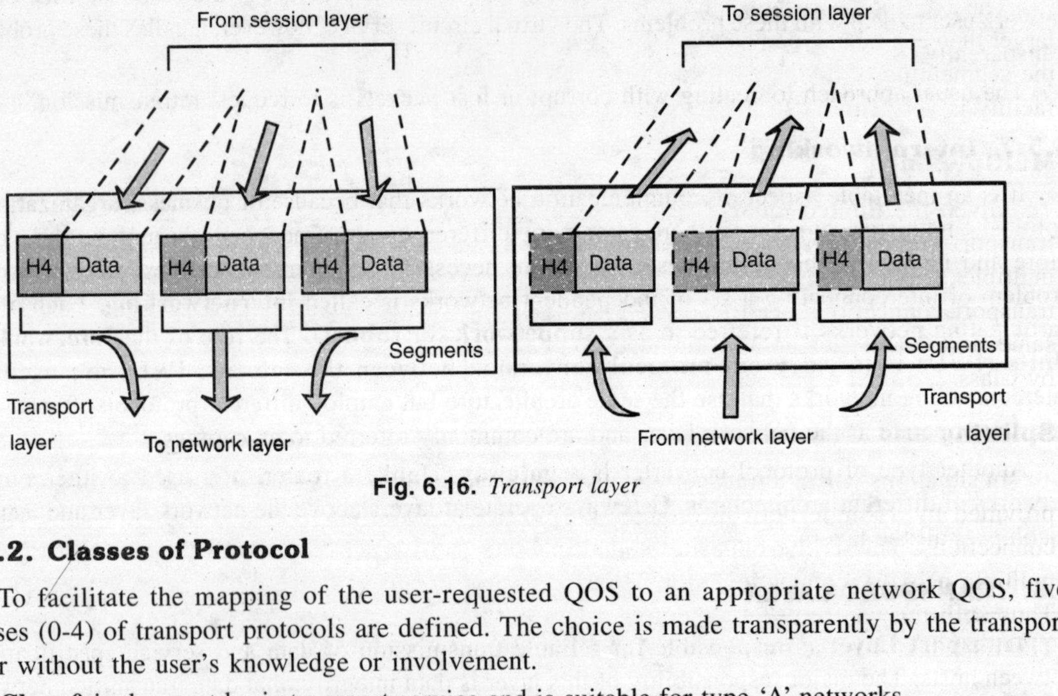

Fig. 6.16. *Transport layer*

6.6.2. Classes of Protocol

To facilitate the mapping of the user-requested QOS to an appropriate network QOS, five classes (0-4) of transport protocols are defined. The choice is made transparently by the transport layer without the user's knowledge or involvement.

Class 0 is the most basic transport service and is suitable for type 'A' networks.

Although Class 0 is capable of detecting data errors (*e.g.*, corrupted, lost, duplicated data) and signaled errors (*e.g.*, failure in switching nodes), it cannot recover from them. These errors therefore result in the connection being terminated. This is the simplest form of transport service and must be supported by all transport layer implementations. (Supporting other classes is optional.)

Class 1 protocol supports recovery from signaled errors. It also supports segmentation, expedited data transfer, and acknowledgment using sequence numbers.

Class 2 is like class 0, except that it also supports multiplexing and flow control.

Class 3 is like class 2, except that it also supports recovery from signaled errors. Finally, Class 4 provides the widest set of features, making its suitable for C type networks. It is the only class that supports the resequencing of TPDUs (Transport protocol data units), and the splitting of the transport connection into multiple network connections to improve throughput.

Figure 6.17 summarizes the five protocol classes and some of their main features, together with the network type suitable for supporting each class.

Class No.	Class Name	Suitable Network
0	Simple class	A
1	Basic error recovery class	B
2	Multiplexing class	A
3	Error recovery multiplexing class	B
4	Error detection and recovery class	C

Fig. 6.17. *Transport protocol classes.*

Segmentation

The Transport Service Data Units (TSDUs) which the transport service users work with may be larger than can be handled by the network layer packets. For this reason, all protocol classes support the segmenting of TSDUs into multiple TPDUs and reassembling them at the receiver end. This facility is very similar to the segmentation facility supported by the internetworking sublayers.

Multiplexing

Given the limited number of ports available on a host and the potentially large number of transport connections usually required by the users, multiple transport connections are often multiplexed onto a single network connection. The price to be paid for this is additional complexity: transport connections need to carry identifiers for distinguishing between connections over the same circuit, and each connection needs to be separately flow controlled. Multiplexing is supported by Class 2, 3, and 4 protocols.

Splitting and Recombining

In situations where a transport connection is needed to have a bandwidth of size that cannot be provided by a single network connection, the connection may be split over multiple network connections. This is the opposite of multiplexing and introduces an additional complexity: with multiple network connections, the TPDUs may arrive out of order and need to be resequenced. Thus splitting can only be supported when the transport protocol can support resequencing of TPDUs. Class 4 is the only protocol that can provide this function.

Addressing

A transport layer source or destination address uniquely identifies a TSAP within a network-wide name space. The TSAP denotes a port on a host to which transport connections are made. Transport addresses are not directly used by TPDUs; **references** are used instead. These are much shorter identifiers that are mapped to the transport addresses (in the variable part of the TPDU header) during connection establishment, allowing the two parties to agree on a common mapping. The source and destination references are used for the remainder of the connection.

Also, for user convenience, a higher level mapping is often provided by a name server, which maps meaningful service **names** for frequently-used transport addresses to the addresses themselves. When wishing to connect to one of these services, the user specifies the service name and the network software looks up the name using the name server to come up with the corresponding transport address, which is then used for initiating a transport connection.

Flow Control

As mentioned earlier, supporting multiplexing in Class 2, 3, and 4 protocols requires transport connections to be separately flow controlled. The flow control protocol used for this purpose is called **credit**. It is similar to the sliding window protocol of the data link and network layers in that a window is used. However, unlike the sliding window protocol where the receiver window size remains fixed, here the window size may vary. The sender is informed of the new window size through a data acknowledgment TPDU which includes a new credit value (the *Credit* field in the TPDU) which denotes the number of TPDUs that can be received.

A suitable credit value is negotiated during connection establishment. This value can be dynamically adjusted subsequently by the receiver. When the receiver is unable to receive further TPDUs, it can suspend flow of further TPDUs by adjusting the credit value to 0.

Error Checking

The transport layer may do additional error checking on top of the ones carried out by lower layers. This may serve one of two reasons: the lower layers might not be performing any error checking, or the network service may be very unreliable and deliver packets that have undetected errors.

To address these needs (especially for the Class 4 protocol) a checksum option is available in the TPDUs. This is a 16-bit checksum that covers the entire TPDU and is designed to be much simpler than CRC so that it can be efficiently computed in software.

6.7. INTERNET PROTOCOL USED IN TRANSPORT LAYER

In the Internet, we have two different transport-layer protocols:

1. UDP (User Data gram Protocol)
2. TCP (Transmission Control Protocol)

UDP is connectionless and therefore unreliable; whereas TCP is connection oriented and reliable. These two can respond to the demand of the application-layer programs. UDP is a convenient transport-layer protocol for applications that provide flow and error control. It is used by multimedia applications.

Important features of TCP/UDP:

- TCP/IP Widely used for network/transport layer (UNIX).
- TCP (Transport Control Protocol): This is a connection oriented protocol.
- UDP (Universal Data gram Protocol): This is a connectionless transport layer protocol.
- Application programs that do not need connection-oriented protocol generally use UDP.

6.8. TCP (TRANSMISSION CONTROL PROTOCOL)

The **Transmission Control Protocol (TCP)** is responsible for reliable end to end delivery of segments of information. A segment is the term which is used to describe the data that is transmitted and received at the **Transport** level of the OSI model where TCP resides. TCP also redirects the data to the appropriate which port (upper level service) that is required.

The reliable end to end delivery of data is accomplished by:

(i) Connection-Oriented Service

Segments are acknowledged to the source when received by the destination. A sliding window s used to enable unacknowledged segments on the "wire" in order to speed up transmission rates

ii) Sequencing of Segments

Data is broken up into segments that are numbered (sequenced) when transmitted. The destination TCP layer keeps track of the received segments and places them in the proper order (resequences).

iii) Requesting Retransmission of Lost Data

If any segment is lost during transmission (missing sequence number) the destination will imeout and request that all segments starting at the lost segment be retransmitted.

iv) Error Checking

Segments are checked for data integrity when received using a 32 bit CRC check.

The redirection of data to the upper level service is accomplished by using Source and Destination Port numbers. Multiple connections to the same service are allowed. For example, you may have nany users (clients) connected to a single web server (http is normally port 80). Each client will ave a unique Port number assigned (typically above 8000) but the web server will only use Port 80.

TCP Header

<table>
<tr><td>0 1 2 3 4 5 6 7</td><td>8 9 10 11 12 13 14 15</td><td>16 17 18 19 20 21 22 23</td><td>24 25 26 27 28 29 30 31</td></tr>
<tr><td colspan="2">Source Port (16 bits)</td><td colspan="2">Destination Port (16 bits)</td></tr>
<tr><td colspan="4">Sequence Number</td></tr>
<tr><td colspan="4">Acknowledgement Number</td></tr>
<tr><td>Offset (1st 4 bits)</td><td>Flags (last 6 bits)</td><td colspan="2">Window</td></tr>
<tr><td colspan="2">Checksum</td><td colspan="2">Urgent pointer</td></tr>
<tr><td colspan="4">Options + Padding</td></tr>
<tr><td colspan="4">Data</td></tr>
</table>

Fig. 6.18.

urce Port

The Source Port is a 16 bit number that indicates the upper level service that the source is ansmitting. For example:

- 21 - ftp servers
- 23 - telnet servers

- 25 - SMTP servers (Internet mail)
- 80 - http services (web)

TCP allows port numbers to be in the range from 0 to 65,535. Clients will have a unique port number assigned to them by the server. Typically the number will be above 8,000.

Destination Port

The Destination Port is a 16 bit number that indicates the upper level service that the source wishes to communicate with at the destination.

Sequence Number

The Sequence Number is a 32 bit number that indicates the first octet of information in the data field. This is used to number each TCP segment transmitted in order to keep track of segments for sequencing of segments and error checking of lost segments. The source numbers the sequence of transmitted segments.

Acknowledgement Number

The Acknowledgement Number is a 32 bit number that is used to acknowledge the receipt of segments by the destination. The acknowledgement is the next sequence number expected. If the sender does not receive an acknowledgement for a segment transmitted, the sender will time-out and retransmit.

Offset (4 bits)

The Offset field consists of the first 4 bits (xxxx0000) of the first byte. The last 4 bits are reserved for future use and are set to 0. The Offset measures the number of **32 bit (4 byte)** words in the TCP header to where the Data field starts. This is necessary because the TCP header has a variable length. The minimum length of the TCP header is 20 bytes which gives an Offset value of 5.

Flags (Last 6 bits)

The Flags Field consists of the last 6 bits (00xxxxxx) of the second byte with the first 2 bits reserved for future use and they are set to 0. The Flags field consists of the following flag bits:

URG (Urgent Flag)

When set indicates that the Urgent Pointer field is being used.

ACK (Acknowledge Flag)

When set indicates that the Acknowledgement Number is being used.

PSH (Push Flag)

An upper level protocol requires immediate data delivery and would use the Push (PSH) flag to immediately forward all of the queued data to the destination.

RST (Reset Flag)

When set the connection is reset. This is typically used when the source has timed out waiting for an acknowledgement and is requesting retransmission starting at a sequence number.

SYN (Synchronize Flag)

When set, it indicates that this segment is the first one in the sequence. The first sequence number assigned is called the Initial Sequence Number (ISN).

FIN (Finish Flag)

When set, it indicates that this is the last data from the sender.

Windows (16 bits)

This contains the number of unacknowledged segments that are allowed on the network at any one time. This is negotiated by the Source and Destination TCP layers.

Checksum

The Checksum field is 16 bits long and calculates a checksum based on the complete TCP Header and what is called the TCP Pseudo header. The TCP Pseudo header consists of the Source IP Address, Destination IP Address, Zero, IP Protocol field and TCP Length. The IP Protocol field value is 6 for TCP.

Urgent Pointer

This field communicates the current value of the urgent pointer as a positive offset from the sequence number in this segment. The urgent pointer points to the sequence number of the octet following the urgent data. This field is only be interpreted in segments with the URG control bit set.

Options

Options may occupy space at the end of the TCP header and are a multiple of 8 bits in length. The allowed options are:

- Kind 0 - End of option list.
- Kind 1 - No Operation.
- Kind 2 - Length 4 Maximum Segment Size. This is used to indicate the maximum segment size allowed.

Padding

The TCP header padding is used to ensure that the TCP header ends and data begins on a 32 bit boundary. The padding is composed of zeros.

6.9. USER DATAGRAM PROTOCOL (UDP)

The User Datagram Protocol (UDP) is a connectionless host to host service that operates at the Transport layer of the OSI model. UDP relies on the upper layer protocol for error correction and reliable service. The protocol is transaction oriented, and delivery and duplicate protection are not guaranteed. The major uses of this protocol are DNS and TFTP.

UDP has a small header and for all intensive purposes adds Port addressing to the IP header. The IP header routes datagram's to the correct host on the network and UDP routes the datagram to the correct application.

UDP Header

0 1 2 3 4 5 4 6 7	8 9 10 11 12 13 14 15	16 17 18 19 20 21 22 23	24 25 26 27 28 29 30 31
Source Port (16 bits)		Destination Port (16 bits)	
Length (16 bits)		Checksum (16 bits)	
Data			

Fig. 6.19.

Source Port

The Source Port is a 16 bit number that indicates the upper level service that the source is transmitting. UDP allows port numbers to be in the range from 0 to 65,535. The Source Port is optional and if not used, a field of 0s is inserted. Clients will have a unique port number assigned to them by the server. Typically the number will be above 8,000.

Destination Port

The Destination Port is a 16 bit number that indicates the upper level service that the source wishes to communicate with at the destination.

Length

The Length field is 16 bits long and indicates the length of the UDP datagram and has a maximum value of 65, 535 bytes and a minimum value of 8 bytes.

Checksum

The Checksum field is 16 bits long and calculates a checksum based on the UDP header, Data field and what is called the UDP Pseudo header. The UDP Pseudo header consists of the Source IP Address, Destination IP Address, Zero, IP Protocol field and UDP Length. The IP Protocol field value is 17 for UDP.

Data

The data field contains the IP header and data. The Data field may be padded with zero octets at the end (if necessary) to make a multiple of two octets.

6.10. SESSION LAYER

The session layer provides a structured means for data exchange between user processes on communicating hosts. Session layer protocols are concerned with the following issues:

1. Negotiating the establishment of a connection (a **session**) between user processes on communicating hosts, and its subsequent termination. This includes the setting of various communication parameters for the session (*e.g.,* synchronization and control).
2. Correct ordering of messages when this function is not performed by the transport layer.
3. Recovery from interrupted transport connections, if necessary.
4. Grouping of messages into a larger message, if necessary, so that the larger message becomes available at the destination only when its constituent messages have all been delivered successfully.

The session layer hides the above details from the higher layers. To the presentation layer, it appears as an **organized** communication service between user processes.

Session layer coordinates dialogue/session/connection between devices over the network. This layer manages communications between connected sessions. Examples of this layer are token management (the session layer manages who has the token) and network time synchronization.

The session layer establishes, manages, and terminates communication sessions. Communication sessions consist of service requests and service responses that occur between applications located in different network devices. These requests and responses are coordinated by protocols implemented at the session layer.

Some examples of session-layer implementations include Zone Information Protocol (ZIP), the AppleTalk protocol that coordinates the name binding process; and Session Control Protocol (SCP), the Decent Phase IV session layer protocol.

6.11. PRESENTATION LAYER

Presentation layer manages the presentation of the information in an ordered and meaningful manner. This layer's primary function is the syntax and semantics of the data transmission. It converts local host computer data representations into a standard network format for transmission on the network. On the receiving side, it changes the network format into the appropriate host computer's format so that data can be utilized independent of the host computer. ASCII and EBCDIC conversions, cryptography, and the like are handled here.

The presentation layer hides the above details from the higher layers. To the application layer, it appears as a **universal** communication service between user processes, regardless of their system-specification, and allowing them to converse in a common **syntax**.

The presentation layer provides a variety of coding and conversion functions that are applied to application layer data. These functions ensure that information sent from the application layer of one system would be readable by the application layer of another system. Some examples of presentation layer coding and conversion schemes include common data representation formats, conversion of character representation formats, common data compression schemes, and common data encryption schemes.

Common data representation formats, or the use of standard image, sound, and video formats, enable the interchange of application data between different types of computer systems. Using different text and data representations, such as EBCDIC and ASCII, uses conversion schemes to exchange information with systems. Standard data compression schemes enable data that is compressed or encrypted at the source device to be properly decompressed, or deciphered at the destination.

Presentation layer protocols are concerned with issues such as the following :

1. Abstract representation of application data.
2. Binary representation of application data.
3. Conversion between the binary representation of application data and a common format for transmission between peer applications.
4. Data compression to better utilize network bandwidth.

Data Encryption as a Security Measure

Presentation layer implementations are not typically associated with a particular protocol stack. Some well-known standards for video include QuickTime and Motion Picture Experts

Group (MPEG). QuickTime is an Apple Computer specification for video and audio, and MPEG is a standard for video compression and coding.

Among the well-known graphic image formats are Graphics Interchange Format (GIF), Joint Photographic Experts Group (JPEG), and Tagged Image File Format (TIFF). GIF is a standard for compressing and coding graphic images. JPEG is another compression and coding standard for graphic images, and TIFF is a standard coding format for graphic images.

6.12. APPLICATION LAYER

Application layer provides a means for the user to access information on the network through an application. This layer is the main interface for the user to interact with the application and therefore the network.

The application layer is the OSI layer closest to the end user, which means that both the OSI application layer and the user interact directly with the software application. This layer interacts with software applications that implement a communicating component. Such application programs fall outside the scope of the OSI model.

Application layer functions typically include identifying communication partners, determining resource availability, and synchronizing communication. When identifying communication partners, the application layer determines the identity and availability of communication partners for an application with data to transmit. When determining resource availability, the application layer must decide whether sufficient network resources for the requested communication exist. In synchronizing communication, all communication between applications requires cooperation that is managed by the application layer.

The application layer is concerned with the **semantics** of data, *i.e.,* what the data means to applications. The application layer provides standards for supporting a variety of application-independent services.

Services include :

1. Virtual terminal standards to allow applications to communicate with different types of terminals in a device-independent manner.
2. Message handling system standards used for electronic mail.
3. File transfer, access, and management standards for exchanging files or parts thereof between different systems.
4. Transaction processing standards to allow different companies with different systems to access each other's on-line databases (for example in banking and Railway & airline reservation).
5. On-line directory standards for storing details of individuals, organizations, and network components.
6. Standards for exchanging formatted documents.

Some examples of application layer implementations include Telnet, File Transfer Protocol (FTP), and Simple Mail Transfer Protocol (SMTP) **DNS, DHCP, FTP, TFTP, SMTP, SNMP, HTTP**.

Application layer services: DNS, DHCP, FTP, TFTP, SMTP, SNMP, HTTP

6.13. INFORMATION EXCHANGE PROCESS

Information Flow

The flow of information is from the top layer (usually application layer) down to the physical layer at information source node; and from the physical layer up to the application later at the destination node as illustrated in the figure below.

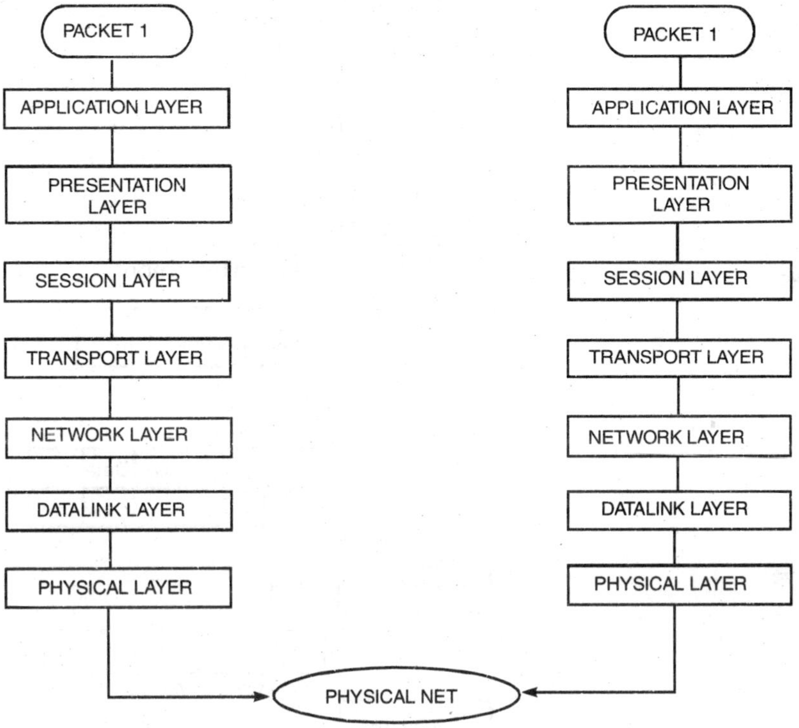

Fig. 6.20. *Information flow.*

The information exchange process occurs between peer OSI layers. Each layer in the source system adds control information to data and each layer in the destination system analyzes and removes the control information from that data.

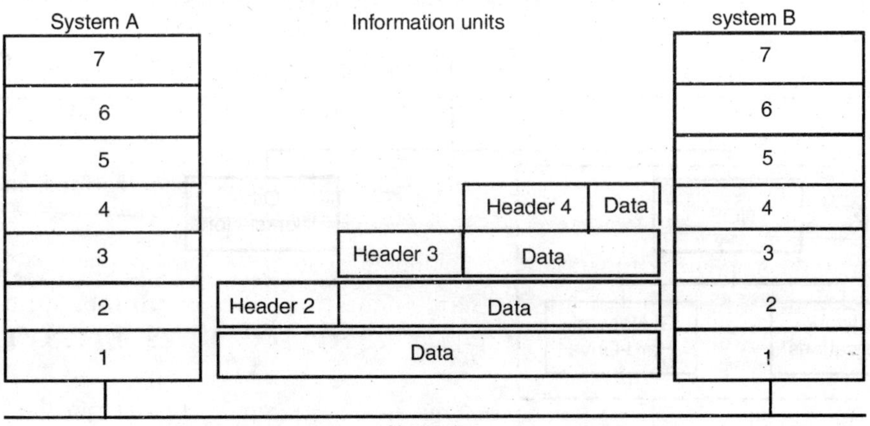

Fig. 6.21. *Information exchange process.*

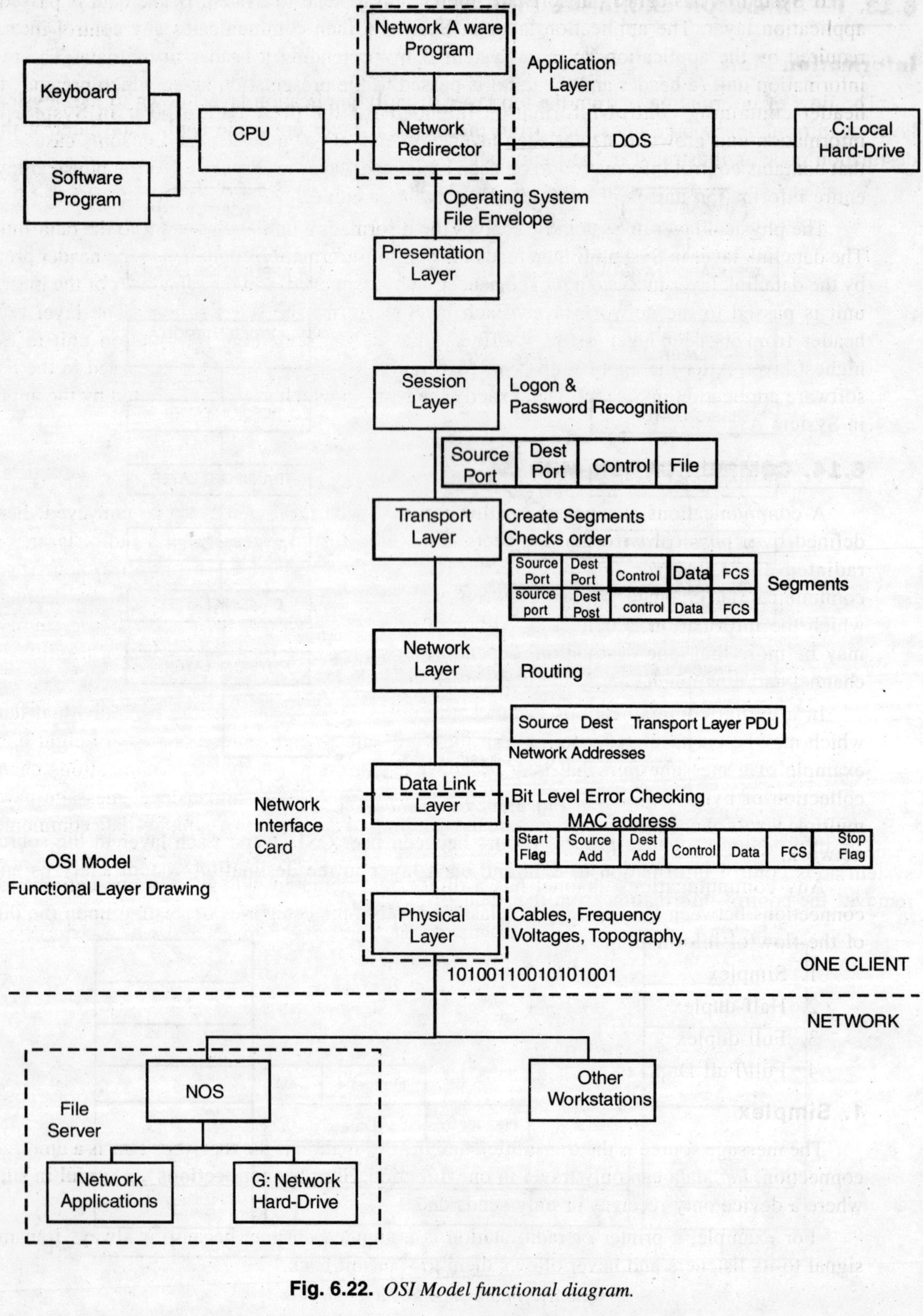

Fig. 6.22. *OSI Model functional diagram.*

If System A has data from software application to send to System B, the data is passed to the application layer. The application layer in System a then communicates any control information required by the application layer in System B by pretending a header to the data. The resulting information unit (a header and the data) is passed to the presentation layer, which pretends its own header containing control information intended for the presentation layer in System B. The information unit grows in size as each layer prep ends its own header (and in some cases a trailer) that contains control information to be used by its peer layer in System B. At the physical layer, the entire information unit is placed onto the network medium.

The physical layer in System B receives the information unit and passes it to the data-link layer. The data link layer in System B then reads the control information contained in the header prepended by the data link layer in System A. The header is then removed, and the remainder of the information unit is passed to the network layer. Each layer performs the same actions: The layer reads the header from its peer layer, strips it off, and passes the remaining information unit to the next highest layer. After the application layer performs these actions, the data is passed to the recipient software application in System B, in exactly the form in which it was transmitted by the application in System A.

6.14. COMMUNICATION MODES

A communications channel is a pathway over which information can be conveyed. It may be defined by a physical wire that connects communicating devices, or by a radio, laser, or other radiated energy source that has no obvious physical presence. Information sent through a communications channel has a source from which the information originates, and a destination to which the information is delivered. Although information originates from a single source, there may be more than one destination, depending upon how many receive stations are linked to the channel and how much energy the transmitted signal possesses.

In a digital communications channel, the information is represented by individual data bits, which may be encapsulated into multibit message units. A byte, which consists of eight bits, is an example of a message unit that may be conveyed through a digital communications channel. A collection of bytes may itself be grouped into a frame or other higher-level message unit. Such multiple levels of encapsulation facilitate the handling of messages in a complex data communications network.

Any communications channel has a direction associated with it, communication modes or connections between devices may be classified into three categories depending upon the direction of the flow of information:

1. Simplex
2. Half-duplex
3. Full-duplex
4. Full/Full Duplex

1. Simplex

The message source is the transmitter, and the destination is the receiver. This is a unidirectional connection, *i.e.,* data can only travel in one direction. **Simplex connections** are useful in situations where a device only receives or only sends data.

For example, a printer or radio station is a simplex channel because it always transmits the signal to its listeners and never allows them to transmit back.

Simplex Transmission

Fig. 6.23. *Simplex transmission.*

2. Half-Duplex

This is a bidirectional connection, with the restriction that data can travel in one direction at a time. A **half-duplex** channel is a single physical channel in which the direction may be reversed. Messages may flow in two directions, but never at the same time, in a half-duplex system. In a telephone call, one party speaks while the other listens. After a pause, the other party speaks and the first party listens. Speaking simultaneously results in garbled sound that cannot be understood. Example : walkie-talkie.

Half-Duplex Transmission

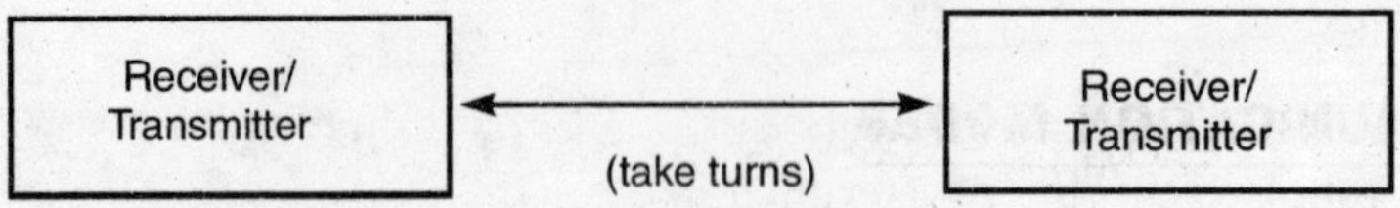

Fig. 6.24. *Half-Duplex transmission.*

3. Full-duplex

This is a bidirectional connection in which data can travel in both directions at once. A **full-duplex connection** is really consists of equivalent to two simplex connections in opposite directions a forward channel and a reverse channel, linking the same points.

A full-duplex channel allows simultaneous message exchange in both directions. The transmission rate of the reverse channel may be slower if it is used only for flow control of the forward channel. Examples- mobile phone, telephone etc.

Full Duplex Transmission

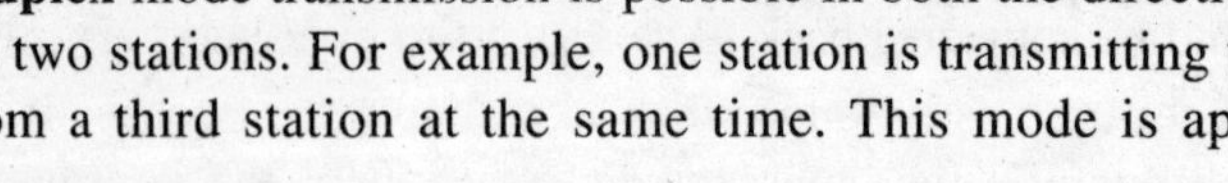

Fig. 6.25. *Full-Duplex Transmission*

4. Full/Full Duplex

In the **full/ full duplex** mode transmission is possible in both the directions simultaneously but not between the same two stations. For example, one station is transmitting to a second station and may be receiving from a third station at the same time. This mode is applicable to multipoint circuits.

6.15. TRANSMISSION MODES

Data communications over a medium can be divided into two basic types :

1. Parallel data communication or Parallel transmission mode
2. Serial data communication or Serial transmission mode

6.15.1. Parallel Transmission Mode

If more than 1 bit of information is transmitted over the data transmission medium at a time then it is considered a **Parallel Communication**.

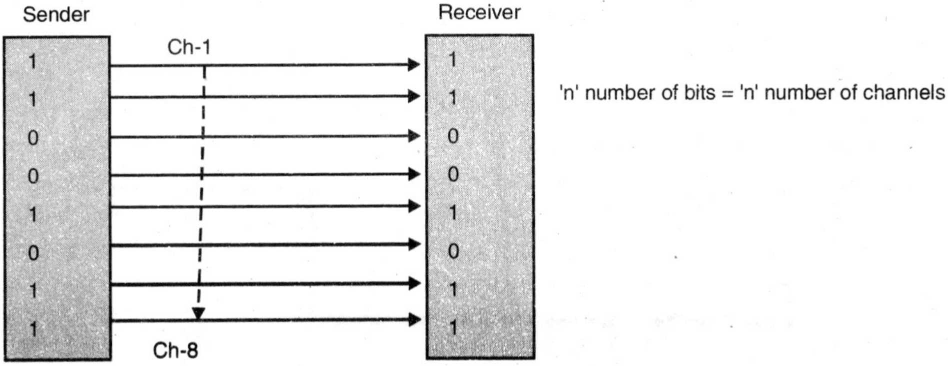

Fig. 6.26. *Parallel transmission mode.*

Byte-serial transmission conveys eight bits at a time through eight parallel channels. Although the raw transfer rate is eight times faster than in bit-serial transmission, eight channels are needed, and the cost may be as much as eight times higher to transmit the message. When distances are short, it may nonetheless be both feasible and economic to use parallel channels in return for high data rates. The popular Centronics printer interface is a case where byte-serial transmission is used as shown in Fig. 6.26. As another example, it is common practice to use a 16-bit-wide data bus to transfer data between a microprocessor and memory chips; this provides the equivalent of 16 parallel channels. On the other hand, when communicating with a timesharing system over a modem, only a single channel is available, and bit-serial transmission is required.

DO YOU KNOW

If more than 1 bit of information is transmitted over the data transmission medium at a time then it is considered a **Parallel Communication**.

6.15.2. Serial Transmission Mode

The physical connection determines how many bits (1's or 0's) can be transmitted at a single instance of time. If only 1 bit of information can be transmitted over the data transmission medium at a time then it is considered a **Serial Communication**. Normally, within the system the data is moving in parallel transmission mode. Now to transfer these data from one system to another system over a single transmission channel, we must convert parallel data into serial data by using the parallel to serial convertor at source end. Similarly at the destination, we receive data in serial mode (1 bit at a time), so it should be converted back to parallel form by using the serial to parallel convertor.

Most digital messages are vastly longer than just a few bits. Because it is neither practical nor economic to transfer all bits of a long message simultaneously, the message is broken into smaller parts and transmitted sequentially. Bit-serial transmission conveys a message one bit at a time through a channel. Each bit represents a part of the message. The individual bits are then reassembled at the destination to compose the message. In general, one channel will pass only one bit at a time. Thus, bit-serial transmission is necessary in data communications if only a single channel is available. Bit-serial transmission is normally just called **serial transmission** and is the chosen communications method in many computer peripherals.

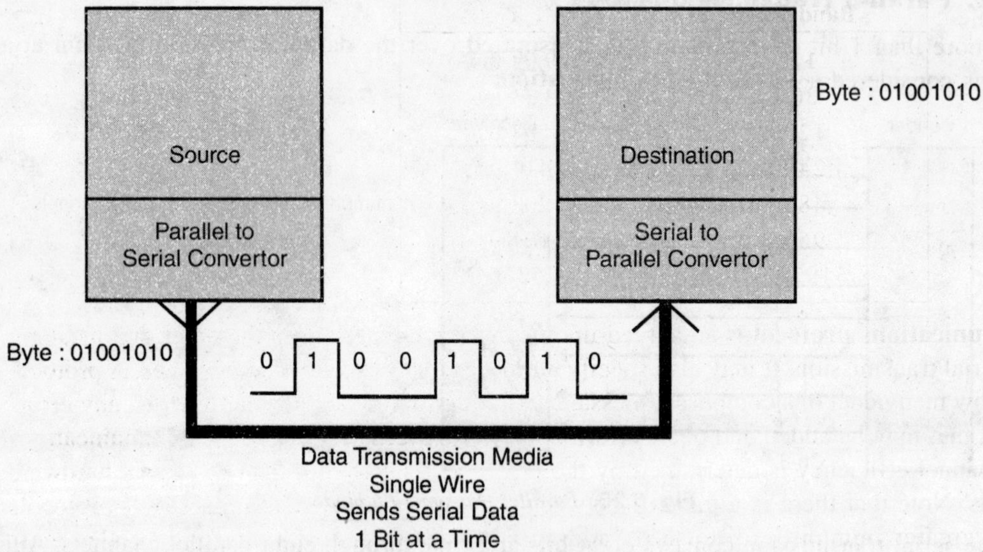

Fig. 6.27. *Serial transmission.*

The baud rate refers to the signaling rate at which data is sent through a channel and is measured in electrical transitions per second. In the EIA232 serial interface standard, one signal transition, at most, occurs per bit, and the baud rate and bit rate are identical. In this case, a rate of 9600 baud corresponds to a transfer of 9,600 data bits per second with a bit period of 104 microseconds (1/9600 sec.). If two electrical transitions were required for each bit, as is the case in non-return-to-zero coding, then at a rate of 9600 baud, only 4800 bits per second could be conveyed. The channel efficiency is the number of bits of useful information passed through the channel per second. It does not include framing, formatting, and error detecting bits that may be added to the information bits before a message is transmitted, and will always be less than one.

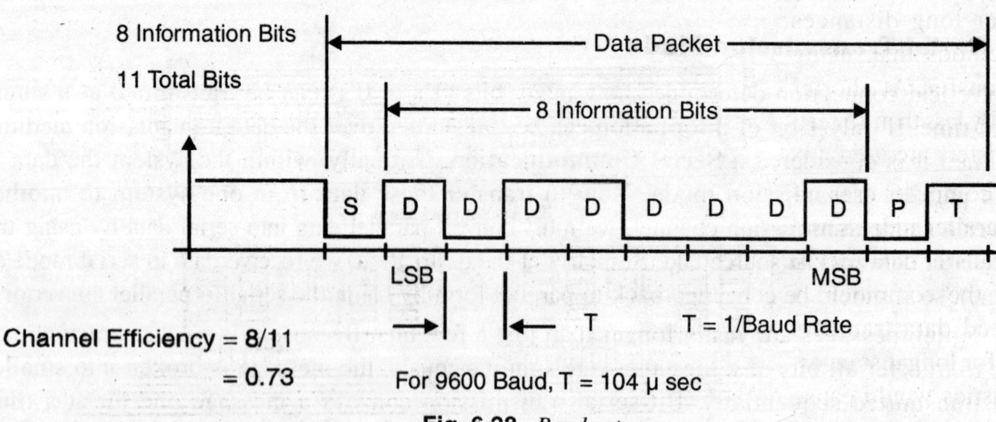

Fig. 6.28. *Baud rate.*

The data rate of a channel is often specified by its bit rate (often thought erroneously to be the same as baud rate). However, an equivalent measure channel capacity is bandwidth. In general, the maximum data rate a channel can support is directly proportional to the channel's bandwidth and inversely proportional to the channel's noise level.

Baud Rate	T
110	9.09 ms
300	3.33 ms
1200	833 μs
2400	417 μs
4800	208 μs
9600	104 μs
19200	52 μs

A **communications protocol** is an agreed-upon convention that defines the order and meaning of bits in a serial transmission. It may also specify a procedure for exchanging messages. A protocol will define how many data bits compose a message unit, the framing and formatting bits, any error-detecting bits that may be added and other information that governs control of the communications hardware. Channel efficiency is determined by the protocol design rather than by digital hardware considerations. Note that there is a tradeoff between channel efficiency and reliability - protocols that provide greater immunity to noise by adding error-detecting and -correcting codes must necessarily become less efficient.

DO YOU KNOW

If only 1 bit of information can be transmitted over the data transmission medium at a time then it is considered a **Serial Communication**.

6.15.3. Comparison of Parallel and Serial Transmission Modes

The transfer rate comparison is relative to serial versus parallel. Serial data transfer is used for high transfer rates over long distances. Data can be transferred at much faster rates over long distances using serial methods than using parallel data transfer. At shorter distances, typically less than 15 feet, parallel data transfers are used.

A personal computer uses 64 bit high speed parallel address and data busses to transfer data within the confines of the computer. If 64 bit high speed data transfers were attempted for long distances, cable characteristics would cause the signals to lead or lag each other and all the bits would not arrive at the exact same time. The bits would arrive staggered and the data would be corrupted.

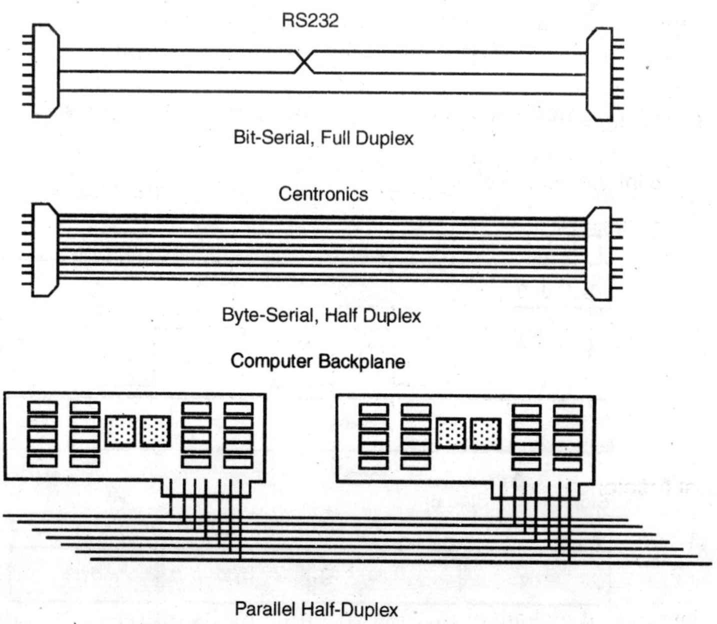

Fig. 6.29. *Comparison of Parallel and Serial Transmission Modes*

Communications	*Advantages*	*Disadvantages*
Parallel	1. Fast Transfer Rates	2. Short distances only 3. More number of channels 4. Complex and costly
Serial	1. Long Distances 2. Single channel 3. Simple	1. Slow transfer rates

Serial data communication only transfers one bit at a time and does not have the timing problems (race conditions) of parallel data communications.

6.16. SYNCHRONIZATIONS

Serialized data is not generally sent at a uniform rate through a channel. Instead, there is usually a burst of regularly spaced binary data bits followed by a pause, after which the data flow resumes. Packets of binary data are sent in this manner, possibly with variable-length pauses between packets, until the message has been fully transmitted. In order for the receiving end to know the proper moment to read individual binary bits from the channel, it must know exactly when a packet begins and how much time elapses between bits. When this timing information is known, the receiver is said to be **synchronized** with the transmitter, and accurate data transfer becomes possible. Failure to remain synchronized throughout a transmission will cause data to be corrupted or lost.

In other words when two devices are about to communicate, the transmitter should somehow notify the receiver as to when to expect to receive data. This allows the receiver to prepare it for receiving the data. Furthermore, such notifications should occur frequently enough so that both devices maintain an agreement about the exact distribution of data over time. This process is called **synchronization**.

When the timing information is known, the receiver is said to be **synchronized** with the transmitter, and accurate data transfer becomes possible.

6.16.1. Asynchronous vs. Synchronous Transmission

There are two basic methods of synchronization: synchronous transmission and asynchronous transmission.

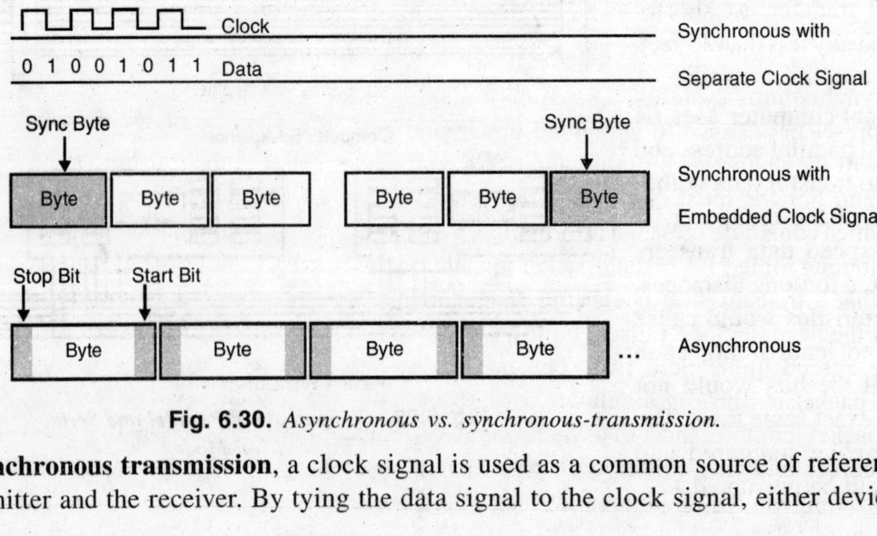

Fig. 6.30. *Asynchronous vs. synchronous-transmission.*

In **synchronous transmission**, a clock signal is used as a common source of reference by both the transmitter and the receiver. By tying the data signal to the clock signal, either device can look at the clock signal to know where data bits may begin or end. The clock signal may be provided on a separate line, or be embedded in the data signal itself (refer Fig. 6.30). Because having a separate clock line increases the costs, it is only used for covering very short distances (*e.g.*, for connecting personal computers).

In synchronous systems, separate channels are used to transmit data and timing information. The timing channel transmits clock pulses to the receiver. Upon receipt of a clock pulse, the receiver reads the data channel and latches the bit value found on the channel at that moment. The data channel is not read again until the next clock pulse arrives. Because the transmitter originates both the data and the timing pulses, the receiver will read the data channel only when told to do so by the transmitter (via the clock pulse), and synchronization is guaranteed.

Techniques exist to merge the timing signal with the data so that only a single channel is required. This is especially useful when synchronous transmissions are to be sent through a modem. Two methods in which a data signal is self-timed are nonreturn-to-zero and biphase Manchester coding. These both refer to methods for encoding a data stream into an electrical waveform for transmission.

In **asynchronous transmission**, the beginning and end of each byte of data is marked by start and stop bits. This enables the receiver to work out the byte boundaries (Figure 6.31). Because of its simplicity, asynchronous transmission is cheaper to implement and is therefore more widely used.

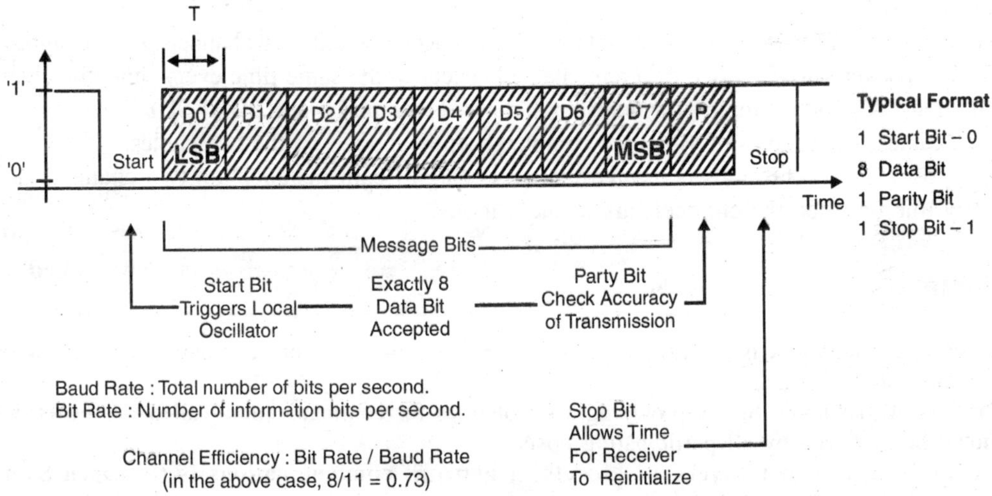

Fig. 6.31. *Asynchronous transmission.*

In **asynchronous systems**, a separate timing channel is not used. The transmitter and receiver must be preset in advance to an agreed-upon baud rate. A very accurate local oscillator within the receiver will then generate an internal clock signal that is equal to the transmitters within a fraction of a percent. For the most common serial protocol, data is sent in small packets of 10 or 11 bits, eight of which constitute message information. When the channel is idle, the signal voltage corresponds to a continuous logic '1'. A data packet always begins with a logic '0' (the start bit) to signal the receiver that a transmission is starting. The start bit triggers an internal timer in the receiver that generates the needed clock pulses. Following the start bit, eight bits of message data are sent bit by bit at the agreed upon baud rate. The packet is concluded with a parity bit and stop bit. One complete packet is illustrated below:

The packet length is short in asynchronous systems to minimize the risk that the local oscillators in the receiver and transmitter will drift apart. When high-quality crystal oscillators are used, synchronization can be guaranteed over an 11-bit period. Every time a new packet is sent, the start bit resets the synchronization, so the pause between packets can be arbitrarily long. Note that the EIA232 standard defines electrical, timing, and mechanical characteristics of a serial interface.

6.16.2. Isochronous and Asynchronous Transmissions

At first glance, the terms *isochronous* and *synchronous*, as applied to network transmissions, describe essentially the same process. Upon further investigation, however, small but significant differences emerge.

An *isochronous* channel has a constant time interval (or integer multiples thereof) between similar "significant instants." This basic definition means that each isochronous packet in a stream begins a constant time interval after the previous packet in that stream. In IEEE 1394, isochronous packets begin 125 μs apart (at least on average). This period is controlled by a clock packet, transmitted by the IEEE 1394 bus master, which signals devices on the network that it is time to send the next isochronous packet. Another example of an isochronous channel is the 270 Mbits/s serial digital video interface, where a line of video begins every 64 μs (625/50) or 63.56 μs (525/60).

The opposite of isochronous is *anisochronous*. An anisochronous channel does not maintain constant time intervals between significant instants. A good example of this is Ethernet, a technology that does not synchronize data transmission from nodes on the network, but instead relies on signaling when conflicts occur.

The term *synchronous* refers to a timing relationship between two different channels. If two "significant instants" (*e.g.*, packet starts) on two different channels occur at the same time every time, the channels are synchronous. A good example of synchronous channels would be a 4:2:24:4 SDI signal carried on two coaxial cables, where lines of video are completely time aligned on the two cables.

The opposite of synchronous is *asynchronous*. If "significant instants" on two channels do not consistently line up, then the channels are asynchronous.

Summary

- A **computer network** is the infrastructure that allows two or more computers (called **hosts**) to communicate with each other.
- The network achieves this by providing a set of rules for communication, called **protocols**, which should be observed by all participating hosts.
- An ISO standard that covers all aspects of network communications is the Open Systems Interconnection (OSI) model.
- The **OSI model** defines internetworking in terms of a vertical stack of seven layers.
- An open system is a set of protocols that allows any two different systems to communicate regardless of their underlying architecture.
- A convenient aid for remembering the OSI layer names is to use the first letter of each word in the phrase: All People Seem To Need Data Processing.
- **Physical layer** handles the bit-level electrical/light communication across the network channel.
- The physical layer is concerned with the transmission of raw data bits over communication lines.
- *The data link layer is responsible for moving frames from one hop (node) to the next.*
- Major duties of Data link layer
  (*a*) Framing (*b*) Physical addressing
  (*c*) Flow control (*d*) Error control (*e*) Access control
- Data link protocols are divided into two basic categories:
  1. Synchronous Protocol 2. Asynchronous Protocol
- Synchronous protocols may be character-oriented or bit-oriented.
- Asynchronous protocols are character-oriented and operate by having the transmitter surround each character with a start and a stop bit.

- The network layer is responsible for the delivery of individual packets from the source host to the destination host.
- Any communications channel has a direction associated with it, communication modes or connections between devices may be classified into three categories depending upon the direction of the flow of information.
- **Simplex connections** are useful in situations where a device only receives or only sends data.
- **Half-Duplex** is a bidirectional connection, with the restriction that data can travel in one direction at a time.
- **Full/Full Duplex :** In the **full/ full duplex** mode transmission is possible in both the directions simultaneously but not between the same two stations.
- If more than 1 bit of information is transmitted over the data transmission medium at a time then it is considered a **Parallel Communication**. Byte-serial transmission conveys eight bits at a time through eight parallel channels.
- If only 1 bit of information can be transmitted over the data transmission medium at a time then it is considered a **Serial Communication.**
- Bit-serial transmission is normally just called **serial transmission** and is the chosen communications method in many computer peripherals.
- Serial data transfer is used for high transfer rates over long distances.
- At shorter distances, typically less than 15 feet, parallel data transfers are used.
- Serial data communication only transfers one bit at a time and does not have the timing problems (race conditions) of parallel data communications.
- When this timing information is known, the receiver is said to be **synchronized** with the transmitter, and accurate data transfer becomes possible.
- There are two basic methods of synchronization: synchronous transmission and asynchronous transmission.
- In **asynchronous transmission**, the beginning and end of each byte of data is marked by start and stop bits.
- In **asynchronous systems**, a separate timing channel is not used.

Review Questions

1. Discuss the objectives of OSI reference model in detail.
2. Explain OSI model in detail. Differentiate between OSI and TCP/IP model. Explain in detail about the various layers of OSI model. **[MDU May 2009, Dec. 2009]**
3. Explain OSI reference model in detail. Explain at which layer the following addressing is used :
   (*a*) IP addressing
   (*b*) Physical addressing
   (*c*) Service point addressing. **[MDU May 2010]**
4. Describe briefly about the token passing ring arrangement of LAN. **[MDU May 2009]**
5. Discuss the functions of physical layer in brief.
6. Discuss the functions of data link layer in brief.
7. Discuss the functions of network layer in brief.
8. Compare different types of network topologies.
9. Compare peer-to-peer and client-server networks.
10. Explain the architecture of the digital subscriber's loop used in the telephone network mentioning its seven major components. What is a DDD network ? **[MDU Dec. 2010]**
11. What is the difference between service point addressing, physical addressing and logical addressing in terms of the functions of the various layers of the OSI model ? **[MDU May 2011]**

❑❑❑

7

CHAPTER

Network Configurations

7.1. INTRODUCTION

INSIDE THIS CHAPTER

This chapter introduces the fundamental concepts of data communication networks. We will first look at constituent network components and various network types.

A network is any collection of independent computers that communicate with one another over a shared network medium. A **computer network** is the infrastructure that allows two or more computers (called **hosts**) to communicate with each other. When these computers are joined in a network, people can share files and peripherals such as modems, printers, or CD-ROM drives. When networks at multiple locations are connected using services available from phone companies, people can send e-mail, share links to the global Internet, or conduct video conferences in real time with other remote users. As companies rely on applications like electronic mail and database management for core business operations, computer networking becomes increasingly more important.

Every computer network includes:

- Minimum two computers Server or Client workstation.
- Networking Interface Card's (NIC)
- A connection medium, usually a wire or cable, although wireless communication between networked computers and peripherals is also possible.
- Network Operating system software, such as Microsoft Windows NT or 2000, Novell NetWare, UNIX and Linux

In this chapter we will be discussing the switching techniques used in computer networks and various types of computer networks such as LAN, MAN, WAN etc.

7.2. PHYSICAL TOPOLOGY

The **physical topology** defines how devices are connected to make a network. For the sake of simplicity, the network wiring is symbolized as a single line between the two devices. In actuality, it may be a twisted pair of wires, a coaxial cable, and an optical fiber. Right now, we're merely focusing on the "shape" of the network, technically known as its *topology*.

If we want to include more devices (sometimes called *nodes*) on this network, we have several options of network configuration to choose from:

Devices can be connected any or combinations of one or more configurations of the following:

(*a*) Star topology (*b*) Ring topology
(*c*) Bus topology (*d*) Mesh topology
(*e*) Hybrid topology (*f*) Point to point topology

(a) Star Topology

The primary feature of the star topology is that each station is radially connected to a central node through a direct point to point connection. With devices known as "gateways" at branching points in the network, data flow can be restricted between nodes, allowing for private communication between specific groups of nodes. This addresses some of the speed and security issues of the simple bus topology. However, those branches could easily be cut off from the rest of the "star" network if one of the gateways were to fail. Figure 7.1 show the star topology.

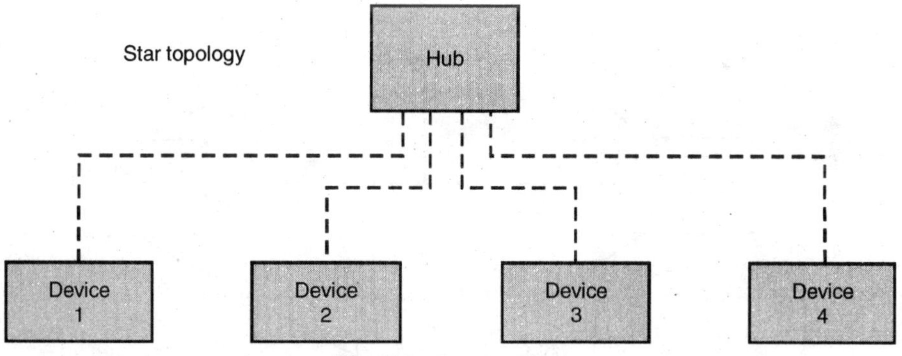

Fig. 7.1. *Star topology.*

(b) Ring Topology

With the ring topology, adjacent devices are interconnected in a closed loop configuration as shown in Fig. 7.2. This topology provides the best reliability with the least amount of wiring. Since each node has two connection points to the ring, a single break in any part of the ring doesn't affect the integrity of the network. The devices, however, must be designed with this topology in mind. Also, the network must be interrupted to install or remove nodes. As with bus topology, ring networks are *broadcast* by nature.

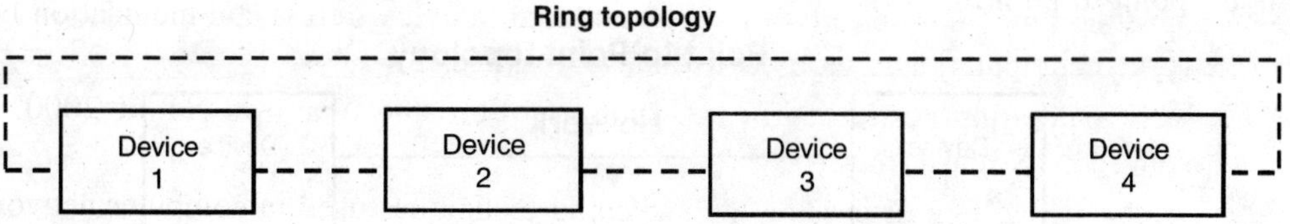

Fig. 7.2. *Ring topology.*

(c) Bus Topology

A linear bus topology is a multipoint circuit configuration where individual devices are connected by a common, shared communication channel (transmission media) as shown in Fig. 7.3. Very simple to install and maintain. Nodes can be easily added or removed with minimal wiring changes. On the other hand, the one bus network must handle *all* communication signals from *all* nodes. This is known as ***broadcast*** networking, and is analogous to a group of people talking to each other over a single telephone connection, where only one person can talk at a time (limiting data exchange rates), and everyone can hear everyone else when they talk (which can be a data security issue). Also, a break in the bus wiring can lead to nodes being isolated in groups.

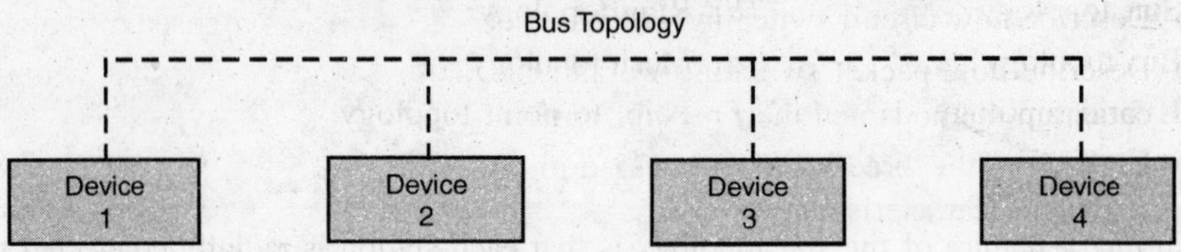

Fig. 7.3. *Bus topology.*

(d) Mesh Topology

Each device is connected to every other device.

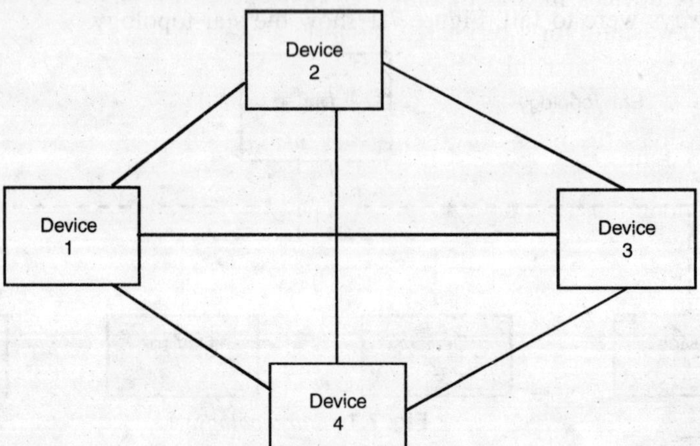

Fig. 7.4. *Mesh topology.*

(e) Hybrid Topology

Hybrid topology is that topology in which combination of two or more topology is used.

(f) Point to Point Topology

If we want to connect two digital devices with a network, we would have a kind of network known as "point-to-point:"

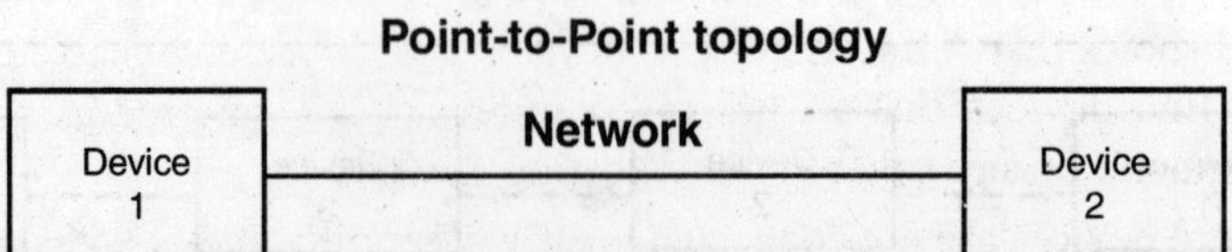

Fig. 7.5. *Point to point topology.*

Many network standards dictate the type of topology which is used, while others are more versatile. Ethernet, for example, is commonly implemented in a "bus" topology but can also be implemented in a "star" or "ring" topology with the appropriate interconnecting equipment. Other networks, such as RS-232C, are almost exclusively point-to-point; and token ring is implemented solely in a ring topology.

7.3. SWITCHING TECHNIQUES

After completing this chapter we should be able to understand the nature of network services and use network primitives to describe network service scenarios.

(*i*) Describe how circuit switching works and appreciate its strengths and weaknesses.

(*ii*) Describe how packet switching works and distinguishes between the virtual circuits and datagram methods and their packet formats.

(*iii*) Understand the three basic routing algorithms (flooding, static routing, and dynamic routing) and their characteristics.

7.4. SWITCHING

We know that a network is set of connected devices; many multiple devices are to be connected to make one-to-one communication. Either we connect each pair of devices in Mesh topology or between central devices in Star topology, to make point to point connection. These methods are impractical and wasteful for a very large network. A better solution is switching. A switched network consists of a series of interlinked nodes, referred as switches.

Switching is the generic method for establishing a path for point-to-point communication in a network. It involves the nodes in the network utilizing their direct communication lines to other nodes so that a path is established in a piecewise fashion. Each node has the capability to 'switch' to a neighboring node that is, a node to which it is directly connected to further stretch the path until it is completed.

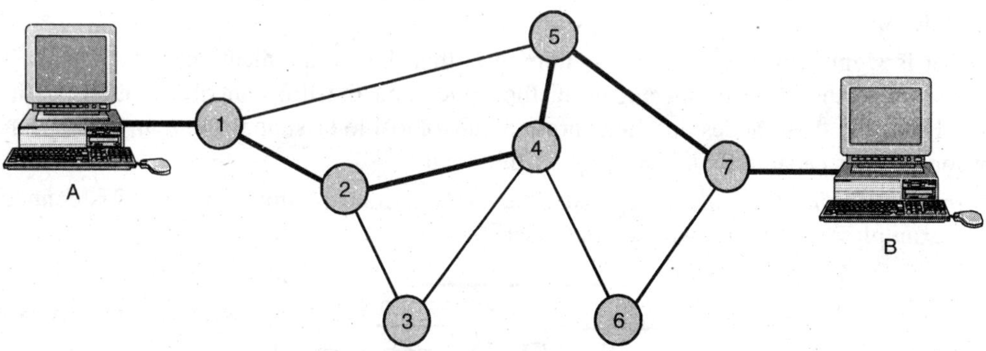

Fig. 7.6. *A switched network.*

Figure 7.6 shows that A and B are end devices and they are connected through the switches or nodes 1 to 7. End devices that wishes to communicate are referred as stations and they may be computer, telephone or any other communicating devices.

In this figure devices are not connected to each other directly but are connected through switches which are used to only route the signal in idle mode. Like in mobile phone, we are always connected to the network but in idle mode. But as we call or receive, the network operator provides a dedicated path to connect to desired user. As soon as we disconnect the call, that path is given to some other user but still you are connected with network by routing signal and this mode is referred as **idle mode**.

Switches are devices capable of making temporary link or connection between two or more devices linked to the switch but not directly to each other. They may be hardware and/or software devices. Switching nodes are not concerned with the content of the data but their purpose is to provide a switching facility that will transfer the data from node to node until they reach their destination.

DO YOU KNOW

Switching is the generic method for establishing a path for point-to-point communication in a network.

One of the most important functions of the network layer is to employ the switching capability of the nodes in order to route messages across the network.

There are traditionally three basic methods of switching, namely :

1. Circuit switching
2. Packet switching.
3. Message switching

These are separately discussed below.

7.5. CIRCUIT SWITCHING

In **circuit switching**, two communicating stations are connected by a *dedicated* communication path which consists of intermediate nodes in the network and the links that connect these nodes. *In circuit switching, an electrical dedicated path is established between the source and the destination before any data transfer take place*. The electrical path may be realized by physical wires or radio or satellite links.

What is significant about circuit switching is that the communication path remains intact for the duration of the connection, engaging the nodes and the links involved in the path for that period. However, these nodes and links are typically capable of supporting many channels, so only a portion of their capacity is taken away by the circuit.

Circuit switching takes place at physical layer. Data transmission using a PSTN connection is a typical example of a circuit switched data transfer.

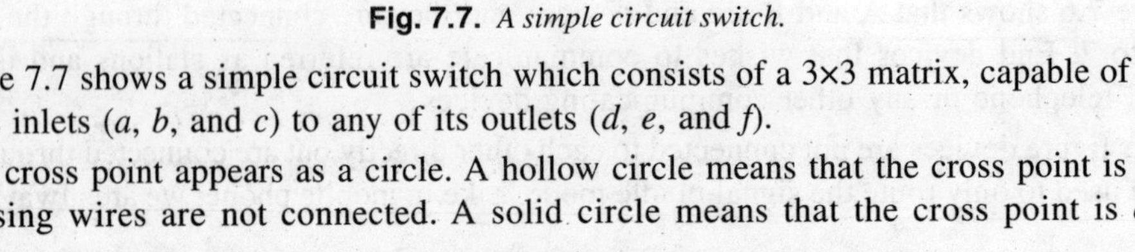

Fig. 7.7. *A simple circuit switch.*

Figure 7.7 shows a simple circuit switch which consists of a 3×3 matrix, capable of connecting any of its inlets (*a*, *b*, and *c*) to any of its outlets (*d*, *e*, and *f*).

Each cross point appears as a circle. A hollow circle means that the cross point is *off i.e.,* the two crossing wires are not connected. A solid circle means that the cross point is *on i.e.,* the crossing wires are connected. The switch can support up to three simultaneous but independent connections. Although we have used an equal number of inlets and outlets here, in general, this need not be the case. Switches may also have more inlets than outlets,

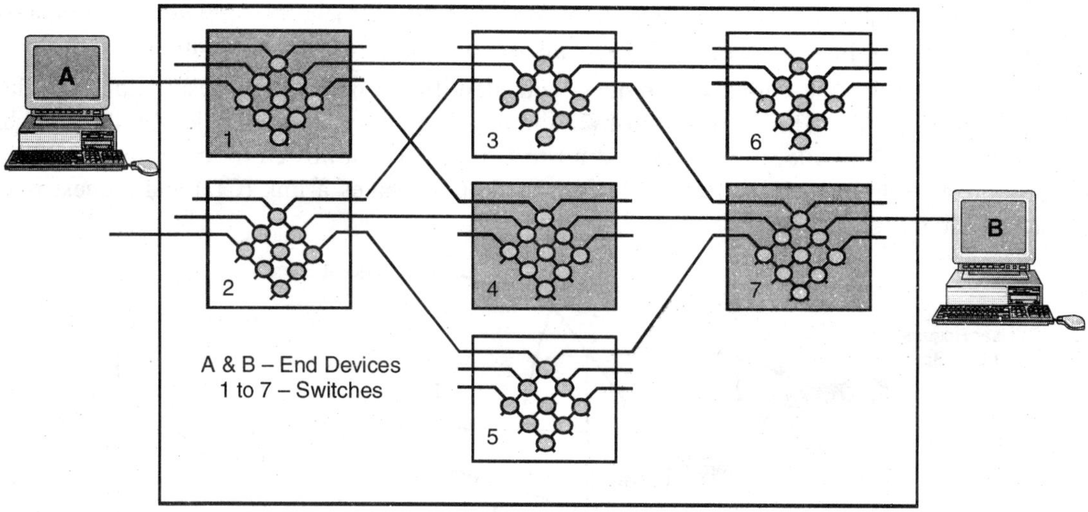

Fig. 7.8. *A simple circuit-switched network.*

or more outlets than inlets. Each link is normally divided in to n-channels by using FDM (Frequency Division Multiplexing) and TDM (Time Division Multiplexing).

Figure 7.8 shows a simple circuit-switched network built using the switch shown in Fig. 8.2. When the two hosts shown in the figure initiate a connection, the network determines a path through the intermediate switches and establishes a circuit which is maintained for the duration of the connection. When the hosts disconnect, the network releases the circuit.

A circuit switched network has three phases which are as follows :

(*i*) Setup phase
(*ii*) Data transfer phase
(*iii*) Tear down phase

Before starting communication, the station must make provision for the resources to use during the communication, such as follows:

(*i*) Channels Bandwidth
(*ii*) Switches buffers
(*iii*) Switch processing time
(*iv*) Switches input/outputs.

The operation of the circuit switched network is explained with the help of an example.

Consider communication between two points *A* and *D* in a network. The connection between *A* and *D* is provided using (shared) links between two other pieces of equipment, *B* and *C*.

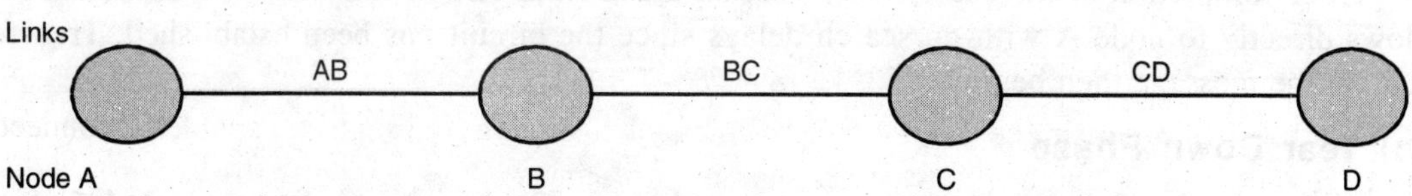

Fig. 7.9. *A Connection between Two systems A and D formed from 3 Links.*

(a) Setup/Connection Phase

Network use is initiated by a connection phase, during which a circuit is set up between source and destination and terminated by a disconnect phase. These phases, with associated timings, are illustrated in the Fig. 7.10.

After a user request a circuit, the desired destination address must be communicated to the local switching node (B). In a telephone network, this is achieved by dialing a number.

Node B receives the connection request and identifies a path to the destination (D) via an intermediate node (C). This is followed by a circuit connection phase handled by the switching nodes and initiated by allocating a free circuit to C (link BC), followed by transmission of a call request signal from node B to node C. In turn, node C allocates a link (CD) and request is then passed to node D after a similar delay.

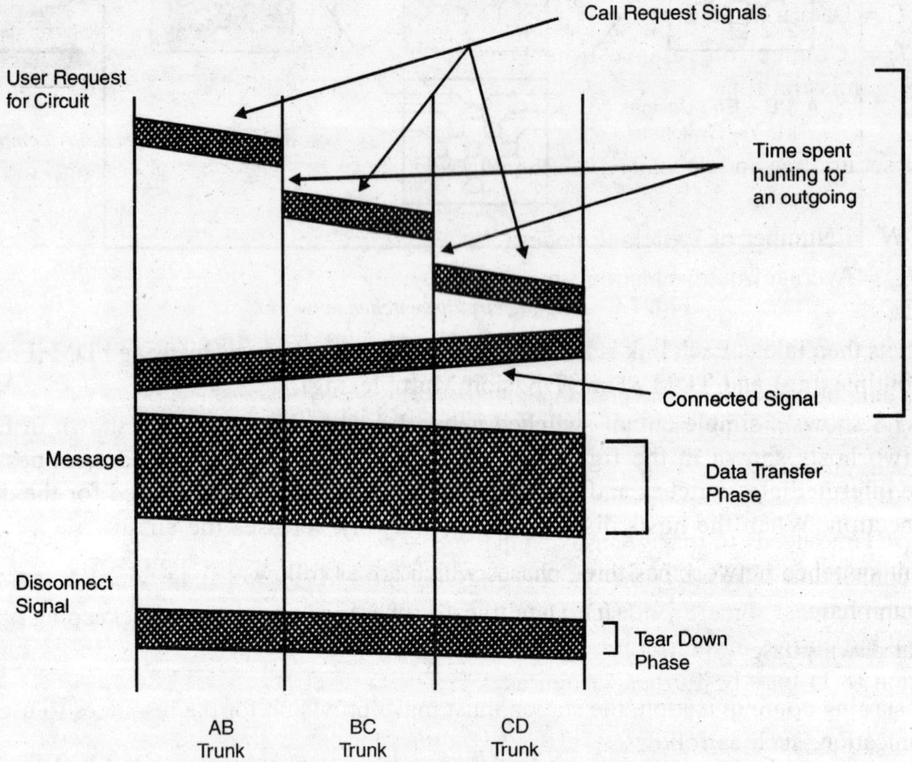

Fig. 7.10. *A Circuit switched connection between A and D.*

(b) Data Transfer Phase

The circuit is then established and may be used. While it is available for use, resources and capacity on the links between the equipment are dedicated to the use of the circuit.

After completion of the connection, a signal confirming circuit establishment is returned. This flows directly to node A with no search delays since the circuit has been established. Transfer of data in the message then begins.

(c) Tear Down Phase

After data transfer, the circuit is disconnected. A simple disconnect phase is included after the end of the data transmission.

7.5.1. Propagation Time in Circuit Switched Network

Propagation time is dependent upon the medium of transmission and the electrical distance between sources to the destination. Free space propagation speed is 3×10^8 m/s as speed of electromagnetic wave and through coaxial cable its speed is 200 m/μs.

Figure 7.11 shows the variations of delay components in circuit switching.

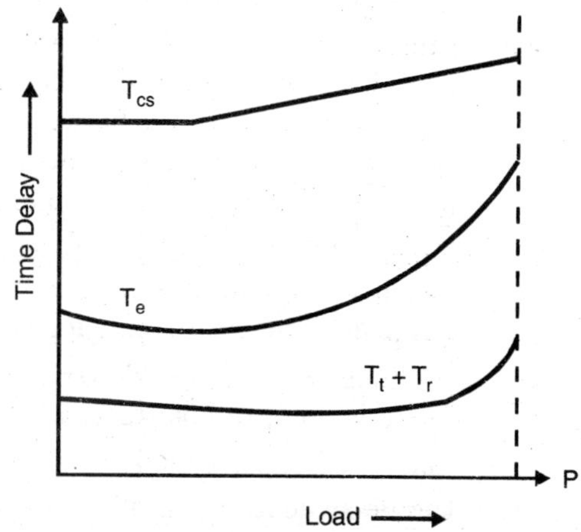

Fig. 7.11. *Variation of delay components in circuit switching.*

$$T_{CS} = T_e + T_t + T_r \qquad ...(7.1)$$

where, T_{CS} = Total circuit switching delay

T_e = Connection establishment time

T_t = Data transmission time

T_r = Connection release time Data transmission time

$$T_e = (N - 1)\, T_m \qquad ...(7.2)$$

T_e is a set up time and depends upon the number of switches.

where, N = Number of switching nodes

T_m = Average route selection time for each node

$$T_t = \left(\frac{M}{R}\right) \qquad ...(7.3)$$

T_t is a data transmission time and depends upon the data rate and size of message.

M = Message length in bits

R = Data rate in bps

$$T_r = (N\, T_n) \qquad ...(7.4)$$

where, T_n = Time needs to make housekeeping entries by each node.

T_r is connection releasing time and depends upon number of switching nodes. On receiving disconnection request signal from each node, the circuit performs certain housekeeping operations, such as making entries in the routing table.

Equation (8.1) may be further written as

$$T_{CS} = (N - 1)\, T_m + \left(\frac{M}{R}\right) + (N\, T_n) \qquad ...(7.5)$$

7.5.2. Advantages/Disadvantages

(*i*) Circuit switching relies on dedicated equipment especially built for the purpose, and is the dominant form of switching in telephone networks. Its main **advantage** lies in its predictable behavior: because it uses a dedicated circuit, it can offer a constant throughput with no noticeable delay in transfer of data. This property is important in telephone networks, where even a short delay in voice traffic can have disruptive effects.

(*ii*) Delays for setting up a circuit connection can be high, especially if ordinary telephone equipment is used. Call setup time with conventional equipment is typically on the order of 5 to 25s after completion of dialing. New fast circuit switching techniques can in theory reduce delays to approximately 140ms. Trade-offs between circuit switching and other types of switching depend on switching times.

(*iii*) Circuit switching main **disadvantage** is its inflexibility in dealing with computer oriented data. A circuit uses a fixed amount of bandwidth, regardless of whether it is used or not. In case of voice traffic, the bandwidth is usually well used because most of the time one of the two parties in a telephone conversation is speaking. However, computers behave differently; they tend to go through long silent periods followed by a sudden burst of data transfer. This leads to significant underutilization of circuit bandwidth.

(*iv*) Another **disadvantage** of circuit switching is that the network is only capable of supporting a limited number of simultaneous circuits. When this limit is reached, the network blocks further attempts for connection until some of the existing circuits are released.

(*v*) There is no error control facilities used in circuit switched network.

(*vi*) The speed of the operation of the circuit is limited by slowest link in circuit.

(*vii*) Circuit switched treats all the transmission equally. There is problem to prioritize messages and select transmission paths.

7.5.3. Efficiency

The circuit switched networks are not as efficient as the other two types of networks because the resources are allocated during the entire duration of connection. Now these resources are not available to other users. Normally, in telephone network, users terminate the communication when not in use whereas in computer networks, a computer can be connected to another computer even if there is no activity for a long period.

DO YOU KNOW

In **circuit switching**, an electrical dedicated path is established between the source and the destination before any data transfer take place.

7.6. PACKET SWITCHING

As we have already discussed about the circuit switched network, in circuit switching, a dedicated channel, called *circuit*, is set up between two stations for the duration of the transmission. The circuit must be established to make a complete the end-to-end connection before any data can be transferred. After some period of data transfer, the connection is terminated.

Packet switching was designed to minimize the shortcomings of circuit switching in dealing with data communication. Unlike circuit switching where communication is continuous along a dedicated circuit, in packet switching, communication is discrete in form of packets. Messages are split at the source host reassembled at the destination host. Each packet is of a limited size and can hold up to certain number of octets of user data. Larger messages are broken into smaller chunks so that they can be fitted into packets. In addition to user data, each packet carries additional information (in form of a header) to enable the network to route it to its final destination, as shown in Fig. 7.12.

In brief in Packet switching, there is no dedicated circuit between any two end nodes. The links can be dynamically shared over time. To provide the line sharing mechanism, user data must be segmented into discrete, variable length unit called **packets.**

Message is broken up into a series of packets and header is used to route the packet through the network.

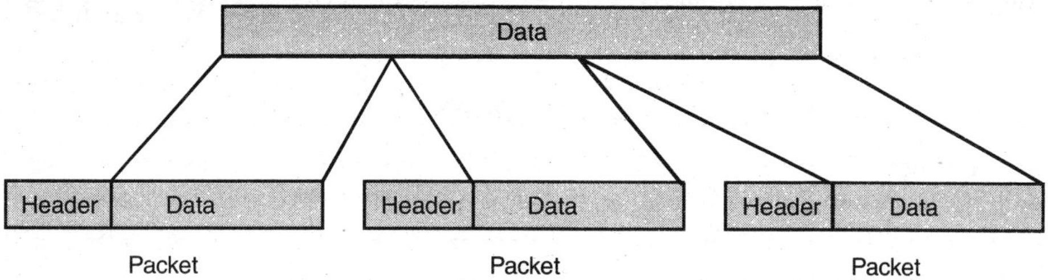

Fig. 7.12. *Packet.*

A packet is handed over from node to node across the network. Each receiving node temporarily stores the packet, until the next node is ready to receive it, and then passes it onto the next node. This technique is called **store-and-forward** and overcomes one of the limitations of circuit switching.

DO YOU KNOW

In **packet switching**, communication is discrete in form of packets and there is no dedicated circuit between any two end nodes.

A packet-switched network has a much higher capacity for accepting further connections. Additional connections are usually not blocked but simply slow down existing connections, because they increase the overall number of packets in the network and hence increase the delivery time of each packet.

Figure 7.13 shows a simple packet switch with six I/O channels (*a* through *f*).

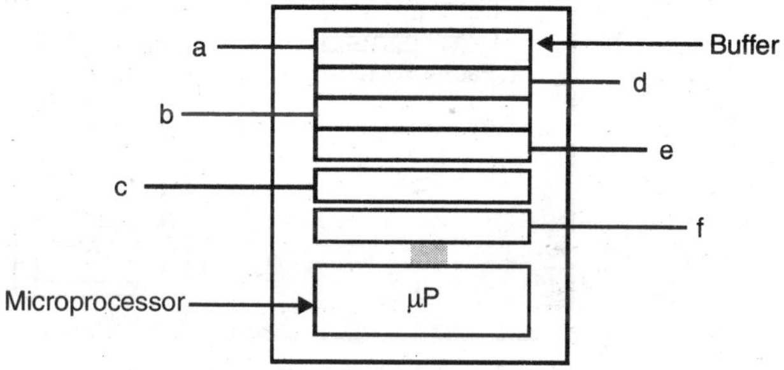

Fig. 7.13. *A simple packet switch with six I/O channels.*

Each channel has an associated buffer which it uses to store packets in transit. The operation of the switch is controlled by a microprocessor. A packet received on any of the channels can be passed onto any of the other channels by the microprocessor moving it to the corresponding buffer.

Packet switching uses two approaches :

1. **The Virtual Circuit** approach (often used in data link layer)
2. **The Datagram** approach (often used in network layer)

7.6.1. The Virtual Circuit

The virtual circuit method (also known as **connection-oriented**) is closer to circuit switching. Here a complete route is worked out prior to sending data packets.

The route is established by sending a connection request packet along the route to the intended destination. This packet informs the intermediate nodes about the connection and the established route so that they will know how to route subsequent packets. The result is a circuit somewhat similar to those in circuit switching, except that it uses packets as its basic unit of communication. Hence it is called a virtual circuit.

A virtual circuit network is normally implemented in the data link layer, while a circuit switched network is implemented in the physical layer and a datagram network in the network layer.

DO YOU KNOW

A virtual circuit network is normally implemented in the data link layer, while a circuit switched network is implemented in the physical layer and a datagram network in the network layer.

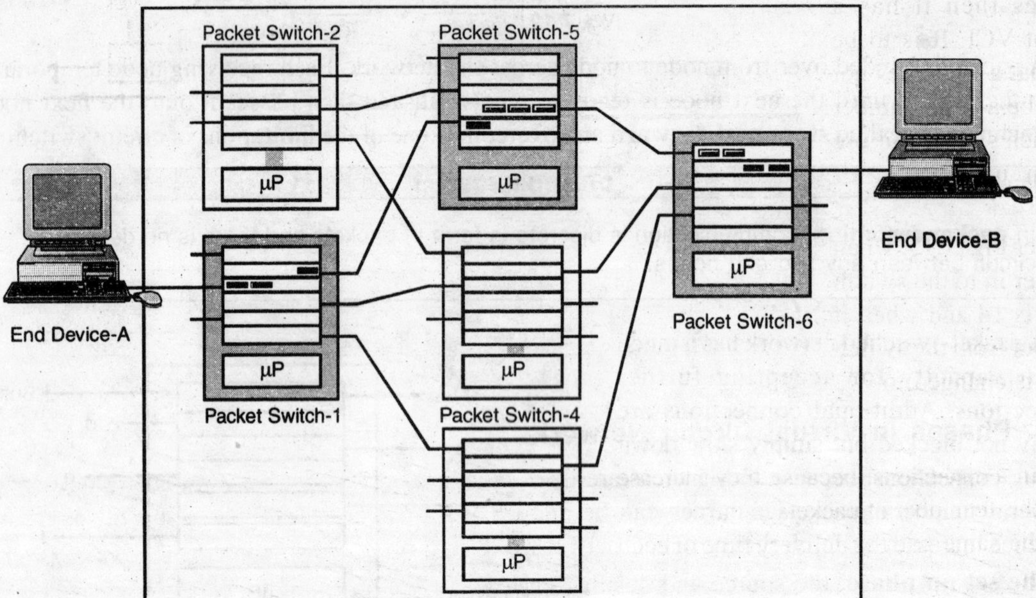

Fig. 7.14. *Packet switched network using virtual circuit*

Figure 7.14 is an example of a virtual circuit wide area network. The network has switches that allow traffic from source to destinations. They may be a computer, router bridges, or any other device that links other networks, such as LANs (local area networks), to the switched WAN.

All communications across a data network are carries out over previously established virtual circuits. These can be set up either on demand by user or permanently by network management. With on-demand connections, the user device sends a request for a *switched virtual connection (SVC)* to be set up to a central control unit known as the *Signaling Control Point (SCP).*

The SCP then determines the availability of both the required destination and the transmission bandwidth appropriate to call across the network.

With *Permanent Virtual Circuits* (PVC), the route is already determined when demand is made. The PVC is a logical predefined path or link through a carrier network.

To understand the operation of virtual circuit network, we should discuss some important terms used in this mechanism.

7.6.1.1. Addressing

There are two types of addressing involves in virtual circuit networks, namely:

(*i*) Global addressing

(*ii*) Local addressing (Virtual circuit identifier)

Global addressing means an address that can be unique in the scope of network, a source or destination required to have global address. However, we will see that global addressing in virtual circuit networks is used only to create a virtual circuit identifier.

Local addressing or Virtual circuit identifier (VCI) means the identifier that is actually used for data transfer from switch to switch. It is used by a packet between two switches, when a packet is arrived at a switch; it has some VCI and when it leaves then it has a different VCI. It is to be noted that a VCI does not need to be a large number since each switch can use its own unique set of VCIs.

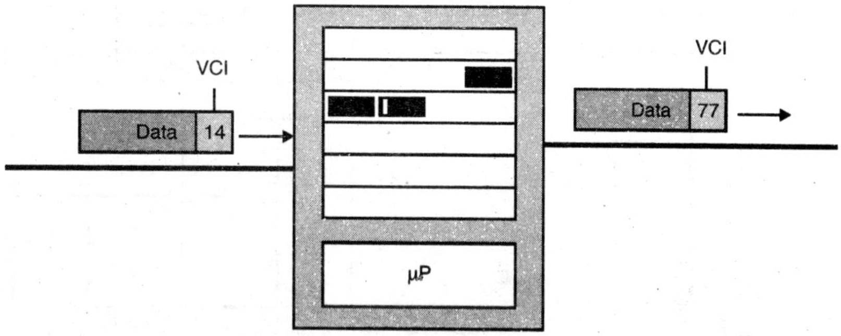

Packet Switch

Fig. 7.15

Refer Fig. 7.15, when data enter in to the switch its VCI is 14 and when it comes out from the switch its VCI is changed to 77.

7.6.1.2. Phases in Virtual Circuit Network

As in a circuit switched network, there are three phases; set-up phase, data transfer phase and finally the tear down phase. Resources can be allocated during the set up phase and all packets follow the same path as in circuit switched network.

In **the set up phase,** the source and destination use their global address to facilitate switches make the table entries for the connection. After the allocated the resources or path, the **data transfer phase** comes. In this phase all the data is transferred and after transferring the data the connection is to be released. In **tear down phase**, the source and destination informs the switches to delete the corresponding entry.

7.6.1.3. Data Transfer Phase

In virtual circuit network, when data has to be transferred from source to the destination, all switches required to have a table entry for this virtual circuit. The table has four columns as shown in Fig. 7.16. The switch holds four pieces of information for each virtual circuit that is already set up. All switches must have a table with entries for all virtual circuits.

Let us take one example, shown in Fig. 7.16, a frame arriving at port-1 with a VCI of 24, when the frame arrives, the switch looks in its table to find port-1, and VCI-24. When it is located, the switch knows that VCI to be changed to 32 and send out the frame from port-2. Another data enter from the same port with a VCI 17 and moves out from port-3 with the changed VCI 55.

Figure 7.17 shows the process of data transfer from source, A, to destination, B, and how its VCI changes from switch to switch. The corresponding table's entries are shown for each switch. The data transfer phase is continuing till the source sends all its frames to the destination. The activities at the switch are the same for each frame of a message. This process creates a virtual circuit between the source and destination.

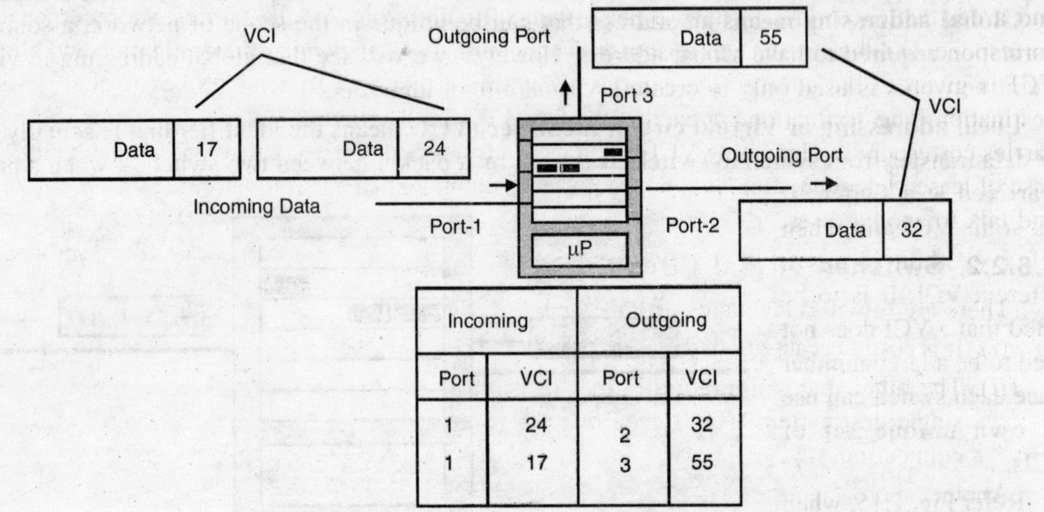

Fig. 7.16. *Switch and Table in virtual circuit.*

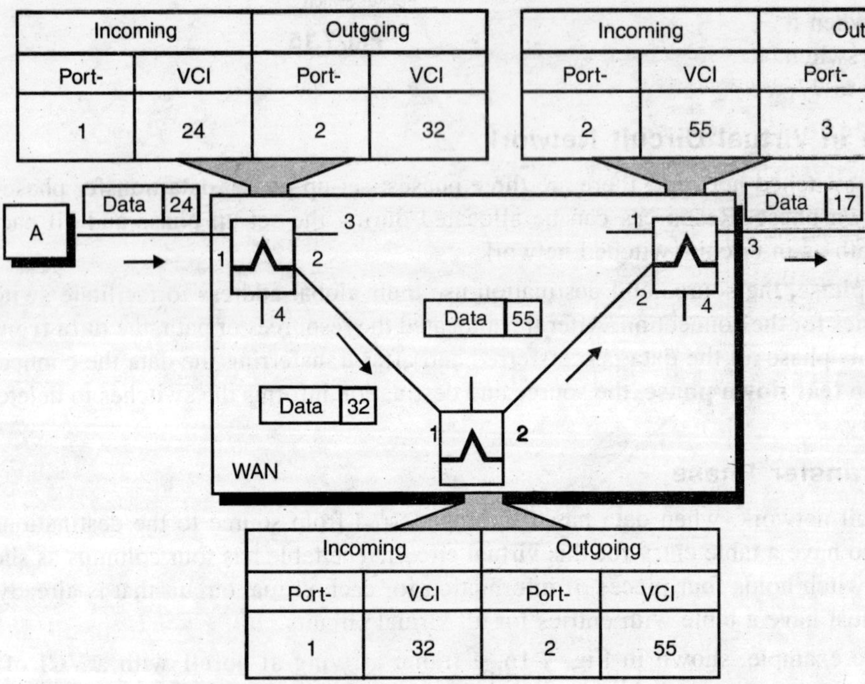

Fig. 7.17. *Source to destination data transfer in virtual circuit.*

7.6.2. Types of Virtual Circuit

During the set up phase a switch create an entry for the virtual circuit. Now there are two approaches to create the virtual circuits.

1. Permanent virtual Circuit 2. Switched Virtual Circuit

7.6.2.1. Permanent Virtual Circuit

With **Permanent Virtual Circuits (PVC),** the route is already determined when demand is made. The PVC is a logical predefined path or link through a carrier network. Sometimes, a source and a destination choose to have a permanent virtual circuit. All the entries are recorded in the corresponding tables by the administrator. An incoming VCI is given to the destination and outgoing VCI is given to the source. The source always uses this VCI to send frames to that particular destination. The destination knows that the frame is coming from that particular source if the frame carries corresponding incoming VCI. Permanent virtual circuit is used in a leased telephone line. In case of leased telephone line, it is not required to dial the number, user can just pick up the receiver and talk to another user.

7.6.2.2. Switched Virtual Circuit

There are few disadvantages of PVC network. Such as

(*i*) It is costly because both the users pay for the connection all the time even when it is not in use.

(*ii*) The other major limitation is that the connection is created from one source to the one single destination. If it is require to have many destinations than it needs PVC for each connection.

Another approach is the **switched virtual circuits** where the circuit creates a temporary connection that exists only when data are being transferred between source and destination. It requires set up phase or connection phase.

Let us suppose X needs to create a virtual circuit to Y. We need to take two steps, the set up request and the acknowledgement.

The Set-up Request

A set-up request frame is sent from the source, X, to the destination, Y, shown in the Fig. 7.18.

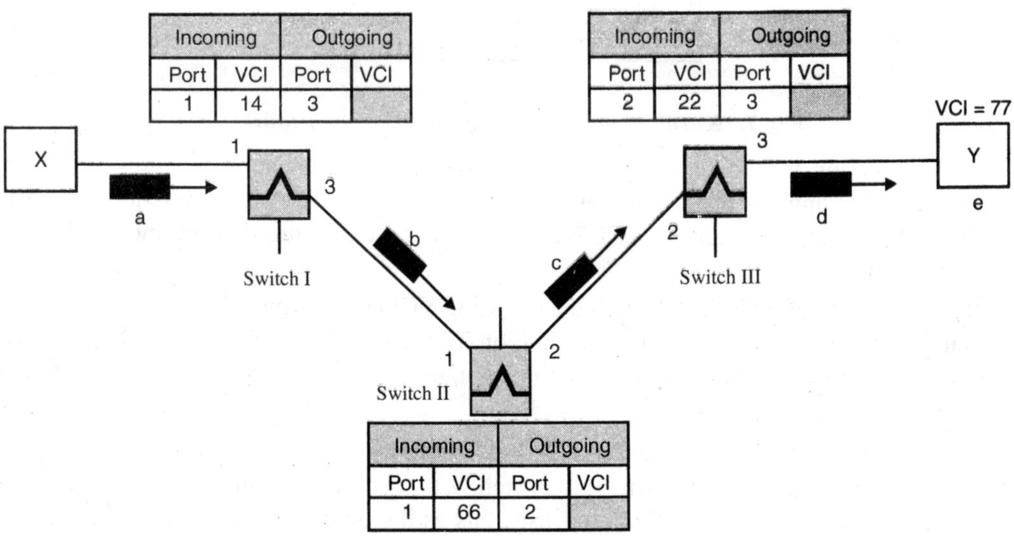

Fig. 7.18. *A set-up request in virtual circuit network.*

1. Source X sends a setup frame to switch-1.
2. Switch-1 receives the setup request frame. This is known that a frame moving from X to Y goes out through port-3.
3. The switch creates an entry in its table for this virtual circuit. But it is only capable to fill only three columns, one column has been left blank.
4. The switch-1 assigns the incoming port-1 and chooses an available incoming VCI-14 and the outgoing port-3. This switch does not yet know the outgoing VCI, which will be found

during the acknowledgment step. Now the switch forwards the frame through port-3 to switch-2.

5. Switch-2 gets the request frame. Now in switch-2, all three columns of the table are filled; in this case, incoming port-1, incoming VCI-66, and outgoing port-2.

Acknowledgment

The acknowledgment frame filled the entries which are unfilled during setup phase, refer Fig.7.19.

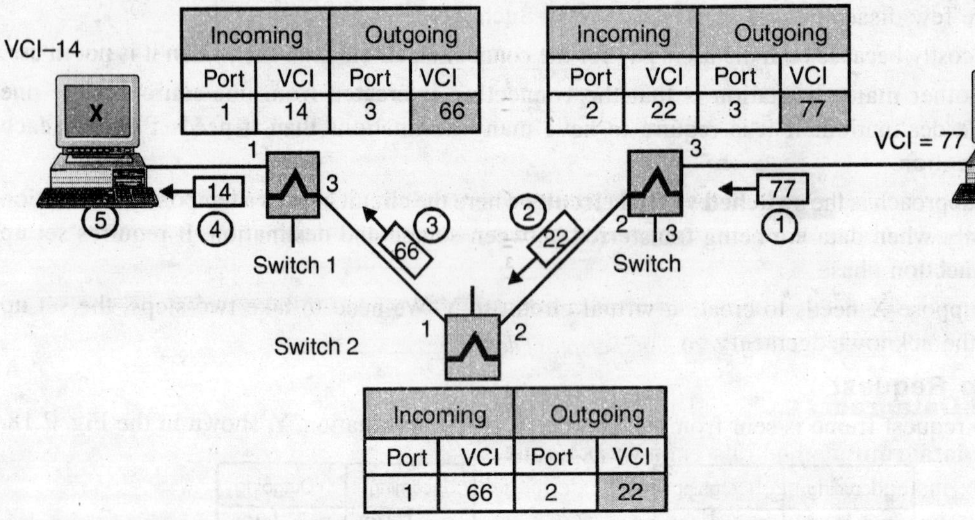

Incoming		Outgoing	
Port	VCI	Port	VCI
1	14	3	66

Incoming		Outgoing	
Port	VCI	Port	VCI
2	22	3	77

Incoming		Outgoing	
Port	VCI	Port	VCI
1	66	2	22

Fig. 7.19. *The acknowledgment in virtual circuit network.*

1. The destination Y sends an acknowledgment to switch-3. The acknowledgment carries the global source and destination addresses so the switch knows which entry in the table to be filled. The frame also carries VCI-77, chosen by the destination as the incoming VCI for frames from X.
2. Switch-3 uses this VCI to complete the outgoing VCI column for this entry. It is to be noted that 77 is the incoming VCI for destination Y, but outgoing VCI for switch-3. Switch-3 sends an acknowledgment to switch-2 that contains its incoming VCI in the table, chosen in the previous step.
3. Switch-2 uses this as the outgoing VCI in the table.
4. Switch-2 sends an acknowledgment to switch -1 that contains its incoming VCI in the table, chosen in the previous step. Switch-1 uses this as the outgoing VCI in the table.
5. At last, finally switch-1 sends an acknowledgment to source X that contains its incoming VCI in the table, chosen in the previous step.
6. The source uses this as the outgoing VCI for the data frames to be sent to destination Y.

Teardowns Phase

When all the frames from the source X are send to the destination Y, sends a special frame referred as teardown request. Destination Y responds with a teardown confirmation frame. All switches erase the corresponding entry from their tables.

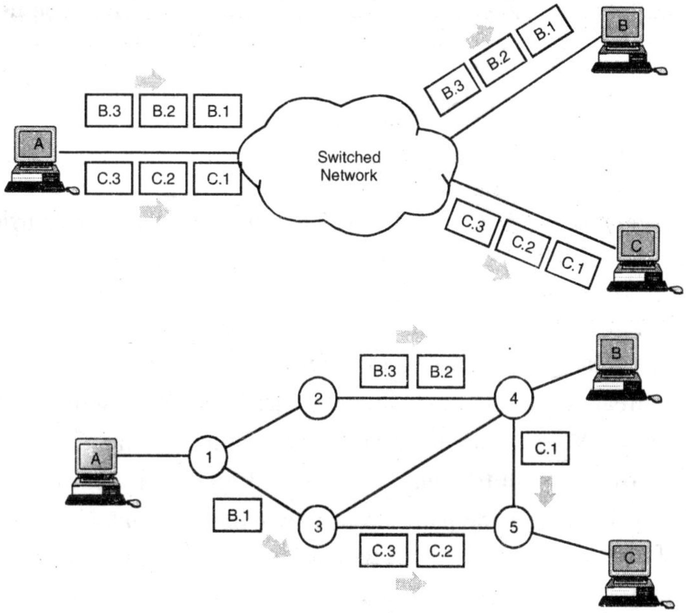

Fig. 7.20. *An example of datagram network.*

7.6.3. The Datagram

(*i*) The **datagram** method (also known as **connectionless**) does not rely on a pre established route; instead each packet is treated independently.

(*ii*) Packet in this approach is referred as datagram. Therefore, it is possible for different packets to travel along different routes in the network to reach the same final destination.

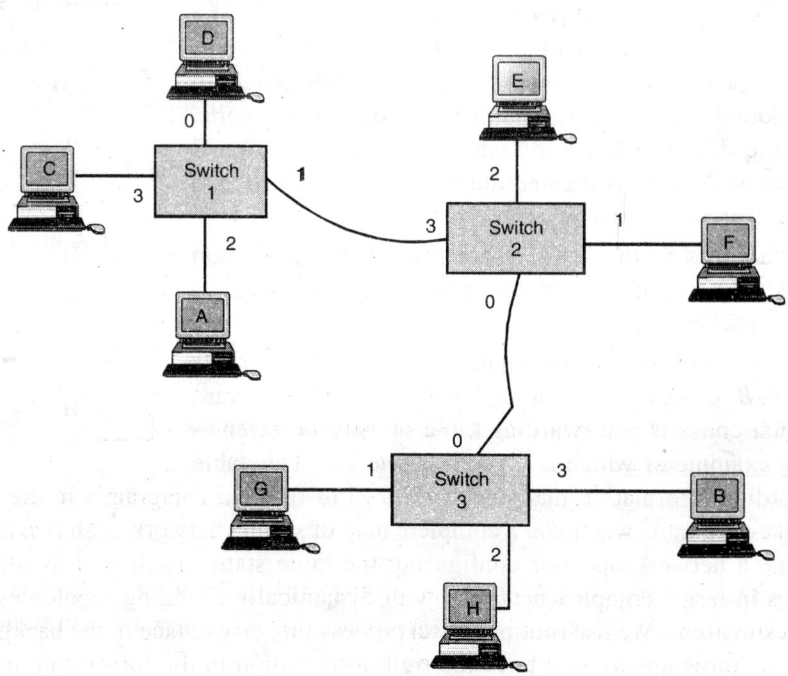

Fig. 7.21. *An example of Datagram network.*

(*iii*) As a result, packets may arrive out of order, or even never arrive (due to node failure). It is up to the network user to deal with lost packets, and to rearrange packets to their original order.

(*iv*) There are no setup and teardown phases.

(*v*) Because of the absence of a pre established circuit, each packet must carry enough information in its header to enable the nodes to route it correctly. That is, every packet contains the complete destination address. This method is analogies to mail (inland letter/ post card / registered letter) which contain complete address to whom it sends. As we drop the mail to the post office, the letter will go through all the way and finally reaches to the addressee. No matter which path it follows.

(*vi*) The switches in a datagram network are referred as **routers**.

(*vii*) Each packet is treated the same by a switch regard less of its source and destination.

Refer Fig. 7.20, source A sends the datagram to destinations B and C.

(*i*) Packet B-1, follow the path through switches 1, 3 and 4 and reaches to destination B.

(*ii*) Packets B-2 and B-3 follow the path through switches 1, 2 and 4; reaches to the destination B. Due to different the paths the packets may reach to destination B, in out of order.

(*iii*) Same way, packets, C-1follows the path through switches 1, 2, 4 and 5; reaches to the destination C. Packets C-2 and C-3 follow the path through switches 1, 3 and 5.

DO YOU KNOW

The **datagram** method (also known as **connectionless**) does not rely on a pre established route; instead each packet is treated independently.

7.6.3.1. Routing Table

In this type of switched network, each switch (router) has a routing table which is based on the destination address. The routing tables are dynamic and updated time to time. The **routing tables** record the destination address and the corresponding forwarding output ports. This table is quite different from the table of a circuit switched network in which each entry is created during the setup phase and deleted when the teardown phase is over.

Destination address in the header of a packet in a datagram network remains the same during the entire journey from source to destination of the packet.

Table 7.1. Forwarding table for switch-2

Destination Address	Output Port
A	3
B	0
C	3
D	3
E	2
F	1
G	0
H	0

Consider the example network shown in Fig. 7.21, in which hosts have addresses, *A*, *B*, *C*, *D*, *E*, *F*, *G* and *H*. To decide how to forward a packet, a switch consults a forwarding table or may be refer as routing table, an example of which is given in Table 7.1. This table shows the forwarding information that switch- 2 need to forward datagram's in the example. It is quite easy to make such table when you a complete map of simple network as shown in the example; we could imagine a network operator configuring the table statically. It is difficult to create the forwarding tables in large, complex networks with dynamically changing topologies and multiple paths between destinations. We use routing, it is a process that takes place in the background so that, when a data packet turns up, we will have the right information in the forwarding table to be able to forward or switch, the packet.

7.6.3.2. Advantages of the Datagram

1. The advantage of the datagram approach is that because there is no circuit, congestion and faulty nodes can be avoided by choosing a different route.
2. Also, connections can be established more quickly because of reduced overheads. This makes datagram's better suited than virtual circuits for brief connections. For example, database transactions in banking systems are of this nature, where each transaction involves only a few packets.

7.6.3.3. Advantages of the Virtual Circuit

1. The advantage of the virtual circuit approach is that because no separate routing is required for each packet, they are likely to reach their destination more quickly; this leads to improved throughput.
2. Packets always arrive in order.
3. Virtual circuits are better suited to long connections that involve the transfer of large amounts of data.

7.6.3.4. Advantages of Packet Switching

There are few important benefits from packet switching.

1. The first and most important benefit is that since packets are short, the communication links between the nodes are only allocated to transferring a single message for a short period of time while transmitting each packet.
2. Longer messages require a series of packets to be sent, but do not require the link to be dedicated between the transmission of each packet. The implication is that packets belonging to other messages may be sent between the packets of the message being sent. This provides a much fairer sharing of the resources of each of the links.
3. Another benefit of packet switching is known as pipelining.
4. The simultaneous use of communications link represents a gain in efficiency; the total delay for transmission across a packet network may be considerably less than for message switching, despite the inclusion of a header in each packet rather than in each message.

7.7. MESSAGE SWITCHING

There is an alternative switching technique known as message switching. There is no pre-established physical path between the source and the destination. In message switching type of switched networks uses a technique which is known as **store-and-forward.** Whenever the sender has blocks of data to be transmitted, it is stored in the first switching office (or router); each block is received in it's entirely inspected for errors and then forwarded to the next switching office. The network uses this technique is referred as store and forward network.

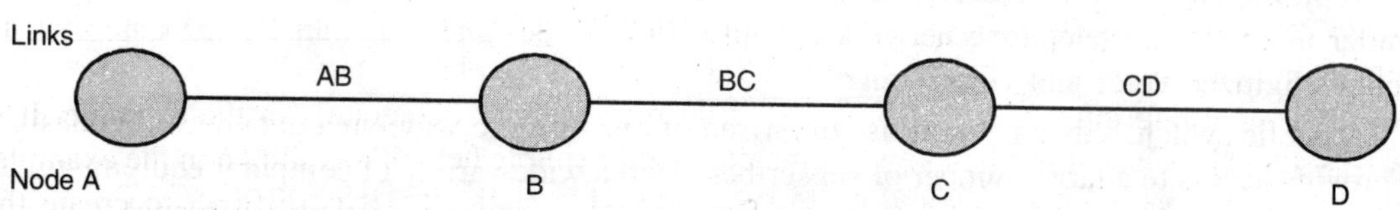

Fig. 7.22. *A Connection between two systems A and D formed from 3 Links.*

Sometimes there is no need for a circuit to be established all the way from the source to the destination. Consider a connection between the users (A and D) in the Fig. 7.22. The connection is represented by a series of links (AB, BC, and CD).

When a message (telex or email) is sent from A to D, it first passes over a local connection (AB). It is then passed at some later time to C (via link BC), and from there to the destination (via link CD). At each message switch, the received message is stored, and a connection is subsequently made to deliver the message to the neighboring message switch. The figure-18below illustrates message switching. The transmission of only one message is illustrated for simplicity. A complete message is sent from node A to node B when the link interconnecting them becomes available. Since the message may be competing with other messages for access to facilities, a queuing delay may be incurred while waiting for the link to become available.

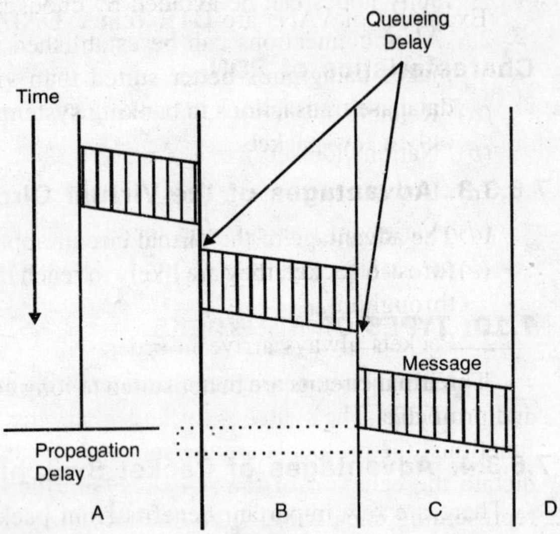

Fig. 7.23. *Message switching.*

The use of message switching to communicate between A and D (see Fig. 7.23).

The message is stored at B until the next link becomes available, with another queuing delay before it can be forwarded. It repeats until reaches its destination. The Circuit setup delays are replaced by queuing delays. Considerable extra delay may result from storage at individual nodes.

This type of network is not so popular nowadays because there is lot of delay at each switch.

7.8. COMPARISONS OF THREE SWITCHING METHODS

Message Switching	Circuit Switching	Packet Switching
(*i*) Long queuing delays and message can take a long time to transmit	(*i*) Long call setup times	(*i*) Break message into packets.
(*ii*) A long message may block system for other messages	(*ii*) The setup times may be negligible compared to data length	(*ii*) Can interpolate packets from other nodes, therefore cannot block system
	(*iii*) Inefficient channel utilisation for burst traffic	(*iii*) Efficiently handles asymmetric traffic.

7.9. PUBLIC DATA NETWORK

A public switched data network (PDN or PSDN) is a switched data communications network similar to the public telephone network except a PDN is designed for transferring data that may include digitized voice and video signals.

A public switched data network is comprised of one or more wide-area data networks designed to provide access to a large number of subscribers with a wide variety of computer equipments. The basic principle behind a PDN is to transport data from a source to a destination through a network of intermediate switching nodes and transmission media. PDN combine the concepts of value-added networks and packet switching networks.

VANs add value to the services provided by common carriers to provide new type of communication services. Examples of added values are error control, enhanced connection reliability, dynamic routing, failure protection, logical multiplexing and data format conversions.

Examples of VANs are GTE Telnet, DATAPAC, TRANSPAC and Tymnet Inc.

Characteristics of PDN

(*a*) Reliable service

(*b*) Nationwide service

(*c*) Lower error rates

(*d*) Variety of transmission speeds

(*e*) Cost-effectiveness

7.10. TYPES OF NETWORKS

Three major components characterize computer communication networks: switches, channels, and protocols. The switches (or nodes) are the hardware entities that house the data communication functions; the protocols are the sets of rules and agreements among the communicating parties that dictate the behavior of the switches, and the channel is the physical medium over which signals, representing data, travel from one switch to another.

We have many types of computer networks, which are briefly discussed here.

Networks may be classified into different categories and types according to various criteria:

1. LANs (Local Area Networks)
2. Metropolitan Area Networks (MANs)
3. WANs (Wide Area Networks)

7.10.1. Local Area Networks (LANs)

LANs (Local Area Networks) have become an important part of most computer installations. Personal computers have been the main driving force behind the LAN proliferation. As personal computers became more widely used in office environments, so it became desirable to interconnect them to achieve two aims: to enable them to exchange information (*e.g.*, e-mail), and to enable them to share scarce and expensive resources (*e.g.*, printers).

This topic is discussed in detail at the end of this chapter.

7.10.2. Metropolitan Area Networks (MANs)

For larger distances, the network is said to be a MANs cover distances of up to a few hundred kilometers and are used for interconnecting hosts spread across a city.

7.10.3. WANs (Wide Area Networks)

Wide area networking combines multiple LANs that are geographically separate. This is accomplished by connecting the different LANs using services such as dedicated leased phone lines, dial-up phone lines (both synchronous and asynchronous), satellite links, and data packet carrier services. Wide area networking can be as simple as a modem and remote access server for employees to dial into, or it can be as complex as hundreds of branch offices globally linked using special routing protocols and filters to minimize the expense of sending data sent over vast distances.

WANs are used to connect hosts spread across a country, a continent, or the globe. LANs, MANs, and WANs usually coexist: closely-located hosts are connected by LANs which can access hosts in other remote LANs via MANs and WANs.

Figure 7.24 shows the example of LANs between WANs.

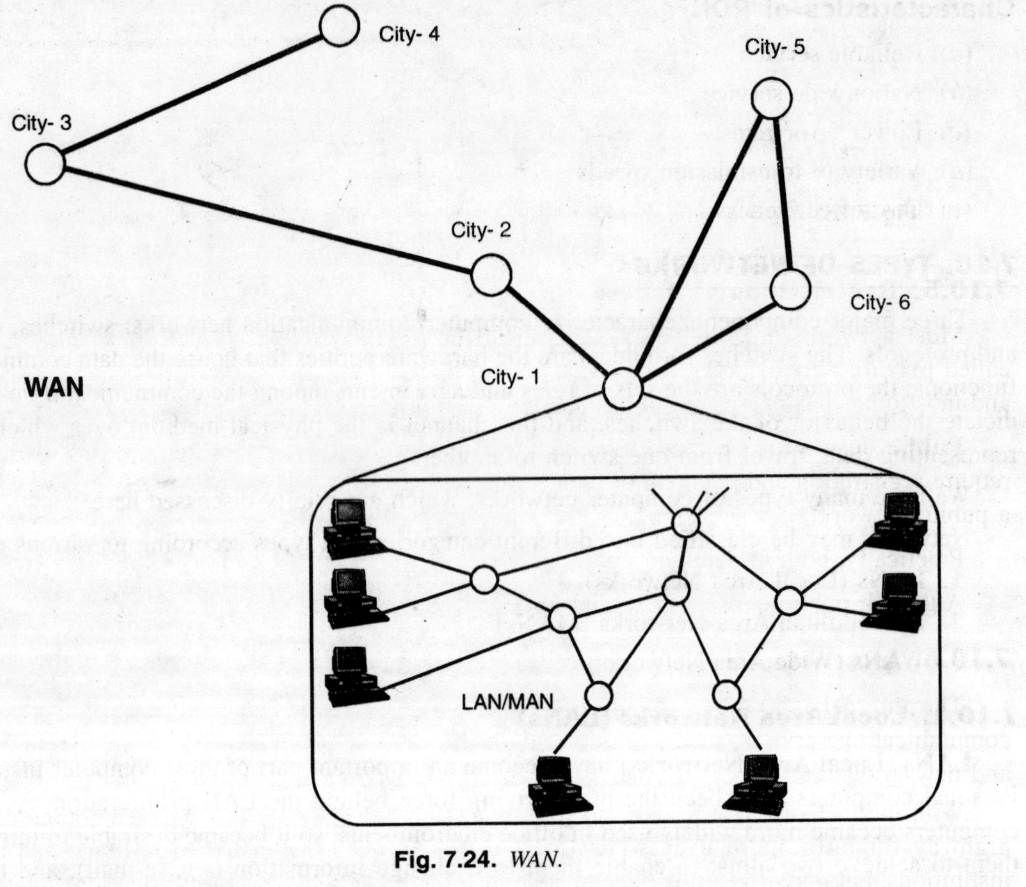

Fig. 7.24. *WAN.*

7.10.4. Communication Model

There are mainly two types of communication models which are discussed below.

(*a*) Point-to-point model

(*b*) Broadcast model

(a) Point-to-Point Model

The communication between the nodes is either based on a Point-to-point model or a Broadcast model. In the point-to-point model, a message follows a specific route across the network in order to get from one node to another.

(b) Broadcast Model

In the broadcast model, on the other hand, all nodes share the same communication medium and, as a result, a message transmitted by any node can be received by all other nodes. A part of the message (an address) indicates for which node the message is intended. All nodes look at this address and ignore the message if it does not match their own address.

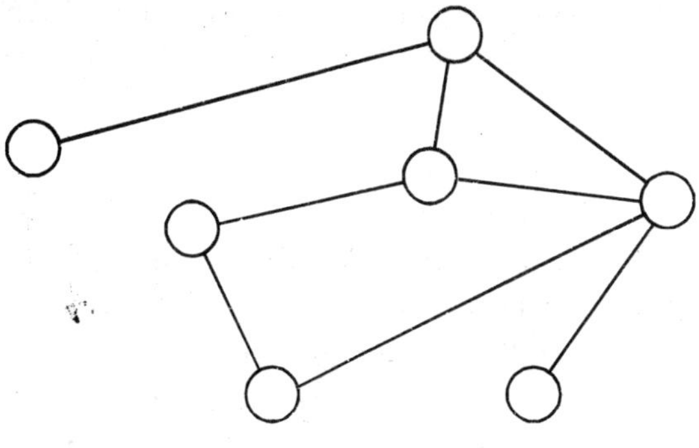

Fig. 7.25. *Point-to-Point model.*

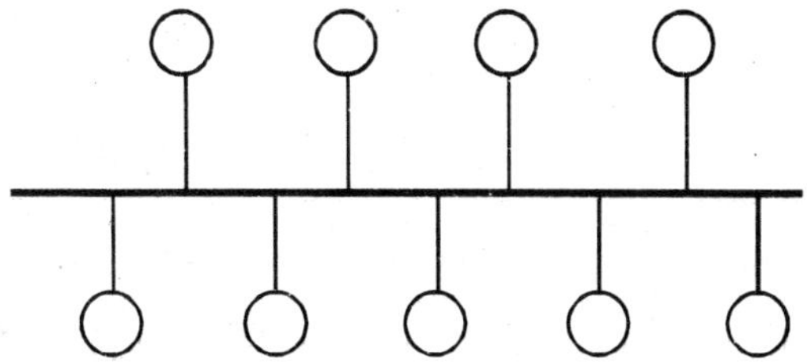

Fig. 7.26. *Broadcast model.*

7.10.5. Restrictions of Access

Most networks are for the private use of the organizations to which they belong; these are called **private networks**. Networks maintained by insurance companies, banks airlines, hospitals, and most other businesses are of this nature.

Public networks, on the other hand, are generally accessible to the average user, but may require registration and payment of connection fees. Internet is the most-widely known example of a public network.

Practically, both private and public networks may be of LAN, MAN, or WAN type.

Although public networks, by their size and nature, tend to WANs.

7.10.6. Internet

The Internet is a system of linked networks that are worldwide in scope and facilitate data communication services such as remote login, file transfer, electronic mail, the World Wide Web and newsgroups.

With the meteoric rise in demand for connectivity, the Internet has become a communications highway for millions of users. The Internet was initially restricted to military and academic institutions, but now it is a full-fledged conduit for any and all forms of information and commerce. Internet websites now provide personal, educational, political and economic resources to every corner of the planet.

7.11. INTRANET

With the advancements made in browser-based software for the Internet, many private organizations are implementing intranets. An intranet is a private network utilizing Internet-type tools, but available only within that organization. For large organizations, an intranet provides an easy access mode to corporate information for employees.

7.12. VPN (VIRTUAL PRIVATE NETWORK)

VPN uses a technique known as tunneling to transfer data securely on the Internet to a remote access server on your workplace network. Using a VPN helps you save money by using the public Internet instead of making long–distance phone calls to connect securely with your private network. There are two ways to create a VPN connection, by dialing an Internet service provider (ISP), or connecting directly to Internet.

Categories of Network

Network can be divided in to two main categories:

(*i*) Peer-to-peer.

(*ii*) Server-based.

In **peer-to-peer** networking there are no dedicated servers or hierarchy among the computers. All of the computers are equal and therefore known as peers. Normally each computer serves as Client/Server and there is no one assigned to be an administrator responsible for the entire network.

Peer-to-peer networks are good choices for needs of small organizations where the users are allocated in the same general area, security is not an issue and the organization and the network will have limited growth within the foreseeable future.

The term Client/server refers to the concept of sharing the work involved in processing data between the client computer and the most powerful server computer.

The client/server network is the most efficient way to provide:

(*i*) Databases and management of applications such as Spreadsheets, Accounting, Communications and Document management.

(*ii*) Network management.

(*iii*) Centralized file storage.

The client/server model is basically an implementation of distributed or cooperative processing. At the heart of the model is the concept of splitting application functions between a client and a server processor. The division of labor between the different processors enables the application designer to place an application function on the processor that is most appropriate for that function. This lets the software designer optimize the use of processors—providing the greatest possible return on investment for the hardware.

Client/server application design also lets the application provider mask the actual location of application function. The user often does not know where a specific operation is executing. The entire function may execute in either the PC or server, or the function may be split between them. This masking of application function locations enables system implementers to upgrade portions of a system over time with a minimum disruption of application operations, while protecting the investment in existing hardware and software.

7.13. LOCAL AREA NETWORKS (LANS)

Local Area Networks (LANs) have become an important part of most computer installations. Personal computers have been the main driving force behind the LAN proliferation. As personal computers became more widely used in office environments, so it became desirable to interconnect them to achieve two aims: to enable them to exchange information (*e.g.*, e-mail), and to enable them to share scarce and expensive resources (*e.g.*, printers).

Figure 7.27 illustrates the main characteristics of LANs.

	Low End	Typically	High End
Number of Users	10s of users	20-100 users	1000s of users
Geographic Coverage	100s of meters	100-1000 meters	10s of km's
Data Rates	10s of kbps	1-10 mbps	100s of mbps
Raw Error Rates	1 bit in 100 million	1 bit in 1-10 billion	1 bit in 100 billion

Fig. 7.27. *The main characteristics of LANs.*

Current LANs are used for interconnecting almost any type of computing devices imaginable, including mainframes, workstations, personal computers, file servers, and numerous types of peripheral devices. Many LANs are further connected to other LANs or WANs via bridges and gateways, hence increasing the reach of their users.

7.13.1. Basic Concepts of LAN

A LAN consists of four general types of components:

(*i*) **User station.** This provides the user with access to the LAN. The most common example is a personal computer. The user station runs special network software (usually in form of a driver) for accessing the LAN.

(*ii*) **LAN protocol stacks.** This implements the LAN protocol layers. This usually takes the form of a hardware card inside the user station, containing a microprocessor and firmware which implements the non-physical protocols.

(*iii*) **Physical Interface Unit.** This directly interfaces the user station-based LAN hardware to the LAN physical medium. The exact form of the PIU is highly dependent on the LAN physical medium. Coaxial cable connectors and cable TV taps are common examples.

(*iv*) **Physical Medium.** This provides a physical path for signals to travel between stations. Coaxial cable, optical fiber, and infra red light are examples.

7.13.2. Topologies and Access Protocols

There are two general categories of LAN topologies:

1. Bus Topology
2. Ring Topology

The **bus topology** uses a broadcast technique; hence only one station at a time can send messages and all other station listen to the message. A listening station examines the recipient address of the message and if it matches its own address, copies the message; otherwise, it ignores the message.

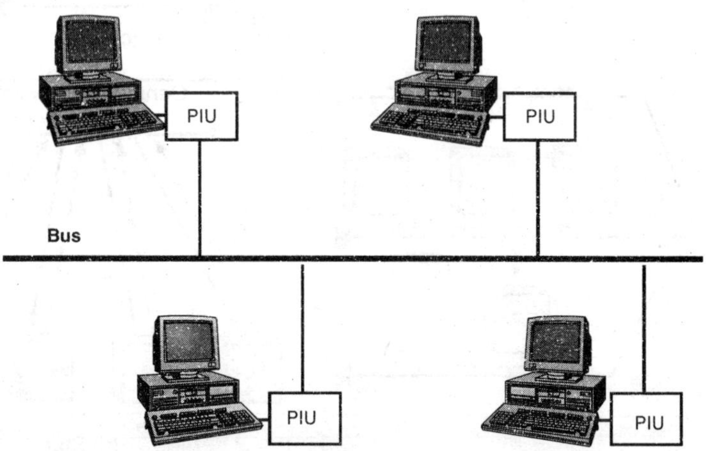

Fig. 7.28. *Bus topology.*

The **ring topology** uses a closed, point-to-point-connected loop of stations.

Data flows in one direction only, from one station to the next.

As with the bus topology, transmission is restricted to one user at a time. When a station gains control and sends a message, the message is sent to the next station in the ring.

Each receiving station in the ring examines the recipient address of the message and if it matches its own address, copies the message. The message is passed around the ring until it reaches the originator which removes the message by not sending it to the next station.

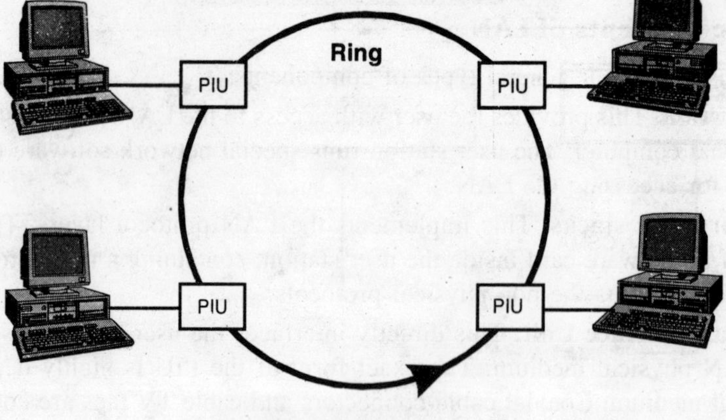

Fig. 7.29. *Ring Topology*

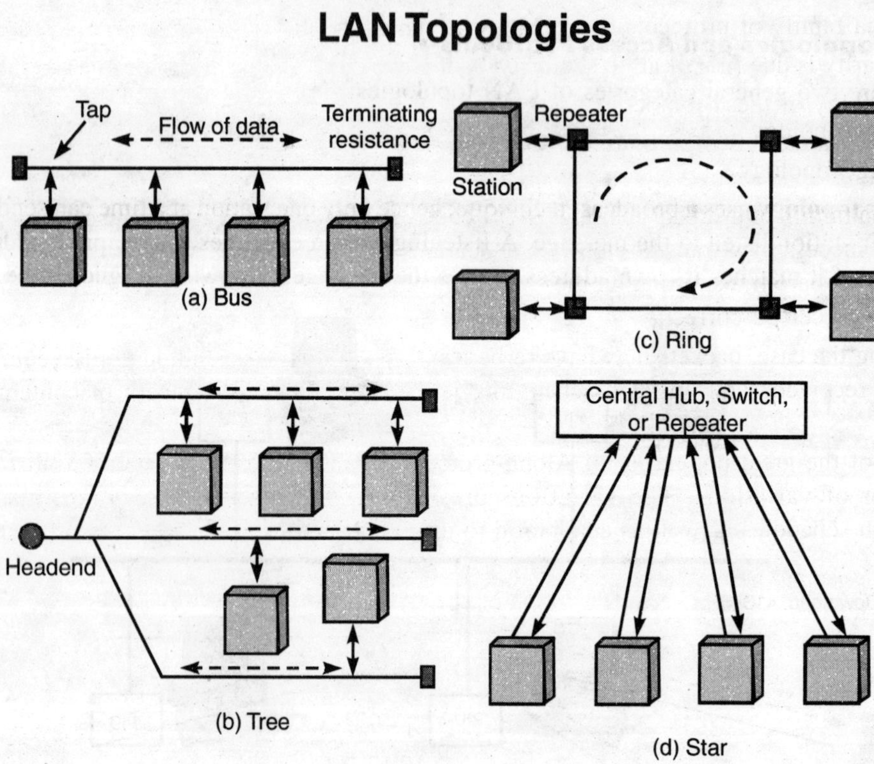

Fig. 7.30. *Various Network Configurations*

7.13.3. Multiple-Access Protocols

Given that access to the bus or ring is restricted to one station at a time, some form of arbitration is needed to ensure equitable access by all stations. Arbitration is imposed by **access protocols.**

A number of such protocols have been devised:

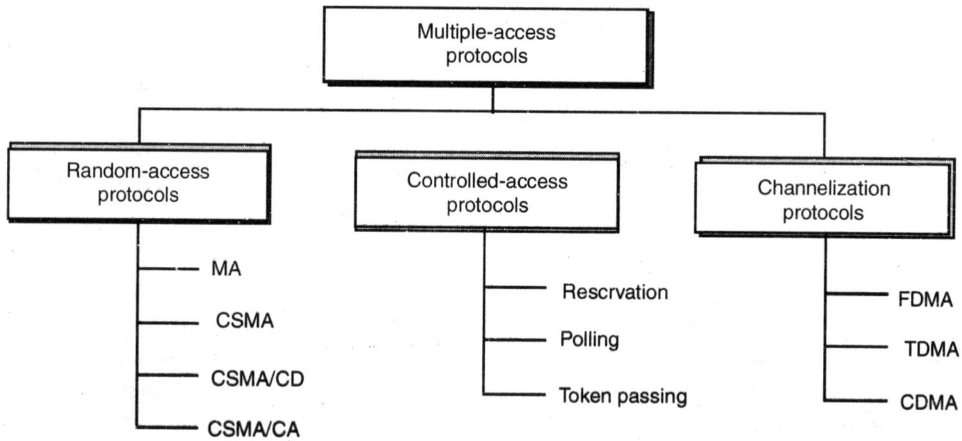

Fig. 7.31. *Multiple access protocols.*

1. Aloha Protocols

The Aloha family of protocols is probably the richest family of multiple access protocols.

Its popularity is due first of all to seniority, as it is the first random access technique introduced.

Second, many of these protocols are so simple that their implementation is straightforward.

Many local area networks of today implement some sophisticated variants of this family's protocols.

The Aloha family of protocols belongs to the contention-type or random retransmission protocols in which the success of a transmission is not guaranteed in advance. The reason is that whenever two or more users are transmitting on the shared channel simultaneously, a collision occurs and the data cannot be received correctly.

This being the case, packets may have to be transmitted and retransmitted until eventually they are correctly received. Transmission scheduling is therefore the focal concern of contention-type protocols.

Because of the great popularity of Aloha protocols, analyses have been carried out for a very large number of variations. The variations present different protocols for transmission and retransmission schedules as well as adaptation to different circumstances and channel features.

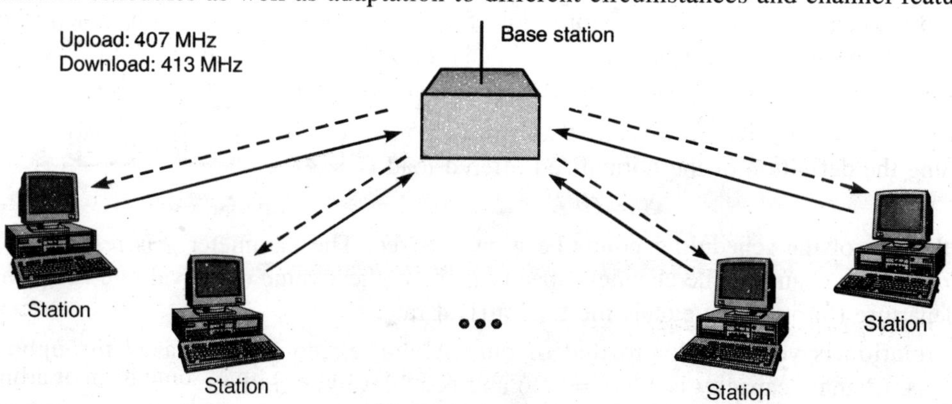

Fig. 7.32. *ALOHA.*

(i) Pure Aloha

The ***pure Aloha*** protocol is the basic protocol in the family of the Aloha protocols. It considers a single-hop system with an infinite population generating packets of equal length T according to a Poisson process with rate g packets/sec. The channel is error-free without capture: whenever a transmission of a packet does not interfere with any other packet transmission, the transmitted packet is received correctly while if two or more packet transmissions overlap in time, a collision is caused and none of the colliding packets is received correctly and they have to be retransmitted. The users whose packets collide with one another are called the *colliding users*. At the end of every transmission each user knows whether its transmission was successful or a collision took place.

The pure Aloha protocol is very simple. It states that a newly generated packet is transmitted immediately hoping for no interference by others. Should the transmission be unsuccessful, every colliding user, independently of the others, schedules its retransmission to a random time in the future. This randomness is required to ensure that the same set of packets does not continue to collide indefinitely. A simple example of the operation of the protocol is depicted in Figure 3.1 where the arrows indicate arrival instants, successful transmissions are indicated by blank rectangles and collided packets are hatched.

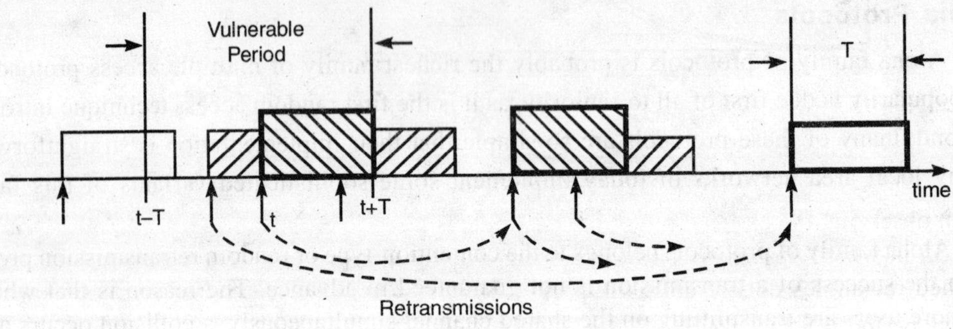

Fig. 7.33. *Pure aloha Packet Timing*

(ii) Slotted Aloha

The *slotted Aloha* variation of the Aloha protocol is simply that of pure Aloha with a slotted channel. The slot size equals T—the duration of packet transmission. Users are restricted to start transmission of packets only at slot boundaries. Thus, the vulnerable period is reduced to a single slot. In other words, a slot will be successful if and only if exactly one packet was scheduled for transmission sometime during the previous slot. The throughput is therefore the fraction of slots (or probability) in which a single packet is scheduled for transmission. Because the process composed of newly generated and retransmitted packets is Poisson we conclude that

$$S = gTe^{-gt}$$

or using the definition of the normalized offered load $G = gT$

$$S = Ge^{-G}$$

Let the rate of the scheduling points be g packets/sec. The parameter g is referred to as the *offered load* to the channel. The channel capacity is the highest value of arrival rate l for which the rate of departure (throughput) equals the total arrival rate.

This relation is very similar to that of pure Aloha, except of increased throughput (see Figure 7.34). Channel capacity is $1 \times e \approx 0.36$ and is achieved at $G=1$.

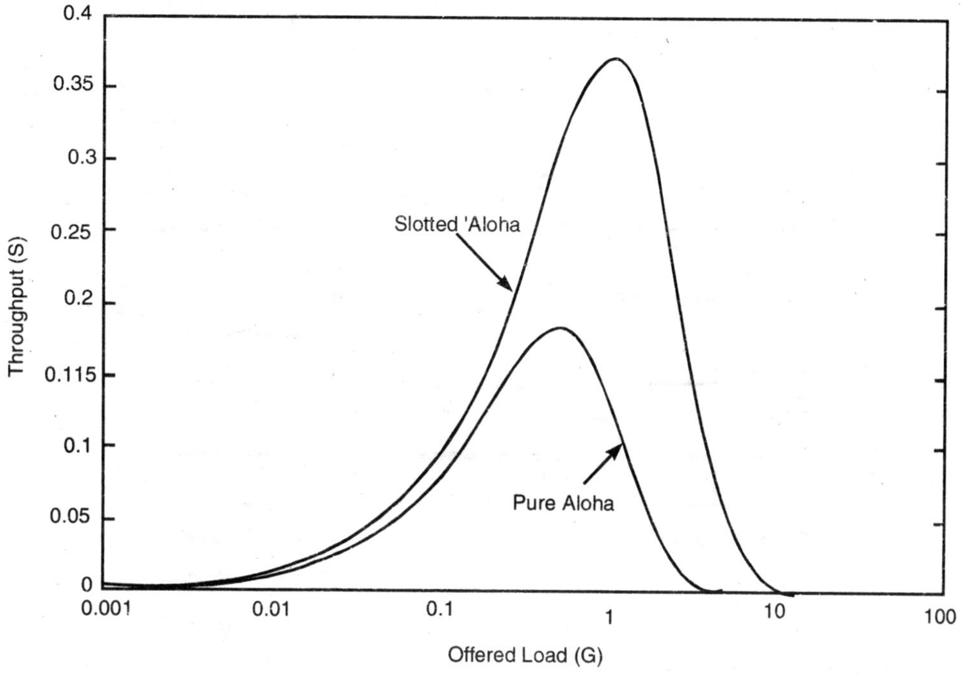

Fig. 7.34. *Throughput-Load of pure and slotted ALOHA.*

When observing the channel over time we notice a cyclic behavior of busy and idle periods (see Figure 7.35 in which up-arrows point at arrival instants, blank rectangles refer to successful slots, and hatched rectangles denote colliding packets, *i.e.*, unsuccessful slots).

A busy period is a succession of slots in which transmission takes place (successful or not). The idle period is defined as the interval between two busy period. The starting times of every cycle (just before the start of the busy period) define renewal points. In fact, these are points of a regenerative process, since the system is memoryless in the sense that its behavior in a given cycle does not depend on the behavior in any previous cycle. As such, the expected fraction of time the system is in a given state equals the expected fraction of time *during a single cycle* that the system is in that state

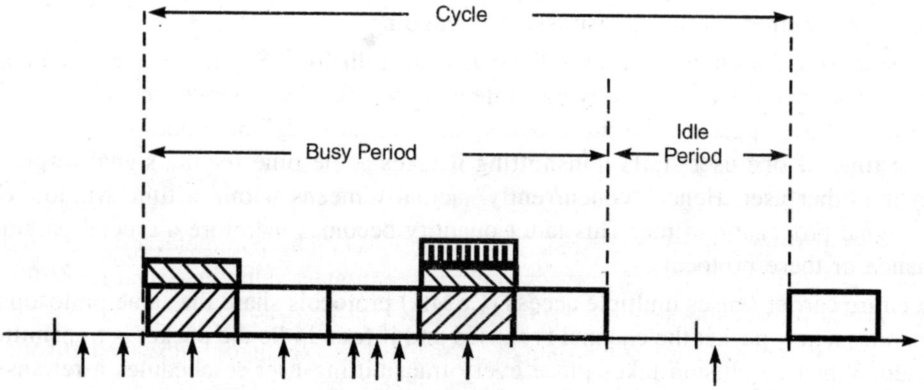

Fig. 7.35. *Slotted-Aloha packet timing.*

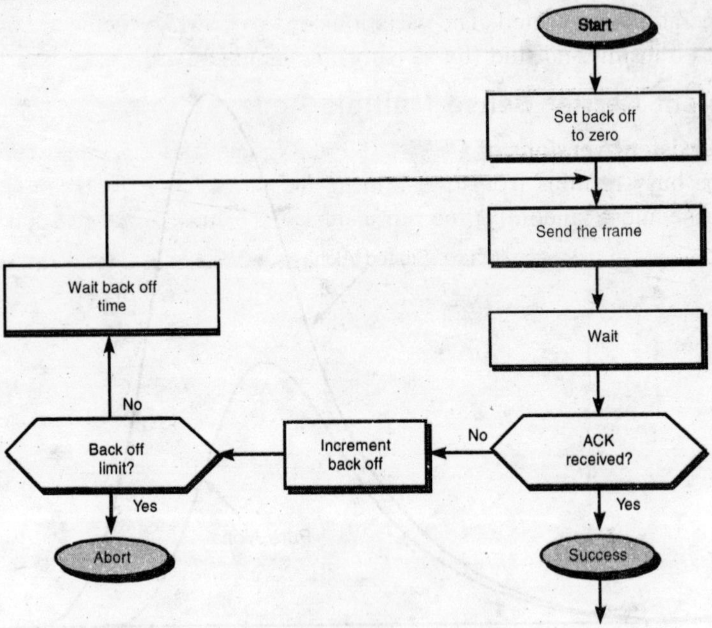

Fig. 7.36. *Procedure for ALOHA.*

2. Carrier Sensing Protocols

The Aloha schemes, described in the previous section, exhibited fairly poor performance which can be attributed to the "impolite" behavior of the users namely, whenever one has a packet to transmit he does so without consideration of others. It does not take much to realize that even little consideration can benefit all. Consider a behavior that we generically characterize as "listen before talk", that is, every user before attempting any transmission listens whether somebody else is already using the channel. If this is the case the user will refrain from transmission to the benefit of all; his packet will clearly not be successful if transmitted and, further, disturbing another user will cause the currently transmitted packet to be retransmitted, possibly disturbing yet another packet.

The process of listening to the channel is not that demanding. Every user is equipped with a receiver anyway. Moreover, to detect another user's transmission does not require receiving the information; it suffices to sense the carrier that is present when signals are transmitted.

The carrier sensing family of protocols is characterized by sensing the carrier and deciding according to it whether another transmission is ongoing.

Carrier sensing does not, however, relieve us from collisions. Suppose the channel has been idle for a while and two users concurrently generate a packet. Each will sense the channel, discover it is idle, and transmit the packet to result in collision. "Concurrently" here does not really mean at the very same time; if one user starts transmitting it takes some time for the signal to propagate and arrive at the other user. Hence "concurrently" actually means within a time window of duration equal to signal propagation time. This latter quantity becomes therefore a crucial parameter in the performance of these protocols.

The entire carrier senses multiple access (CSMA) protocols share the same philosophy: when a user generates a new packet the channel is sensed and if found idle the packet is transmitted without further ado. When a collision takes place every transmitting user reschedules a retransmission of the collided packet to some other time in the future (chosen with some randomization) at which time the same operation is repeated. The variations on the CSMA scheme are due to the behavior of users that wish to transmit and find (by sensing) the channel busy.

(i) Nonpersistent Carrier Sense Multiple Access

In the **nonpersistent versions of CSMA** (NP-CSMA) a user that generated a packet and found the channel to be busy refrains from transmitting the packet and behaves exactly as if its packet collided, i.e, it schedules (randomly) the retransmission of the packet to sometime in the future.

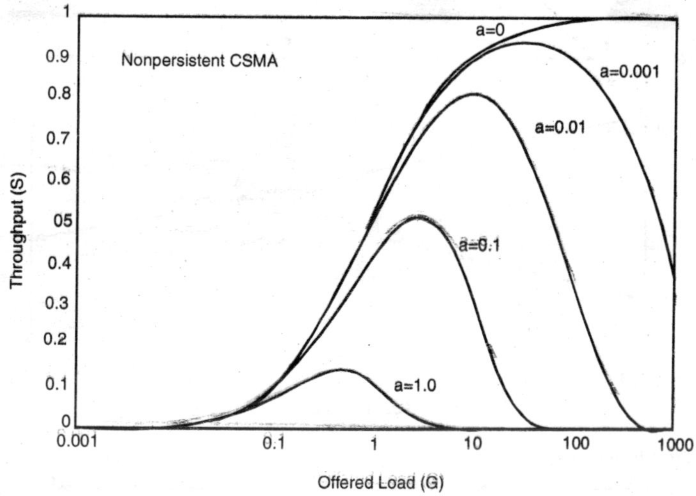

Fig. 7.37: *Throughput-Load of Nonpersistent CSMA.*

(ii) 1-Persistent Carrier Sense Multiple Access

With nonpersistent CSMA there are situations in which the channel is idle although one or more users have packets to transmit. The 1-persistent CSMA (1P-CSMA) is an alternative to nonpersistent CSMA that avoids such situations. This is achieved by applying the following rule: A user that senses the channel and finds it busy, persists to wait and transmits as soon as the channel becomes idle. Consequently, the channel is always used if there is a user with a packet.

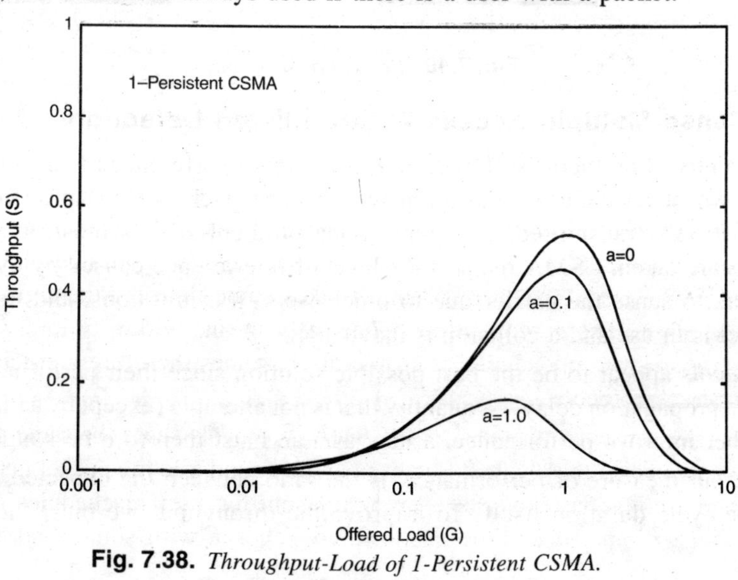

Fig. 7.38. *Throughput-Load of 1-Persistent CSMA.*

This protocol is applicable to a bus topology. Before a station can transmit, it listens to the channel to see if any other station is already transmitting. If the station finds the channel idle, it attempt to transmit; otherwise, it waits for the channel to become idle. Because of an unavoidable delay in a station's transmission to reach other stations, it is possible that two or more stations find the channel idle and simultaneously attempt to transmit. This is called a **collision.**

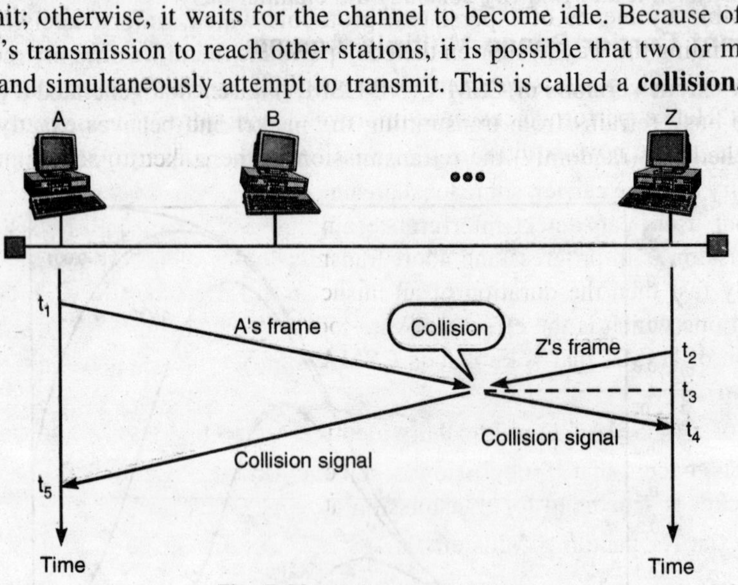

Fig. 7.39. *Collision in CSMA.*

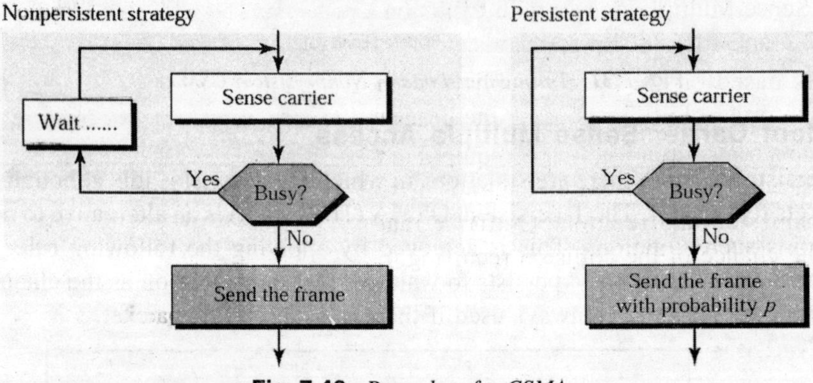

Fig. 7.40. *Procedure for CSMA.*

(iii) Carrier Sense Multiple Access With Collision Detection

The Aloha family of protocols suffers from the inherent interference of concurrently transmitted packets, that is, whenever the transmission of two or more packets overlap in time, even a bit, all are lost and must be retransmitted. The pure Aloha protocol suffers most, as no precautions to reduce collisions are taken. CSMA reduces the level of interference caused by overlapping packets by allowing users to sense the carrier due to other users' transmissions, and inhibit transmission when the channel is in use and a collision is inevitable.

CSMA protocols appear to be the best possible solution since their performance depends only on the end-to-end propagation delay—a quantity that is not alterable (except by a different topological design). To further improve performance, a new avenue must therefore be sought.

Throughput, our measure of performance, is the ratio between the expected useful times spent in a cycle to the cycle duration itself. To improve the throughput we must therefore reduce the cycle length, an observation that is the foundation of the protocols described in this section. As we have seen, a cycle is composed of a transmission period followed by an idle period. Shortening the idle period is possible by means of 1-persistent protocols which, unfortunately, perform poorly under most loads. Finding the way to shorten the busy period is therefore our only recourse. Clearly, the duration of successful transmission periods should not be changed for this is the time the channel is used best. Hence, performance can be improved by shortening the duration of unsuccessful transmission periods, as we now explain.

Beside the ability to sense carrier, some local area networks (such as Ethernet) have an additional feature, namely that users can detect interference among several transmissions (including their own) while transmission is in progress and abort transmission of their collided packets. If this can be done sufficiently fast then the duration of an unsuccessful transmission would be shorter than that of a successful one which is the effect we were looking for. Together with carrier sensing this produces a variation of CSMA that is known **as CSMA/CD (Carrier Sense Multiple Access with Collision Detection).**

The operation of all CSMA/CD protocols is identical to the operation of the corresponding CSMA protocols, except that if a collision is detected during transmission, the transmission is aborted and the packet is scheduled for transmission at some later time.

Two schemes exist for handling collisions:

Collision Detection

The Carrier Sense Multiple Access with Collision Detection (CSMA/CD) protocol is specified by the IEEE 802.3 and ISO 8802.3 standards.

CSMA/CD is based on the widely-publicized and well-adopted Ethernet specification, and offers data rates in order of 10 mbps using the baseband or the broadband technique.

In this scheme a transmitting station is required to also listen to the channel, so that it can detect a collision by observing discrepancies in the transmission voltage levels. Upon detecting a collision, it suspends transmission and re-attempts after a random period of time. Use of a random wait period reduces the chance of the collision recurring.

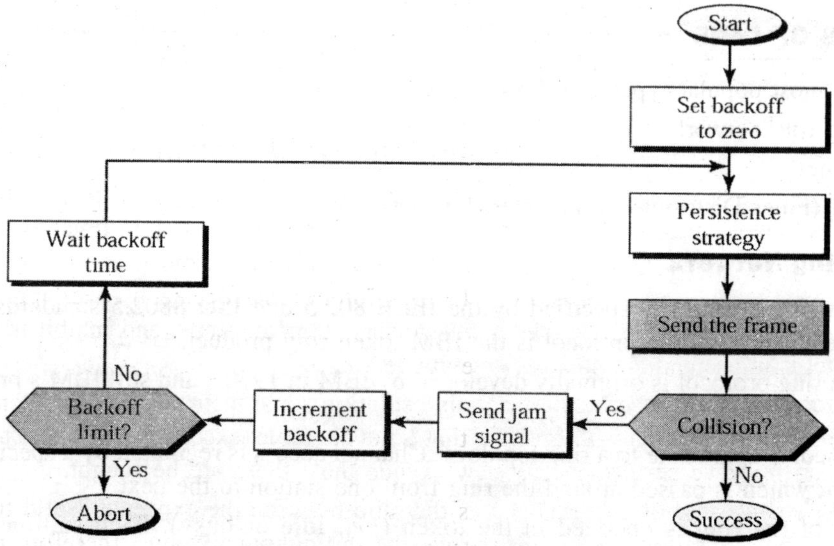

Fig. 7.41. *CSMA-CD procedure.*

Collision Free

This scheme avoids collisions occurring in the first place. Each station has a predetermined time slot assigned to it which indicates when it can transmit without a collision occurring. The distribution of time slots between stations also makes it possible to assign priorities.

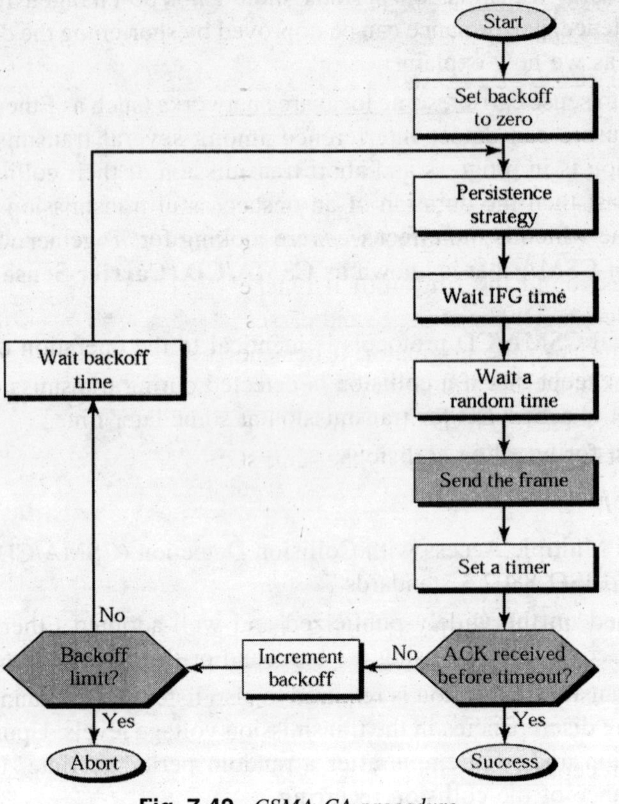

Fig. 7.42. *CSMA-CA procedure.*

7.14. TYPES OF LANS

The three most popular types of LANs are:

1. Token ring network
2. Ethernet
3. FDDI (Fiber Distributed Data Interface) network

1. Token Ring Network

The token ring protocol is specified by the IEEE 802.5 and ISO 8802.5 standards. The most well-known realization of this protocol is the IBM token ring product.

The token ring protocol is originally developed by IBM in 1970's and still IBM's primary LAN technology.

This protocol is applicable to a ring topology. Channel access is regulated by a special message, called a **token,** which is passed around the ring from one station to the next.

The state of the ring is encoded in the token (*i.e.,* idle or busy). Each station wishing to transmit needs to get hold of the idle token first.

When a station gets hold of the idle token, it marks it as busy, appends to it the message it wishes to transmit, and sends the whole thing to the next station.

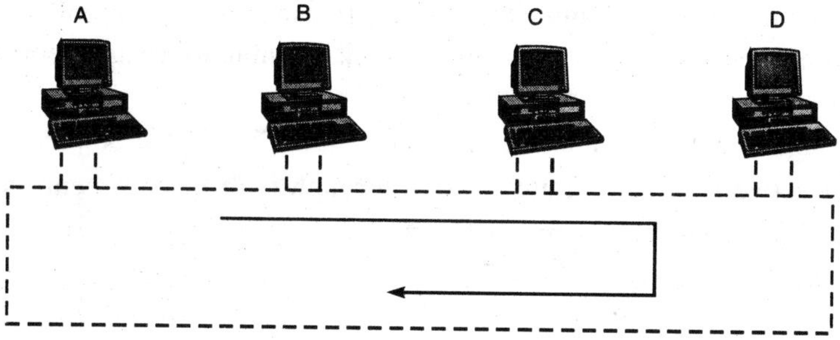

Fig. 7.43. *Token ring network.*

The message goes round the ring until it reaches the intended recipient which copies the message and passes it on. When the message returns to the originator, it detaches the message, marks the token as idle and passes it on.

To ensure fair access, the token should go round the ring, unused, at least once before it can be used by the same station again.

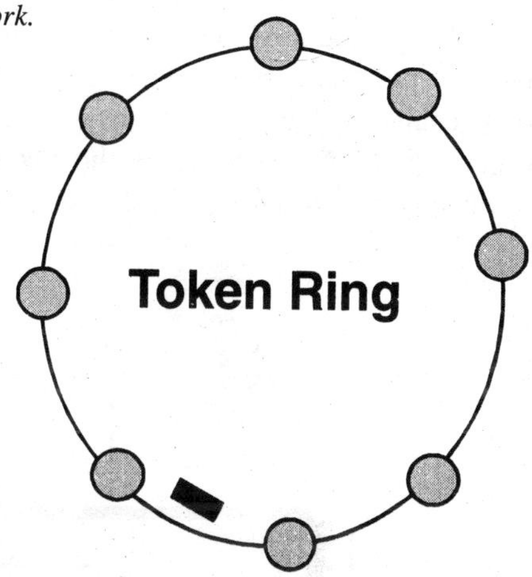

Fig. 7.44.

Release of Token

1. Release after transmission
   - Nod replaces token on ring as soon as it is done transmitting the packet
   - Next node can use token after short propagation delay
2. Release after reception

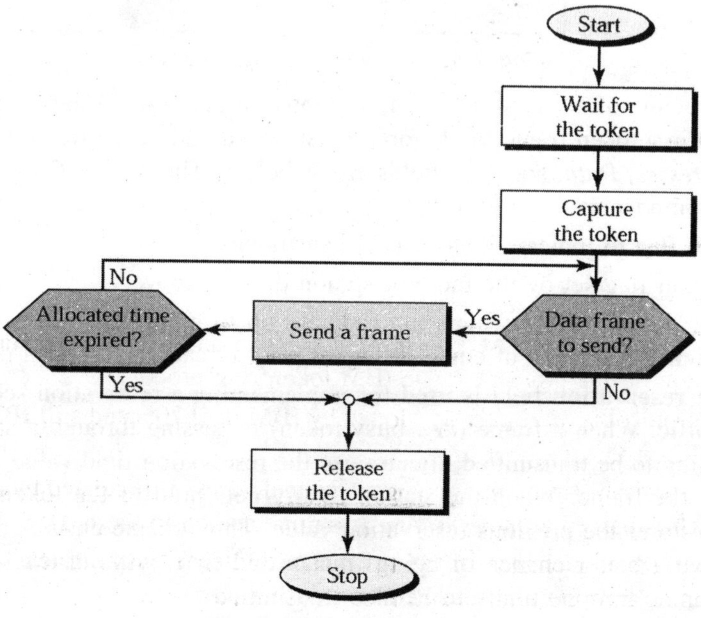

Fig. 7.45. *Token Ring in LAN.*

Node releases token only after its own packet has returned to it, serves as a simple acknowledgement mechanism.

Few important points to remember:

1. In cases of heavy traffic, the token ring network has higher throughput than Ethernet due to the deterministic (non-random) nature of the medium access.
2. Is used in applications in which delay when sending data must be predictable
3. Is a robust network *i.e.,* it is fault tolerant through fault management mechanisms
4. Can support data rates of around 16 Mbps
5. Typically uses twisted pair

Token Bus

This protocol is applicable to a bus topology but makes it behave as a ring. Each station on the bus has two other stations designated as its logical predecessor and its logical successor, in a way that results in a logical ring arrangement. A special message is provided which plays the role of a token. Each station receives the token from its predecessor, readdresses it to its successor, and retransmits it on the bus. The rest of the protocol is as in a token ring.

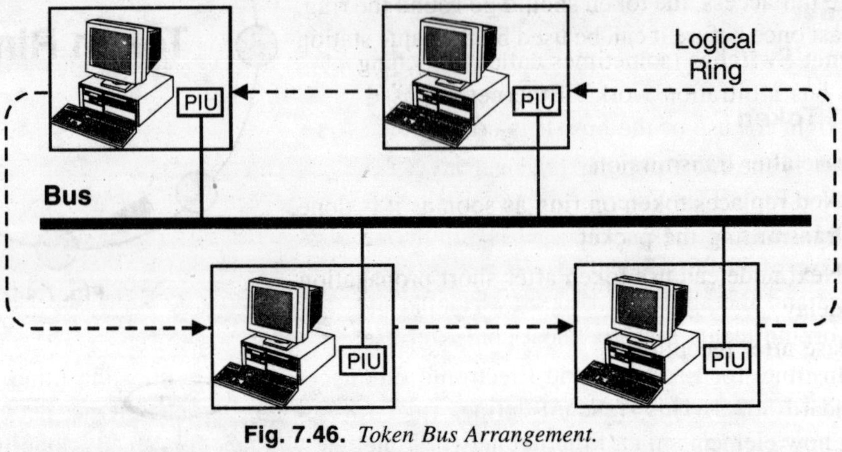

Fig. 7.46. *Token Bus Arrangement.*

The general structure of a token ring frame is shown in Fig. 7.46. The shaded fields appear in data frames but not in a token frame. As before, the start and end of the frame is delimited by two special octets. *Addresses*, *Data*, and *FCS* fields are as before. The *Access Control* field is an octet divided into four components:

(i) A **token** bit flag to indicate if this is a token frame.

(ii) A **monitor** bit flag set by the monitor station during recovery.

(iii) A three-bit **priority** field which can indicate up to eight token priority levels. Only a station which has a frame of equal or higher priority can gain control of the ring.

(iv) A three-bit **reservation** field is used for implementing a reservation scheme for handling priority traffic. When a frame (*i.e.,* busy token) is passing through a station which has a frame waiting to be transmitted, it can raise the reservation field value to the value of the priority of the frame. When the station finally gets hold of the token and transmits its frame, it restores the previous reservation value. This scheme ensures that higher priority frames have a better chance of being transmitted first, without totally denying lower-priority frames an opportunity to be also transmitted.

The *Frame Control* field specifies the frame type (*i.e.*, whether it is an LLC PDU or some kind of MAC control frame). The *Frame Status* field is used by the receiving station to indicate that it has successfully received the frame and extracted its data, so that when the frame reaches the originator, it can be disposed of.

Field	Description
Start Delimiter	Marks the beginning of frame
Access Control	Various access-related flags and values.
Frame Control	Denotes the frame type.
Addresses	Source and destination addresses
Data	Actual user data (e.g., LLC PDU)
FCS	Frame check sequence.
End Delimiter	Marks the end of frame.
Frame Status	Used by receiving station to record receipt status.

Fig. 7.47. *Token ring MAC frame structure.*

2. Ethernet

Ethernet Switches (sometimes called switching hubs) are devices that have changed the way Ethernet's bus arbitration works. Ethernet is based on CSMA/CD, a shared bus where all nodes listen and fight for use of the bus. It is a first come first serve protocol.

Ethernet is the most popular physical layer LAN technology in use today. Other LAN types include Token Ring, Fast Ethernet, Fiber Distributed Data Interface (FDDI), Asynchronous Transfer Mode (ATM) and Local Talk. Ethernet is popular because it strikes a good balance between speed, cost and ease of installation. These benefits, combined with wide acceptance in the computer marketplace and the ability to support virtually all popular network protocols, make Ethernet an ideal networking technology for most computer users today.

The Institute for Electrical and Electronic Engineers (IEEE) defines the Ethernet standard as IEEE Standard 802.3. This standard defines rules for configuring an Ethernet network as well as specifying how elements in an Ethernet network interact with one another. By adhering to the IEEE standard, network equipment and network protocols can communicate efficiently.

Fast Ethernet

For Ethernet networks that need higher transmission speeds, the Fast Ethernet standard (IEEE 802.3u) has been established. This standard raises the Ethernet speed limit from 10 Mbps to 100 Mbps with only minimal changes to the existing cable structure.

There are three types of Fast Ethernet:

1. 100BASE-TX for use with level 5 UTP cable,
2. 100BASE-FX for use with fiber-optic cable, and
3. 100BASE-T4 which utilizes an extra two wires for use with level 3 UTP cable.

The 100BASE-TX standard has become the most popular due to its close compatibility with the 10BASE-T Ethernet standard. For the network administrator, the incorporation of Fast Ethernet into an existing configuration presents a host of decisions. Network administrator must determine the number of users in each site on the network that need the higher throughput, decide which segments of the backbone need to be reconfigured specifically for 100BASE-T and then choose the necessary hardware to connect the 100BASE-T segments with existing 10BASE-T segments.

Gigabit Ethernet is a future technology that promises a migration path beyond Fast Ethernet so the next generation of networks will support even higher data transfer speeds.

Ethernet switches are based on circuit switching technology. When a node on a port wants to talk to another node on a separate port, a private circuit is made between the two ports. No one else is allowed to use the connection. It is not a shared line anymore. The complete bandwidth is available for the conversation.

In addition most if not all Ethernet switches are full duplex which means that they can transmit data and receive data at the same time. This effectively doubles the bandwidth between the two ports. Modern Ethernet switches run at 100 Mbps and 1 Gbps full duplex.

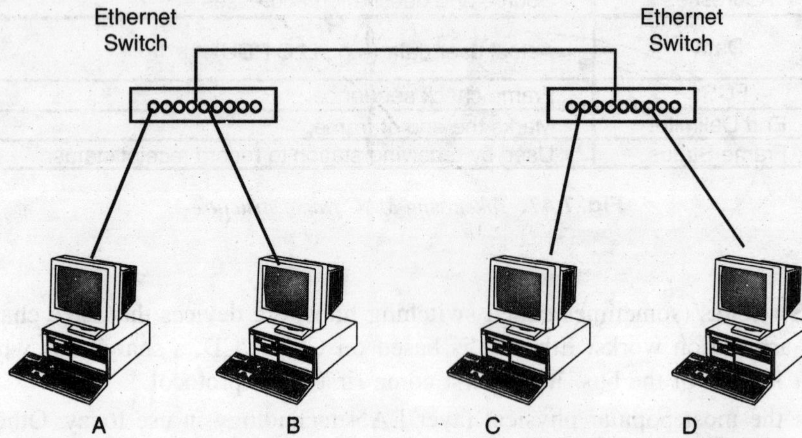

Fig. 7.48.

If suppose device-A wanted to communicate with device-B, a dedicated 100 Mbps connection would be established between the two. If device-C wanted to communicate with device-D, another dedicated 100 Mbps connection would be established.

The Ethernet switch examines the MAC address of the Ethernet frame to determine which port to send the frame. They are similar to bridges in a many respects because they work at the Data Link layer and operate similarly to bridges. Ethernet switches which work only at the Data Link layer is called **layer 2 switches**.

New generations of Ethernet switches examine frame information at the Network layer and above layers. They can make decisions based on the Network layer information which used to be the domain of routers which is covered later. Ethernet switches which can make decisions at the Network layer are called **layer 3 switches**.

3. Fiber Distributed Data Interface (FDDI)

FDDI (Fiber-Distributed Data Interface) is a standard for data transmission on fiber optic lines in a local area network that can extend in range up to 200 km (124 miles). The FDDI protocol is based on the token ring protocol. In addition to being large geographically, an FDDI local area network can support thousands of users.

The Fiber Distributed Data Interface (FDDI) specifies a 100-Mbps token-passing, dual-ring LAN using fiber-optic cable. FDDI is frequently used as high-speed backbone technology because of its support for high bandwidth and greater distances than copper. It should be noted that relatively recently, a related copper specification, called Copper Distributed Data Interface (CDDI) has emerged to provide 100-Mbps service over copper. CDDI is the implementation of FDDI protocols over twisted-pair copper wire. This section focuses mainly on FDDI specifications and operations, but it also provides a high-level overview of CDDI.

FDDI uses dual-ring architecture with traffic on each ring flowing in opposite directions (called *counter-rotating*). The dual-rings consist of a primary and a secondary ring. During normal operation, the primary ring is used for data transmission, and the secondary ring remains idle. The primary purpose of the dual rings, as will be discussed in detail later in this chapter, is to provide superior reliability and robustness. Figure 7.49 shows the counter-rotating primary and secondary FDDI rings.

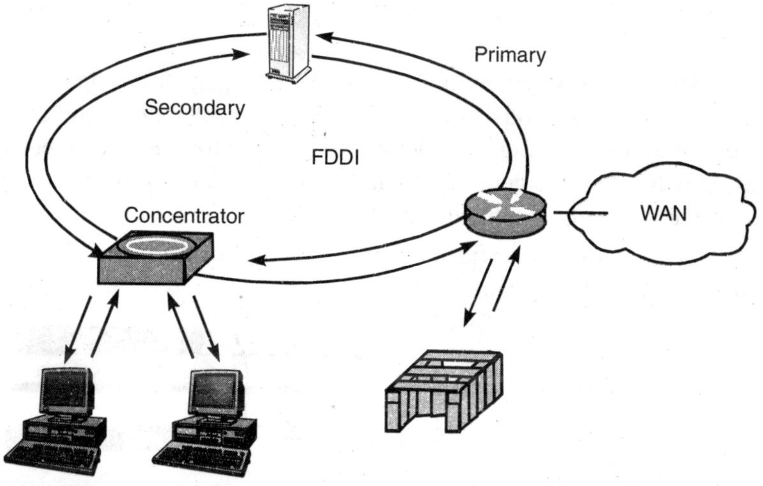

Fig. 7.49. *FDDI uses counter-rotating primary and secondary rings.*

Standards

FDDI was developed by the American National Standards Institute (ANSI) X3T9.5 standards committee in the mid-1980s. At the time, high-speed engineering workstations were beginning to tax the bandwidth of existing local area networks (LANs) based on Ethernet and Token Ring). A new LAN media was needed that could easily support these workstations and their new distributed applications. At the same time, network reliability had become an increasingly important issue as system managers migrated mission-critical applications from large computers to networks. FDDI was developed to fill these needs. After completing the FDDI specification, ANSI submitted FDDI to the International Organization for Standardization (ISO), which created an international version of FDDI that is completely compatible with the ANSI standard version.

FDDI Transmission Media

FDDI uses optical fiber as the primary transmission medium, but it also can run over copper cabling. As mentioned earlier, FDDI over copper is referred to as *Copper-Distributed Data Interface* (CDDI). Optical fiber has several advantages over copper media. In particular, security, reliability, and performance all are enhanced with optical fiber media because fiber does not emit electrical signals. A physical medium that does emit electrical signals (copper) can be tapped and therefore would permit unauthorized access to the data that is transiting the medium. In addition, fiber is immune to electrical interference from radio frequency interference (RFI) and electromagnetic interference (EMI). Fiber historically has supported much higher bandwidth (throughput potential) than copper, although recent technological advances have made copper capable of transmitting at

100 Mbps. Finally, FDDI allows two kilometers between stations using multi-mode fiber, and even longer distances using a single mode.

FDDI defines two types of optical fiber: single-mode and multi-mode. A *mode* is a ray of light that enters the fiber at a particular angle. Multi-mode fiber uses LED as the light-generating devices, while single-mode fiber generally uses lasers.

Multi-mode fiber allows multiple modes of light to propagate through the fiber. Because these modes of light enter the fiber at different angles, they will arrive at the end of the fiber at different times. This characteristic is known as *modal dispersion*. Modal dispersion limits the bandwidth and distances that can be accomplished using multi-mode fibers. For this reason, multi-mode fiber is generally used for connectivity within a building or within a relatively geographically contained environment.

Single-mode fiber allows only one mode of light to propagate through the fiber. Because only a single mode of light is used, modal dispersion is not present with single-mode fiber. Therefore, single-mode is capable of delivering considerably higher performance connectivity and over much larger distances, which is why it generally is used for connectivity between buildings and within environments that are more geographically dispersed.

Figure 7.50 depicts single-mode fiber using a laser light source and multi-mode fiber using a light-emitting diode (LED) light source.

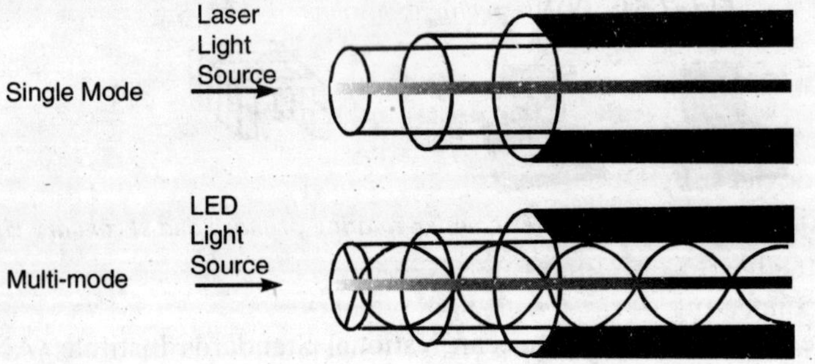

Fig. 7.50. *Light sources differ for single-mode and multi-mode fibers.*

FDDI Specifications

FDDI specifies the physical and media-access portions of the OSI reference model. FDDI is not actually a single specification, but it is a collection of four separate specifications each with a specific function. Combined, these specifications have the capability to provide high-speed connectivity between upper-layer protocols such as TCP/IP and IPX, and media such as fiber-optic cabling.

FDDI's four specifications are the Media Access Control (MAC), Physical Layer Protocol (PHY), Physical-Medium Dependent (PMD), and Station Management (SMT). The MAC specification defines how the medium is accessed, including frame format, token handling, addressing, algorithms for calculating cyclic redundancy check (CRC) value, and error-recovery mechanisms. The PHY specification defines data encoding/decoding procedures, clocking requirements, and framing, among other functions. The PMD specification defines the characteristics of the transmission medium, including fiber-optic links, power levels, bit-error rates, optical components, and connectors. The SMT specification defines FDDI station configuration, ring configuration, and ring control features, including station insertion and removal, initialization, fault isolation and recovery, scheduling, and statistics collection.

FDDI is similar to IEEE 802.3 Ethernet and IEEE 802.5 Token Ring in its relationship with the OSI model. Its primary purpose is to provide connectivity between upper OSI layers of common protocols and the media used to connect network devices. Figure 7.51 illustrates the four FDDI specifications and their relationship to each other and to the IEEE-defined Logical-Link Control (LLC) sublayer. The LLC sublayer is a component of Layer 2, the MAC layer, of the OSI reference model.

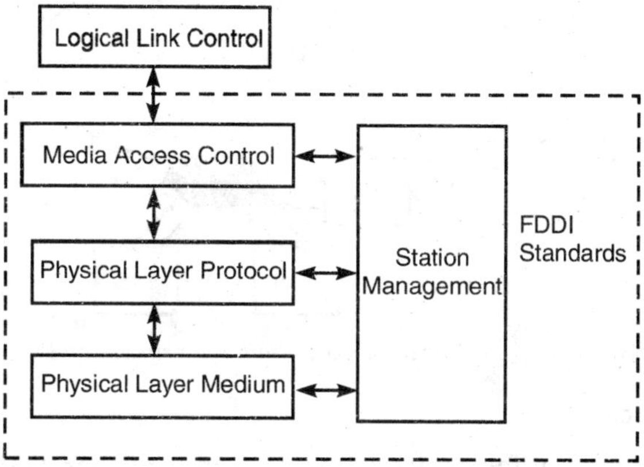

Fig. 7.51. *FDDI specifications map to the OSI hierarchical model.*

FDDI Station-Attachment Types

One of the unique characteristics of FDDI is that multiple ways actually exist by which to connect FDDI devices. FDDI defines three types of devices: single-attachment station (SAS), dual-attachment station (DAS), and a concentrator.

An SAS attaches to only one ring (the primary) through a concentrator. One of the primary advantages of connecting devices with SAS attachments is that the devices will not have any effect on the FDDI ring if they are disconnected or powered off. Concentrators will be discussed in more detail in the following discussion.

Each FDDI DAS has two ports, designated A and B. These ports connect the DAS to the dual FDDI ring. Therefore, each port provides a connection for both the primary and the secondary ring. As you will see in the next section, devices using DAS connections will affect the ring if they are disconnected or powered off. Figure 7.52 shows FDDI DAS A and B ports with attachments to the primary and secondary rings.

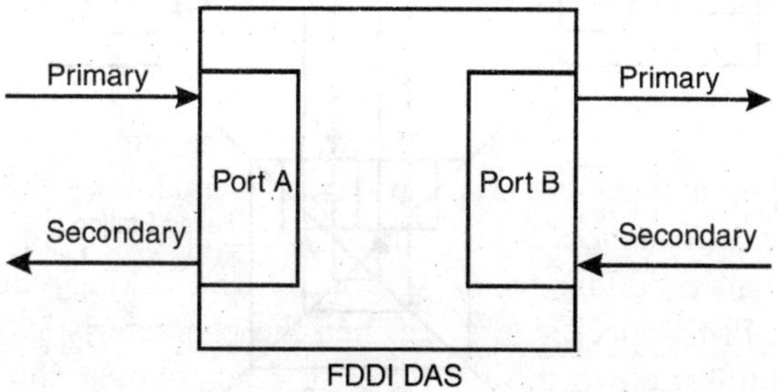

Fig. 7.52. *FDDI DAS ports attach to the primary and secondary rings.*

An FDDI concentrator (also called a *dual-attachment concentrator* [DAC]) is the building block of an FDDI network. It attaches directly to both the primary and secondary rings and ensures that the failure or power-down of any SAS does not bring down the ring. This is particularly useful when PCs, or similar devices that are frequently powered on and off, connect to the ring. Figure 7.53 shows the ring attachments of an FDDI SAS, DAS, and concentrator.

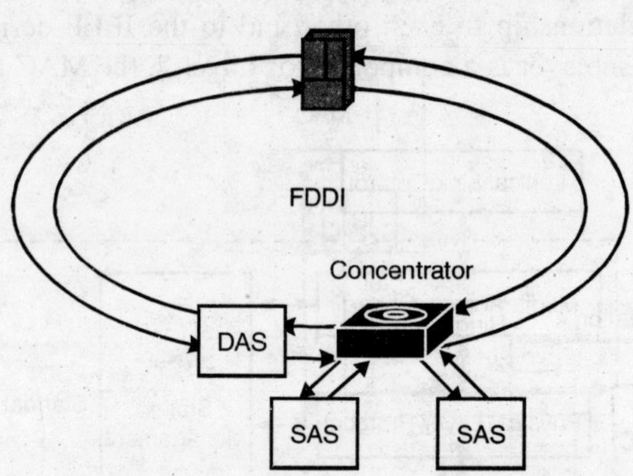

Fig. 7.53. *A concentrator attaches to both the primary and secondary rings.*

FDDI Fault Tolerance

FDDI provides a number of fault-tolerant features. In particular, FDDI's dual-ring environment, the implementation of the optical bypass switch, and dual-homing support make FDDI a resilient media technology.

Dual Ring

FDDI's primary fault-tolerant feature is the dual ring. If a station on the dual ring fails or is

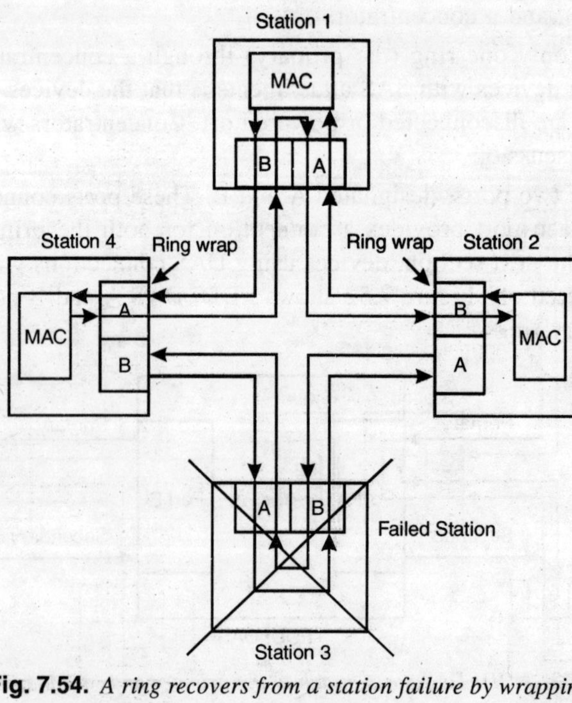

Fig. 7.54. *A ring recovers from a station failure by wrapping.*

powered down, or if the cable is damaged, the dual ring is automatically *wrapped* (doubled back onto itself) into a single ring. When the ring is wrapped, the dual-ring topology becomes a single-ring topology. Data continues to be transmitted on the FDDI ring without performance impact during the wrap condition. Figure 7.54 and Figure 7.55 illustrate the effect of a ring wrapping in FDDI.

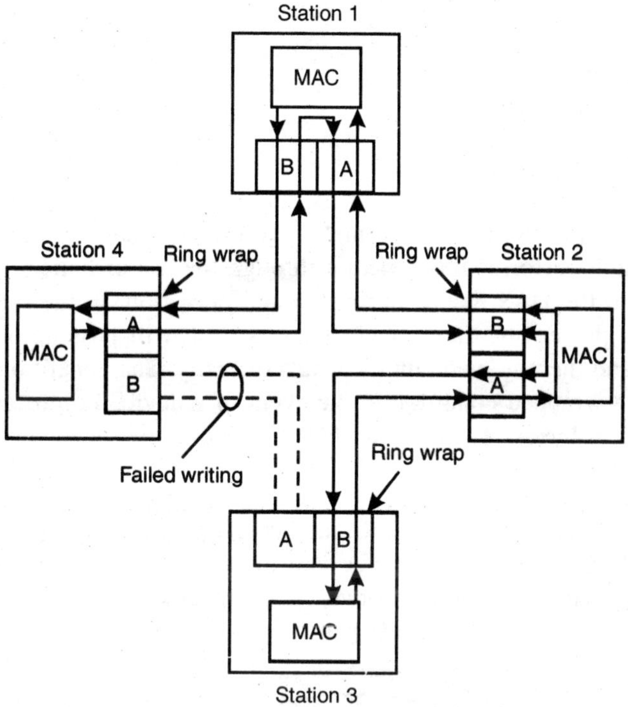

Fig. 7.55. *A ring also wraps to withstand a cable failure.*

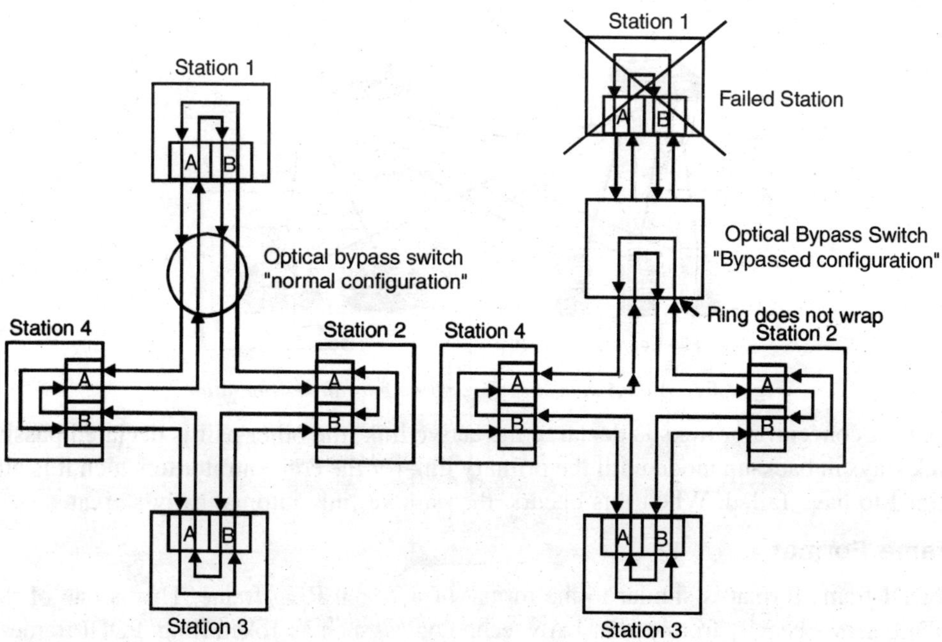

Fig. 7.56. *The optical bypass switch uses internal mirrors to maintain a network.*

When a single station fails, as shown in Figure 7.55, devices on either side of the failed (or powered down) station wrap, forming a single ring. Network operation continues for the remaining stations on the ring. When a cable failure occurs, as shown in Figure 7.55, devices on either side of the cable fault wrap. Network operation continues for all stations.

It should be noted that FDDI truly provides fault-tolerance against a single failure only. When two or more failures occur, the FDDI ring segments into two or more independent rings that are unable to communicate with each other.

Optical Bypass Switch

An optical bypass switch provides continuous dual-ring operation if a device on the dual ring fails. This is used both to prevent ring segmentation and to eliminate failed stations from the ring. The optical bypass switch performs this function through the use of optical mirrors that pass light from the ring directly to the DAS device during normal operation. In the event of a failure of the DAS device, such as a power-off, the optical bypass switch will pass the light through itself by using internal mirrors and thereby maintain the ring's integrity. The benefit of this capability is that the ring will not enter a wrapped condition in the event of a device failure. Figure 7.56 shows the functionality of an optical bypass switch in an FDDI network.

Dual Homing

Critical devices, such as routers or mainframe hosts, can use a fault-tolerant technique called *dual homing* to provide additional redundancy and to help guarantee operation. In dual-homing situations, the critical device is attached to two concentrators. Figure 7.57 shows a dual-homed configuration for devices such as file servers and routers.

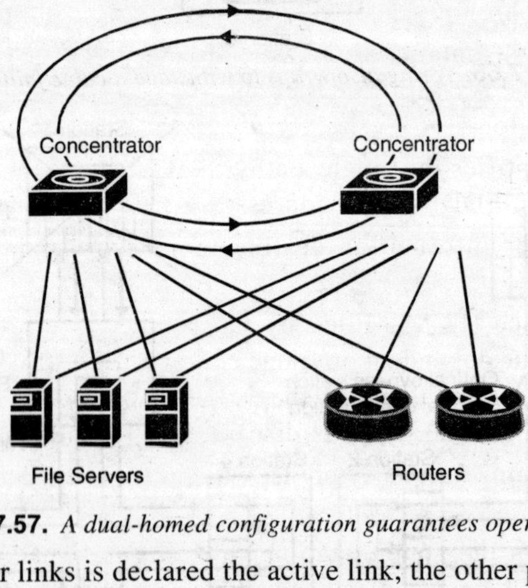

Fig. 7.57. *A dual-homed configuration guarantees operation.*

One pair of concentrator links is declared the active link; the other pair is declared passive. The passive link stays in back-up mode until the primary link (or the concentrator to which it is attached) is determined to have failed. When this occurs, the passive link automatically activates.

FDDI Frame Format

The FDDI frame format is similar to the format of a Token Ring frame. This is one of the areas where FDDI borrows heavily from earlier LAN technologies, such as Token Ring. FDDI frames can be as large as 4,500 bytes. Figure 7.58 shows the frame format of an FDDI data frame and token.

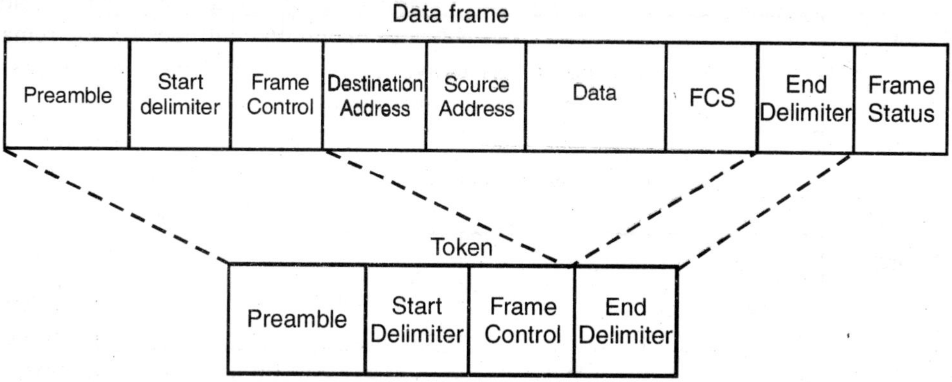

Fig. 7.58. *The FDDI frame is similar to that of a Token Ring frame.*

FDDI Frame Fields

The following descriptions summarize the FDDI data frame and token fields illustrated in Fig. 7.59.

(*i*) Preamble—A unique sequence that prepares each station for an upcoming frame.

(*ii*) Start Delimiter—Indicates the beginning of a frame by employing a signaling pattern that differentiates it from the rest of the frame.

(*iii*) Frame Control—Indicates the size of the address fields and whether the frame contains asynchronous or synchronous data, among other control information.

(*iv*) Destination Address—Contains a unicast (singular), multicast (group), or broadcast (every station) address. As with Ethernet and Token Ring addresses, FDDI destination addresses are 6 bytes long.

(*v*) Source Address—Identifies the single station that sent the frame. As with Ethernet and Token Ring addresses, FDDI source addresses are 6 bytes long.

(*vi*) Data—Contains either information destined for an upper-layer protocol or control information.

(*vii*) Frame Check Sequence (FCS)—Filed by the source station with a calculated cyclic redundancy check value dependent on frame contents (as with Token Ring and Ethernet). The destination address recalculates the value to determine whether the frame was damaged in transit. If so, the frame is discarded.

(*viii*) End Delimiter—Contains unique symbols, which cannot be data symbols, that indicate the end of the frame

(*ix*) Frame Status—Allows the source station to determine whether an error occurred and whether the frame was recognized and copied by a receiving station.

Copper Distributed Data Interface (CDDI)

Copper Distributed Data Interface (CDDI) is the implementation of FDDI protocols over twisted-pair copper wire. Like FDDI, CDDI provides data rates of 100 Mbps and uses a dual-ring architecture to provide redundancy. CDDI supports distances of about 100 meters from desktop to concentrator.

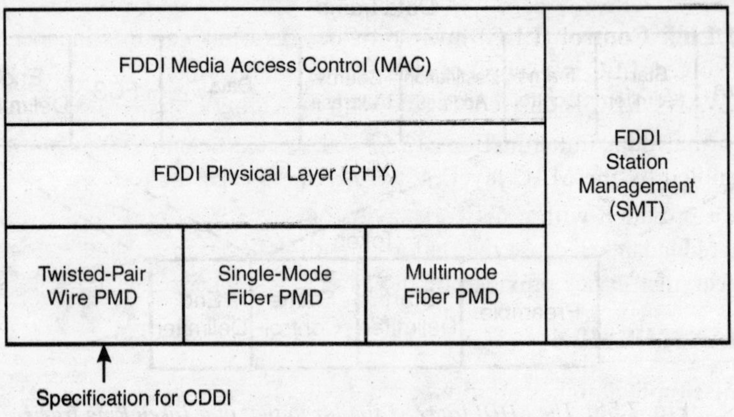

Fig. 7.59. *CDDI TP-PMD and FDDI specifications adhere to different standards.*

CDDI is defined by the ANSI X3T9.5 Committee. The CDDI standard is officially named the *Twisted-Pair Physical Medium Dependent* (TP-PMD) standard. It is also referred to as the *Twisted-Pair Distributed Data Interface* (TP-DDI), consistent with the term *Fiber-Distributed Data Interface* (FDDI). CDDI is consistent with the physical and media-access control layers defined by the ANSI standard.

The ANSI standard recognizes only two types of cables for CDDI: shielded twisted pair (STP) and unshielded twisted pair (UTP). STP cabling has a 150-ohm impedance and adheres to EIA/TIA 568 (IBM Type 1) specifications. UTP is data-grade cabling (Category 5) consisting of four unshielded pairs using tight-pair twists and specially developed insulating polymers in plastic jackets adhering to EIA/TIA 568B specifications.

Figure 7.59 illustrates the CDDI TP-PMD specification in relation to the remaining FDDI specifications.

7.15. LAN ARCHITECTURE

Figure 7.60 depicts the LAN protocol layers in relation to the OSI model. The role of the physical layer is the same as in the OSI model. It includes the connectors used for connecting the PIU to the LAN and the signaling circuitry provided by the PIU.

The OSI data link layer is broken into two sublayers. The **Media Access Control (MAC)** layer is responsible for implementing a specific LAN access protocol, like the ones described earlier.

This layer is therefore highly dependent on the type of the LAN. Its aim is to hide hardware and access protocol dependencies from the next layer.

As we will see shortly, a number of MAC standards have been devised, one for each popular type of access protocol.

OSI Layer	LAN Layer	Purpose
Higher layers	Undefined	Application dependent
Data Link	Logical Link control	Provides generic data link services to higher layers.
	Media Access control	Implements the protocol for accessing the LAN.
Physical	Physical	Transmission of data bits over the channel.

Fig. 7.60. *LAN protocol architecture.*

7.16. LOGICAL LINK CONTROL (LLC)

The **Logical Link Control (LLC) layer** provides data link services independent of the specific MAC protocol involved. LLC is a subset of HDLC and is largely compatible with the data link layer of OSI-compatible WANs. LLC is only concerned with providing Link Service Access Points (LSAPs).

All other normal data link functions (*i.e.,* link management, frame management, and error handling) are handled by the MAC layer.

LANs are not provided with a network layer (or any other higher layer) because such a layer would be largely redundant. Because the stations are directly connected, there is no need for switching or routing. In effect, the service provided by the LLC is equivalent to the OSI network layer service.

7.17. LAN TRANSMISSION

LAN transmission techniques are divided into two categories:

(*i*) Baseband (*ii*) Broadband.

In the **baseband technique,** the digital signal from a transmitting device is directly introduced into the transmission medium (possibly after some conditioning).

In the **broadband technique,** a modem is used to transform the digital signal from a transmitting device into a high frequency analog signal. This signal is typically frequency multiplexed to provide multiple FDM channels over the same transmission medium.

7.18. COMPARISON OF BASEBAND AND BROADBAND

Baseband is a simple and inexpensive digital technique. By comparison, broadband has additional costs: each device requires its own modem; also, because transmission is possible in one direction only, two channels typically need to be provided, one for either direction.

Broadband, however, has the advantages of offering a higher channel capacity which can be used for multiplexing data from a variety of sources (*e.g.,* video, voice, fax), not just digital data. It is also capable of covering longer distances, typically tens of kilometers compared to up to a kilometer for baseband.

7.19. IEEE 802 STANDARDS

The IEEE 802 series of recommendations provide a widely-accepted set of LAN standards. These recommendations are formulated by nine subcommittees.

Table 7.2. IEEE 802 subcommittees

Subcommittee	Title	Purpose
802.1	High-level interface	Specification of standards for LAN architecture, interconnection, management
802.2	Logical Link Control	Specification of standards for the LLC layer
802.3	CSMA/CD	Specification of standards for CSMA/CD architectures
802.4	Token Bus	Specification of standards for token bus architectures
802.5	Token Ring	Specification of standards for token ring architectures
802.6	Metropolitan Area Networks	Specification of standards for MANs
802.7	Broadband Technical Advisory group	Provision of guidance to other groups working on broadband LANs
802.8	Fiber Optical Technical Advisory Group	Provision of guidance to other groups working on fiber optic-based LANs
802.9	Integrated Data and Voice Networks	Specification of standards for interfaces to ISDN

Table 7.3. Main IEEE LAN Recommendations.

<table>
<tr><td colspan="3">LLC
(IEEE 802.2 / ISO 8802.2)</td></tr>
<tr><td>CSMA/CD
(IEEE 802.3 / ISO 8802.3)</td><td>Token Bus
(IEEE 802.4 / ISO 8802.4)</td><td>Token Ring
(IEEE 802.5 / ISO 8802.5)</td></tr>
<tr><td colspan="3">Physical</td></tr>
</table>

7.20. ADVANTAGES AND DISADVANTAGES OF LOCAL AREA NETWORKS

Local area networks have several advantages,

1. Hardware and software sharing,
2. Workstation survival during network failure,
3. Component and system evolution,
4. Heterogeneous mix of hardware and software, and
5. Access to other LANs, WANs, and mainframe computers.

Disadvantages include

1. Complexity,
2. Maintenance costs, and
3. The network is only as strong as the weakest link.

7.21. RFID

7.21.1. Introduction

RFID, short for **R**adio **F**requency **ID**entification, is a technology that enables identification of a tag (that is normally attached with any physical item) by using electromagnetic waves. The function served by RFID is similar to bar code identification, but line of sight signals are not required for operation of RFID.

Important components of an RFID system are:

(*i*) An RFID reader (also called transceiver) with an antenna and a transceiver,

(*ii*) A transponder (Also called a tag) that includes an antenna and a chip).

Radio Frequency Identification technologies provide a wireless means of communication between objects and readers. Radio frequency identification (RFID) involves the use of tags, or transponders, that collect data and manage it in a portable, changeable database. Unlike bar codes, RFID has the ability to identify and track products and equipment in real-time without contact or line-of-sight and the tags can withstand harsh, rugged environments.

Radio frequency identification (RFID) is a general term that is used to describe a system that transmits the identity (in the form of a unique serial number) of an object wirelessly, using radio waves. RFID technologies are grouped under the more generic Automatic Identification (Auto ID) technologies.

The barcode labels that triggered a revolution in identification systems long time ago are inadequate in an increasing number of cases. They are cheap but the stumbling block is their low storage capacity and the fact that they cannot be reprogrammed.

A feasible solution was putting the data on silicon chips. The ideal situation is contactless transfer of data between the data carrying device and its reader. The power required to operate the electronic data carrying device would also be transferred from the reader using contactless technology. These procedures give RFID its name.

7.21.2. RFID Standards

RFID standards deal with the following :

- **Air Interface Protocol**: The way tags and readers communicate.
- **Data Content**: Organizing of data.
- **Conformance**: Tests that products meet the standard.
- **Applications**: How applications are used.

There are two major and somewhat conflicting organizations into the business—ISO and Auto-ID Centre (now handled by EPC Global). We will review the standards proposed by both these organizations.

Tags are required to be disposable (manufacturer may not get the tags back from the retailer to reuse it). Hence, the primary mission for any standard developer is to make the tags low cost. It should operate in UHF, as only UHF delivers read range needed for supply chain applications. And since the goods are needed to be tracked as they move across the globe, the standards must be open and globally accepted. There should also be an accompanying network architecture, which would enable anyone to look up information associated with a serial number stored on a tag. The network too needs to be based on open standards.

As per EPC standards there are two classes of the tags, the class 0 and class 1 tags:

- **Class 1**: A simple, passive, read-only backscatter tag with one-time, field-programmable non-volatile memory.
- **Class 0**: read-only tag that was programmed at the time the microchip was made.

Class 1 and Class 0 have a couple of drawbacks, in addition to the fact that they are not interoperable. One issue is that they are incompatible with ISO standards. The new EPC standard ~V Gen2 is designed to be fast tracked with ISO standards but for some disagreements over the 8 bit Application Family Identifier (AFI).

ISO has developed RFID standards for automatic identification and item management. This standard, known as the ISO 18000 series, covers the air interface protocol for systems likely to be used to track goods in the supply chain. They cover the major frequencies used in RFID systems around the world.

The seven parts are:

1. 18000~V1: Generic parameters for air interfaces for globally accepted frequencies
2. 18000~V2: Air interface for 135 KHz
3. 18000~V3: Air interface for 13.56 MHz
4. 18000~V4: Air interface for 2.45 GHz
5. 18000~V5: Air interface for 5.8 GHz
6. 18000~V6: Air interface for 860 MHz to 930 MHz
7. 18000~V7: Air interface at 433.92 MHz

It is not so far in future, RFID enabled stores will monitor the consumption in real time. Shelf will signal the inventory when it needs more stuff and inventory will pull supplies from the manufacturer based on its level of stock.

7.21.3. Compare RFID with Bar Code Technology

Given below are the brief differences between the Barcode technology and RFID:

Table 7.4. Compare RFID with Bar Code Technology

S. No.	Parameter	RFID	Barcode
1	Frequencies used for tag reading	Optical frequencies	Radio frequencies
2	Type of communication	Line of sight communication	Non-Line of sight communication
3	Range of data readability	Very limited range, less than feet or two.	Can be read up to several feet.
4	Data Volume	Physical limitation exists. It is very difficult to read a very long barcode.	Can carry relatively large volume of data.
5	Cost	Cheap	Expensive, but likely to cost less as more industries adopt the technology.

7.21.4. Advantages and Limitations of RFID

The following are the **Advantages of RFID** Systems:

(*i*) Ability to identify moving elements that have tags embedded.

(*ii*) Larger area of coverage. Up to several feet.

(*iii*) Automatic integration with back end software solutions provide end to end integration of data in real time.

(*iv*) Non-line of sight identification of tags.

(*v*) Unattended operations are possible, minimizing human errors and high cost.

(*vi*) Can be used in diverse environments, including live stock, military, and scientific areas.

(*vii*) RFID can be used in addition to Bar Code. These two technologies can be complementing each other.

The following are the **Limitations of RFID** Systems:

(*i*) Expensive with barcode technique.

(*ii*) It is prone to physical/electrical damage due to environmental conditions. Tags that are subjected to space exploration may encounter extreme temperatures. The tags are required to be designed for a given application, and may be costly when designed for use under extreme environmental conditions.

(*iii*) RFID are bulky due to embedding of electronics components in the tag. However, with advanced techniques, it is possible to reduce the size, and weight of the tags to a large extent.

7.21.5. Components of RFID

An RFID system is composed of *tags*, which carry the data in suitable transponders, and an RFID *reader*, which retrieves the data from the tags (CII, 2002). Products that contain RFID tags embedded in them or fastened to them enable stored information to be transferred from an RFID tag to a remote reader through radio frequency waves of a specific wavelength. There currently is not a definite industry standard for wavelength, but the most common applications around the world use wavelengths of 125 kHz and 13.56 MHz.

Initially, data is written to the RFID tag enabling it to identify and characterize a product as a particular manufactured good with a determined application. At some later point, a RFID remote reader will scan and acknowledge the information once the tag is within range of an electromagnetic field activating the tag to perform a user defined function. Some of the components may be combined into one hand-held unit to allow user mobility. Also, many passive RFID transponders have antennas sealed with the tags to give them greater read-write abilities (see Figure 7.61).

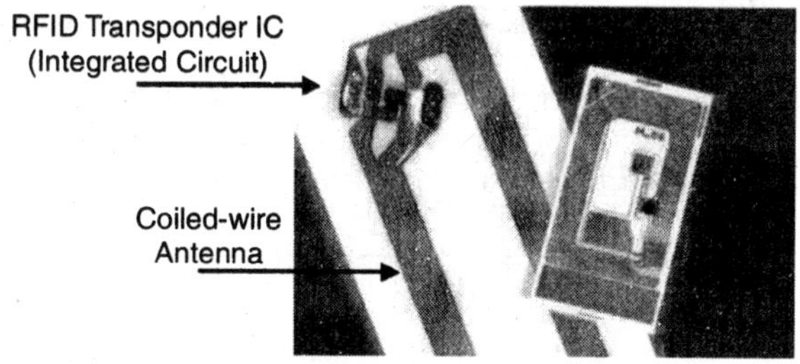

Fig. 7.61. *Antenna sealed with RFID tag*

RFID Tags

There are two classifications of RFID tags:

1. Passive
2. Active.

The means in which they receive power for transmission determines their classification. Passive tags depend on a power source provided by the RFID reader's energy field and may have read-write or read-only capabilities, whereas, the active tags have an internal power source and are rewritable.

Passive tags generally have shorter read ranges but have a life that usually outlasts the object that it is identifying. Active tags have longer reading ranges, high memory, and better noise protection. However, these tags are larger and heavier, more expensive, and have a shorter life (3 – 10 years) than passive tags. Read-only tags are used for simple identification purposes because they can only store a limited amount of information that cannot be altered. Such tags may be used to identify a package of nails or screws because they have many applications and are not designated to a particular item or activity.

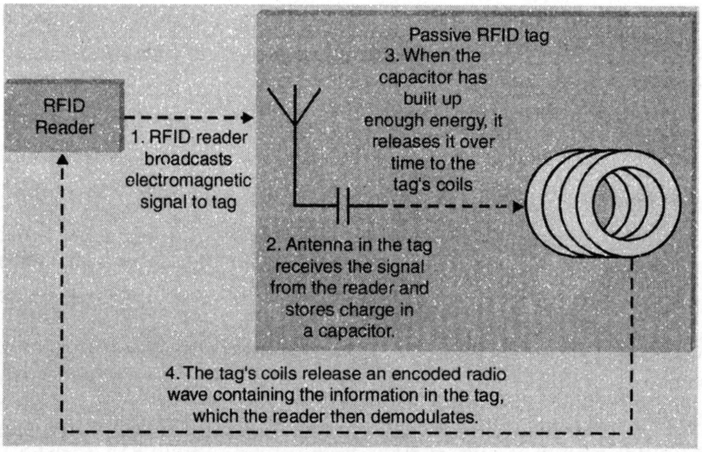

Fig. 7.62. *Passive RFID Principle*

Fig. 7.63. *An RF Tag*

Given below are the primary differences between a Passive and Active RFID tags:

Table 7.5. The primary differences between a Passive and Active RFID tags.

S. No.	Parameter	Active RFID	Passive RFID
1.	Power Source	Internal (Battery)	External (Reader provided)
2.	Energization	An active tag is always energized.	A passive tag is energized only when there is a reader present.
3.	Tag Readability	Can provide signals over an extended range, typically up to 100 meters	Only within the area covered by the reader, typically up to 3 meters.
4.	Magnetic Field Strength	Low, since the tag emits signals using internal battery source	High, since the tag draws power from the electromagnetic field provided by the reader.
5.	Shelf Life	Limited to about 5 years, the life of a battery.	Very high, ideally does not expire over a life time.
6.	Data storage	Can store larger amounts of data.	Limited data storage, typically 128 bytes.

Antenna

The function of the antenna attached to a reader is to transmit an electromagnetic field that activates a passive tag when it is within reading range. Once a passive tag is activated it can transmit information from its antenna to that of the reader where it is processed. During rewriting applications the antenna of the reader acts as a relay device in the reverse direction, the reader communicates a message through its antenna, which transfers and stores the new data to the activated transducer via its antenna. The RFID tag's antenna is practically maintenance free and can be configured in a variety of shapes and sizes ranging in size from a grain of rice to the size of a brick (Zebra Technologies, 2002). However, it is very common to see transponders and tag antennas packaged as smart labels (sealed RFID tags) consisting of an integrated circuit (IC) attached to an antenna in the shape of a coil of wires as in **Figure 7.61**.

Readers

Readers may be integrated into handheld computers or they may be stationary and positioned at strategic points, such as a facility entrance or on an assembly line (Zebra Technologies, 2002) (see **Fig. 7.64**).

The handheld readers offer portability, however, the stationary devices offer a larger reading range. As stated above, readers have an antenna for sending and receiving signals and a processor for decoding them. The reader receives instructions and information from the antenna through the scanner, which is a part of the reader that examines analog output from the antenna. The scanner's information is then converted into a digital format by the reader, which the computer or processor can then use for data analysis, recording, and reporting (CII, 2001). There are readers today that can simultaneously read 100 to 2000 tags per second.

Fig. 7.64. *Handheld and Stationary Readers*

Read Ranges and Tag Frequency

Reading range may be determined by the power available or the frequency of the tag. Generally, active tags that have power supplies embedded in them have a larger reading range than those of passive tags; however, they do come at a cost.

Passive tags on the other hand have a read range of up to 2 meters. Other factors affecting reading distances include the frequency at which the tags communicate.

Table 7.6 shows the classification of four frequency bands used in RFID tags, their frequencies, and their reading ranges.

The most commonly used tags are classified as low frequency because they more easily readable through materials and are not as orientation sensitive as higher frequency tags. Generally speaking, higher frequencies have greater reading ranges and are less sensitive to noise than the lower frequency tags. Conversely, RFID tags with microwave frequency do have greater read ranges and higher reading speeds than lower frequency tags, but they tend to be line of sight dependent, orientation sensitive, and require more power (Liard, 2003).

Table 7.6. The classification of four frequency bands used in RFID tags

S. No.	Frequency Classification	Frequency Band (Hz)	Reading Range	Typical Applications
1.	Low Frequency	125 kHz	0.3 meters	(Inductive coupling) Access Control & Security Widgets identification through manufacturing processes Ranch animal identification OEM applications
2.	High Frequency	13.56 MHz	1 meter	(Inductive coupling) Access Control Library books Laundry identification OEM applications
3.	Ultra High Frequency (UHF)	433 MHZ to 2.45 GHz	1-3 meters	(Backscatter coupling) Supply chain tracking
4.	Microwave Frequency	2.45 GHz to 300 GHz	20+ meters	(Backscatter coupling) Asset tracking Highway toll tags Vehicle tracking

7.21.6. RFID Block Diagram

A simplified block diagram of an RFID tag (also called transponder) is shown in the Fig. 7.65. Various components of the tag are as shown. Normally, the antenna is external to the tag chip, and large in size.

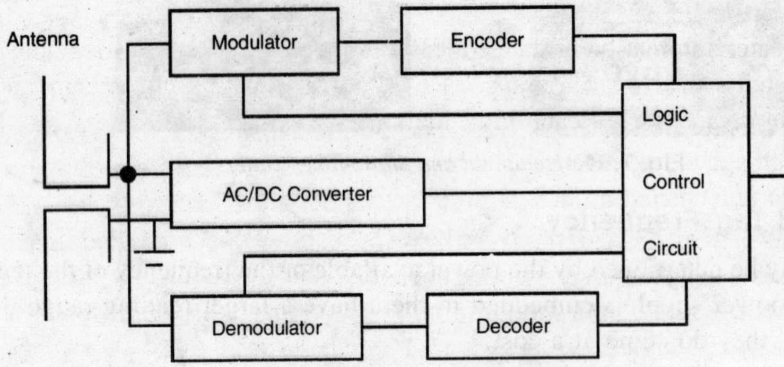

Fig. 7.65. *Block Diagram of RFID*

The operation of the RFID tag is described below:

Handshaking With the Reader (Interrogator)

(*i*) The reader continuously emits RF carrier signals, and keeps observing the received RF signals for data.

(*ii*) The presence of a tag (for our discussion, we consider only passive tag) modulates the rf field, and the same is detected by the reader.

(*iii*) The passive tag absorbs a small portion of the energy emitted by the reader, and starts sending modulated information when sufficient energy is acquired from the rf field generated by the reader. Note that the data modulation (modulation for 0s and 1s) is accomplished by either direct modulation or FSK or Phase modulation.

(*iv*) The reader demodulates the signals received from the tag antenna, and decodes the same for further processing.

Backscatter Modulation

Backscatter is one of the most widely used modulation schemes for modulating data on to rf carrier. In this method of modulation, the tag coil (load) is shunted depending on the bit sequence received. This in turn modulates the rf carrier amplitude as shown in the Fig. 7.66. The reader detects the changes in the modulated carrier and recovers the data.

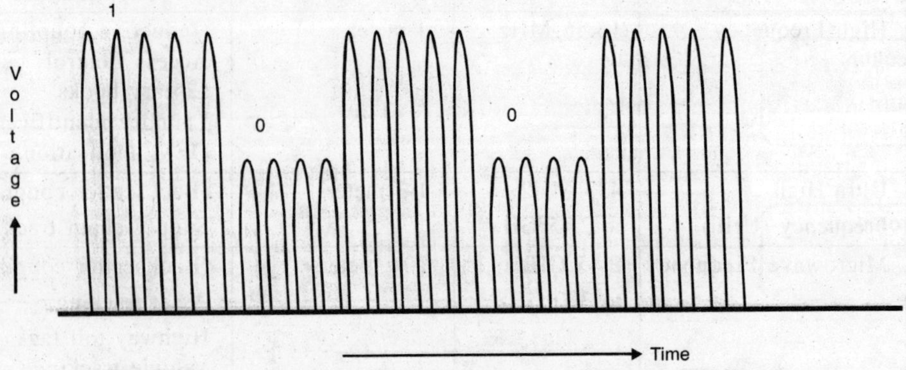

Fig. 7.66. *A simplified modulated carrier signals from the RF ID tags.*

7.21.7. Applications of RFID Technology

There are two main area of applications, defined broadly as proximity (short range) and vicinity (long range). Long range or vicinity applications can generally be described as track and trace applications, but the technology provides additional functionality and benefits for product authentication.

RFID enables greater automation of data collection process. Most companies spend considerable effort in knowing what's in their warehouse. RFID will help them dig deeper and much more easily, tracking to the detail of even each unit, long after it has left the factory or warehouse.

RFID allows all this data to be transferred securely. Companies use independent suppliers, data from each of them can be carried on tags and uploaded to the Company's central system.

Imagine the control that the Company will have on a product's life cycle. The creation of successes and defeats can be better understood. There have been numerous instances when companies had to recall the entire product due to a fault in a minor component. Imagine the costs involved in recalling a whole car for a mistake in the AC system! RFIDs can make such recalls much more focused.

While the potential for RFID applications appears virtually limitless there are only a few applications that have been implemented and are continually being used.

The most common uses for radio frequency identification technologies included:

1. Attendance
2. Security and access control
3. Library management
4. Transportation,
5. Materials management,
6. Agriculture,
7. construction, and
8. Athletics, etc.

Fig. 7.67. *Applications of RFID*

7.21.7.1. Library Automation Using RFID

RFID is a technology that is sparking interest in the library community because of its applications that promise to increase efficiency, productivity and enhance user satisfaction.

A library stacked with books and other information dissemination processes, has a physical presence. A library is an institution of knowledge acquisition and learning; it provides invaluable service to its members, patrons and to a wider local community.

Current library management systems use barcode technology and security strips. Using barcodes, a library management system can keep records of lending, borrowing and shelving status of items such as books, audio or video tapes, CDs, DVDs, etc. Security strips on library items tag their movements.

But barcodes and security strips (electronic article surveillance or EAS) have their limitations. They are slow to read and are prone to sabotaging by thieves. All these lead to irreparable loss to a library and its valuable inventory stock. This is where RFID technology can come to the aid of library managers and users.

Many libraries are switching over to RFID applications, for example, the Vatican Library. With its priceless, ancient collections of 2 million books and manuscripts, the Vatican Library is now using RFID to track, manage and secure its assets. The main problem these ancient libraries face are thefts, non-returns and mis-filed items. It is expected that by adopting an RFID solution the Vatican Library will be able to control misuse of its library and at the same time provide its users the best possible facilities and access to rare manuscripts.

In the ensuing discussions we will see how RFID technology is recommended for library automation. RFID technology is not just there to tags books and other library assets; it will provide a comprehensive route for enhancing all library services and upgrade operations for everyone concerned with the library.

Issues: During a day, library managers have to supervise many activities within their libraries. These activities have to be performed smoothly and simultaneously for the benefit of all persons concerned.

Some of the important functions in a library are:

- Check-in/check-out of items
- Shelve items
- Prevent thefts
- Check inventory

If each of these functions is done by conventional methods, they will take time and lead to inefficiencies and unsatisfactory services to the library patrons.

Let us see how check-out of items is done with barcodes and EAS.

Scan Patron Card → Receive Materials → Locate/Scan Barcode → Deactivate Security → Return Materials → Date Due Receipt

It is a six step process. This is tedious and slow. If by using RFID technology this process is made for time-efficient, then it could be recommended.

As far as shelving is concerned, in a conventional library, it is done manually. In most busy libraries, many books, DVDs, audio cassettes and videos must be shelved every day. This is a time consuming task. Misplacement of items, mis-shelving of books, is quite common. Not only this, repetitive re-shelving task can be health risk to library staff. If automated, all these problems can be avoided.

In each and every activity within a library, RFID technology will provide a greater amount of efficiency and error free functioning. It will ensure that there is:

I. Quick check-in/check-out of items for convenience of members
II. Quick and correct shelving of items
III. Complete prevention of thefts
IV. Quick inventory check

RFID tags contain a memory chip and RF antenna that can send and receive several bits of data. Such RFID tags are known as smart labels or digital identification tags. These tags provide benefits of electronic article surveillance (EAS) as well as barcodes.

For a library, smart labels have several added advantages over EAS and barcodes. One of the major benefits of an RFID system in a library is the ease of check-in and check-out of library items. Patrons can self check-in and check-out library items, saving themselves valuable time. To understand this in details, let us see how a smart label on an item communicates.

In a library, an RFID system consists of:

- A smart label
- A reader or a hardware for interrogating the smart label
- A software for controlling the hardware and decoding the responses from smart labels

Since RFID tags do not have line of sight requirements:

- Many items can be read simultaneously even whilst stacked
- Items do not have to be opened and scanned one at a time
- Items in multiple formats like books, CDs, etc. can be read at the same time

Automated counters, which read RFID signals, can help patrons:

- To self check-in/out of items from a height adjustable counter (may be placed anywhere suitable within the library)
- To quickly renew their issued out items
- To obtain a receipt of the transaction.

These automated self check-in/out counters are suitable for children and persons with physical disabilities.

After a check-out operation, the same RFID system disables the security requirements so that patrons can take the books, CDs, videos, etc. outside the library.

Many libraries, where RFID technology is operational, have an external chute, mounted on the exterior walls of the library, for returns of books and other borrowed materials. This is operational 24 x 7. (In principle, the return drop boxes may be placed anywhere – at airports, at town centers, at railway stations, etc.) These return chutes have RFID readers installed in them.

The external return chutes can be opened by member's card (secured access). On-board receipt printer can print out the details of the items returned and automatically check items back to the library management system.

7.21.7.2. Security

Security and personal identification applications are a major and broad application of RFID.A common use of RFID is in identification cards to control building access.

Many organizations use RFID tags embedded in ID cards, which are readable at a building entrance. On a similar note, some credit cards and other payment devices, such as ExxonMobil's Speedpass (http://www.speedpass.com), use RFID tags. Other cards use tags for automatic fare payment in mass-transit systems, such as the SmarTrip card for the Washington DC area subway and bus system.

Figure 7.68 shows samples of such cards. Essentially, these are a replacement for identification cards with magnetic stripes, providing a more reliable way to store identification information—magnetic stripes tend to wear out and lose information over time. RFID tags also have a higher memory capacity than magnetic stripes.

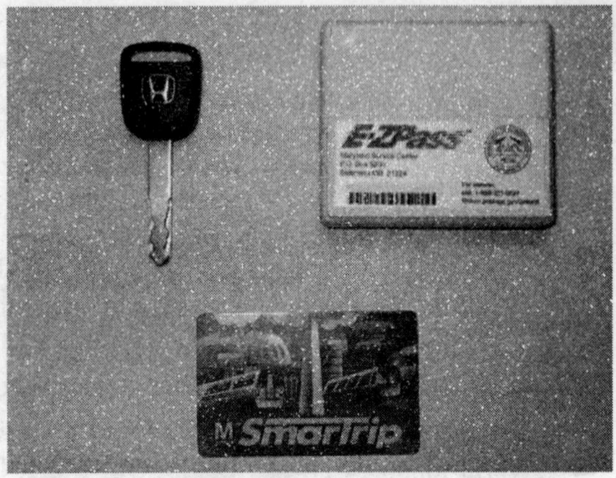

Fig. 7.68. *(i) Some car keys (upper left) contain a passive tag. (ii) An EzPass (upper right) helps collect highway tolls. (iii) A SmarTrip card (bottom) collects fares in a publictransit system.*

Keys for new cars also incorporate passive RFID tags that work with a reader near the car's ignition switch. The reader will only accept codes stored in certain keys. If the code in a key does not match the reader in the car, the car will not start, making it more difficult to steal vehicles by copying keys.

7.21.7.3. Supply Chain Management

In supply chain management, RFID tags are used to track products throughout the supply chain—from supplier delivery, to warehouse stock and point of sale. New applications target tracking from checkout through customer billing. A central database records product movement, which manufacturers or retailers can later query for location, delivery confirmation, or theft prevention.

For this application, RFID basically serves as a replacement for the bar code scanners used to track products and shipments in similar ways. Gao and colleagues outlines four ways in which RFID is superior to bar codes for tracking inventory flow over the supply chain ("An Approach to Security and Privacy of RFID System for Supply

Chain," X. Gao and colleagues, *Proc. IEEE Int'l Conf. E-Commerce Technology for Dynamic E-Business*, 2004):

- The read range of RFID is larger than that of a bar code reader.
- RFID does not require line-of-sight access to read the tag.
- Readers can simultaneously communicate with multiple RFID tags. Because of this capability, an RFID reader can capture the contents of an entire shipment as it is loaded into a warehouse or shipping container.
- A reader collects detail information in one pass, without having to scan each product.
- Tags can store more data than bar codes.

The last capability has several interesting applications in supply chain management. For example, read-write tags can store information about their environment. They can physically store their position and time throughout their movement in the supply chain.

7.21.7.4. Movement tracking

Because moving objects can easily carry RFID tags, a common use is to track the movement of people and the information associated with them. Some hospitals now use tags on newborns, to ensure identification and to alert hospital staff should someone attempt to take the baby outside of the hospital without authorization.

Some schools are requiring children to wear tag-embedded bracelets or wrist bands while on school grounds, to monitor attendance and to locate lost children.

Summary

- A **computer network** is the infrastructure that allows two or more computers (called **hosts**) to communicate with each other. When these computers are joined in a network, people can share files and peripherals such as modems, printers, or CD-ROM drives.
- The **physical topology** defines how devices are connected to make a network.
- The primary feature of the star topology is that each station is radially connected to a central node through a direct point to point connection.
- With the ring topology, adjacent devices are interconnected in a closed loop configuration.
- A linear bus topology is a multipoint circuit configuration where individual devices are connected by a common, shared communication channel (transmission media).
- Each device is connected to every other device.
- Hybrid topology is that topology in which combination of two or more topology is used.
- If we want to connect two digital devices with a network, we would have a kind of network known as "point-to-point:"
- **Switching** is the generic method for establishing a path for point-to-point communication in a network.
- **Switches** are devices capable of making temporary link or connection between two or more devices linked to the switch but not directly to each other.
- There are traditionally three basic methods of switching, namely :
  1. Circuit switching
  2. Packet switching.
  3. Message switching
- *In circuit switching, an electrical dedicated path is established between the source and the destination before any data transfer take place.*
- **Packet switching** was designed to minimize the shortcomings of circuit switching in dealing with data communication.
- In brief in Packet switching, there is no dedicated circuit between any two end nodes. The links can be dynamically shared over time.
- Message is broken up into a series of packets and header is used to route the packet through the network.
- A packet-switched network has a much higher capacity for accepting further connections.
- The virtual circuit method (also known as **connection-oriented**) is closer to circuit switching.
- A virtual circuit network is normally implemented in the data link layer, while a circuit switched network is implemented in the physical layer and a datagram network in the network layer.
- **Global addressing** means an address that can be unique in the scope of network, a source or destination required to have global address.
- **Local addressing or Virtual circuit identifier (VCI)** means the identifier that is actually used for data transfer from switch to switch.
- The **datagram** method (also known as **connectionless**) does not rely on a pre established route; instead each packet is treated independently.
- The switches in a datagram network are referred as **routers**.
- The **routing tables** record the destination address and the corresponding forwarding output ports.
- In message switching type of switched networks uses a technique which is known as **store-and-forward.**
- **A public switched data network** (PDN or PSDN) is a switched data communications network similar to the public telephone network except a PDN is designed for transferring data that may include digitized voice and video signals.
- Three major components characterize computer communication networks: switches, channels, and protocols.

- The Internet is a system of linked networks that are worldwide in scope and facilitate data communication services such as remote login, file transfer, electronic mail, the World Wide Web and newsgroups.
- With the advancements made in browser-based software for the Internet, many private organizations are implementing intranets. An intranet is a private network utilizing Internet-type tools, but available only within that organization.
- VPN uses a technique known as tunneling to transfer data securely on the Internet to a remote access server on your workplace network.
- In **peer-to-peer** networking there are no dedicated servers or hierarchy among the computers.
- The *slotted Aloha* variation of the Aloha protocol is simply that of pure Aloha with a slotted channel.
- The carrier sensing family of protocols is characterized by sensing the carrier and deciding according to it whether another transmission is ongoing.
- CSMA protocols appear to be the best possible solution since their performance depends only on the end-to-end propagation delay.
- The Carrier Sense Multiple Access with Collision Detection (CSMA/CD) protocol is specified by the IEEE 802.3 and ISO 8802.3 standards.
- This protocol is applicable to a bus topology but makes it behave as a ring.
- **Ethernet Switches** (sometimes called switching hubs) are devices that have changed the way Ethernet's bus arbitration works. Ethernet is based on CSMA/CD, a shared bus where all nodes listen and fight for use of the bus.
- For Ethernet networks that need higher transmission speeds, the Fast Ethernet standard (IEEE 802.3u) has been established.
- Gigabit Ethernet is a future technology that promises a migration path beyond Fast Ethernet so the next generation of networks will support even higher data transfer speeds.
- FDDI (Fiber-Distributed Data Interface) is a standard for data transmission on fiber optic lines in a local area network that can extend in range up to 200 km (124 miles).

Review Questions

1. What is meant by topology ? Differentiate between various topologies of LAN. **[MDU Dec. 2008]**
2. Describe briefly about the token passing ring arrangement of LAN ? **[MDU May 2009]**
3. What is LAN ? Differentiate between star and mesh topology in detail. **[MDU May 2010]**
4. In a 16 Mbps token ring network the length of token is 3 bytes. How long does it take for a station to produce a token ? **[MDU Dec. 2010]**
5. Write short note on Ethernet. **[MDU Dec. 2010]**
6. What is the difference between a public telephone network and a public switched data network ? A message is broken up into three piece. Discuss the transmission of packets in a Public Data Network using :
   (*a*) Datagram approach
   (*b*) Permanent virtual circuit
   (*c*) Switched virtual circuit **[MDU May 2011]**
7. Draw and discuss the frame formats of all the three types of frames in a token ring network. **[MDU May 2011]**
8. Explain the working principle of RFID technology. Give the applications of RFID.
9. Briefly explain the process of library automations using RFID.
10. Write short notes on
    (*i*) RF tag
    (*ii*) Physical topology
    (*iii*) Datagram
    (*iv*) Packet switching
    (*v*) PDN
    (*vi*) LAN

8

CHAPTER

Integrated Service Digital Network (ISDN)

INSIDE THIS CHAPTER

8.1. INTRODUCTION: INTEGRATED SERVICE DIGITAL NETWORK (ISDN)

Current trends in telecommunication are toward integration of voice and data services. So far these services have been available separately, requiring separate subscription, communication links, and equipment. It has long been acknowledged that the integration of these services will result in significant flexibility and cost benefits to both service users and service providers. The integrated service digital network (ISDN) is a major attempt to realize these objectives.

ISDN is the integrating of a wide range of services in to a single multipurpose network. ISDN is proposed network designed by the major telephone companies in conjunctions with ITU-T (International Telecommunication Union—Telephony) with the intent of providing worldwide telecommunication support of voice, data, video and facsimile information within the same network.

ISDN or integrated services digital network is an international standard for end to end digital transmission of voice, data and signaling. It can operate over copper based systems and allows the transmission of digital data over the telecommunications networks, typically ordinary copper based systems and providing higher data speeds and better quality than analogue transmission. The ISDN specifications provide a set of protocols that enable the set up, maintenance and completion of calls.

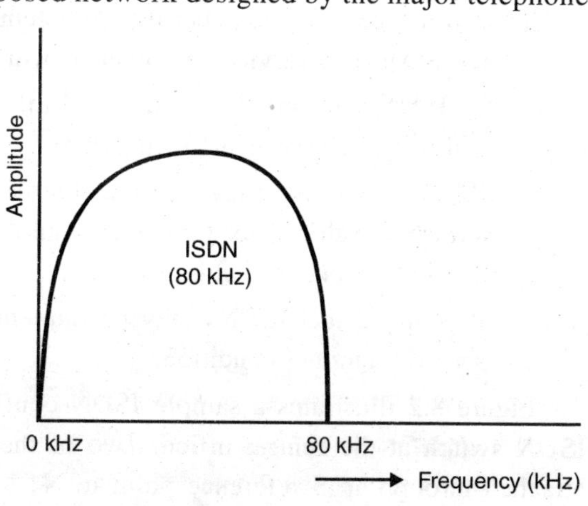

Bandwidth allocation on local loop

Fig. 8.1. *ISDN bandwidth.*

The basic principles and evolution of ISDN have been framed by ITU-T in its recommendation ITU-T1.120 (1984).

The original specifications for the integrated services digital network (ISDN) were based around voice and non-voice telephone-type services: telephony, data, telex, facsimile, as it was hoped that the ISDN would evolve from the emerging digital telephone networks.

The Integrated Services Digital Network (ISDN) is a method used to bridge the local loop (the last mile) between the Central Office and the premise connection (home). ISDN uses the existing wiring so no new cabling is required. It is a digital service which replaces the analog POTS (plain old telephone set) phone.

DO YOU KNOW

ISDN or Integrated Services Digital Network is an international standard for end to end digital transmission of voice, data and signaling.

In North America and Europe, ISDN has a bandwidth of 80 kHz. In Japan, ISDN has a bandwidth of 320 kHz.

8.2. PRINCIPLES OF ISDN

Conceptual principles of ISDN are as follows :

(*i*) ISDN concept should support a wide range of voice or telephone and nonvoice or digital data applications in the same network using minimum standard facilities.

(*ii*) ISDNs must support a large variety of applications, including switched or dedicated connections.

(*iii*) Whenever new services are introduced into an ISDN should be compatible with 64 kbps switched digital connections.

(*iv*) The transition from the existing networks to a comprehensive ISDN may take a period of time extending over one or two decades.

(*v*) During the transitions period, arrangement must be made for the interworking of service on ISDNs and services on other networks.

(*vi*) An ISDN will contain intelligence for the purpose of providing service features, maintenance and network management functions.

(*vii*) ISDN should use a layer protocol structure that specifies the access procedures. Standard already developed for OSI related applications can be for ISDN, such as X.25 level 3 for access to packet switching services.

(*viii*) It is stated that ISDNs may be implemented in a variety of configurations according to specific national condition.

Figure 8.2 illustrates a sample ISDN configuration and shows three devices attached to an ISDN switch at the central office. Two of these devices are ISDN-compatible, so they can be attached through an S reference point to NT2 devices. The third device (a standard, non-ISDN telephone) attaches through the reference point to a TA. Any of these devices also could attach to an NT1/2 device, which would replace both the NT1 and the NT2. In addition, although they are not shown, similar user stations are attached to the far right ISDN switch.

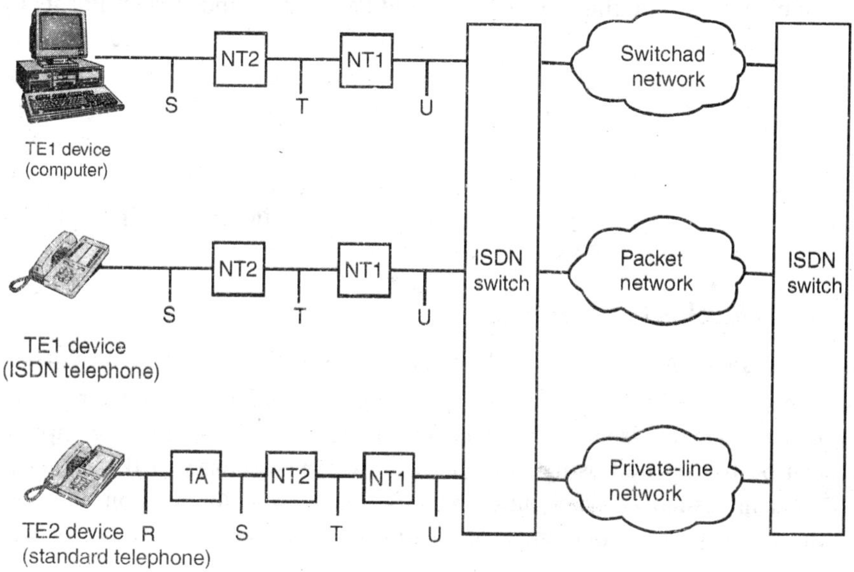

Fig. 8.2. *Sample ISDN configuration illustrates relationships between devices and reference points.*

8.3. BASIC CONCEPTS

ISDN provides a fully integrated digital network for voice and data communication.

It supports both circuit and packet switching. **Figure 8.3 illustrates** the overall arrangement of a typical ISDN network, which brings together a mix of old and new technologies and devices.

Each ISDN switch consists of an **exchange termination** part, which performs the necessary circuit switching functions, and a **packet handler**, which performs the necessary packet switching functions. The packet handlers implement X.25 and are connected to a public packet switched network via X.75. The exchange terminations are interconnected via tandem exchanges. STPs and SCPs provide network intelligence. Network access is facilitated through a set of **service switching points** (SSPs). The signaling link between the switches is controlled through a set of **signaling transfer points** (STPs). A **service control**

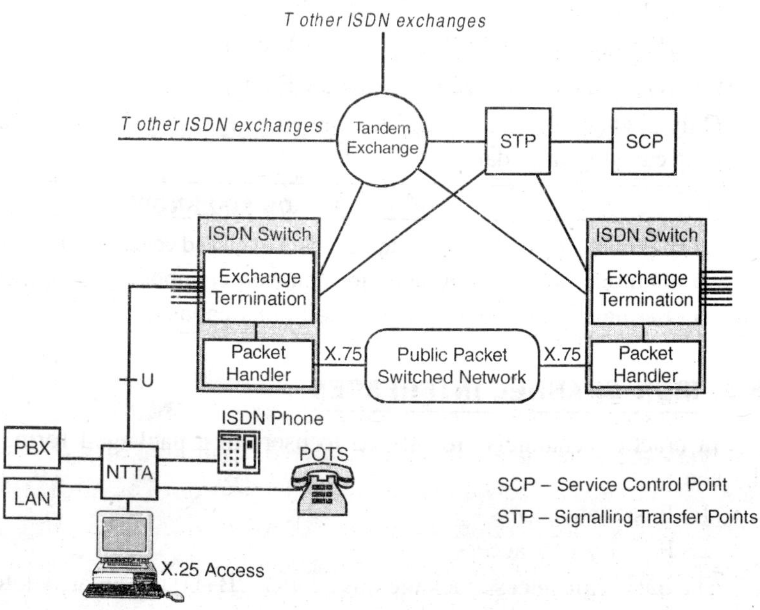

Fig. 8.3. *A typical ISDN network.*

point (SCP) is in charge of dictating how calls should be handled and routed. For its operation, the SCP uses a service database which contains service profile definitions provided by the customer.

Service profiles use a number of network-provided parameters (*e.g.*,, time and date of the call, origin of the call, network-related conditions) to determine how to handle calls. A **support management system** (SMS) provides network management and customer control capabilities.

Subscriber access is provided via a network termination and/or terminal adapter (NT/TA). This provides the connectivity for a variety of user devices, including ISDN phones, Plain Old Telephone Sets (POTS), LANs, PBXs, and X.25 terminals.

8.4. ISDN TRANSMISSION CHANNELS

Subscriber access to ISDN is via digital channels, of which there are three types :

(*i*) **B channels** are used for carrying user data (digitized voice or computer generated data) at 64 kbps. This data rate is more than necessary in many situations (*e.g.*,, compressed digitized voice can be transmitted using less bandwidth). Consequently, a B channel is sometimes subdivided into smaller subchannel. Whether there is a subdivision or not, the network treats the whole thing as one channel. All subchannels therefore are between the same two endpoints and follow the same route.

(*ii*) **D channels** are primarily used for common channel signaling purposes. They are typically associated with B channels and carry the control signals for B channel calls. D channels are also used for packet-switched data communication. D channel may operate at 16 or 64 kbps.

(*iii*) **H channels** are used in a high-speed trunk capacity. They are suitable for applications that require higher than 64 kbps data rates. Multi-media applications (*e.g.*,, audio, video, and graphics multiplexed over the same channel) are examples. H channels are divided into three categories depending on their speed:

(*iv*) H0 operates at 384 kbps (= 6 B channels)

(*v*) H11 operates at 1536 kbps (= 23 B channels)

(*vi*) H12 operates at 1920 kbps (= 30 B channels)

Only D channels can be used for carrying signaling information. B and H channels can only be used for carrying user data.

DO YOU KNOW

B channels are used for carrying user data (digitized voice or computer generated data) at 64 kbps.

D channels are primarily used for common channel signaling purposes.

H channels are used in a high-speed trunk capacity.

8.5. ISDN CHANNEL INTERFACES

In practice, channels are offered to users in a packaged form. Two such packages have been defined :

1. Basic rate access
2. Primary rate access.

The **basic rate access** package (also called 2B+D) is primarily intended for residential subscribers and consists of the following:

(*i*) Two B channels

(*ii*) One 16 kbps D channel

(*iii*) Overhead of 48 kbps for framing, synchronization, etc.

This produces a total bit rate of 192 kbps. The channels may be used for a variety of purposes. For example, the two B channels can be used for two independent voice services, or one of them can be used for voice and the other for a data service such as fax, teletext, or remote LAN access. Modest data communication requirements (*e.g.*,, remote banking transactions) may be met by the D channel alone. Other permitted combinations for basic access are: B+D or just D.

The **Primary Rate Access (PRA)** package is aimed at business users with greater bandwidth requirements. Primary access comes in two configurations.

At a bit rate of 1.544 Mbps (North America and Japan) and consisting of:

(*i*) 23 B channels

(*ii*) One 64 kbps D channel

(*iii*) Overhead of 8 kbps

At a bit rate of 2.048 Mbps (Europe) and consisting of :

(*i*) 30 B channels

(*ii*) One 64 kbps D channel

(*iii*) Overhead of 64 kbps

As with the basic access, lower configurations are also possible, depending on requirements. Primary access can also support H channels.

8.6. ISDN SERVICES

ISDN provides three types of services :

1. Bearer services
2. Teleservices
3. Supplementary services

Tele and supplementary services represent the type of features and functions which are visible to end-users, while bearer services represent the parts of the network which remain hidden from end-users.

Bearer services facilitate the real-time communication of digital information between end-users. These services mainly relate to network functions and account for OSI layers 1-3. An example of a bearer services is the *64 kbps, 8 kHz structured, speech* service. This service uses a data rate of 64 kbps together with 8 kHz timing information (which structures the data into octet intervals) for transmitting a pulse code modulated (PCM) speech signal. The fact that the signal represents speech is known to the network, allowing it to employ transformations which may not preserve bit integrity but will result in good quality audio reproduction.

By contrast, the *64 kbps, 8 kHz Structured, Unrestricted* service makes no assumptions about the signal's data content and is a general purpose service. There are also a number of other unrestricted services, offering successively higher bandwidths (*e.g.*,, *384, 1536, and 1920 kbps, 8 kHz structured, unrestricted* services).

In addition to the above circuit-switched bearer services, there are a number of packet-switched bearer services. For example, the *User Signaling* service is a packet-switched bearer service suitable for exchanging signaling information between end-users and is defined by recommendation I.451.

Another example is the *connectionless on a D channel* service which provides a datagram service on a D channel and is suitable for control-oriented or transaction-oriented applications.

Teleservices provide a set of higher-level functions on top of bearer services.

These services account for OSI layers 4-7. Examples of teleservices are:

(*i*) Telephony services which provide speech communication over a B channel with control signaling over the D channel.

(*ii*) Facsimile services which facilitate the communication of bitmap images over a B channel with control signaling over the D channel.

(*iii*) Teletex services which facilitate the interchange and communication of textual as well as formatted documents over a B channel with control signaling over the D channel.

Supplementary services enhance bearer and teleservices in an independent fashion. Examples of supplementary services are:

(*i*) The *Centrex* service emulates a private network and provides specialized features to a set of subscribers.

(*ii*) The *Call Transfer* service allows a user to transfer an active call to a third-party.

(*iii*) The *Call Waiting* service allows a user already engaged in a call to be informed of another incoming call.

(*iv*) The *Calling Line ID* service provides the calling party's address information to the called party.

Although these services all appear geared toward circuit-switched telephone calls, they are equally applicable to packet-switched data calls.

8.7. ISDN - OSI LAYERS

The D channel uses up to the OSI network layer.

The B channel uses both the data link and physical layers.

The line encoding is used by ISDN is standard telecommunication 2B1Q which stands for 2 Binary elements encoded in 1 quaternary. A digital bit is represents two binary elements for each voltage change. The following table illustrates the 2B1Q encoding used by ISDN:

<table>
<tr><td>OSI Model</td><td colspan="2">ISDN</td></tr>
<tr><td>Network</td><td rowspan="4">D Channel</td><td></td></tr>
<tr><td>Data Linkl</td><td rowspan="3">B Channel</td></tr>
<tr><td>Physical</td></tr>
</table>

Fig. 8.4. *ISDN - OSI Model.*

Digital bit	Voltage
10	+3
11	+1
01	−1
00	−3

8.8. FUNCTIONAL GROUPINGS AND REFERENCE POINTS

User access to ISDN is provided at a number of different levels of abstraction.

These levels are defined by **functional groupings**, which encompass functions equivalent to those denoted by one or more OSI layers. The interfaces between the functional groupings are called **reference points** (see **Fig. 8.5**). The U (User) interface is a 2-wire physical interface to the network.

The **Network Termination 1** (NT1) functional grouping provides OSI layer 1 capabilities and deals with signal transmission and physical connectors for interfacing customer premises equipment (CPE) to ISDN.

The NT1 transforms the U interface into a 4-wire subscriber S/T interface which supports 2B+D channels (in case of basic access) or T interface which supports 23B+D or 30B+D (in case of primary access). NT1 multiplexes these channels using TDM into a continuous bit stream for transmission over the U interface. NT1 also supports up to eight CPEs connected in a multidrop line arrangement to basic access. The NT1 device may be owned and operated by the service provider, baring the customer from direct access to the U interface, or it may be a CPE. The **network termination 2** (NT2) functional grouping provides additional OSI layer 2 and 3 capabilities on top of NT1. NT2 is a CPE which transforms the T (Terminal) interface into an S (System) interface. The S interface supports 2B+D channels. NT2 may perform switching and concentration functions. A typical NT2 device would be a digital PBX, serving a set of digital phones, or a LAN, serving a set of personal computers.

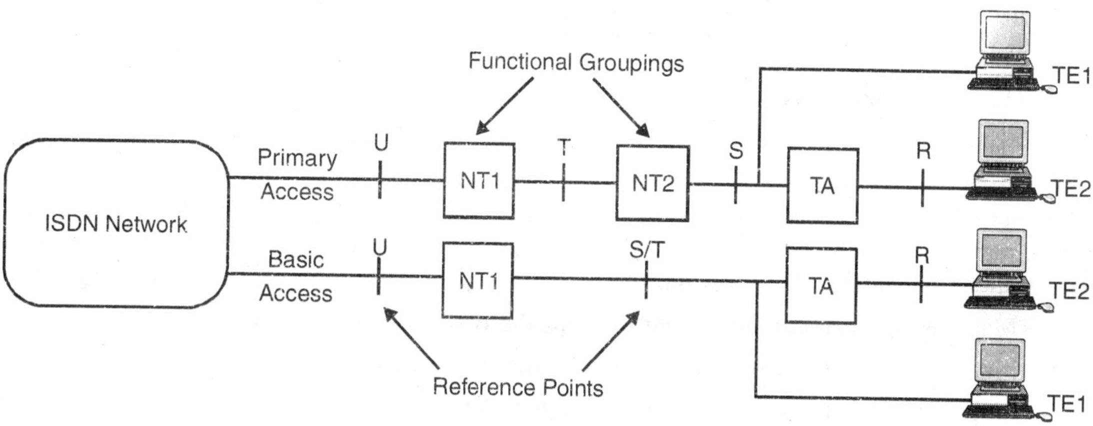

Fig. 8.5. *ISDN functional grouping and ISDN reference points.*

Two types of terminal equipment may be used for ISDN access.

Terminal Equipment 1 (TE1) denotes ISDN terminals which use a 4-wire physical link to the S or S/T interface. TE1 devices conform to ISDN standards and protocols and are especially designed for use with ISDN. A digital ISDN telephone and a PC with an ISDN card are examples.

Terminal Equipment 2 (TE2) denotes non-ISDN terminal equipment. Ordinary terminals and personal computers are examples. These devices can be connected to ISDN at the R (Rate) reference point. RS-232 and V.21 are examples of the type of standards that may be employed for the R reference point. The mapping between the R interface and the S or S/T interface is performed by a **terminal adapter** (TA), which performs the necessary protocol conversions and data rate adaptations between the two interfaces.

It is worth pointing out that although NT1, NT2, and TAs may be offered as separate devices, in practice this is not always the case. For example, some CPE manufacturers produce TAs that has NT1 and NT2 capabilities, as well as additional interfaces for other devices (*e.g.,*, analog telephones).

Figure 8.6 illustrates an example of such a device: an ISDN point-of-sale terminal adapter designed as a retail environment CPE which provides a variety of interfaces.

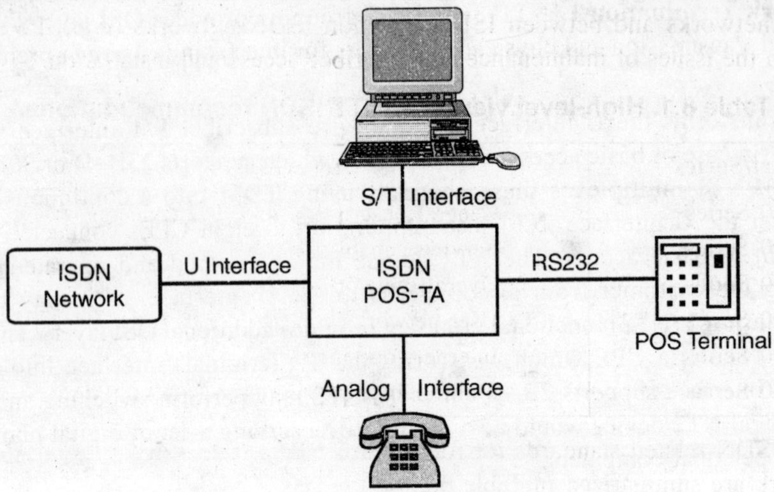

Fig. 8.6. *ISDN point-of-sale terminal adapter.*

8.9. INTERNETWORKING

For economic and practical reasons, the introduction of ISDN involves a rather long transitory phase, during which time ISDN and earlier networks have to coexist. For this to work, some means of internetworking is needed. For examples, subscribers who are served by ISDN may need to contact subscribers who are served by PSTN. Such internetworking involves a number of considerations, including the following:

1. Mapping between the numbering plans of the two networks.
2. Mapping between the services offered by the two networks.
3. Mapping between the control signals employed by the two networks.
4. The physical interfacing of the two networks.

An additional set of reference points have been defined by CCITT to address the above. These reference points define interfaces between ISDN and other networks:

(*i*) The K reference point is used for interfacing ISDN to an existing, non-ISDN voice or data network, where the internetworking mappings are performed by ISDN.

(*ii*) The L reference point is used for interfacing ISDN to an existing, non-ISDN voice or data network, where the internetworking mappings are performed by the non-ISDN network.

(*iii*) The M reference point is used for interfacing ISDN to a specialized network (*e.g.*,, MHS), where the internetworking mappings are performed by the latter.

(*iv*) The N reference point is used for interfacing two ISDN networks, where both networks may need to do internetworking mappings to provide a common set of services.

8.10. ISDN STANDARDS

A high-level view of CCITT ISDN recommendations are given in following Table 8.1. Each of the series consists of many parts. The I.100 series provides a general introduction to ISDN, its principles, objectives, and envisaged evolution path. The I.200 series defines the ISDN end-user services and a set of attributes for each service. The I.300 series defines the ISDN network and its functions and protocols in relation to ISDN services. The I.400 series defines the physical and logical interfaces between ISDN and its users. The I.500 series deals with the issues of internetworking between ISDN networks and between ISDN and non-ISDN networks (*e.g.*,, PSTN). The I.600 series deals with the issues of maintenance of subscriber access and installation.

Table 8.1. High-level view of CCITT ISDN recommendations.

Part/Series	Description
I. I.100 Series	General description of ISDN concepts
II. I.200 Series	Service capabilities
III. I.300 Series	Overall network aspects and functions
IV. I.400 Series	User-network interfaces
V. I.500 Series	Internetwork interfaces
VI. I.600 Series	Maintenance principles

The major ISDN-related standards for four areas of basic rate, primary rate, SS7, and packet-switched network are summarized in Table 8.2.

Table 8.2. Major ISDN-related standards.

	Layer	Standard	Description
Basic Rate	1	I.430	2B + D channels (D = 16 kbps)
	2	I.441	LAP-D (with multiple TEs)
	3	I.451	User-network interface
Primary Rate	1	I.431	23B + D or 30B + D channels (D = 64 kbps)
	2	I.441	LAP-D (point-to-point)
	3	I.451	User-network interface
Signaling System 7	1-3	Q.701-Q.708	Message Transfer Part (MTP)
	4	Q.711-Q.714	Signaling Connection Control Part (SCCP)
	4+	Q.761-Q.766	ISDN User Part (IUP)
	4+	Q.791-Q.795	Operation, Admin and Maintenance (OA&M)
Packet Network	1	X.21	Physical interface
	2	LAP-B	Link access
	3	X.25, X.75	Packet interface

ISDN Advantages

(*i*) ISDN is a mature technology; it has been around since the late 1980s. It is has been tried, tested and works.

(*ii*) It is governed by a world-wide set of standards.

(*iii*) It provides symmetrical transfer rates: the transmit rate is the same as the receive rate.

(*iv*) It has consistent transfer rates. If you have a 64 kbps bearer channel then that's the speed that you transfer at.

(*v*) It is competitive priced compared to other technologies.

ISDN Disadvantages

(*i*) An external power supply is required. The telephone companies don't supply power for ISDN lines. If the power fails, the phones won't work.

(*ii*) Special digital phones are required or a Terminal Adapter to talk to the existing POTS devices.

(*iii*) It is very expensive to upgrade a central office switch ($500,000+) to ISDN

(*iv*) If the ISDN fails—the phone fails.

8.11. BROADBAND ISDN

The original specifications for the **integrated services digital network (ISDN)**, were based around voice and non-voice telephone-type services: telephony, data, telex, facsimile, as it was hoped that the ISDN would evolve from the (then) emerging digital telephone networks. Indeed, this is one of the reasons that the fundamental element of an ISDN link is the 64 kbps **B-Channel**. However, the planning for ISDN was started around 1976, and as technology evolved, so did the requirements of the users that wanted to use this technology. In 1988, the CCITT released a document that described a new set of **Broadband ISDN (B-ISDN)** services. To distinguish this new concept from the original ISDN service, we now refer to the latter as **Narrowband ISDN (N-ISDN)**.

Since then the CCITT has become the ITU. The B-ISDN work is far from complete, and some of the factors influencing the emergence of the B-ISDN from ITU are:

(*i*) **Demand :** Users (both commercial and residential) are showing interest in receiving high speed, reliable services.

(*ii*) **Technology :** This has been one of the biggest factors. Advances in technology have increased demand, as well as the ability to supply it. As data processing technologies available to the user have become more sophisticated so has his/her demands, while high speed transmission (based on the use of optical fibre), high speed switching and increased processing power make the realization of these demands possible.

(*iii*) **Service integration.** There is a need to integrate both circuit switched and packet switched services into one network that can provide interactive and distribution services.

(*iv*) **Flexibility.** The resulting network must be able to satisfy the needs of the users as well as the network operators in terms of its functionality and usability.

DO YOU KNOW

In 1988, the CCITT released a document that described a new set of **broadband ISDN (B-ISDN)** services.

8.11.1. Broadband ISDN Services

The need for a broadband ISDN service sprung from the growing needs of the customers. The planned broadband ISDN services can broadly be categorized as follows :

(*i*) **Interactive services :** These are services allowing information flow between two end users of the network, or between the user and the service provider. Such services can be subdivided:

(*a*) **Conversational services :** These are basically end-to-end, real-time communications, between users or between a user and a service provider, *e.g.*, telephone-like services. Indeed, B-ISDN will support N-ISDN type services. (Note also that the user-to-user signaling, user-to-network signaling, and inter-exchange signaling are also provided but outside our scope.) Also the additional bandwidth offered will allow such services as video telephony, video conferencing and high volume, high speed data transfer.

(*b*) **Messaging services :** This differs from conversational services in that it is mainly a store-and-forward type of service. Applications could include voice and video mail, as well as multi-media mail and traditional electronic mail.

(*c*) **Retrieval services :** This service provides access to (public) information stores, and information is sent to the user on demand only. This includes things like tele-shopping, videotext services, still and moving pictures, telesoftware and entertainment.

(*ii*) **Distribution services :** These are mainly broadcast services, are intended for mainly one way interaction from a service provider to a user:

(*a*) **User control of presentation :** This would be for instance, a TV broadcast, where the user can choose simply either to view or not. It is expected that cable TV companies will become interested in Broadband ISDN as a carrier for the high definition TV (HDTV) services that are foreseen for the future.

(*b*) **User controlled presentation :** This would apply to broadcast information that the user can partially control, in that the user can decide which part of it he/she accesses, *e.g.*, teletext and news retrieval services.

8.11.2. B-ISDN User-Network Interface

In addition to the narrowband channels defined for ISDN, B-ISDN supports the following three User-Network Interfaces (UNIs):

(*i*) **Symmetric full-duplex 155.52 Mbps :** This interface provides the basis for interactive services. This will therefore be the most common interface offered. It is commonly referred to as the 150 Mbps interface. The effective payload of this interface is 149.76 Mbps.

(*ii*) **Symmetric full-duplex 622.08 Mbps :** This interface is suitable for situations which involve very high traffic volumes. It is commonly referred to as the 600 Mbps interface. The effective payload of this interface is 599.04 Mbps or 600.768 Mbps, depending on how it is structured.

(*iii*) **Asymmetric full-duplex 622.08/155.52 Mbps :** This is a hybrid interface which combines the earlier two. It uses bit rates of 622.08 in the network-to-subscriber direction and 155.52 Mbps in the subscriber-to-network direction. It is suitable for situations where the traffic volume in the network-to-subscriber direction is much higher than in the opposite direction.

The functional groupings and reference points for B-ISDN UNI closely follow those of the narrowband, as illustrated by Fig. 8.4. Their role remains identical to those of narrowband ISDN. To highlight the broadband nature of the functional groupings and reference points, the letter 'B' is added (*e.g.*,, B-TE1). The R reference point is an exception because it may or may not have broadband capabilities. Accordingly, it may support a B-TE2 or TE2.

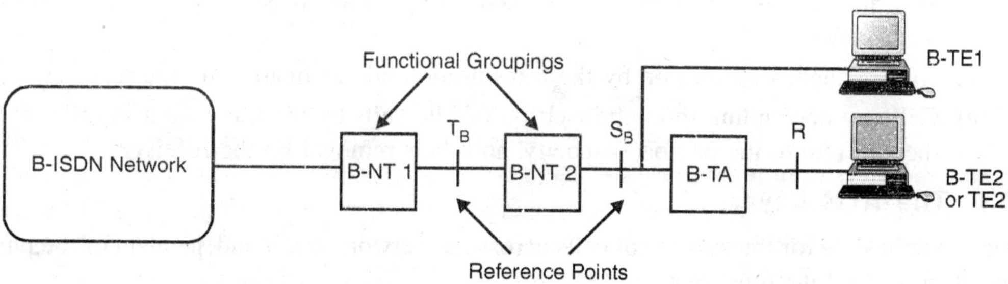

Fig. 8.7. *B-ISDN functional groupings and reference points.*

8.11.3. B-ISDN Protocol Architecture

The reference model for the B-ISDN protocol architecture is shown in **Fig. 8.8.** The architecture is partitioned by three planes.

The **user plane** facilitates the transfer of user information and associated control signals. The **control plane** is responsible for connection control functions such as call setup and release.

The **management plane** is responsible for management functions relating to the system as a whole and coordination between the planes (Plane Management), and for management functions related to resources and parameters residing in its protocol entities (Layer Management).

8.11.3.1. The physical layer is responsible for the physical transmission of ATM cells

To achieve some degree of physical medium independency, this layer is divided into two sub layers.

(*i*) The **physical medium sub layer** covers the medium-dependent aspects of the physical layer and provides bit transmission and bit timing functions.

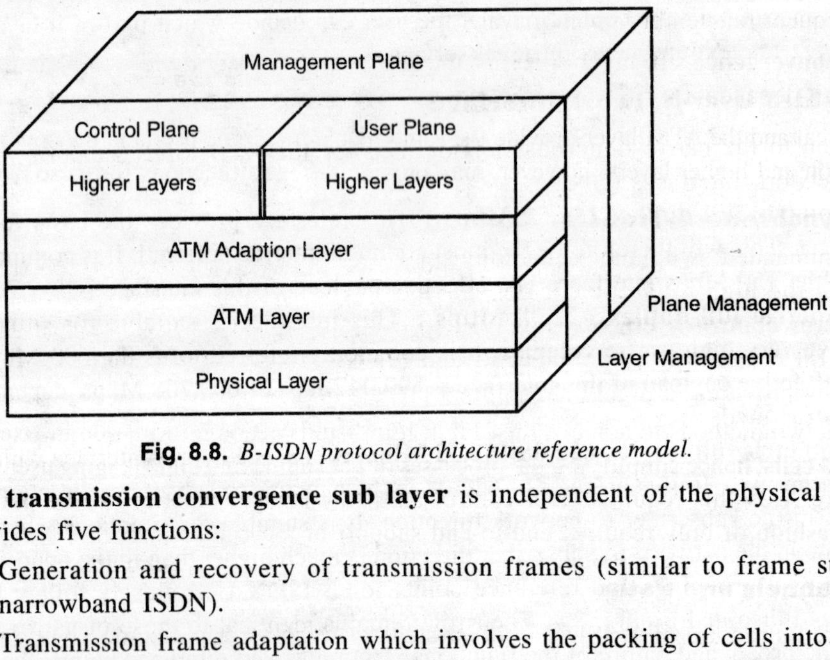

Fig. 8.8. *B-ISDN protocol architecture reference model.*

(*ii*) The **transmission convergence sub layer** is independent of the physical medium and provides five functions:

(*a*) Generation and recovery of transmission frames (similar to frame structures for narrowband ISDN).

(*b*) Transmission frame adaptation which involves the packing of cells into transmission frames and their extraction upon arrival.

(*c*) Cell delineation to enable the receiving-end identifies the boundaries of cells whose information fields have been scrambled prior to transmission and are descrambled by the receiver.

(*d*) HEC sequence generation by the transmitter, and verification by the receiver.

(*e*) Cell rate decoupling through insertion of idle cells by the transmitter in order to adapt the cell rate to the payload capacity, and their removal by the receiver.

8.11.3.2. The ATM Layer

This is responsible for the transfer of cells across the network and is independent of the physical medium. It provides four functions:

(*i*) Cell multiplexing by the transmitter in order to support multiple logical channels, and their demultiplexing by the receiver.

(*ii*) Virtual path and virtual channel identifier translation.

(*iii*) Cell header generation by the transmitter, and its extraction by the receiver.

(*iv*) Generic flow control for the control of traffic in a customer network by placing flow control information in cell headers.

8.11.3.3. The ATM Adaptation Layer (AAL)

Packages information from higher layers into ATM cells to be transported by the ATM layer at the transmitter end, and extracts information from ATM cells for passing to higher layers at the receiver end.

It consists of two sub layers :

(*i*) The **segmentation and reassembly sub layer** handles the segmentation of information from higher layers into smaller units so that they can be transported by ATM cells, and their subsequent reassembly upon arrival.

(*ii*) The **convergence sub layer** specifies the services that AAL provides to higher layers and is service-dependent.

The physical and the ATM layer provide the same functions for the user and the control plane. The ATM adaptation and higher layers, however, may provide different functions for these two planes.

8.11.4. Asynchronous Transfer Mode

The recommended switching technology for B-ISDN is the **Asynchronous Transfer Mode** (ATM). Given the high reliability of optical fiber for transmission of digital information, the significant error control overheads involved in earlier protocols (such as X.25) become very questionable. Like frame relay, ATM takes advantage of this increased reliability to improve network performance by providing a highly streamlined protocol stack.

ATM uses two methods to achieve this. First, it transmits all its information in fixed-size small packets, called **cells**, hence simplifying the processes of packaging and unpackaging user information. Second, unlike X.25 this requires error control and flow control functions to be performed in a link-by-link fashion, it only requires end-to-end support of these functions.

8.11.5. Channels and Paths

The layered structure of B-ISDN is illustrated in **Fig. 8.9.** The **higher layer functions** are application-dependent and beyond the scope of this chapter. The **ATM transport network** is divided into two layers, both of which are hierarchically organized.

The **ATM layer transport functions** are divided into two path levels :

(*i*) Virtual channel level

(*ii*) Virtual path level.

Virtual Channel Level

A **virtual channel** (VC) denotes the transport of ATM cells which have the same unique identifier, called the **virtual channel identifier (VCI)**.

This identifier is encoded in the cell header. A virtual channel represents the basic means of communication between two end-points, and is analogous to an X.25 virtual circuit.

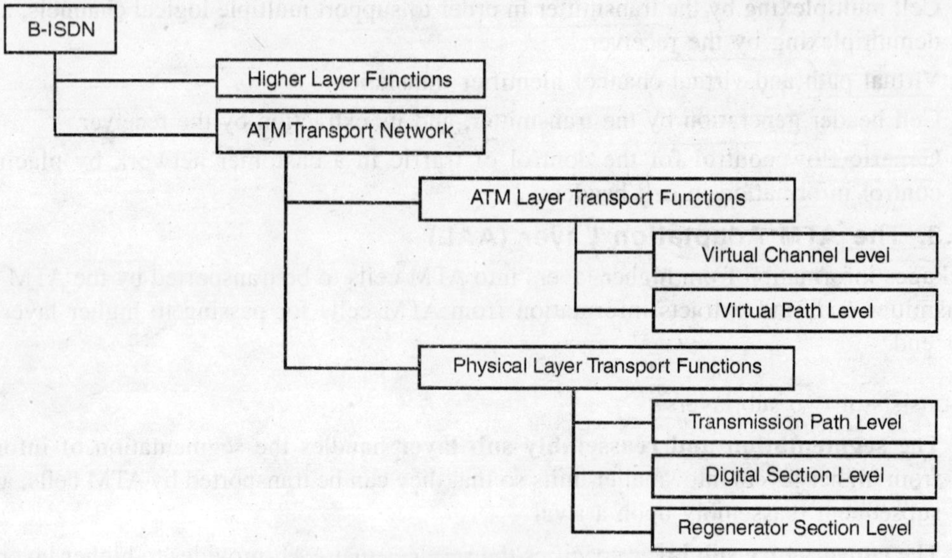

Fig. 8.9. *Layered structure of B-ISDN.*

Virtual Path Level

A **Virtual Path** (VP) denotes the transport of ATM cells belonging to virtual channels which share a common identifier, called the virtual path identifier (VPI), which is also encoded in the cell header. A virtual path, in other words, is a grouping of virtual channels which connect the same end-points. This two layer approach results in improved network performance. Once a virtual path is setup, the addition/removal of virtual channels is straightforward.

8.11.6. Physical Layer Transport Functions

The **physical layer transport functions** are divided into three levels of functionality:

(*i*) The **transmission path** connects network elements that assemble and disassemble the transmission system payload. This payload may contain user or signaling information to which transmission overheads are added.

(*ii*) The **digital section** connects network elements (such as switches) that assemble and disassemble continuous bit/byte streams.

(*iii*) The **regenerator section** is simply a portion of a digital section which connects two adjacent repeaters along a transmission path which is otherwise too long to sustain the signal.

Figure 8.10 illustrates the relationship between virtual channel, virtual path, and transmission path.

Two types of switching are possible:

1. VP switching
2. VC switching.

A **VP switch** connects a set of incoming VP terminations to a set of outgoing VP terminations. The mapped VPs will have different VPIs, but the VCI will remain unchanged. A **VC switch**, on the other hand, connects a set of incoming VC terminations to a set of outgoing VC termination. VCIs are not preserved. Because VCs are embedded in VPs, VC switching also involves VP switching.

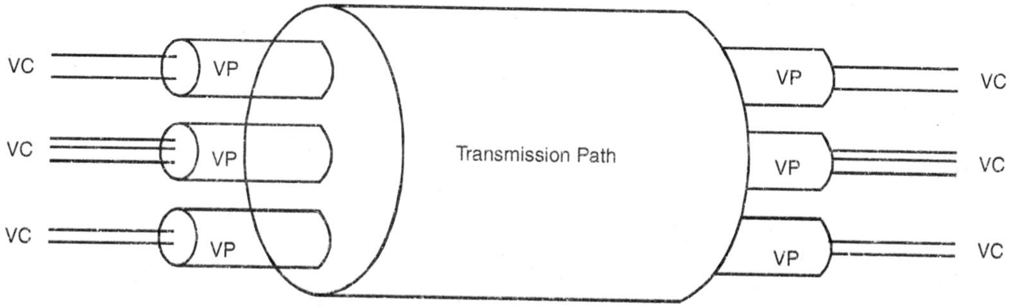

Fig. 8.10. *Virtual channel, virtual path, and transmission path*

8.11.7. ATM Cells

An ATM cell consists of 53 consecutive octets, of which 48 octets are used to represent user information and the remaining 5 octets form a header for the private use of the ATM protocols (see Table 8.3).

Table 8.3. ATM cell structure

	Field	Description
Header (5 octets)	GFC	Generic Flow Control field for end-to-end flow control
	VPI	Virtual Path Identifier for routing
	VCI	Virtual Channel Identifier for routing
	PT	Payload Type for the information field
	CLP	Cell Loss Priority for discarding of cells in congestions
	HEC	Header Error Control Code
Body (48 octets)	Information	User information or inband signaling information

The **GFC** field is intended for the user-network interface only. (It does not appear in the header format of cells used for the internal operation of the network. Its 4 bits are instead used by the VPI.) It can be used by applications to provide end-to-end flow control.

The **VPI** and **VCI** fields collectively support the routing of a cell by the network. The **PT** field denotes the type of information stored in the information field.

Summary

- ISDN is the integrating of a wide range of services in to a single multipurpose network.
- ISDN or integrated services digital network is an international standard for end to end digital transmission of voice, data and signaling.
- In North America and Europe, ISDN has a bandwidth of 80 kHz. In Japan, ISDN has a bandwidth of 320 kHz.
- **B channels** are used for carrying user data (digitized voice or computer generated data) at 64 kbps.
- **D channels** are primarily used for common channel signaling purposes.
- **H channels** are used in a high-speed trunk capacity.
- The **basic rate access** package (also called 2B+D) is primarily intended for residential subscribers.
- The **primary rate access** (PRA) package is aimed at business users with greater bandwidth requirements.

- ISDN provides three types of services :
  1. Bearer services
  2. Teleservices
  3. Supplementary services
- In 1988, the CCITT released a document that described a new set of **broadband ISDN (B-ISDN)** services.

Review Questions

1. Explain ISDN.
2. Explain ISDN layer model in detail. **[MDU Dec. 2008]**
3. Describe briefly about the following in context with the ISDN network. Explain its model ? Give
   (*a*) BRI and PRI
   (*b*) NT 1, NT 2
   (*c*) TE 1, TE 2.
   Brief description about various layers.
4. What is the difference between In band and Out band signaling in context with ISDN. Explain all ISDN service and channels including its functional groupings ? **[MDU Dec. 2010]**
5. Give a comparison of PSTN AND ISDN. Also give their advantages and disadvantages. **[MDU May 2007]**
6. Explain ISDN and compare it with PSTN and ADSL. **[MDU Dec. 2008]**
7. Write short notes on following:
   (*i*) Components of PSTN
   (*ii*) PSTN **[MDU Dec. 2010]**
   (*iii*) ISDN
   (*iv*) Broad ISDN

❑❑❑

9
CHAPTER

Multiplexing

9.1. INTRODUCTION : MULTIPLEXING

INSIDE THIS CHAPTER

One requirement of a data communications system is for it to transmit as many intelligent signals as possible in a fixed period using a single-communications channel. **Multiplexing** is a technique which makes it possible to cram a number of logical channels (each capable of supporting an independent connection) into the same physical channel or line. Multiplexing may be defined as any process of sending of a number of separate signals together, over the same cable or bearer, simultaneously and without interference.

As we know that usage of data and telecommunication is increasing day by day, so the traffic also. We can accommodate to add individual lines each time for a new channel or we may use transmissions lines, with the higher bandwidth (fiber optics or satellite microwaves), so that it can carry more numbers of signals.

In other words, we can define multiplexing is the set of techniques that allows the simultaneous transmission of multiple signals across a single data link.

The objective of multiplexing should be obvious:

(*i*) To reduce costs by better utilizing the capacity of a line.

(*ii*) To reduce complexity.

Multiplexing is an operation in which a number of independent signals are combined into a composite signal for transmission over a communication channel. The signals must be combined in a way so that they can be separated at the receiving end without interfering one another. The separation of the composite signal into the original signals is called **demultiplexing.** The transmission media may be metallic wire, a coaxial cable, optical fiber, a terrestrial microwave radio system or a satellite system, refer Fig. 9.1.

The rate of data transmission is measured in the number of bits per second (bps) transmitted. The bps rate is often confused with the baud rate. Baud refers to the rate at which a modulated signal between two devices changes in 1 second. For example, if the signal between two modems changes frequency or phase at a rate of 2,400 times per second, the baud is 2,400. If we use a modulation method in which a single modulation change carries one bit, the 2,400 baud is also 2,400 bits per second. Using more sophisticated modulation methods, several bits of information can bits of data are transmitted with each modulation change, the data transfer rate is 4,800 bits per second at 2,400 baud. The data signals being transmitted are normally multiplexed to increase the transmission rate of data over the communications channel or to increase the efficiency of the channel by allowing multiple users of the same channel.

In a multiplexed system, many lines (say *n* lines) share the bandwidth of one link (Physical path) as shown in Fig. 9.1.

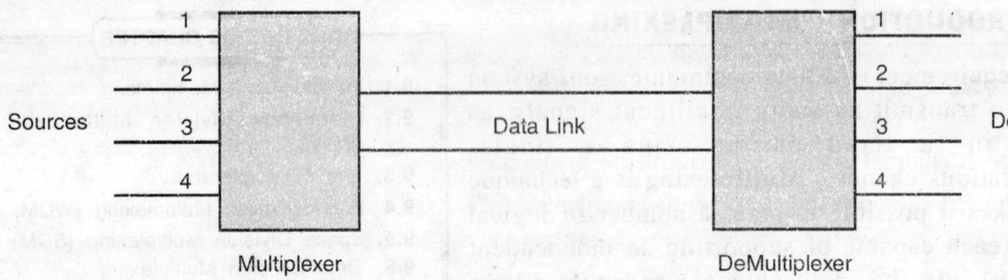

Fig. 9.1. *Concept of Multiplexing/Demultiplexing.*

> **TIPS**
>
> Multiplexing is an operation in which a number of independent signals are combined into a composite signal for transmission over a communication channel.

In telecommunication, the multiplexing means the use of one telecommunication line to handle several channels of voice or data. The best example of multiplexing is our TV cable. We select a particular channel using remote control from the cable, which carries many channels.

The individual channels entering and leaving the terminal station is called baseband channels. A link, which carries multiplex signal, is called broad band channel.

Types of Multiplexing

There are many domains in which multiplexing can be achieved. They are time, space, phase, frequency and wavelength. However, the most common techniques of multiplexing are as follows:

1. Frequency division multiplexing (FDM) (mainly used for Analog communication).
2. Time division multiplexing (TDM)
3. Wave division multiplexing (WDM) (mainly used for Analog communication).
4. Space Division Multiplexing (SDM).

9.2. FREQUENCY DIVISION MULTIPLEXING (FDM)

FDM is a broad band analog transmission technique in which multiple signals are transmitted over a single cable simultaneously.

Frequency Division Multiplexing (FDM) is an analog technique where each communications channel is assigned a carrier frequency. To separate the channels, a guard-band would be used. This is to ensure that the channels do not interfere with each other.

For example, if we had our 3 terminals each requiring a bandwidth of 3 kHz and a 300 Hz guard-band, Terminal 1 would be assigned the lowest frequency channel 0-3 kHz, Terminal 2 would be assigned the next frequency channel 3.3 kHz-6.3 kHz and Terminal 3 would be assigned the final frequency channel 6.6 kHz-9.6 kHz.

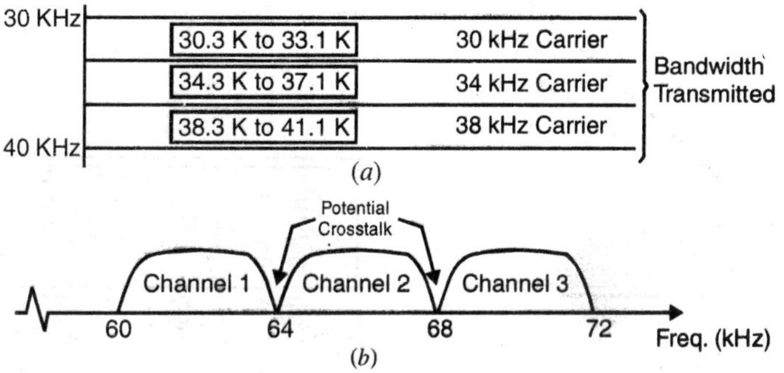

Fig. 9.2. *FDM in Telephone Transmission.*

The frequencies are stacked on top of each other and many frequencies can be sent at once. The downside is that the overall line bandwidth increases. Individual terminal requirement were 3 kHz bandwidth each, in the above example: the bandwidth to transmit all 3 terminals is now 9.6 kHz.

Each speech channel occupies 4 kHz of the available bandwidth.

Figure 9.2 (*b*) shows FDM in telephone transmission.

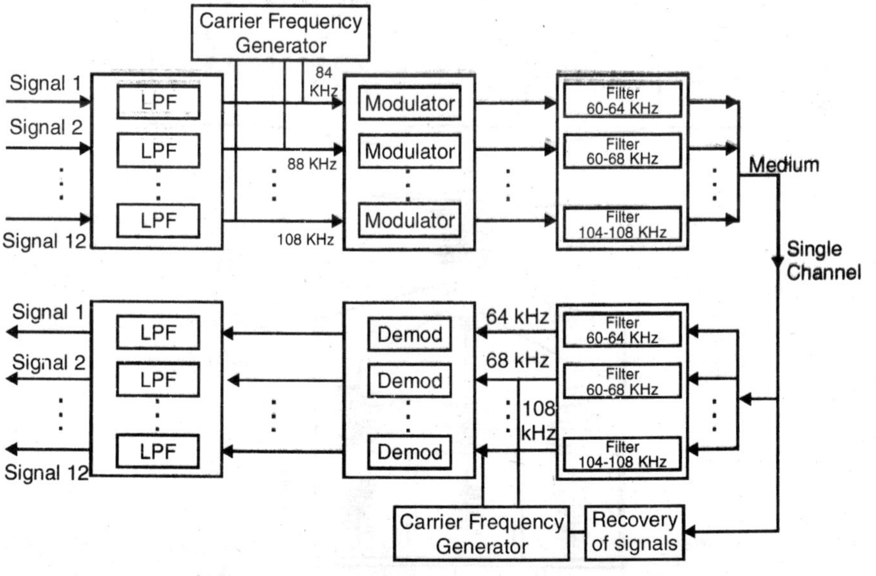

Fig. 9.3(*a*)

A typical FDM system is shown in Fig. 9.3 (*a* and *b*). The message signals are assumed to be low-pass signals. The LPF's ensure that the signals entering the modulators are bandlimited to some frequency, say, W. Each signal is then modulated with a different carrier. The frequencies of the carriers differ by at least so that the modulated signals are separated in frequency domain. The BPF's are used to restrict the modulated signals to their prescribed ranges. The outputs of the BPF's are added into a composite signal and sent through the communication channel.

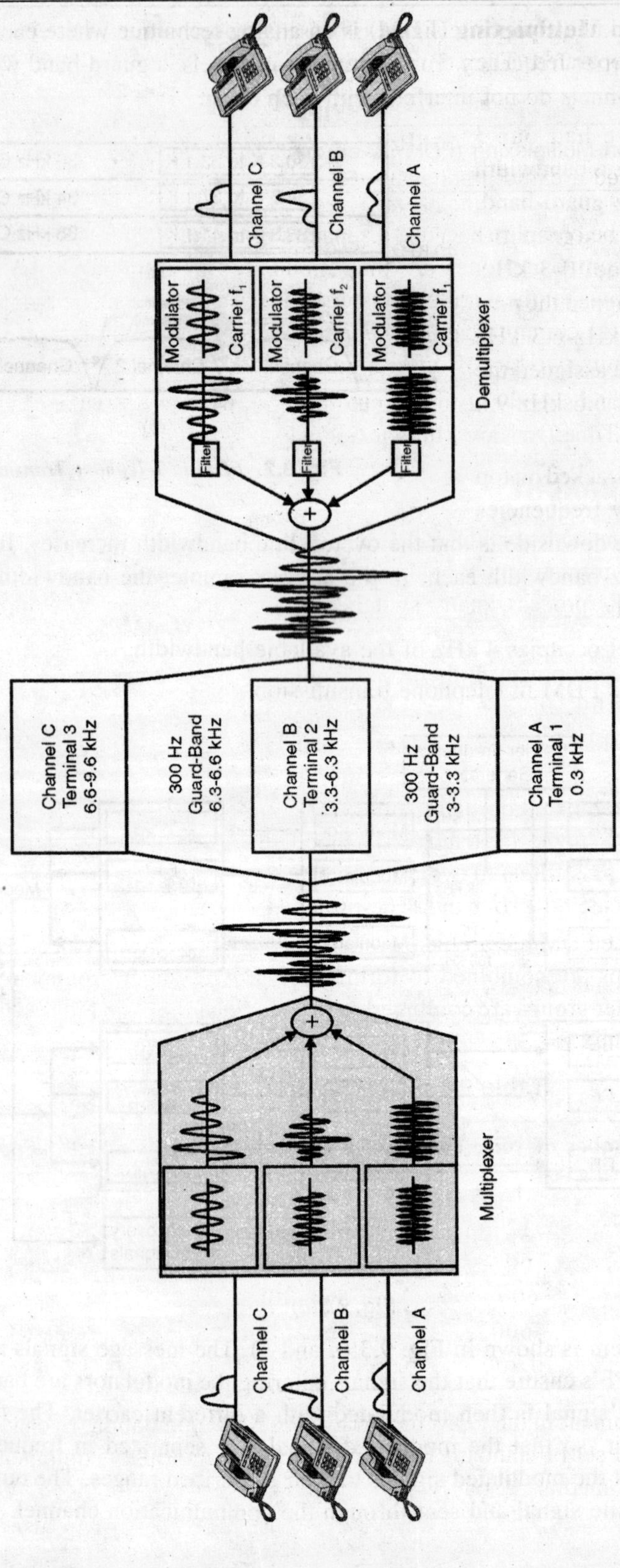

Fig. 9.3 *(b) A typical FDM System*

At the receiver end, the composite signal is passed through different BPF's so that the modulated message signals are separated. Then demodulation follows.

TIPS

Frequency Division Multiplexing (FDM) is an analog technique where each communications channel is assigned a carrier frequency.

FDM devices can be complex because a separate modulator/demodulator circuit is required for each tone used. The composite tones are then modulated to a single carrier frequency for radio transmission. FDM allows for the parallel transmission of data over a single communications channel. For example, the Link-11 communications system uses 15 audio tones to transmit 30 bits of parallel data. Each tone transmits two bits of differential quadrature phase-shift keyed data.

FDM is an analog and slightly historical multiplexing technique. It is prone to noise problems and has been overtaken by Time Division Multiplexing which is better suited for digital data.

9.3. THE FDM HIERARCHY

We know that for analog lines, FDM is used. Telephone companies traditionally used to maximize the efficiency of their infrastructure, multiplexed signals from lower bandwidth lines onto higher bandwidth lines. In this process, many switched or leased lines can be combined into fewer but bigger channels.

To standardize the equipment in the broad band transmission system CCITT recommended the following FDM hierarchy.

FDM principle : FDM for 12 telephone channels (group multiplex level) is considered for example. The signals pass through 12 low pass filters (LPF) to remove any high frequency components. The LPF outputs are modulated on 12 separate carrier signals separated by 4 kHz. The modulations in all FDM hierarchy are single side band (SSB) modulation. The output of each of the 12 modulations must be filtered (band pass filter) to avoid the interference with each other. BPF are used to restrict each signal to the allocated 4 kHz band. The principle of operation is shown in Fig. 9.3 (*a*).

Figure 9.4 shows the FDM hierarchical system. As the figure shows, voice channels are combined to forms groups, groups are combined to form super groups, super groups are combined to form master groups and master groups are combined to form jumbo groups and jumbo groups are combined to form jumbo group mix.

Table 9.1. FDM Hierarchical System.

Multiplex level	Number of voice circuits	Frequency band (kHz)	Formation
Voice channel	1	0–4	—
Group	12	60–108	12 voice circuits
Super group	60	312–552	5 group (12 × 5 = 60)
Master group	600	564–3084	10 super group (60 × 10 = 600)
Jumbo group	3600	564–17548	6 master group
Jumbo group mix	10,800	3000–6000	3 Jumbo groups

All multiplex equipment in the FDM hierarchy uses SSB modulation. Each level of the hierarchy is implemented using a set of standard FDM modules. The multiplex equipment is independent of particular transmission media.

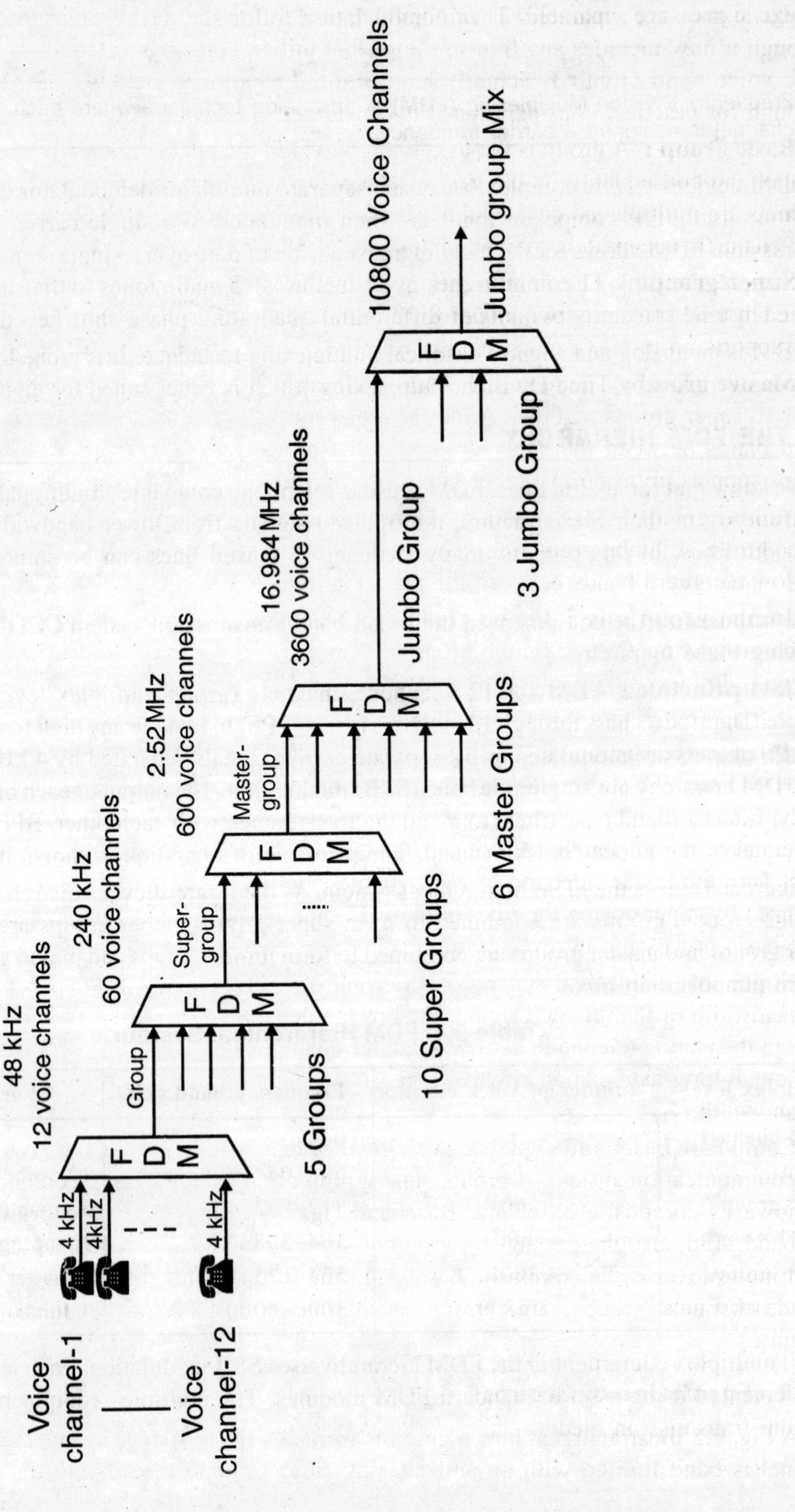

Fig. 9.4. *The FDM Hierarchical System.*

Message channel : The **message channel** is the basic building block of the FDM hierarchy. The basic message channel was originally intended for the analog transmission of voice signals, although it now includes any transmissions that utilize voice band frequency from 0 to 4 kHz. The basic voice band circuit is actually band limited to approximately a 300 Hz to 3000 Hz band, although for practical consideration it is considered a 4 kHz.

Basic group : A group is the next higher level in the analog hierarchy above the basic message channel and consequently is the first multiplexing step for combining message channels. 12 voice channels are multiplexed onto a higher bandwidth line to create a basic group. Twelve 4 kHz voice band channels occupy a combined bandwidth of 48 kHz.

Super groups : The next higher level in the analog hierarchy is the super group, which is formed by FDM five groups containing 12 channels each. A super group has a bandwidth of 240 kHz and supports up to 60 voice channels.

Master group : The next higher level of multiplexing is the master group, which is formed by FDM 10 super groups together. A master group must have 2.40 kHz of bandwidth, but the need for guard bands between the super groups increases the necessary bandwidth to 2.52 MHz Master groups support up to 600 voice channels.

Jumbo group : Master group can be further multiplexed in master group banks to create a jumbo group. A jumbo group must have 15.12 MHz (6×2.52 MHz) but is augmented to 16.984 MHz to allow for guard bands between the master groups.

Jumbo group mix : Jumbo group can be further multiplexed, which is formed by 3 FDM Jumbo groups together. A Jumbo group mix support up to 10800 voice channels.

All multiplex equipment in the FDM hierarchy used SSB modulation. Thus, every voice circuit required approximately 4 kHz of bandwidth. The lowest level building block in the hierarchy is a channel group consisting of 12 voice channels. A channel group multiplex uses a total bandwidth of 48 kHz. Figure 9.5 shows a block diagram of an A5 channel group multiplexer, the most common A –type channel bank used for first' level multiplexing. Twelve modulators using 12 separate carriers generate 12 double-sidebands signals as indicated. Each channel is then bandpass filtered to select only the lower sideband of each double-sideband signal. The composite multiplex signal is produced by super posing the filter outputs. Demultiplex equipment in a receiving terminal uses the same basic processing in reverse order.

Notice that a sideband separation filter not only removes the upper sideband but also restricts the bandwidth of the retained signal: the lower sideband. These filters therefore represented a basic point in the analog telephone network that defined the bandwidth of a voice circuit. Since FDM was used on all long-haul analog circuits, long-distance connections provided somewhat less than 4 kHz of bandwidth.

Composite baseband signals : A single message channel is baseband i.e the modulating signal in a communication system. A group, super group or signals prior to modulation of the final carrier is known as composite baseband. Baseband signal in Fig. 9.3, the output of a channel bank is baseband. And, also the output of a group or super group bank is baseband. The final output of the FDM multiplexer is the composite baseband. The formation of the composite base band signal can include channels, groups, super group and master group banks, depending on the capacity of the systems.

9.3.1. Formation of Groups

A Fig. 9.5 illustrated that how a group is formed with an A-type channel bank. Each voice band channel is band limited with an anti aliasing filter prior to modulating the channel carrier. The

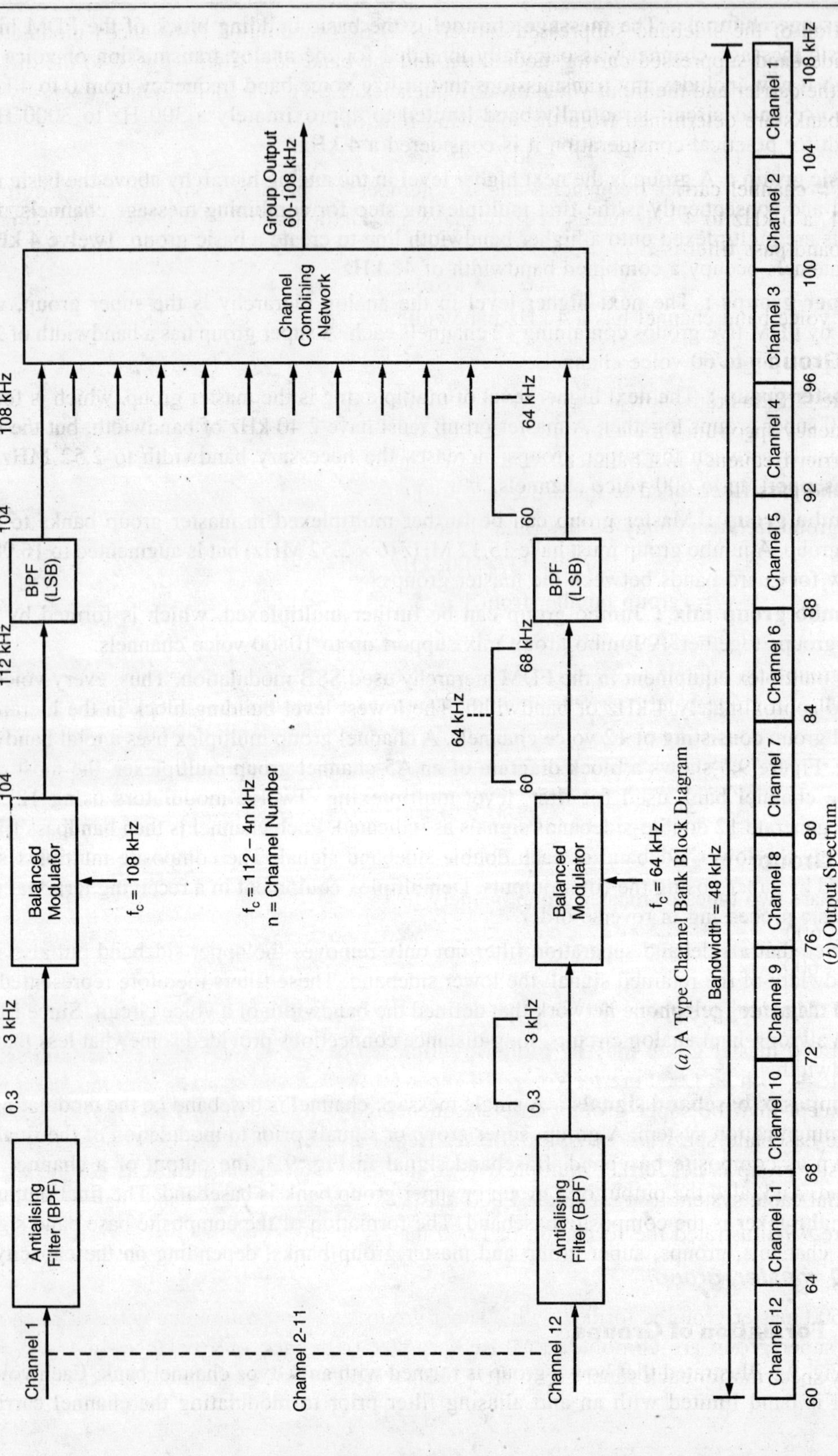

Fig. 9.5. *A block diagram of A-Type channel group multiplexer.*

combination of the sideband suppressed carrier (SSB-SC) modulator. A balanced modulator is a double side band suppressed carrier modulation and the band pass filter is tuned to the difference between the carrier and the input voice band frequencies (LSB). The ideal input frequencies for the channel banks are determined from the following relation,

$$Fc = 112 - 4n \text{ kHz}$$

where, f_c = channel carrier frequencies and n = number of channels from 1 to 12. Therefore, for channel 1, a 0 kHz to 4 kHz bands of frequencies modulation a 108 kHz carrier. The output of a channel band pass filter is

$$F_{out} = (f_c - 4 \text{ kHz}) \text{ to } f_c$$

Earn voice band channel has a 4 kHz bandwidth.

Super Group

Figure 9.6 illustrates how a super group is formed with a group bank and combining network. The frequency spectrum for each group is 60 kHz to 108 kHz. Each group is mixed with a different group carrier frequency in a balanced modulator. The output of balanced modulator is band limited with a band pass filter tuned to the difference frequency band (LSB) to produce a SSB-SC signal.

The group frequency may be determined from the following relation:

$$F_c = 327 + 48n \text{ kHz}$$

where, n = group number from 1 to 5

F_c = group carrier frequencies

For group-1, a 60 kHz to 80 kHz group signal modulator a 420 kHz group carrier frequency. The output of a group band pass filter is given as

$$F_{out} = (f_c - 108 \text{ kHz}) \text{ to } (f_c - 60 \text{ kHz})$$

$$= (327 + 48n - 108 \text{ kHz}) \text{ to } (327 + 48n - 60 \text{ kHz})$$

$$F_{out} = (48n + 264 \text{ kHz}) \text{ to } (48n + 312 \text{ kHz})$$

Master Group

There are two types of master group, namely :

1. L 600
2. U 600

1. L600 Master group

The L600 master group may be further multiplexed and used for higher capacity microwave radio systems with an L600 master group: 10 super groups are combined as with the U600 master group, except that the super group carrier frequencies are lower. In an L600 master group, the composite baseband spectrum occupies a lower frequency band than the U600 master group. An L600 master group is not further multiplexed. So the maximum channel capacity for a microwave or a coaxial cable system using a single L600 master group is 600 voice band channels.

Figure 9.7 illustrated the formation of L600 master group.

2. U600 master group

A U600 master group is formed with a super group bank and combining network as shown in Fig. 9.8 super group are combined to form a carrier frequency in a balanced modulator. The frequency spectrum for each super group is 312 kHz to 552 kHz. The output of balanced modulator

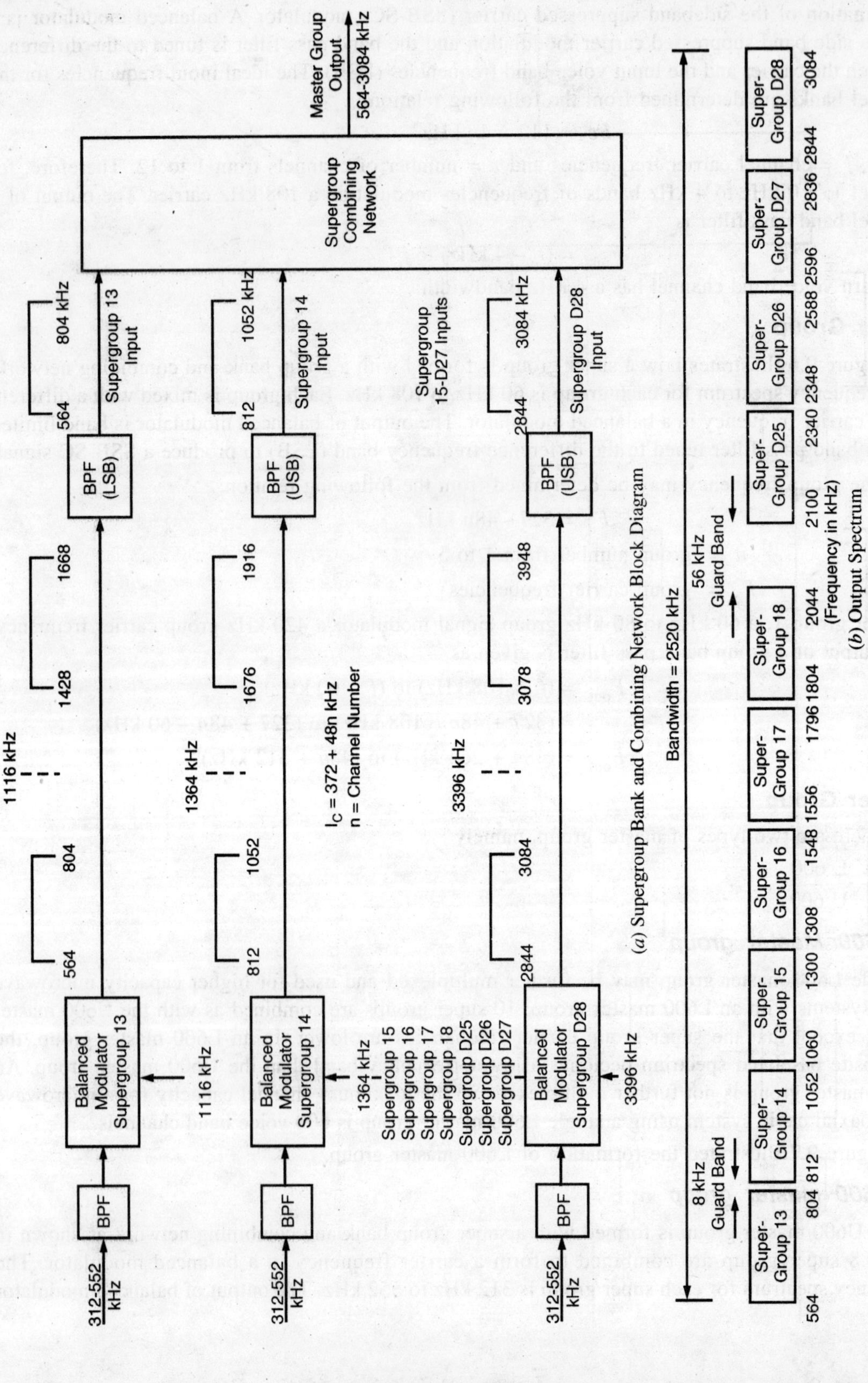

Fig. 9.6. *Formation of L600 Master Group.*

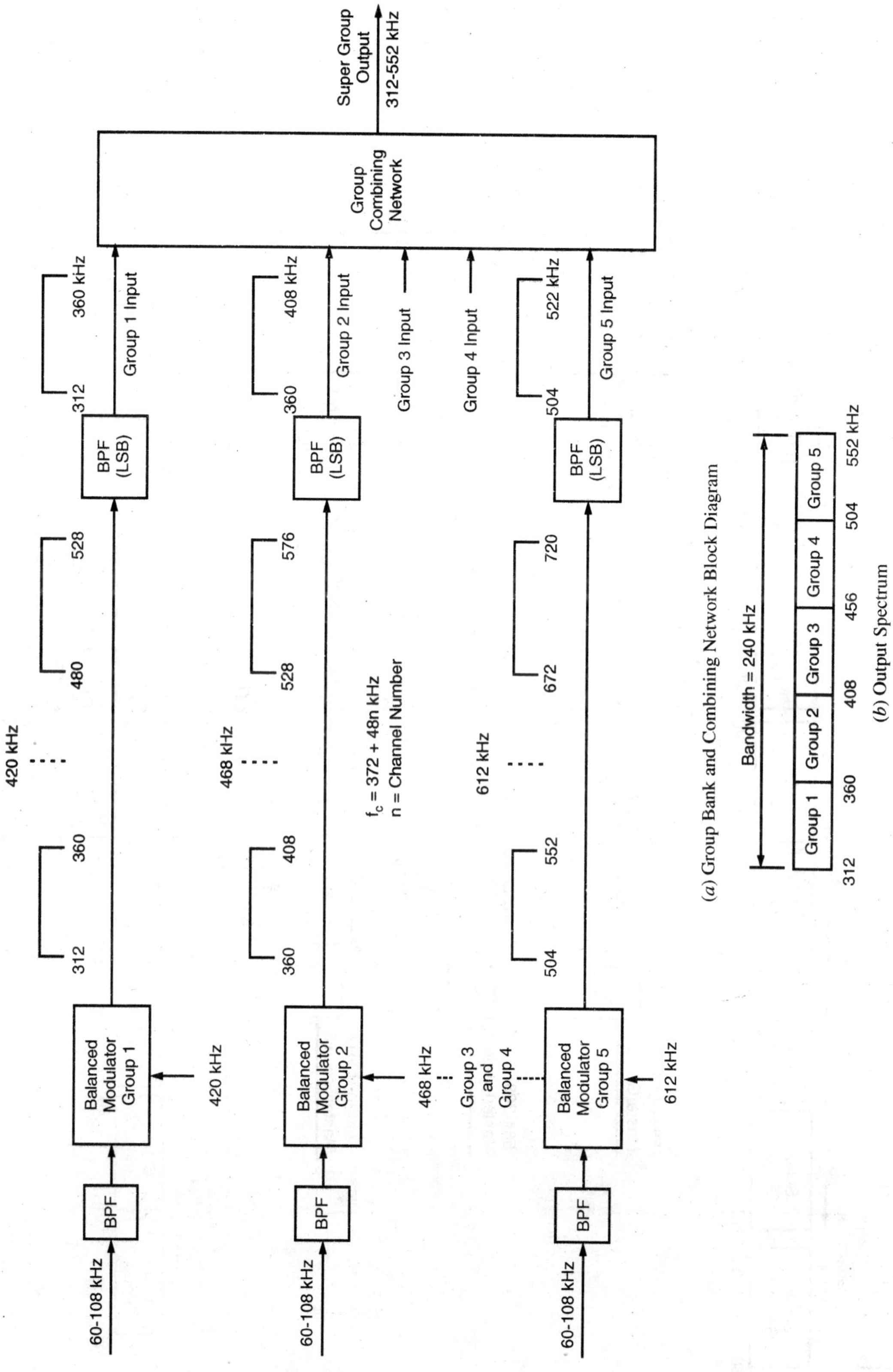

(*a*) Group Bank and Combining Network Block Diagram

(*b*) Output Spectrum

Fig. 9.7.

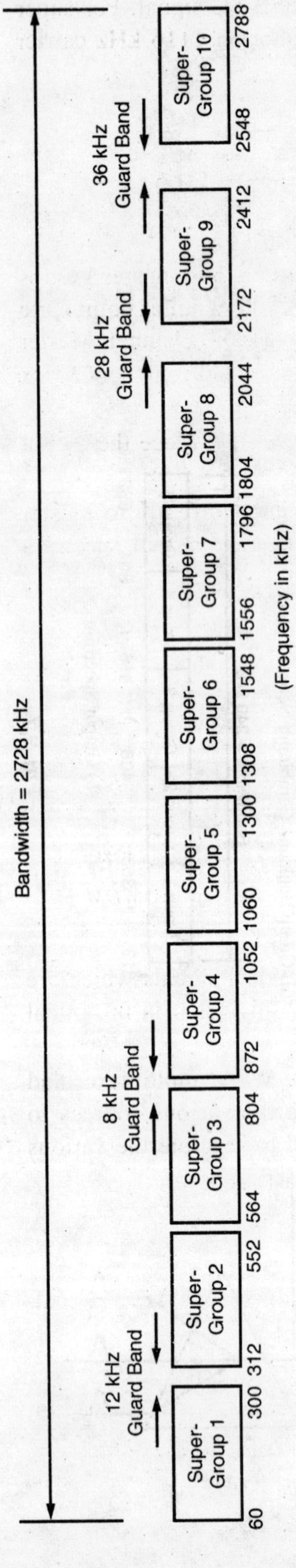

Fig. 9.8.

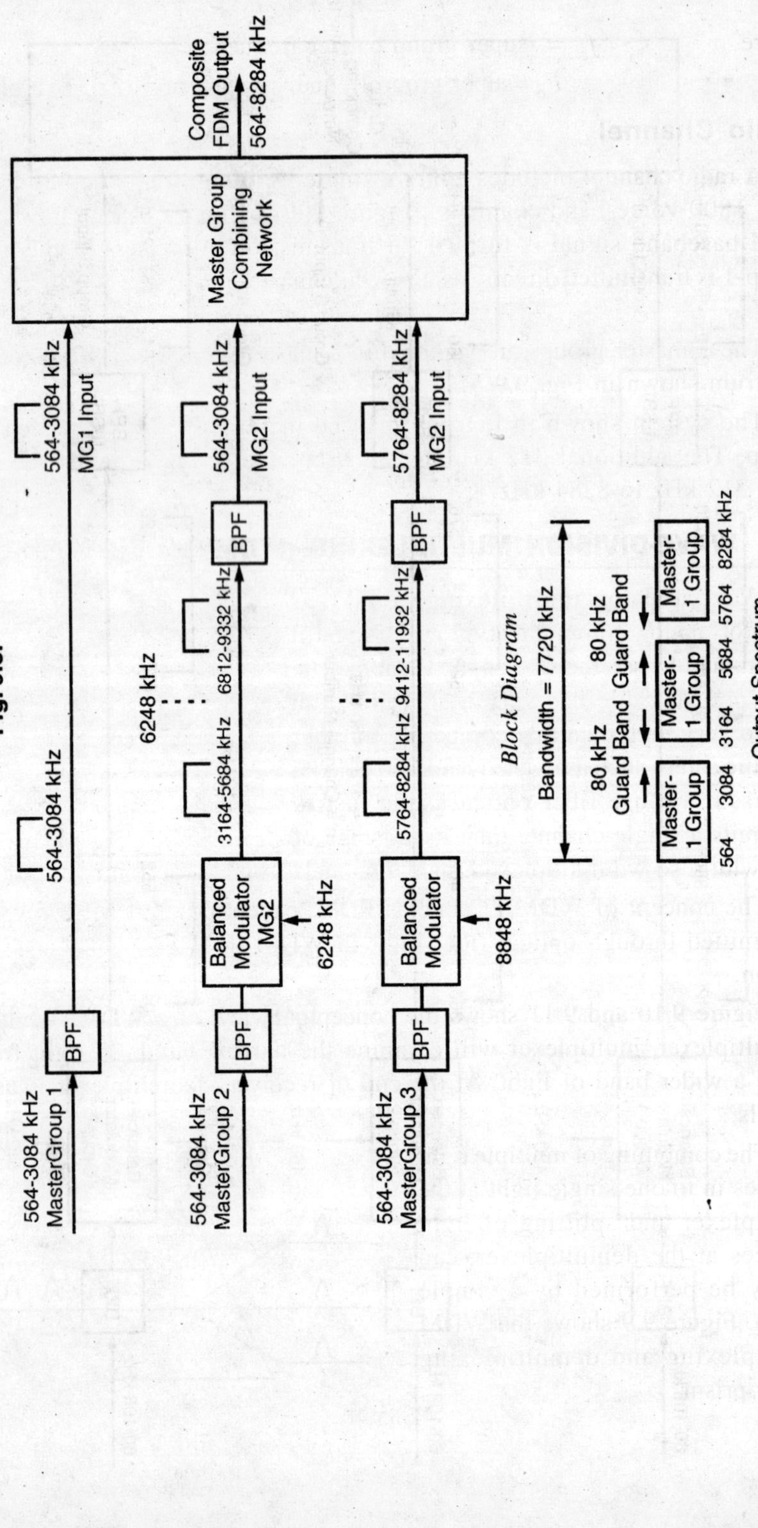

Fig. 9.9. *Three-Mastergroup Radio Channel*

is then band limited to the difference frequency band (LSB) to form a SSB-SC signal. For super group 13, a 312 kHz to 552 kHz super group band of frequency modulator an 1116 kHz carrier frequency. The output from a super group band pass filter is

$$F_{out} = f_c - f_s \text{ to } f_c$$

where f_c = super group carrier frequencies

F_c =super group frequency spectrum (*i.e.*, 312 kHz to 552 kHz)

Radio Channel

A radio channel includes either a single L 600 master group up to three U600 master groups (*i.e.*, 1800 voice-band channels). Figure 9.9(*a*) illustrates that how an 1800 channel composite FDM baseband signal is formed for transmission over a single microwave radio channel. Master group-1 is transmitted directly as it is, while master groups 2 and 3 undergoes as additional multiplex step.

The 3 master groups are summed in a master group combining network to produce the output spectrum shown in Fig. 9.9(*b*).

The system shown in Fig. 9.9 may be increased from 180 voice channel directly to master group. The additional 312 kHz to 552 kHz super group extends the composite output spectrum from 312 kHz to 8284 kHz.

9.4. WAVE-DIVISION MULTIPLEXING (WDM)

Wave division multiplexing (WDM) is used for analog communication system and it is designed to use the fiber optical cable. The fiber optical cable has very high bandwidth and so very high data rate comparing to other metallic cables such as twisted pairs and coaxial cables. If we use fiber optical cable just to transmits a single channel than its wastage of bandwidth, so we multiplexed many channels to make optimum use of data link.

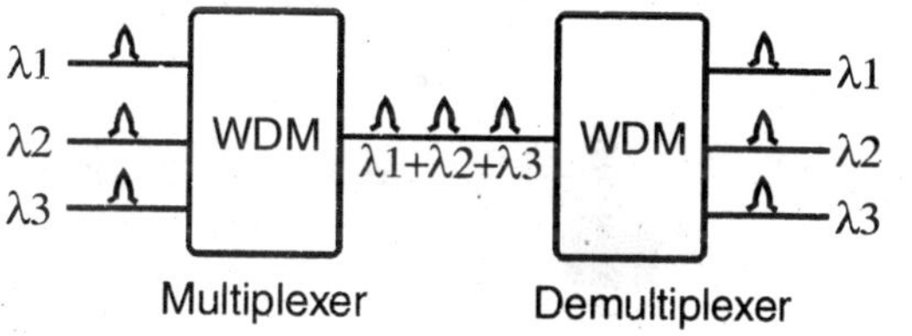

Fig. 9.10. *The Conceptual View of a WDM.*

The concept of WDM is same as FDM, except that WDM involves optical signals which are transmitted through optical fiber link. In WDM the frequencies are very high that is in optical region.

Figure 9.10 and 9.11 shows the conceptual view of a WDM. We have WDM multiplexer and demultiplexer, multiplexer will combine the narrow bands of light from the various sources to make a wider band of light. At the end of receiver, demultiplexer is used to separate the various signals.

The combining of multiple light sources in to one single light at the multiplexer and splitting of light sources at the demultiplexer, can easily be performed by a simple prism. Figure 9.9 shows the WDM multiplexing and demultiplexing using prism.

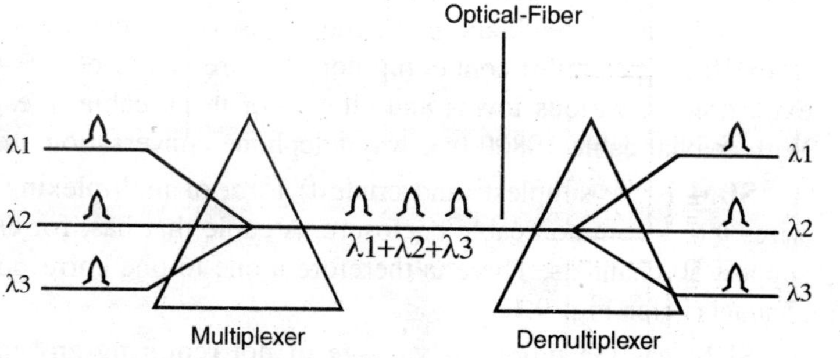

Fig. 9.11. *Prisms used in WDM.*

Using a new method, called **Density Wave-Division Multiplexing (DWDM)**, the spacing between adjacent frequencies is considerably less, can multiplex a very large number of channels.

9.5. SPACE DIVISION MULTIPLEXING (SDM)

In space divisions multiplexing more than one physical transmission path are grouped together. A telephone cable consisting of hundreds (or thousands) of twisted pair constitutes a space division multiplexed system, wire pair cables are constructed containing many hundreds of wire pairs. Several coaxial tubes bound together in one cable is also an example of space division switching.

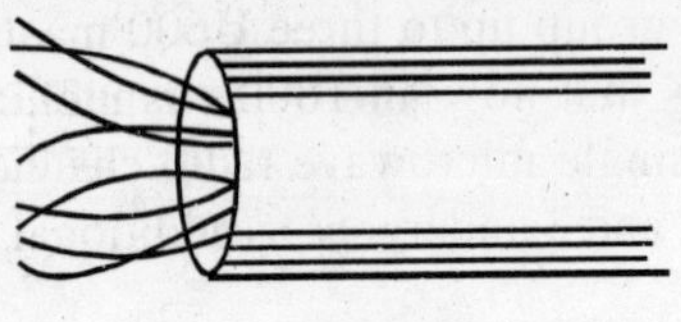

(*a*) Unshielded Twisted-pair (UTP) cable

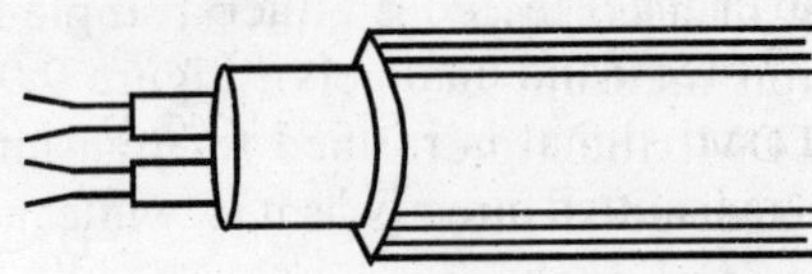

(*b*) Shielded Twisted-pair (STP) cable

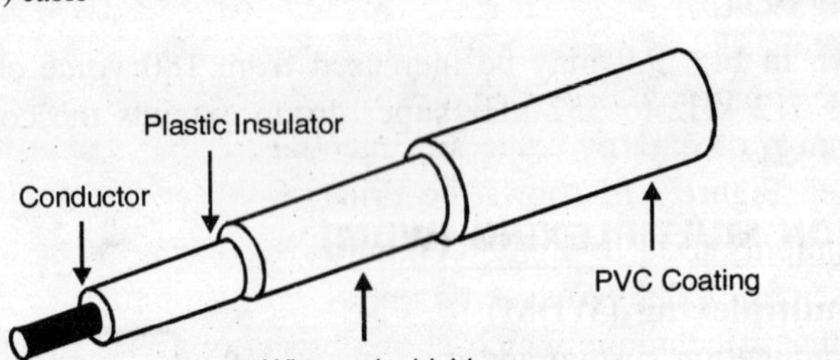

(*c*) Coaxial cable

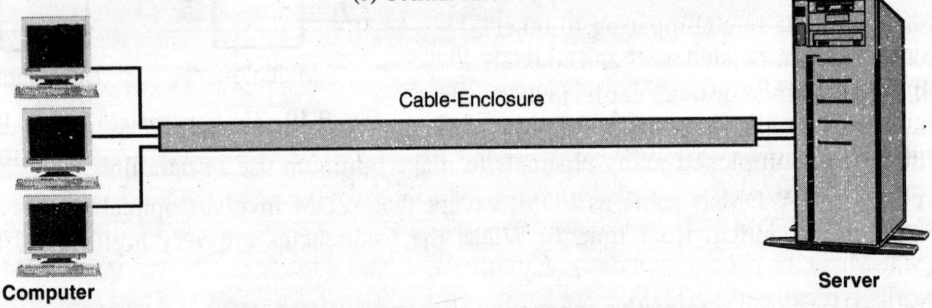

Fig. 9.12. *Space Division Multiplexing (SDM).*

A very large number of separate telephone calls can be transmitted together down a coaxial system. Single wire pair commonly carries 12 or 24 voice channels. But one single coaxial tube commonly carries 3600 and the higher capacity one can carry 10800. In India, the telephone cables (5 or 10 lbs per mile) containing copper wires pairs of 20, 40, 60, 80, 100, 1000 and 5 per cent extra used in various towns and cities. For trunk cable lines, 70 – 300 lb copper wires are used. With coaxial cable 10800 two way telephone conversation are possible.

SDM is the simplest (and crudest) form of multiplexing. It involves grouping many separate wires into a common cable enclosure. A cable that has, for example, 50 twisted pairs inside it can support 50 channels. There is therefore a one-to-one correspondence between physical and logical channels refer Fig. 9.12.

SDM has the unique advantage of not requiring any multiplexing equipment. It is usually combined with other multiplexing techniques to better utilize the individual physical channels.

The space division switching is not limited to voice frequency circuits alone. Many high capacity transmission systems using either frequency or time division multiplexing can also be space division multiplexed.

9.6. TIME DIVISION MULTIPLEXING

In 1970's TDM was introduced and now favored over FDM. TDM is the sharing of a common transmission medium in time. In TDM, the time available is divided into small slots, and each of them occupied by a piece of one of the signals to be sent. Thus the multiplexing device should scan the input signal in round-robin fashion. TDM is a base band technology in which individual channels of data or voice are interleaved into a stream of framed bits across a communication channel. Figure 9.13 shows the Principle of TDM.

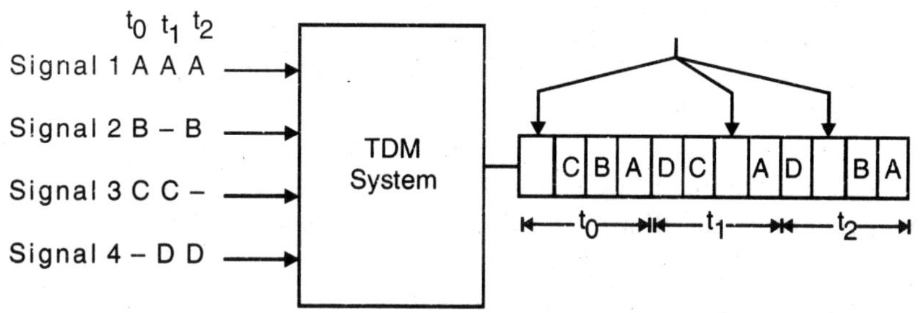

Fig. 9.13. *Principle of TDM.*

TDM is more commonly used in digital communication systems. **Time-division multiplexing (TDM)** grants each user full channel capacity, but assigns time slots to each user. In simple term TDM can be defined that a technique which allows all the users to share the time period, use the available data link with the full capacity for the predefined period, say T. Transmissions from various sources are interleaved in the time domain.

There are two basic modes of operation for TDM: those that repeatedly assign a portion of the transmission capacity to each source and those that assign capacity only as it is needed. The first form of operation is referred to as synchronous time division multiplexing (STDM) when necessary to; distinguish it from the "as-needed" mode of operation. Otherwise, TDM is generally understood to imply the synchronous variety. The as-needed form of TDM is referred to as a statistical time division multiplexing.

Multiplexing (ATDM), Asynchronous Transfer Mode (ATM), statistical time division multiplexing (Stat-Mux), or packet switching. Circuit-switched telephone networks use STDM, whereas data networks typically use ATDM.

If analog signals are to be multiplexed, the signals should be sampled. The analog signals are digitized by CODEC (coder/decoder) device. It produces 7 or 8 bit number. Sampling is 8000 per sec (125 μ sec/sample) to capture all the information from the 4 kHz telephone channel bandwidth. Pulse cone modulation (PCM) technique is used in TDM. T1 carrier is the widespread method (in North America and Japan) T1 consists of 24 voice channels multiplexed together. Each channel has 8 bits, 7 for information and 1 for signalling,

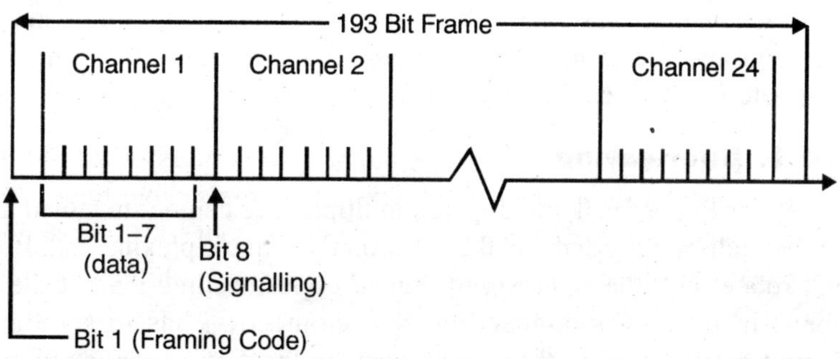

Fig. 9.14.

as shown in Fig. 9.14. The output frame of 125 micro sec consists of 24 × 8 = 192 bits plus one extra bit for framing.

Pulse Code Modulation is the most prevalent encoding technique used for TDM digital signals. Two or more voice channels are sampled and converted to PCM codes and then time multiplexed onto a single transmission line. Each user is connected to a time-division multiplexer. Data signals from the user are fed to the time-division multiplexer buffer, and the time slots are rotated among the users and scanned for data. The data from each user can be in the form of bits, bytes, or blocks. The data signals from all users are compiled into frames for transmission on a single, high-speed communications channel. Transmit and receive frames are used for half-duplex communications. Transmit frames are sent and a receive time slot is enabled for return information. In this manner, a single carrier frequency and modem may be used to transmit and receive information at a fairly high rate of speed. Since time slots are preset and assigned, if a user has no data to transmit, the time slot is wasted.

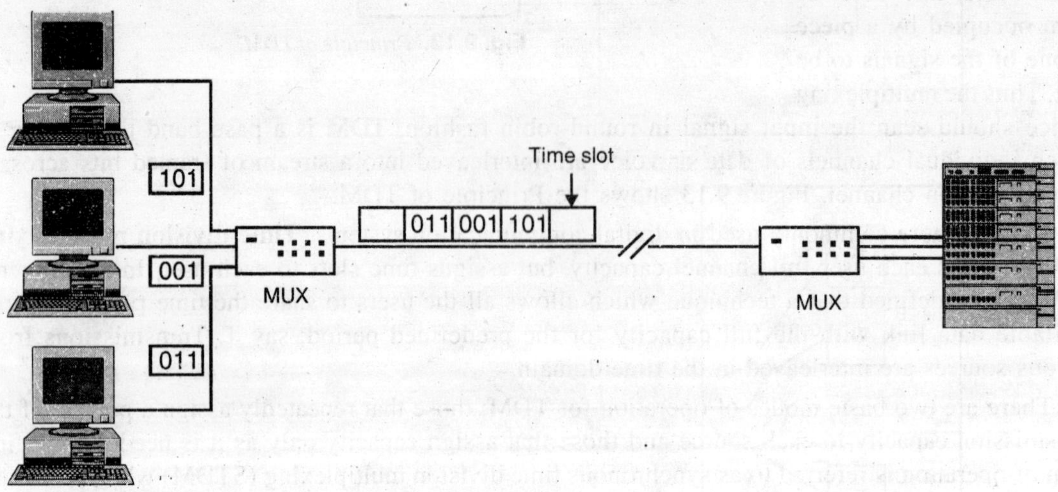

Fig. 9.15. *A TDM system.*

TDM requires special multiplexing/demultiplexing hardware (MUX) at either end of the line (see Fig. 9.15). Because the channels are spread across time, some means of initial synchronization is also needed. Basically, the receiving end needs to know which time slot belongs to the first channel when the connection is established, and can work everything else out from this reference point.

Each channel has a time slice assigned to it whether the terminal is being used or not. Again, to the send and receiving stations, it appears as if there is a single line connecting them. All lines originate in one location and end in one location.

9.6.1. Interleaving

Refer Fig. 9.16, Time division multiplexing can be imagined as two fast rotating switches, one on the multiplexing end and the other on the demultiplexing end. Both the switches are synchronized and rotate with the same speed, but in opposite direction. At the multiplexing end, as the switch opens in front of a connection, that connection has to send a unit onto the path and on the demultiplexing end, as the switch opens in front of a connection, now the time has come to receive a unit from the path. This process is referred as **Interleaving.**

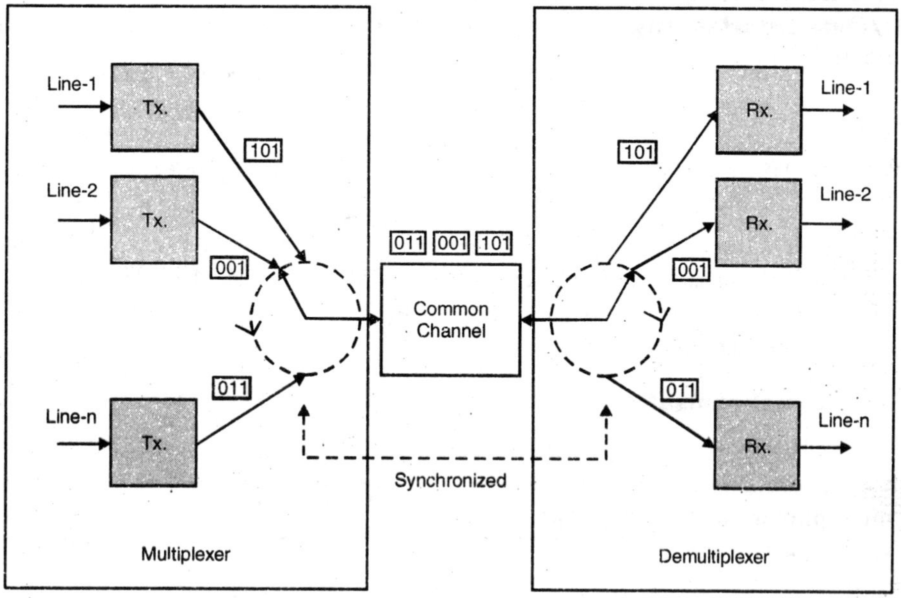

Fig. 9.16. *A TDM System Shows Interleaving.*

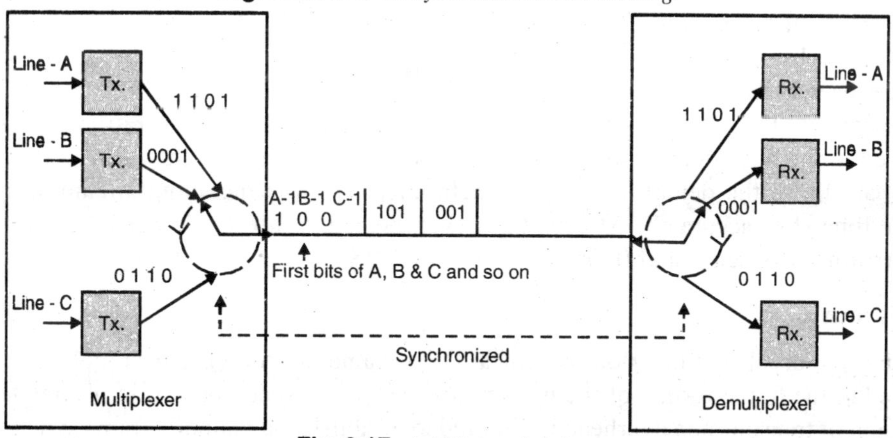

Fig. 9.17. *(a) Bit interleaving.*

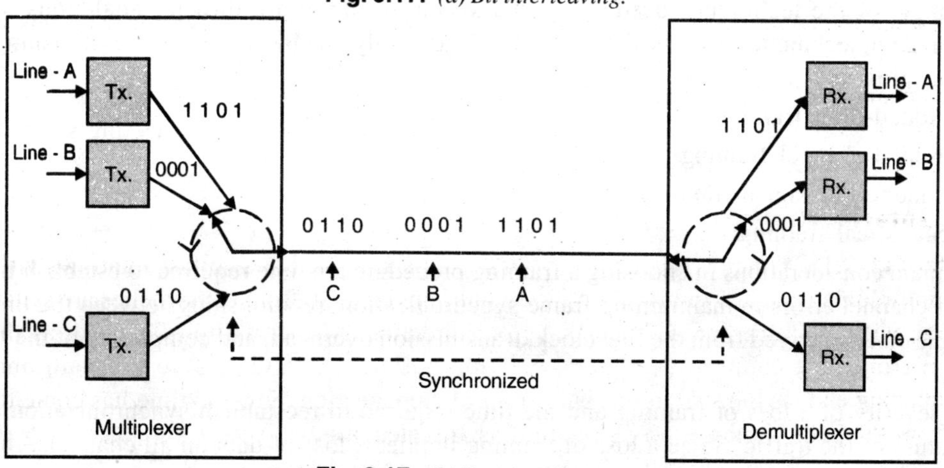

Fig. 9.17. *(b) Byte interleaving.*

9.6.2. Bit/Byte Interleaving

Time division multiplexing can be either bit by bit or byte by byte basis, known as bit interleaving and byte interleaving respectively. Bit interleaving also known as digit interleaving, used in North American Digital Hierarchy (T-1 system). Byte interleaving also known as word interleaving, used in Digital services (DS) and SONET (synchronous optical network system).

Figures 9.16 and 9.17 shows these two schemes. It can be seen from the example shown in Fig. 9.17, that we have three data, A, B and C. When one bit of each data is being is sent then it is referred as bit interleaving and when we sent 4 bits (word/byte) of A, 4 bits of B and 4 bits of C then it is referred as byte/word interleaving shown in Fig. 9.17. Reverse process is done at the receiver end to recover the data.

9.6.3. Frame Synchronization

As we have already discussed that synchronization between multiplexer and demultiplexer is must in the interleaving process. If the synchronization between multiplexer and demultiplexer is not there than a bit of one channel may be received by the wrong channel. To avoid this problem, we use one or more synchronization bits, which are usually added to the beginning of the each frame. These synchronizing bits follow a pattern, frame to frame, that allows the demultiplexer to synchronize with the incoming stream so that it can separate the time slots accurately. Mostly, this synchronization information consists of one bit per frame, alternating between 1 and 0, as shown in Fig. 9.18.

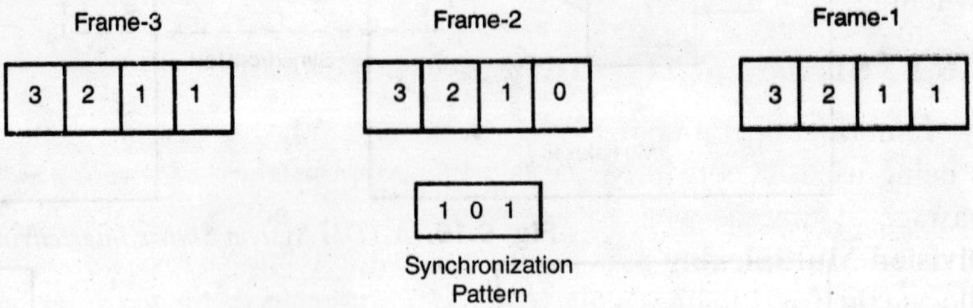

Fig. 9.18. *Synchronization Bits Pattern.*

9.6.4. Framing

To identify individual time slots within a TDM frame, a receiving terminal uses a counter synchronized to the frame format of the transmitter. Just as for synchronization of sample clocks, a certain amount of transmission overhead is required to establish and maintain frame synchronization. In fact, most of the techniques used for frame synchronization are directly analogous to clock synchronization techniques discussed previously. Specifically, the basic means of establishing frame synchronization are;

1. Added-digit framing
2. Added-channel framing
3. Unique line signal framing
4. Statistical framing

The main considerations in choosing a framing procedure are time required to establish framing, effects of channel errors in maintaining frame synchronization, relationships between the line clock and sample clocks derived from the line clock, transmission overhead, and complexity of the framing circuitry.

The severity of a loss of framing and the time required to reestablish synchronization depend on the nature of the traffic. Since a loss of framing implies a loss of data on all channels, the mean time between misframes must be as long as possible. For voice traffic in frequent misframes can be tolerated if frame synchronization is reestablished rapidly enough to minimize the duration of the "glitch" in the output speech.

For data traffic the duration of reframe times is not as critical as the frequency of occurrence since most data communications protocols initiate recovery procedures and retransmit a message no matter how many data bits are lost.

A critical requirement for reframe time in the telephone network comes from the possibility that various in-channel control signals may be lost and interpreted as disconnects. Thus the maximum reframe time on a particular digital transmission link is often determined from analog network signalling conventions.

A loss of framing is also used as a performance monitor for the transmission link and usually sets alarm conditions, which in turn cause automatic switching to spare terminals or transmission facilities. Generally speaking, terminals serving 600 or more channels require automatic protection switching.

9.6.5. Statistical Time Division Multiplexing (STDM)

In normal operation of TDM, each channel has a time slice assigned to it whether the terminal is being used or not. In case when there is no data to transmit in assigned slot of time than it is wastage of time as well as bandwidth. So an another type of TDM, known as **Statistical Time Division Multiplexing** uses intelligent devices capable of identifying when a terminal is idle. They allocate time only to lines when required. This means that more lines can be connected to a transmission medium as this device statistically compensates for normal idle time in data communication lines. Newer STDM units provide additional capabilities such as data compression, line priority, mixed speed lines, host port sharing, network port control, automatic speed detection and much more.

> **DO YOU KNOW**
>
> **Statistical Time Division Multiplexing** uses intelligent devices capable of identifying when a terminal is idle.

9.6.6. Advantages of a TDM System

1. TDM is more efficient,
2. More easier to operate,
3. Less complex and
4. Less expensive than FDM
5. Its ability to handle devices with varying speeds, its effectiveness when used with devices that transmit data almost continuously, and its simple implementation.

9.6.7. Time Division Multiplex Loops and Rings

It is mentioned that TDM is not as acquiescent to applications with distributed sources and sinks of traffic as is FDM. In this section a particular form of a TDM network is described that is quite useful in interconnecting distributed nodes. The basic structure of interest is referred to as a TDM loop or TDM ring and is shown in Fig. 9.19.

Basically, a *TDM ring* is configured as a series of unidirectional (two-wire) links arranged to form a closed circuit or loop. Each node of the network is implemented with two fundamental operational features. First, each node acts as a regenerative repeater merely to recover the incoming bit stream and retransmit it. Second, the network nodes recognize the TDM frame structure and communicate on the loop by removing and inserting data into specific time slots assigned to each node. As indicated in Fig. 9.19, a full-duplex connection can be established between any two nodes by assigning a single time slot or channel for a connection. One node insets information into the assigned time slot that propagates around the loop to the second node (all intervening nodes merely repeat the data in the particular time slot). The destination node removes data as the assigned time slot passes by and inserts return data in the process. The return data propagates around the loop to the original node where it is removed and replaced by new data, and so forth.

Since other time slots are not involved with the particular connection shown, they are free to be used for other connections involving arbitrary pairs of nodes. Hence a TDM Loop with C time slots per frame can support C simultaneous full-duplex connections. If, as channels become available, they are reassigned to different pairs of nodes, the transmission facilities can be highly utilized with high concentration factors and provide low blocking probabilities between all nodes. Thus a fundamental attraction of a loop network is that the transmission capacity can be assigned dynamically to meet changing traffic patterns. In contrast, if a star network with a centralized switching node is used to interconnect the nodes with four-wire links, many of the links to particular nodes would be underutilized since they cannot be shared as in a loop configuration.

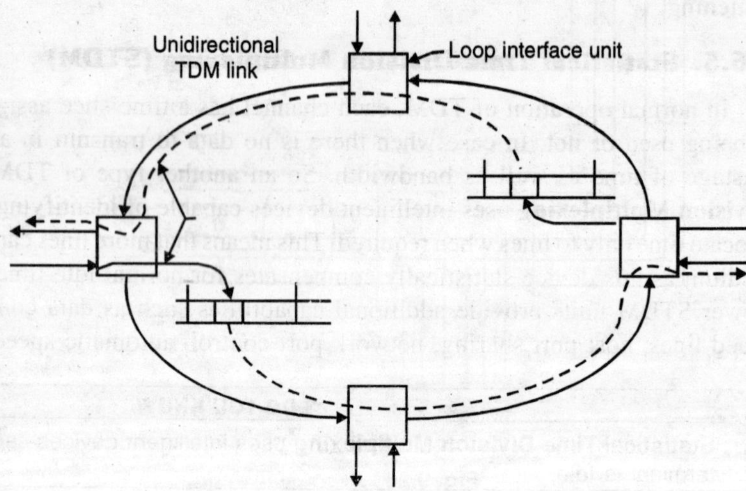

Fig. 9.19. *TDM Loop.*

Another feature of the loop-connected network is the ease with which it can be reconfigured to accommodate new nodes in the network. A new access node is merely inserted into the nearest link of the network and the new node has complete connectivity to all other nodes by way of the TDM channels. In contrast, a star structured network requires transmission to the central node and expansion of the centralized switching facilities.

The ability to reassign channels to arbitrary pairs of nodes in a TDM loop implies that the loop is much more than a multiplexer. It is, in fact, a distributed transmission and switching system. The switching capabilities come about almost as a by-product of TDM transmission. TDM loops represent the epitome of integrated transmission and switching.

DO YOU KNOW

TDM loops have been used within computer complexes to provide high capacity and high interconnectivity between processors, memories and peripherals.

The loop structure in this application is sometimes more attractive than more conventional bus structures since all transmission is unidirectional and therefore avoids timing problems on bidirectional uses that limit their physical length. Furthermore, as more nodes are added to a bus, the electrical loading increases, causing a limitation on the number of nodes that can be connected to a bus. Loops, on the other hand, have no inherent limits of transmission length or numbers of nodes.

The loop structure of Fig. 9.19 is topologically identical to the token-passing ring developed by IBM and standardized by the IEEE as an 802.5 local area network.

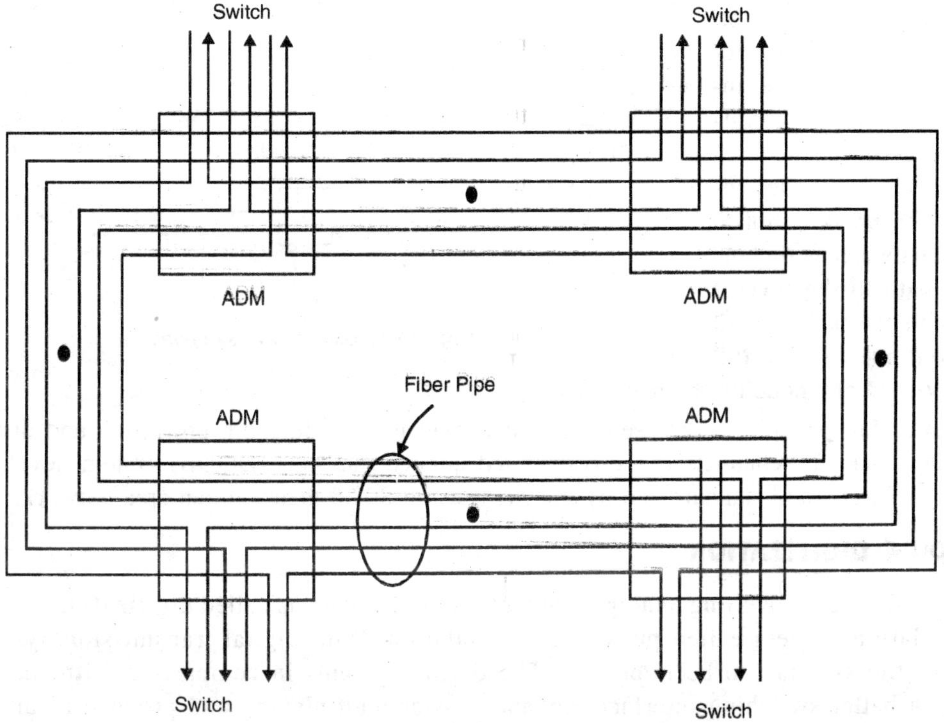

Fig. 9.20. *Functional Mesh, Fiber Loop and ADM's.*

However a token-passing ring operates differently than a TDM loop in that there is only one channel. When a node on a ring becomes active, it uses the entire capacity of the outgoing link until it is through sending its message. In contrast, a node on a loop uses only specific time slots in the TDM structure, allowing other nodes to be simultaneously connected, using other time slots. In essence, a TDM loop is a distributed-circuit switch and an 802.5 ring is a distributed-packet switch. A particularly attractive use of a loop with high-bandwidth links is shown in Fig. 9.19. This figure illustrates the use of add-drop multiplexers (ADMs) that access whatever bandwidth is needed at a local node but pass the rest on to other nodes.

In typical applications the amount of bandwidth allocated to each node is quasi-static. It is changed only in response to macroscopic changes in traffic patterns, possibly as a function of the time of day. This basic operation is generally often referred to as a cross-connect function as opposed to a switching function, which involves call-by-call reconfigurations.

An important point to note about Fig. 9.20 is the ability to utilize a general-purpose physical topology but define an arbitrary functional topology on top of it.

One obvious limitation of a loop is its vulnerability to failures of any link or node.

The effect of a node failure can be minimized by having bypass capabilities included in each node. When bypassed, a node becomes merely a regenerative repeater, as on T-carrier transmission links. Link failures can be circumvented by providing alternate facilities. Figure 9.21 shows one particular structure using a second, reverse-direction loop to provide backup capabilities in the case of failures. When fully operational, the network can use the reverse loop as a separate, independent network for traffic as needed. Whenever a failure occurs, the nodes adjacent to the break establish a new loop by connecting the forward path to the reverse path at both places. Hence all nodes continue to have full connectivity to any node on the new loop.

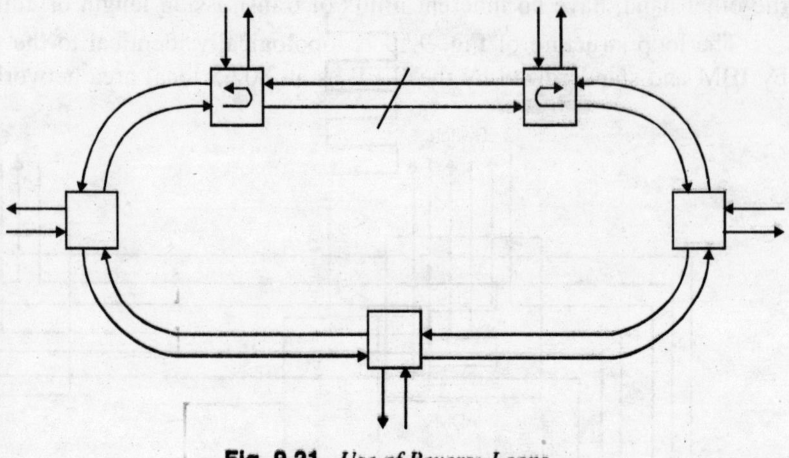

Fig. 9.21. *Use of Reverse Loops.*

A particular example of the use of the dual reverse loops for both protection and distributed queued access to the channels is the distributed queued dual-bus (DQDB) system developed by QPSX in Australia and standardized by the IEEE as an 802.6 metropolitan area network.

9.7. VOICE DIGITIZATION

The process of converting analog signal into digital signals is called **digitization**. It involves the modulation process called pulse code modulation. With digital transmission system, the quality of the system can be improved. The digital systems in comparison with the analog, provides a better switching interface, enhance easier multiplexing and produce clear signals. To convert analog signals to digital signals, a coding system called pulse code modulation (PCM) is used.

The process of digitization generally involves four steps as shown in Fig. 9.22.

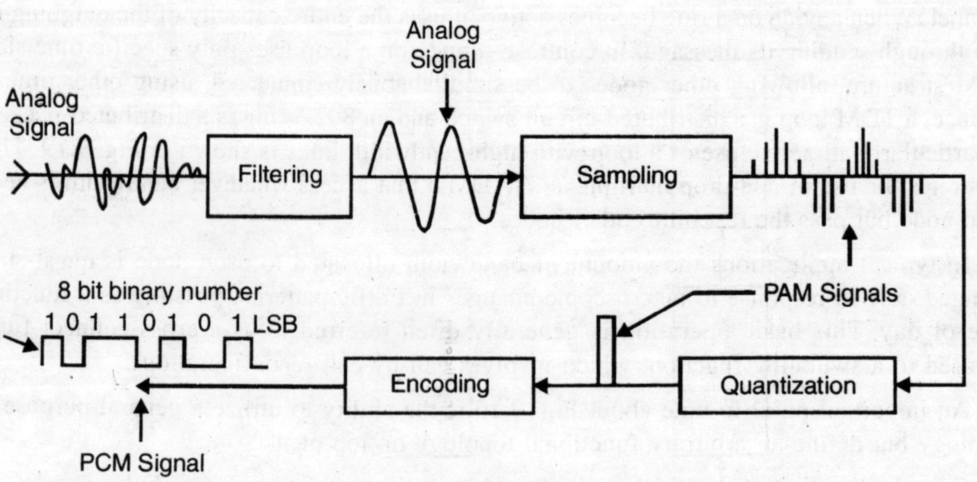

Fig. 9.22. *A Simple Block Diagram of PCM Encoder.*

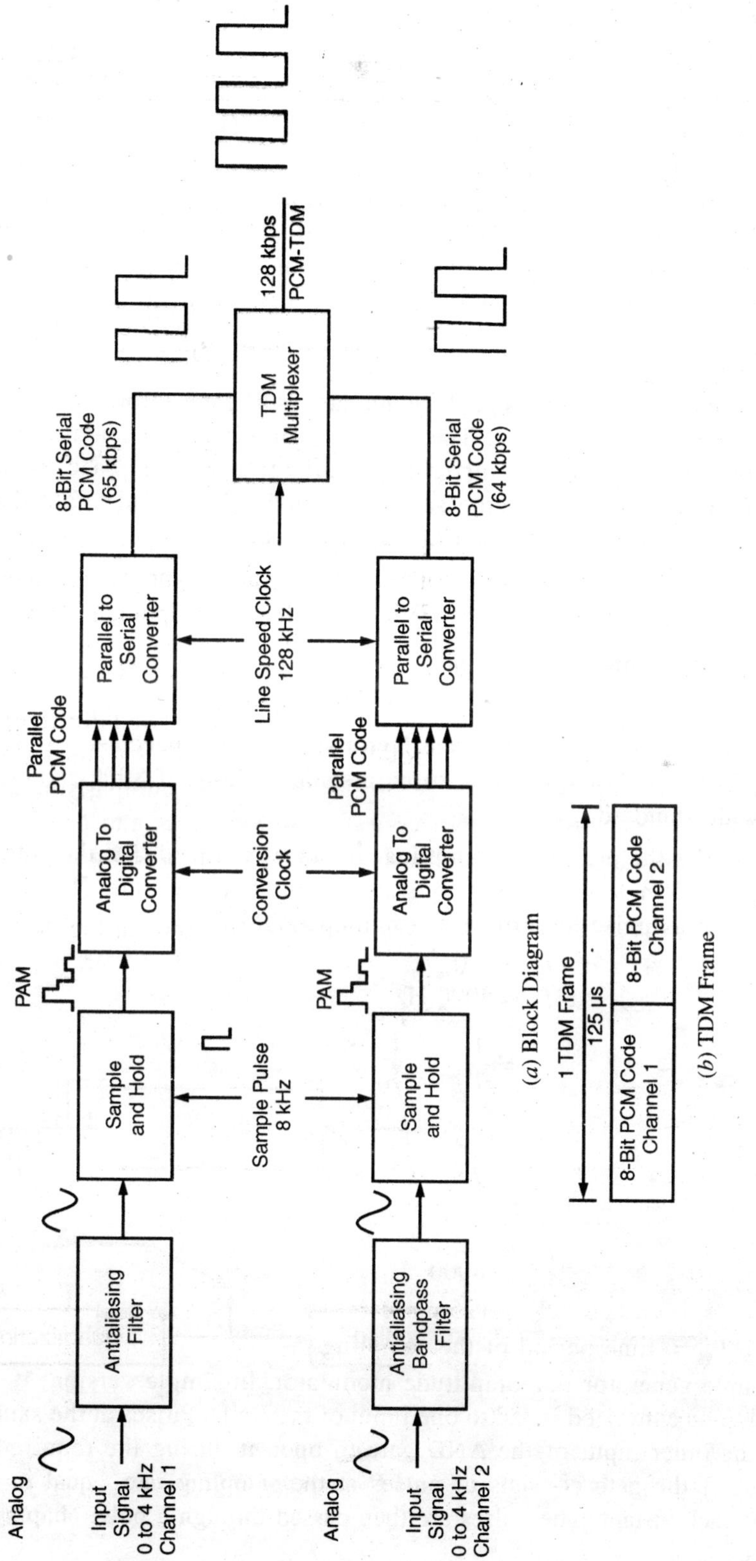

Fig. 9.23.. *Two Channel PCM-TDM System*

9.7.1. Filtering

When signal is transmitted over wires from subscriber to local exchange, electrical noise and crosstalk disturbances from neighboring wire paths also will be picked up.

The signal also gets attenuated as the distance increases. Thus the speech signal transmitted is combined with noise and attenuated. By filtering process, the lower frequencies are filtered out to remove electrical noise which may also be induced from power lines. The higher upper frequencies are filtered out because they require additional bits and add to the cost of a digital transmission system. Frequencies below 300 Hz and above 3400 Hz (voice frequency range) are filtered from the analog signal. Thus the actual bandwidth of the filtered signal is 3100 Hz.

9.7.2. Sampling

The sampled signal from analog signal should carry sufficient information so as to receive at the receiver with minimum distortion. The reduced sampling rate contains less information. The higher sampling rate contains sufficient information but increase in rate extends the bandwidth required for transmitting the sampled signal. The Nyquist rate derived by Harry Nyquist in 1933 gave engineering compromise for the sampling rate.

By principle, the sampled signal can be recovered exactly when

$$f_s \geq (2)\ (f_m) \qquad \text{...(9.1)}$$

where f_s = sampling frequency

f_m = highest frequency of the band limited signal.

This minimum sampling rate is known as **Nyquist rate**. The speech signal is in the frequency range of 300 to 3400 Hz. For this application, a sampling rate of 8000 samples per second is almost a worldwide standard.

Figure 9.23 shows the process of sampling. It can be realized that the sampling process is equivalent to amplitude modulation of a constant amplitude pulse train. Hence this technique is normally referred to as pulse amplitude modulation (PAM).

The sampling frequency

$$f_s = 2\ (f_m) \qquad \text{...(9.2)}$$

$$f_s \approx 8000 \text{ Hz} \qquad \text{...(9.3)}$$

As

$$f_s = \frac{1}{T} \qquad \text{...(9.4)}$$

$$T_s = \frac{1}{f_s}$$

$$T_s = \frac{1}{8000}$$

$$T_s = 125\ \mu \text{ sec} \qquad \text{...(9.5)}$$

where, T_s = time period of the sampling.

In PAM sample generator is a amplitude modulator. In simple version, it contains a AND gate. The signal to be converted is fed to one input of the AND. Pulses at the sampling frequency are applied to the other input of the AND gate to open it during the required time intervals. Thus, the output of the gate consists of pulses at the sampling rate, equal in amplitude to the signal voltage at each instant. The pulses are then passed through a pulse shaping network which gives the flat tops.

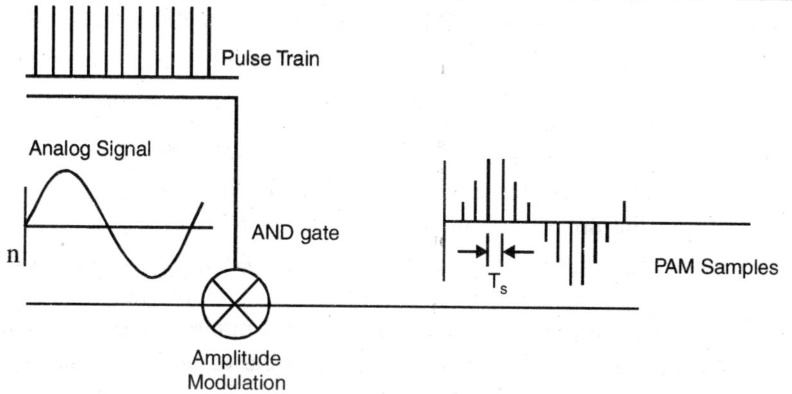

Fig. 9.24. *Generation of Pulse Amplitude Modulation Samples.*

9.7.3. Quantization and Encoding

Instead of sending a pulse train capable of continuously varying one of the parameters, the PCM technique produces a series of binary code which represents the approximate amplitude of the signal sample at that instant. This process of approximation is called signal quantization. Thus, the combined operations of sampling and quantizing generate a quantized PAM waveform.

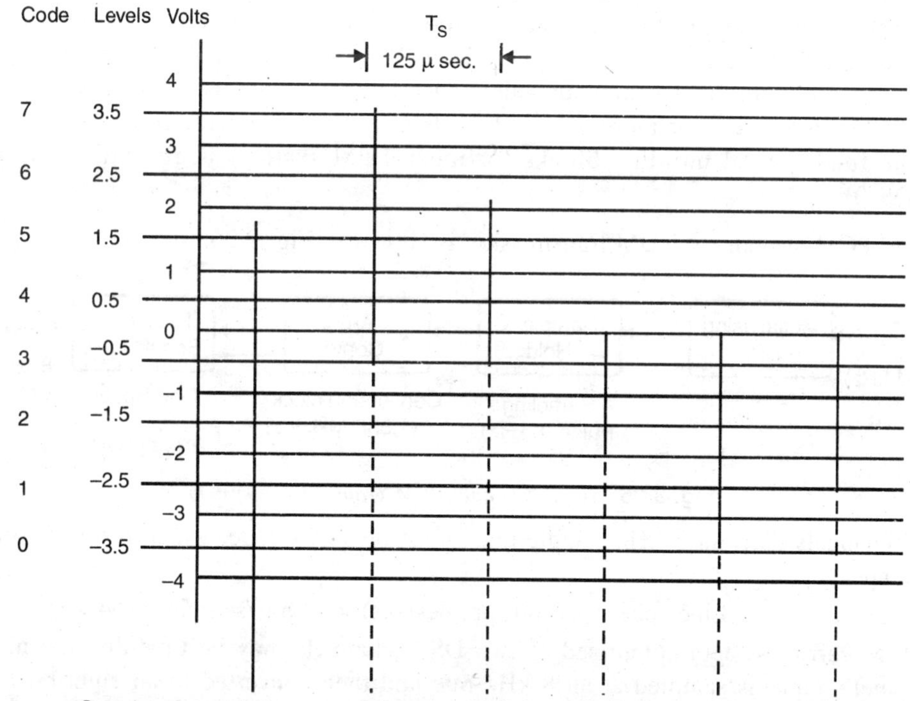

Sample value	1.8	3.6	2.2	–1.7	–3.3	–2.2
Nearest level	1.5	3.5	2.5	–1.5	–3.5	–2.5
Code	5	7	6	2	0	1
Binary representation	101	111	110	010	000	001

Fig. 9.25. *Quantization With Number of Levels 8.*

The quantizing process is as follows. Let the signal $m(t)$ is to be quantized. Identify the signal excursion (negative peak to positive peak) which is ranges from V_L to V_H. Divide the total range into M equal intervals. In the center of each of these steps, we locate quantization levels. If the sampled amplitude is in a particular interval, the nearest value is chosen and the corresponding binary digit is approximated as equivalent to the sampled amplitude. The quality of the approximation may be improved by reducing the size of the steps, thereby increasing the number of allowable levels. Figure 9.25 shows the process of quantization and encoding.

In the fourth step of the digitization process, the quantized samples are encoded into a digital bit stream (series of electrical pulses). In practical system, 256 levels can be used to obtain quality of signal. This results in the output of the encoding as 8 bit number. This binary number (or 8 bit word) is transmitted over the network as a series of electrical or optical pulses.

In the case of 256 (0 to 255) levels, the digital encoder recognizes the 255 different voltage levels of the quantized samples and converts each into a string of eight bits (1*s* and 0*s*).

The series of pulses is called a digital bit stream. The PCM process require a 64 kbps channel to encode a 4 kHz audio input signal because 8000 samples/sec × 8 bit/words = 64000 bps.

9.8. TIME DIVISION MULTIPLEXING OF PCM SIGNALS

A PAM or PPM can be employed for TDM. These methods are now obsolete as the transmitted pulses dispersed due to attenuation and delay distortion. They spread in time and interfere with the pulses of adjacent channels, thus causing inter-channel crosstalk. The PCM technique used in TDM eliminates this problem. PCM is most prevalent encoding technique used for TDM digital signals. In TDM-PCM system, two or more voice channels are sampled, quantized and encoded to PCM codes. The fundamental building blocks for most TDM systems begin with a digital signal level 0 (DS-0).

Typical PCM system for TDM transmission is shown in Fig. 9.26.

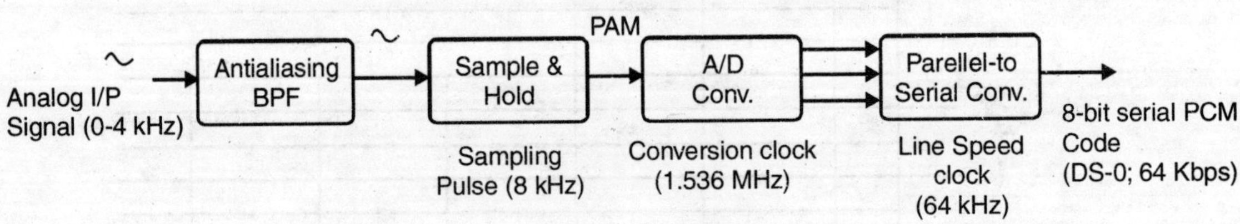

Fig. 9.26. *Single Channel PCM Transmission system.*

DS-0 channels used an 8 kHz sample rate and 8-bit PCM code, which resulted to a 64 kbps PCM line speed,

$$\text{Line speed} = (800 \text{ samples/sec}) \times (8 \text{ bits/sample}) = 64 \text{ kbps.}$$

A PCM carrier system comprised of two DS-0 channels may be time-division multiplexed. Each channel's input is sampled at an 8 kHz rate and then converted to an eight-bit PCM code. While the PCM code for channel first is being transmitted, second channel's is sampled and converted to a PCM code. Now while PCM code for second channel is being transmitted, the next sample is taken from first channel and converted to a PCM code. This process will continue and samples are taken alternatively from each channel, converted to PCM code and transmitted.

The PCM code for each channel occupies a fixed time slots within the total TDM time frame. One 8-bit PCM code from each channel (total 16 bits) is called a *TDM frame*, and the time it takes to transmit one TDM frame is called the *frame time*. With a two channel system, one sample is taken from each channel during each frame; time allocated to transmit the PCM bits from each channel is equal to one-half the total fame time. Thus the line speed at the output of multiplexer is

Line speed = (2-channels/frame) × (8000 samples/ sec) × (8 bits/sample) ...(9.6)

Line speed = 2 × 64 kbps = 128 kbps

9.9. DIGITAL CARRIER SYSTEMS

Telephone companies make uses of multiplexing long distance calls onto high- speed trunks. International standards bodies issue multiplexing rates that from the digital transmission hierarchy. Digital carrier is a digital signalling represents a telecommunications service. The base level in the hierarchy is the equivalent of a single 64 kbps digitized voice. The hierarchy differs from region to region.

Digital service in North America defines a four level transmission hierarchy called T-carrier, range from T_1, T_2, T_3 and T_4. In Europe and South America, there is five-level transmission hierarchy called **E-carrier**, range from E_1, E_2, E_3, E_4 and E_5. Both systems are using PCM to encode an analog signal in digital form. The signal is sampled 8000 times per second, and each sample value is encoded in an 8-bit value. The signal transmission uses TDM. SONET and SDH are intended to provide a specification for taking advantage of high-speed digital transmission capability of optical fiber.

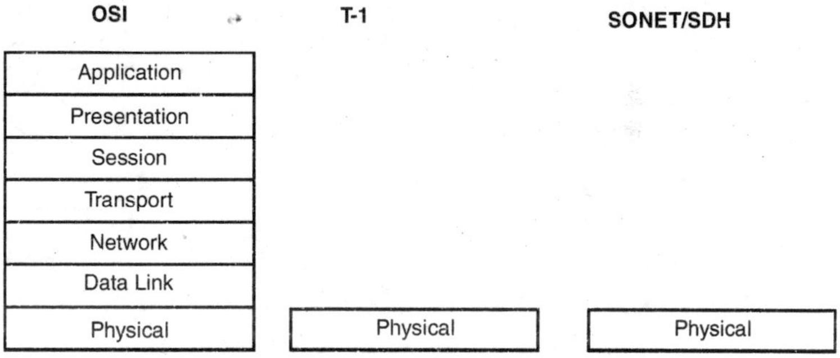

Fig. 9.27. *Comparison of the layer.*

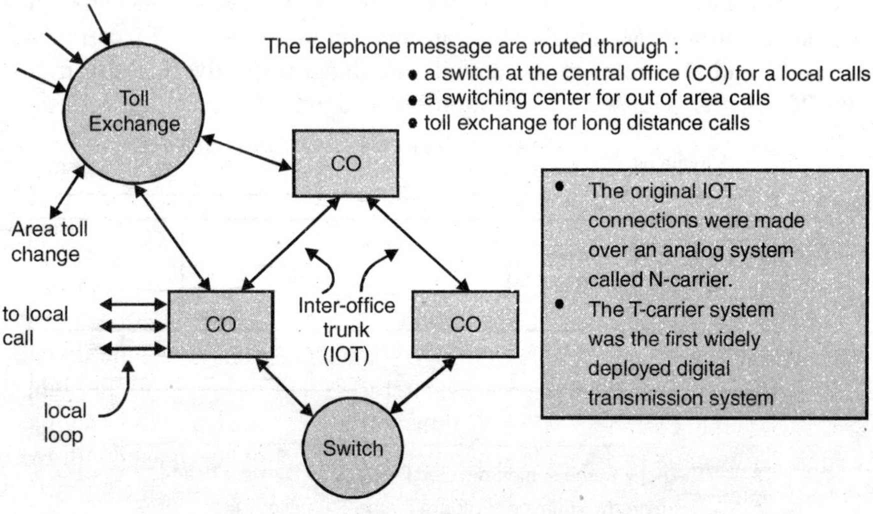

Fig. 9.28. *Organization of telephone services.*

T-1 and SONET/SDH standard define the rates and formats, the physical layer, network and the network operational criteria.

The non-local calls are routed between central offices through inter-office trunks (IOT). The original IOT connections were made an analog system called the N-carrier. The N-carrier system deployed an analog transmission, allowing 12 voice calls to share the same line. The L-carrier system was designed in the late 1960s for handling much higher capacity. The system can carry up to 6000 calls on the coaxial cable. However, the repeater was needed to improve signal periodically along the line. The system was very expensive for investment.

The new digital transmission called T-carrier system was later developed. The T-1 version can handle 24 calls on two pairs of copper wire.

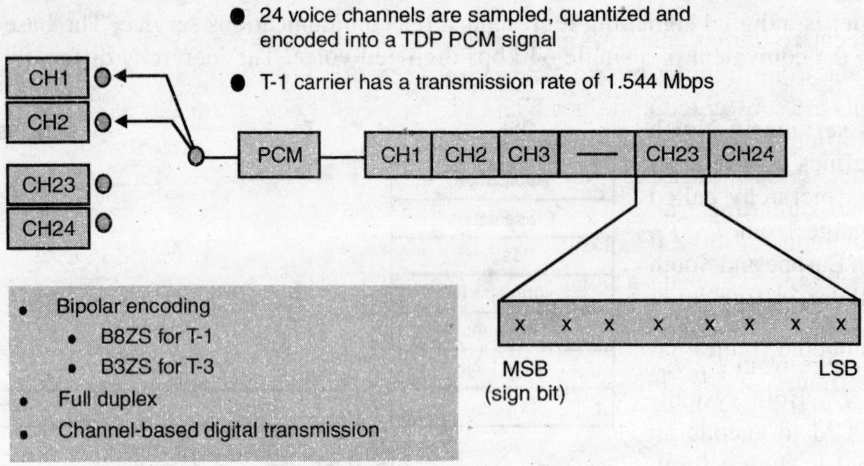

Fig. 9.29. *T-1 carrier system.*

Frame

The Bell system has established time division multiplexed PCM systems in North America. The transmission system is known as the T-carrier systems. The T1 carrier is the fundamental building block of the multiplexing hierarchy. The T1 carrier consists of 24 voice channels that are sampled, digitized and encoded to a PCM signal. The transmission rate onto the T1 carrier is 1.544 Mbps. The destructions T1 and DS1 refer to the circuit and signal respectively. Pulse trains in T carrier rely on the B8ZS (Bipolar with 8 Zero Substitution) coding.

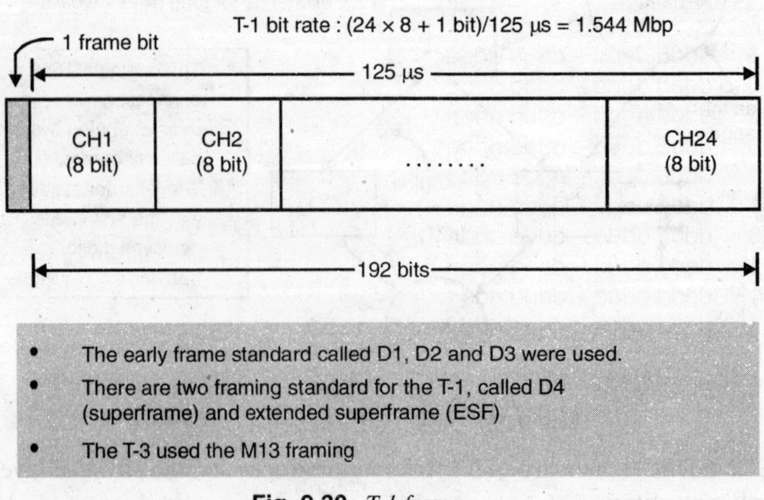

Fig. 9.30. *T-1 frame.*

A DS1 frame contains 193 bits divided into 24 slots (one for each voice channel) of 8 bits and one extra bit called a framing bit (f-bit), which is used for synchronizing, as illustrated in Fig. 9.29.

Eight-bit voice samples are taken each of 24 channels at a rate of 8000 times per second (64 Kbps rate).

To support this speed, T1 must transmit a DS1 frame every 1/8000 of a second (125 μs). This yields the bit rate of $(24 \times 8 + 1)$ bits/125 μs= 1.544 Mbps.

Superframe

12 consecutive frames comprise a superframe (also called D4 framing). The 12 framing bits in this superframe (one framing bit per frame) goes through the 12-bit pattern 100011011100. The signalling bits are substituted in frames 6 and 12, the most significant bit into frame 6, and the least significant bit into frame 12. Frames 1 to 6 are called the A-highway; with frame 6 designated the A-channel signalling frame. Frames 7 to 12 are called B-highway; with frame 12 designated the B-channel signalling frame. Therefore, in addition to identifying the signalling frame, the 6th and 12th frame must also be positively identified.

The synchronization pattern occurring in the odd-numbered frames is 101010. These are the synchronization bits for the channel banks. The voice clock synchronization pattern occurring in the even-numbered frames has a pattern of 001110. The combined Fs bit pattern is 100011011100 for the superframes. It is call the S bit is shared between framing and signalling.

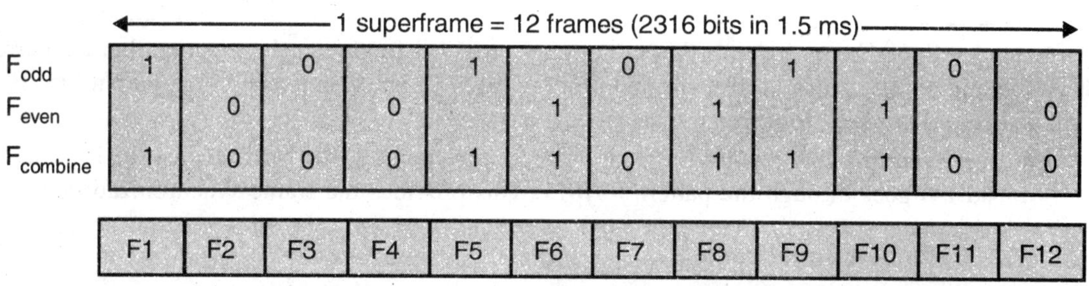

Frame #	F	data (192 bits)
1	1	dddd dddd – dddd dddd
2	0	dddd dddd – dddd dddd
3	0	dddd dddd – dddd dddd
4	0	dddd dddd – dddd dddd
5	1	dddd dddx – dddd dddd
6	1	dddd dddd – dddd dddd
7	0	dddd dddd – dddd dddd
8	1	dddd dddd – dddd dddd
9	1	dddd dddd – dddd dddd
10	1	dddd dddd – dddd dddd
11	0	dddd dddd – dddd dddd
12	0	dddd dddx – dddd dddd

- T-1 carrier frames are transmitted in groups of 12 called *superframes*
- F-bit in even-numbered frame has a pattern of 101010 for synchronization
- Signalling information is accomplished by robbing the LSB position of each channel. This is performed only in the 6th and 12th frame to keep distortion minimum

Fig. 9.31. *Frame and superframe.*

Extended Superframe

The extended superframe format consists of 24 frames of 193 bit, totalling 4632 bits, of which 24 are framing bits. The extended superframe occupies 3 ms.

$$\left[\frac{1}{1.544\ \text{Mbits/s}}\right]\left[\frac{193\ \text{bits}}{\text{Frame}}\right](24\ \text{frames}) = 3\ \text{ms}$$

Table 9.1. Extended Superframe Format

Frame number	Framing bit	Frame number	Framing bit
1	C	13	C
2	F	14	F
3	F	15	F
4	S = 0	16	S = 0
5	C	17	C
6	F	18	F
7	F	19	F
8	S = 0	20	S = 1
9	C	21	C
10	F	22	F
11	F	23	F
12	S = 1	24	S = 1

Extended Superframes extends this to not only provided frame synchronization, but also error detection and a data channel, all using the framing bit. The value in every 193rd bit (in bits 193, 386, and so on) are used for three purposes :

- Every fourth bit of this 24-bit cycle (that is, the framing bits for frames 4, 8, 12, 16, 20, and 24) goes through the pattern 001011. This provides the frame synchronization.

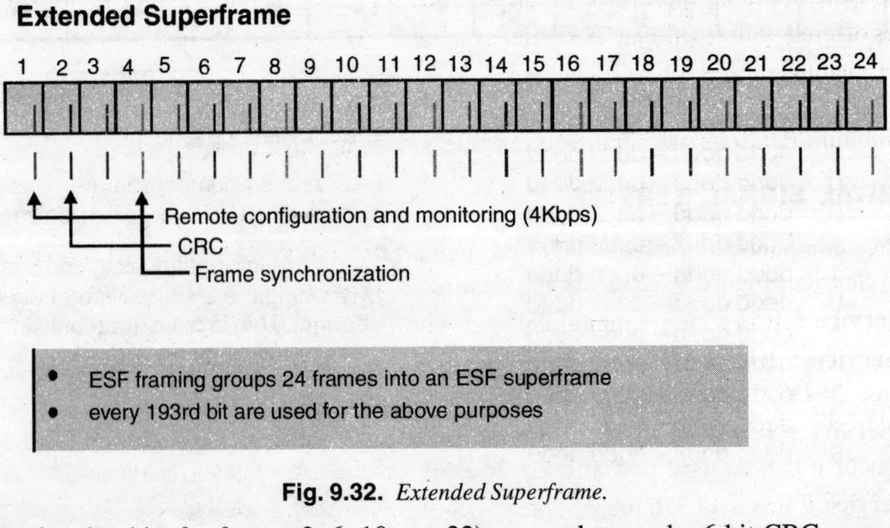

Fig. 9.32. *Extended Superframe.*

- The framing bits for frames 2, 6, 10, ..., 22).are used to send a 6 bit CRC, generated from the data in the previous 24 frames. This provides error detection. The receiving CSU can then track the error rate and generate an alarm if it gets too high. This error checking is done constantly while the links is in service and for any type of data.

- The remaining framing bits (for frames 1, 3, 5, 7,..., 23) provide a 4kbps supervisory data channel that is used for other functions such as remote configuration and monitoring of CSUs.

The eighth bit in every channel of frames 6, 12, 18 and 24 is used for signalling between central offices.

9.10. FRACTIONAL T CARRIER SERVICE

A fractional T1 is one or more channels bundled together and sold to a customer as a set. This allows a consumer to purchase less than a full T1's bandwidth at a lower cost. Like the channelized T1, individual channels can be voice or data and a CSU/DSU is used to split the channels, however less than the full set of 24 channels is available to the customer for use.

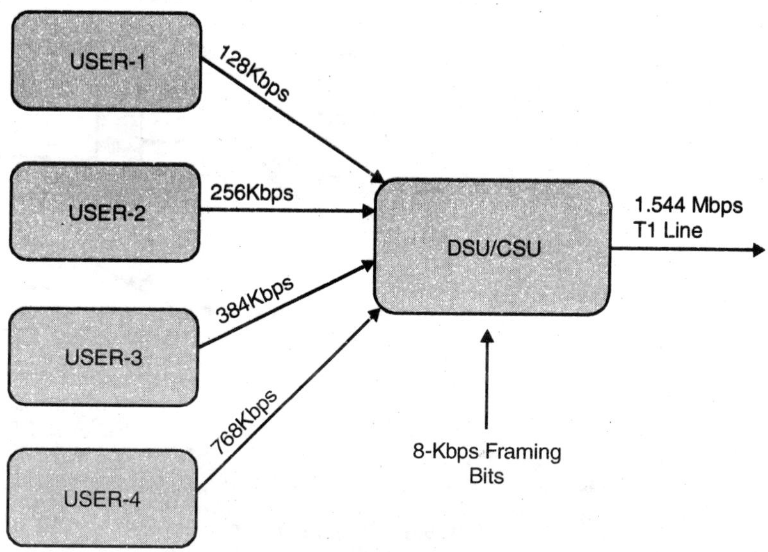

Fig. 9.33. *Fractional T1 Carrier System.*

Fractional T-carrier emerged because standard T1 carriers provide a higher capacity (higher bit rate) than most users require. Fractional T1 carrier system distributes the channels (*i.e.,* bits) in a standard T1 system among more than one user, allowing several subscribers to share one T1 line.

Bit rates offered with fractional T1 carrier system are 64 kbps (1-channel), 128 kbps (2-channels) 256 kbps (4-channels), 384 kbps (6-channels), 512 kbps (8-channels) and 768 kbps (12-channels) with 384 kbps (1/4 T1) and 768 kbps (1/2 T1) being the most common.

The minimum data rate required to propagate video information is 384 kbps.

9.11. DIGITAL SIGNAL SERVICE

Telephone companies implement TDM through a hierarchy of Digital signal, known as Digital Signal (DS) Service. Figure 9.32 shows the data rates supported by at each level.

DS-0 Service : It is a single digital channel with the data rate of 64 kbps.

DS-1 Service : It is a 1.544 Mbps service. DS-1 service as single service for 1.544 Mbps transmissions. 24 DS-0 channels also can be multiplexed.

DS-2: Service : It is a 6.312 Mbps service. It can be used as a single service for 6.312 Mbps transmissions or it can be used to multiplex 96 DS-0 channels or 4 DS-1 channels.

DS-3 Service : It is a 44.376 mbps service. DS-3 can be used as a single service for 44.376 Mbps transmissions. It may be used to multiplex 7 DS-2 channels, 28 DS-1 channels, or 672 DS-0 channels.

DS-4 Service : It is a 274.176 Mbps service. DS-4 may be used to multiplex 6 dDS-3 channels, 42 DS-2 channels or 4032 dS-0 channels.

DS-5 Service : It is a 560.160 Mbps service. DS-5 may be used to multiplex 2 DS-4 channels, 12 DS-3 channels, 88 DS-2 channels, 362 DS-1 channels or 875 DS-0 channels.

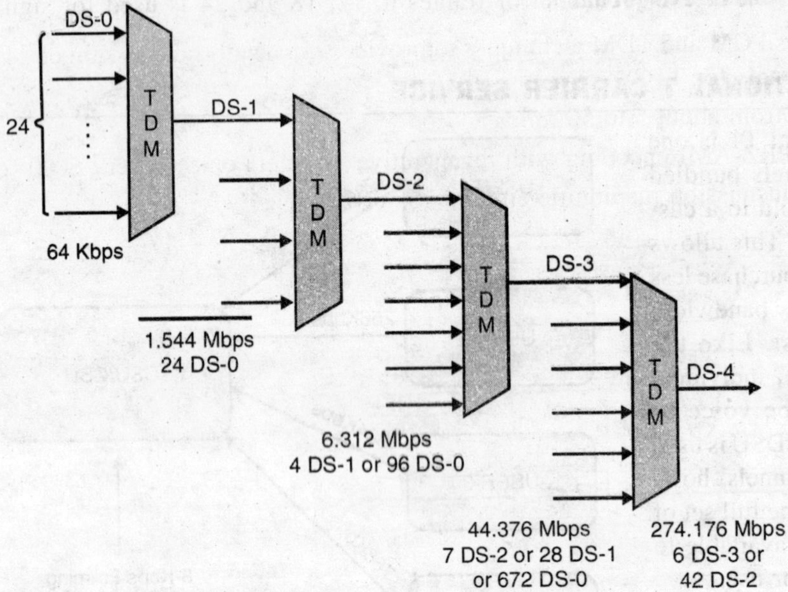

Fig. 9.34. *Digital Signal Service Using TDM Hierarchy.*

Telephone companies implement digital services (DS-0 to DS-5) by using t lines. The capacities of these lines matched to the data rates of the digital services. DS-0 has been defined as a basis for reference purposes, it actually does not offer as a service.

DS-1, DS-2, DS-3, DS-4 and DS-5 digital signals may be placed on T-1, T-2, T-3, T-4, and T-5 lines respectively. Table 9.2, lists the digital signals, their bit rates, channel capacities and services offered for the line types.

Table 9.2. The Digital Signals, Their Bit Rates, Channel Capacities and Services Offered.

DS	*Line*	*Rate (Mbps)*	*Voice Channels*	Services
DS-1	T-1	1.544	24	Voice or Data
DS-2	T-2	6.312	96	Voice or Data Picture phone
Ds-3	T-3	44.736	672	Voice or Data, Picture Phone, Broadcast
DS-4	T-4	274.176	4032	Same as T-3
DS-5	T-5	560.160	8064	Same as T-3

TIPS

Telephone companies implement TDM through a hierarchy of digital signals, known as digital signal (DS) service.

9.12. T-CARRIERS

T-Carriers are used for the transmission of PCM-encoded time-division multiplexed digital signals. It utilized special line-encoded signals and metallic cables that have been conditioned to meet the relatively high bandwidths required for high-speed digital transmission

T1 and T1C Carrier System

1. Utilizes PCM and TDM techniques to provide short-haul transmission of 24 VB signals (48 VB signals for T1C)
2. Range from about 5 to 50 miles
3. Use BPRZ-AMI encoding with regenerative repeaters placed every 6000 ft
4. The transmission medium is either a 19- or 22-gauge wire pair.

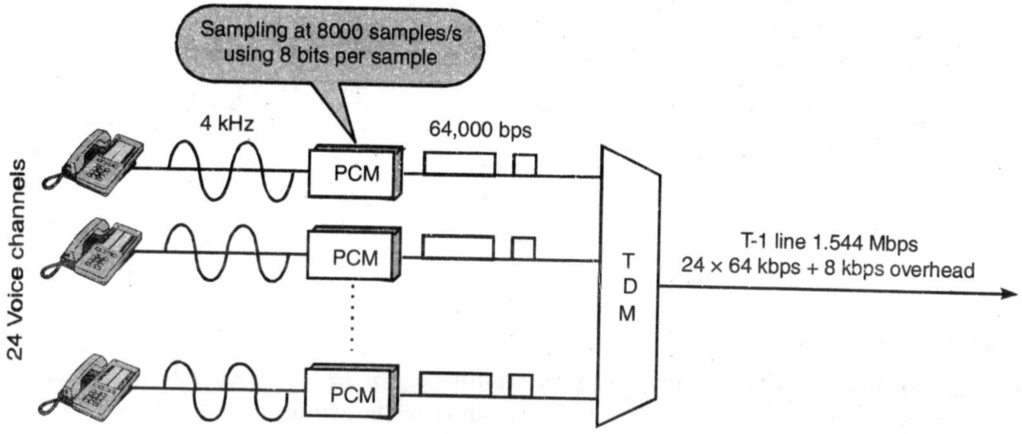

Fig. 9.35. *T-1 Line for Multiplexing telephone lines.*

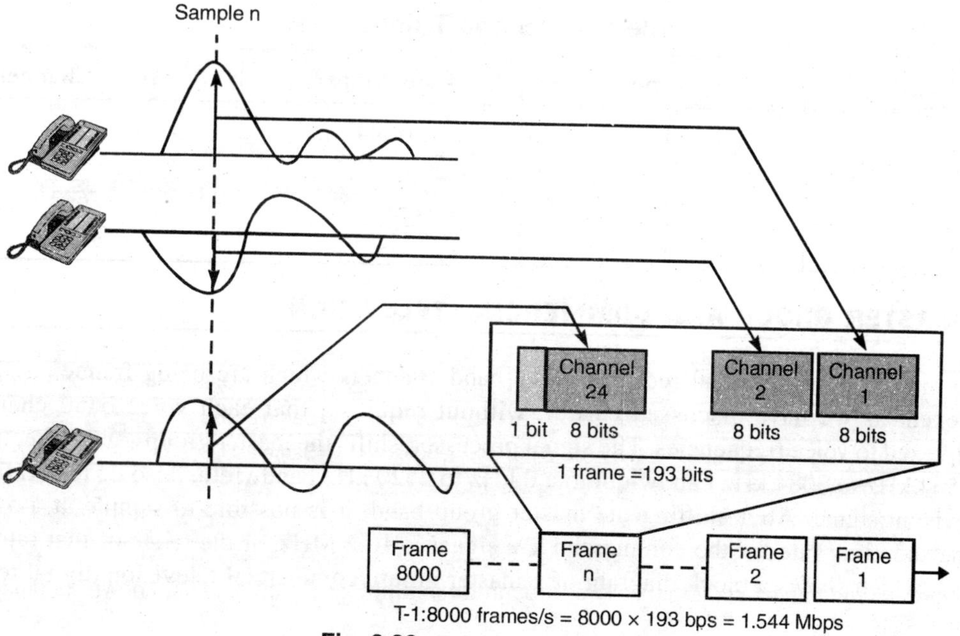

Fig. 9.36. *T-1 frame structure.*

Line Rate of the T1 Carrier

$$\left(\frac{8 \text{ bits}}{\text{channel}} \times \frac{24 \text{ channels}}{\text{frame}}\right) + \frac{1 \text{ framing bit}}{\text{frame}} = 193\frac{\text{bits}}{\text{frame}}$$

$$193\frac{\text{bits}}{\text{frame}} \times 8000\frac{\text{frames}}{\text{seconds}} = 1.544 \text{ Mbps}$$

T2 Carrier System

T2 Carrier System utilizes PCM to time division multiplex 96 VB channels into a single 6.312 Mbps data signal for transmission up to 500miles over a special LOCAP (low capacitance) cable. T2 Carrier System uses an alternative BPRZ-AMI method of ensuring that ample transitions occur in the data. This method is called binary six zero substitution (B6ZS). With B6ZS, whenever six consecutive 0's occur, one of the following codes is substituted in its place: 0 - + 0 + - or 0 + - 0 - +. The + and – represents positive and negative logic 1's.

E Line	Rate (Mbps)	Voice Channels
E-1	2.048	30
E-2	8.448	120
E-3	34.368	480
E-4	139.264	1920

T3 Carrier System

T3 Carrier System utilizes Time-division multiplexes 672 PCM-encoded voice channels for transmission over a single metallic cable. The encoding technique used with T3 carriers is binary three zero substitution (B3ZS). There are four substitution patterns used with B3ZS: 00-, -0-, 00+, and +0+. The pattern chosen should cause a bipolar error in the third substitute bit.

Table 9.3. DS and T lines rates

Service	Line	Rate (Mbps)	Voice Channels
DS-1	T-1	1.544	24
DS-2	T-2	6.312	96
DS-3	T-3	44.736	672
DS-4	T-4	274.176	4032

9.13. MASTER GROUP AND COMMERCIAL TELEVISION

The master group terminal receives voice band channels which are being frequency division multiplexed, as we have discussed earlier, without requiring that each voice band channel be demultiplexed to voice frequencies. The signal processor, shifts the master group signals frequencies, from a 564 kHz to 3084 kHz bandwidth to a 0 kHz to 2520 kHz bandwidth. Also dc restorations for the television signal. After shifting the master group band, it is possible to sample at a 5.1 MHz rate. The sampling rate for the commercial TV signal is 10.2 MHz, at the twice of that rate. Refer Fig. 9.37, which shows a block diagram of a master group /commercial television digital terminal.

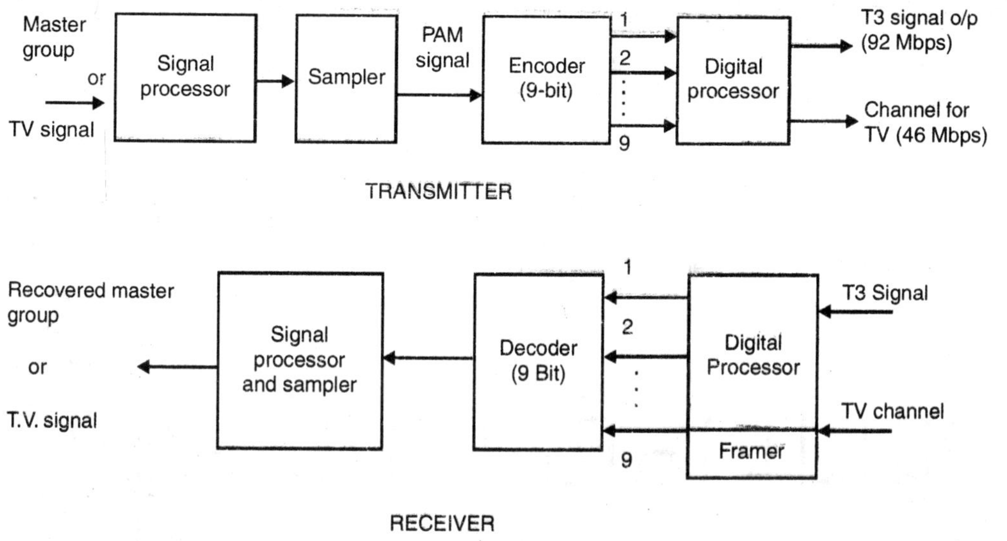

Fig. 9.37. *Block Diagram of a Master Group/Commercial TV Digital Terminal.*

A 9-bit PCM code is used to digitize each sample of the master group or television signal, to meet the transmission requirements. The output of the digital processor is approximately 46 Mbps for the master group and 92 Mbps for the TV signal.

The digital terminal shown in Fig. 9.36 has three main functions, which are as follows :

1. Conversion of the parallel data from the output of the encode to the serial data,
2. Flame synchronizing bits are inserted, and
3. It converts the serial binary signal to form more suitable for transmission.

As there is no 92 Mbps line speed in the digital hierarchy, the 92 Mbps digital must be split into two 46 Mbps signals for the commercial TV terminal.

9.14. PICTURE PHONE TERMINAL

A picture phone is used to transmit the low quality video signals for the use of non-dedicated subscribers. This picture phone signal is encoded into T2 capacity of 6.312 Mbps, which is substantially less than that for commercial network broadcast signals, so that cost may be reduced and makes the service affordable. At the same time, it allows the transmission of adequate detail and contrast resolution to satisfy the average picture phone subscriber.

Picture phone service is ideally suited to a differential PCM code in which only difference of the magnitude between the present sample and the previous sample is encoded and transmitted. By using differential PCM, we need to transmit substantially fewer bits than encoding the actual sample, so that result in saving of bandwidth.

9.15. DATA TERMINAL

Data terminal are designed to transmit the signals other than voice. In most of the cases the data rates generated by each individual subscriber are substantially less than the data rate capacities of digital lines. Therefore, it seems only logical that terminals be designed that transmit data signal from several sources over the same digital line.

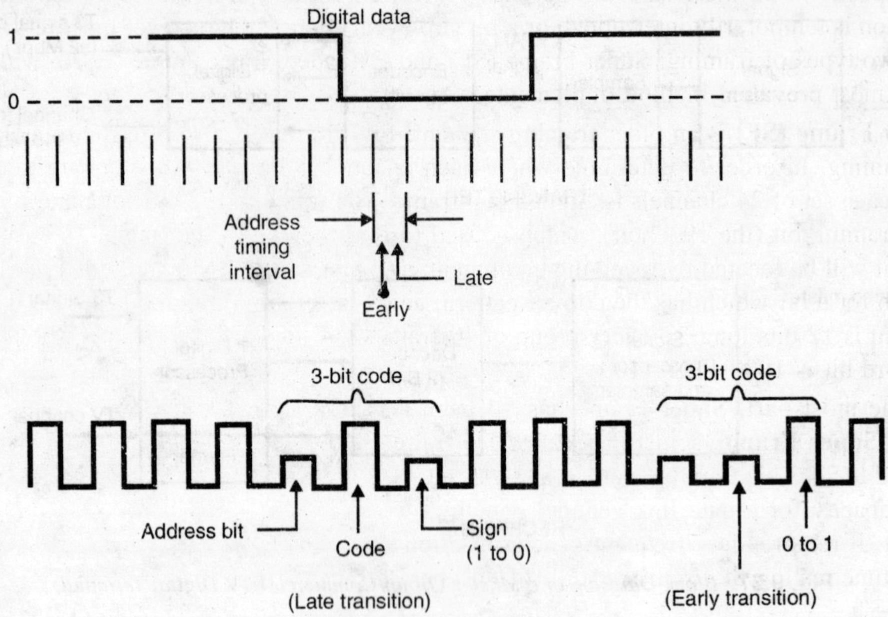

Fig. 9.38. *Data Coding Format.*

A method of coding the data is shown in Fig. 9.38. This method is more efficient, it codes the transition time. In the coding format shown in Fig. 9.38 is a uses three bit code to identify when transition occurs on the data and whether the transition is from a '1' to '0' or vice-versa. The first bit is referred as address bit. When there is no transition it indicates logic '1' and when transition did occur, it indicates logic '0'. The second bit indicates whether the transition occurred during the first half (0) or during the second half (1) of bit period. The third bit indicates the direction of the transition; a '1' for this bit indicates a 0-1 transition; and a '0' indicates a 1-0 transition.

Consequently; where there are no transition in the data, a second of all '1' is transmitted. Although the transmissions of only address bit would be sufficient, but however, the sign bit provides a degree of error protection and limit error propagations.

The advantage of using a coded format rather than the original data is that coded data are more efficiently substituted for voice in analog system. Without this coding format, transmitting a 250 kbps data signal requires the same bandwidth as would be required to transmit 60 voice channels with analog multiplexing. When we use this coded format, a 50 kbps data signal displaces three 64 kbps PCM-encoded channels and a 250 kbps data stream displaces only 12 voice-band channels.

9.16. FRAME SYNCHRONIZATION

Framing, or **frame synchronization**, is the process by which incoming frame alignment signals, *i.e.,* distinctive bit sequences, are distinguished from data bits, permitting the data bits within the frame to be extracted for decoding or retransmission. A T1 time slot is an 8-bit segment for each DS0. A frame consists of 24 time slots (one for each DS0 in the T1) and an additional framing bit, for a total of 193 bits.

The framing bit does not contain any user data; however, it is used by the transmitting and receiving equipment for the actual synchronization of the incoming data with the receiver. Within a bit stream, framing bits are used to determine the beginning or end of a frame. They occur at specified positions in the frame, do not carry information, and are commonly repetitive. If the transmission is temporarily interrupted, or a bit slip event occurs, the receiver must re-synchronize. T1s use two types of framing: Super Frame (SF) and Extended Super Frame (ESF), with the latter being the most prevalent method implemented.

Super Frame (SF) is an older framing standard for T1s that is also referred to as D4 or even D3/D4 framing. In order to determine where each channel is located in the stream of data being received, each set of 24 channels is aligned in a frame. The frame is 192 bits long and is terminated with the framing bit (the 193rd bit), which is used to find the end of the frame. To ensure that the framing bit will be located by receiving equipment, a specific pattern is sent on this bit. Equipment will search for a bit which has the correct pattern, and will align its framing based on that bit. The pattern sent is 12 bits long, so every group of 12 frames is called a Super Frame. The pattern used in the 193rd bit is 1000 1101 1100.

For the most part, Super Frame has all been replaced by Extended Super Frame, or ESF. **Extended Super Frame**, sometimes called D5 framing, doubles the number of the frames in the SF from 12 to 24. This framing method includes a Cyclic Redundancy Check, or CRC, and a 4 kbps channel capacity for a data link channel, which is used to pass out-of-band information between equipment. It requires less frequent synchronization than the earlier SF format, and provides on-line, real-time testing of circuit capability and operating condition.

9.17. E-CARRIER CIRCUITS

Outside of the United States, E-carrier circuits are used to carry digital transmissions. The most common E-carrier circuit is the E1, which has its electrical interference and framing defined in ITU-T G.703 and G.704. E1 time-slots are numbered TS0 to TS31, where TS1 through TS15 and TS17 through TS31 are used to carry voice, which is encoded with PCM, or to carry 64 kbps data. Of these 32 timeslots, 1 is used for framing, 1 is used for telephony signalling, and 30 are used for voice and data transport for an aggregate capacity of 2048 kbps or 2.048 Mbps.

Timeslot 0 is used for framing, alarming, and international bits; while timeslot 16 is used to carry signalling information, which may be CAS, ISDN, SS7, or proprietary signalling. CAS, ISDN, and SS7 signalling will be described later in this chapter. There is a 16-frame multi-frame structure that allows a single 8 bit time slot to handle the line signalling for all 30 data channels.

The E1 standard also allows for optional CRC-4 error-checking and is most commonly used in Europe, parts of Asia, South and Central America and Africa.

9.18. CODECS

Codec is a generic term that refers to the coding and decoding function performed by a device. It converts analog signals to digital codes and digital codes to analog signals.

A codec is a large-scale integration (LSI) chip designed for use in the telecommunication industry for central office switches, digital handsets; private branch exchanges (PBXs), voice store-forward systems and digital echo suppressors. The codec is applicable for any purpose that needs the digitizing of analog signals, such as in a PCM-TDM carrier system.

Functions of Codec

The codec performs following functions:

1. Analog sampling.

2. Analog to digital and digital to analog conversion.
3. Digital companding.

Recently developed codecs are referred as combo-chips because they combine codec and filter functions in the same LSI package. The filters (input and output) perform the following functions :

1. Noise rejection
2. Band limiting,
3. Antialiasing
4. Reconstruction of analog audio waveforms after decoding.

9.19. COMBO-CHIPS

Essentially, a combo chip replaces the older codec and filter chip combination. A combo-chip can provide the analog to digital conversion and transmit and receive filtering necessary to interface a full-duplex *i.e.,* four wire voice telephone circuit to the PCM highway of a TDM carrier system.

Functions of Combo-Chip

A combo- chip performs the following functions :

1. Band pass filtering of the analog signals before encoding and after decoding.
2. Encoding and decoding of voice signals.
3. Encoding and decoding of signalling and supervision information.
4. Digital companding.

Figure 9.38 shows the simple block diagram of combo-chip. The frequency response curves for transmit band pass filter and receive band pass filter shown in Fig. 9.40 (*a*) and (*b*).

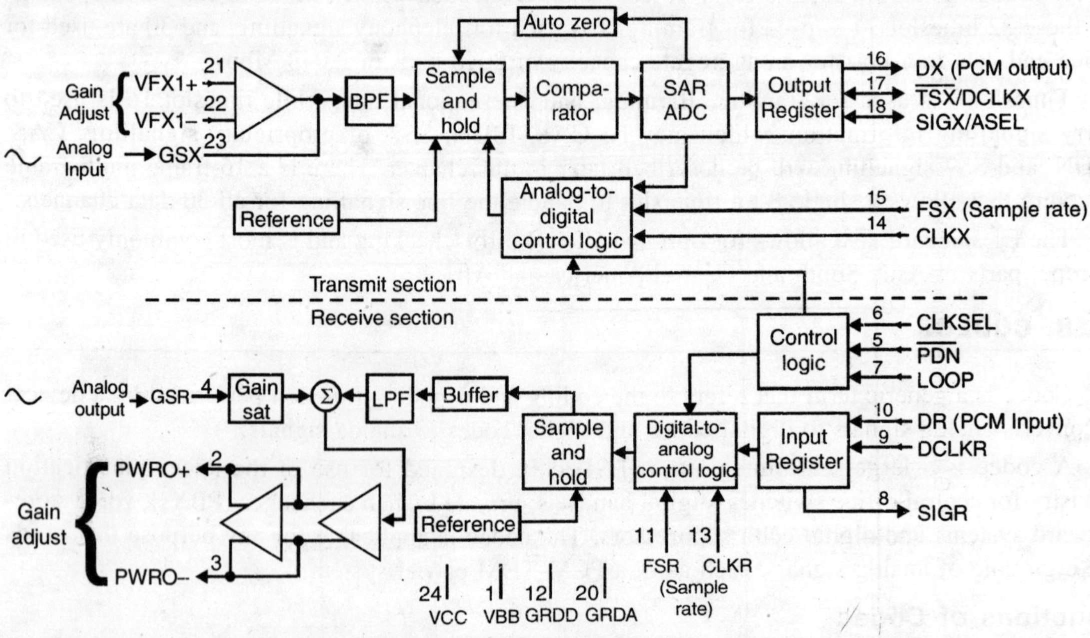

Fig. 9.39. *A Simple Block Diagram of Combo-Chip.*

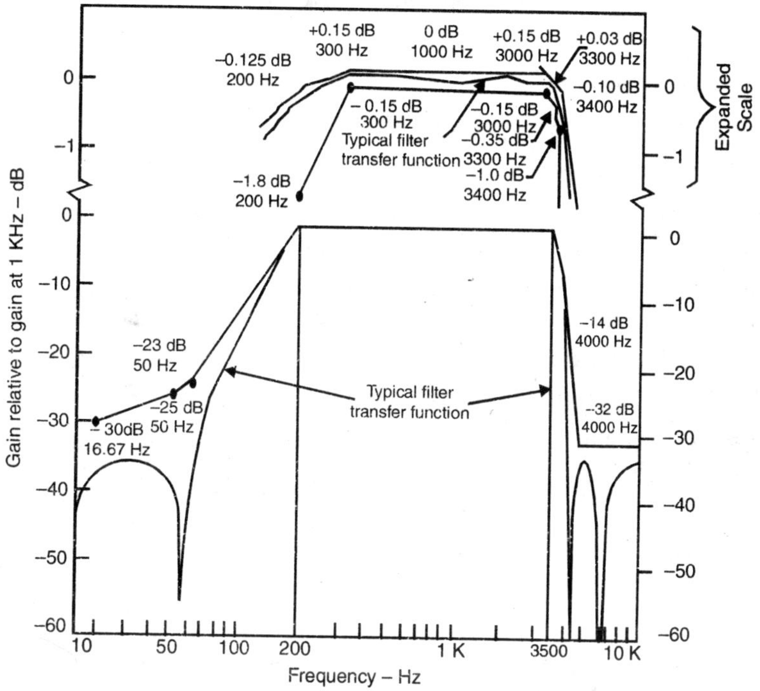

Fig. 9.40. *(a) Transmit BPF Response Curves.*

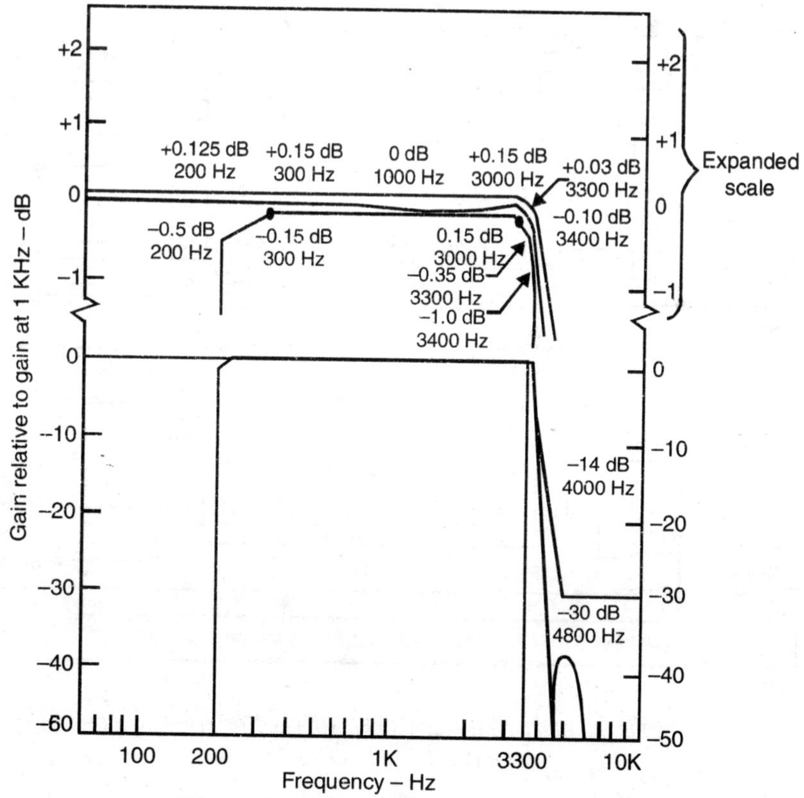

Fig. 9.40. *(b) Receive LPF Response Curves.*

There are two types of modes of operation performed in combo-chip. They are as follows :

1. Fixed-data-rate mode
2. Variable-data-rate mode

Both the modes of operation in combo-chip are discussed in brief here :

1. Fixed-Data-Rate Mode

In the fixed-data-rate mode, transmit and receive data must be either 1.536 Mbps or 2.048 Mbps— the same as the master clock rate. In this mode, the master transmits and receives clocks on a combo-chip performs the following functions:

- It provides the master clock for the on board switched capacity filter.
- It provides the clock for the analog to digital and digital to analog convertors.
- It determines the input and output data rate between the codec and the PCM highway.

A simple block diagram of a single channel PCM system using a combo-chip in the fixed- data-rate mode shown in Fig. 9.41.

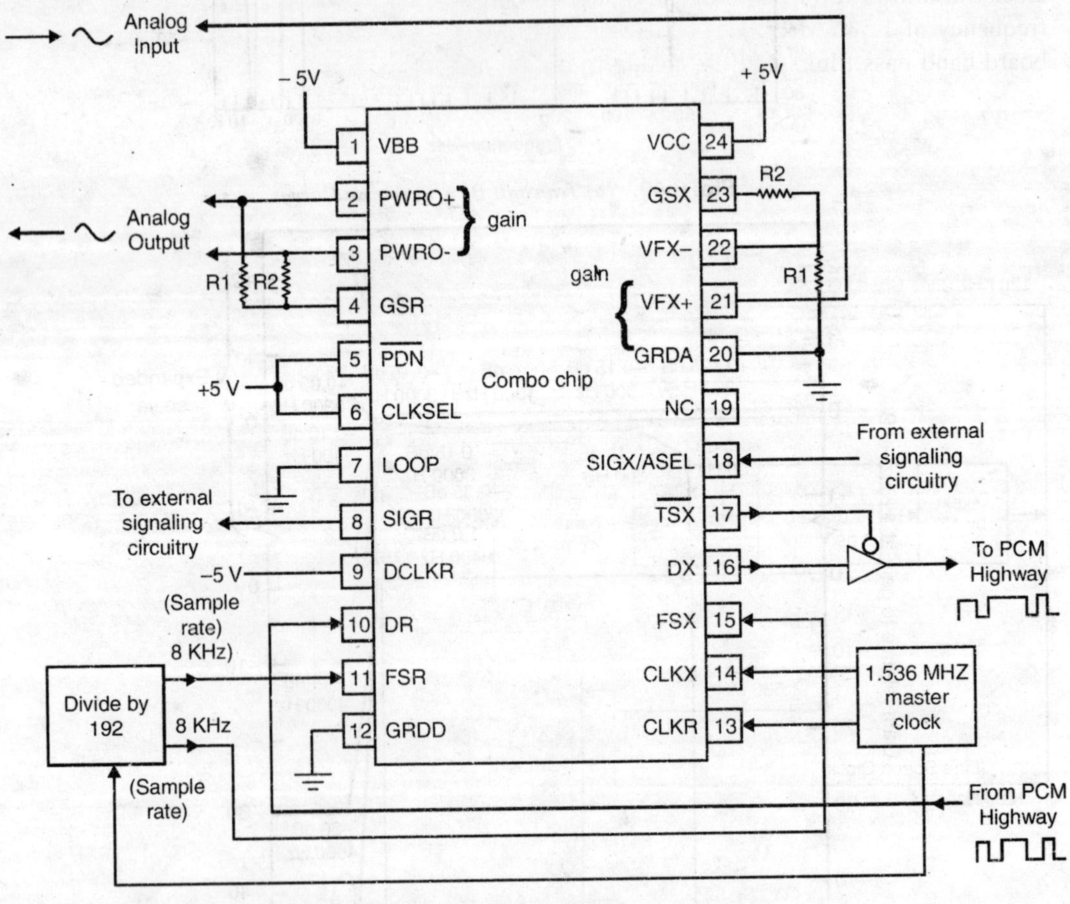

Fig. 9.41. *A Simple Block Diagram a Single Channel PCM Using a Combo-Chip in the Fixed Data Rate.*

Transmit and receive frame synchronizing pulses are 8 kHz input set transmit and receive sampling rates and distinguish between signalling and non-signalling frames.

TSX	Time slot strobe buffer enable output
CLK X	Clock transmit
CLK R	Clock receive
DX	Data o/p
DR	Data input

In the fixed data rate mode, data are input and output for a single channel in short bursts and that's why, it is known as *burst mode*.

- The inputs and output bits rates from the codec are equal to the master clock frequency *i.e.*, 1.536 Mbps.
- The DX and DR are enabled only 1/24 of the total frame time *i.e.*, 125 μs.

2. Variable-Data-Rate Mode

Variable-data-rate mode provides the ability to vary the frequency of transmit and receive bit clocks. It allows for a flexible data input and output clock frequency. In this mode, a master clock frequency of 1.536 MHz, 1.544 MHz or 2.048 MHz is still needed for proper operations of the on board band pass filters and the analog to digital and digital to analog convertor.

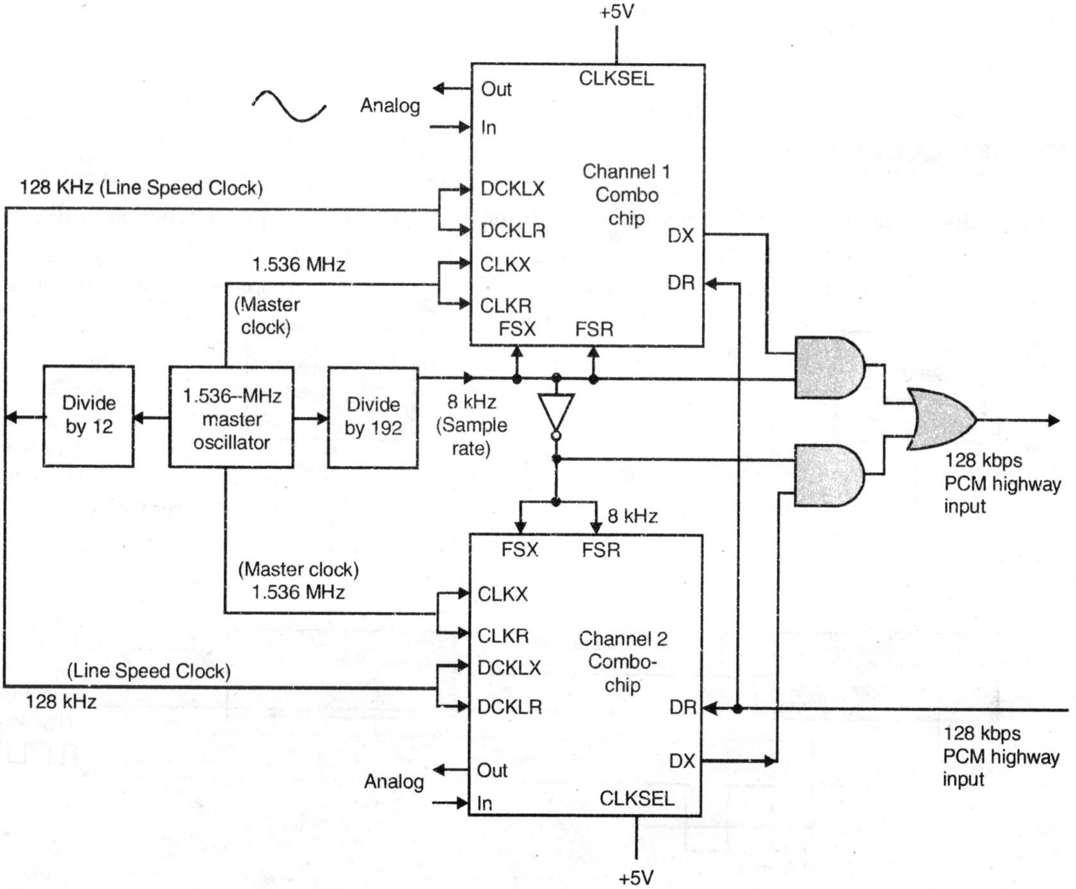

Fig. 9.42. *A Simple Block Diagram for a Two-Channel-PCM-TDM System.*

However in the variable data rate mode, DCL KR and DCL KX becomes the data clock for receive and transmit PCM highways, respectively.

This mode of operation also known as the *shift register mode*. When FSX (Frame synchronizing pulses transmit) is high, data rate are transmitted on to the PCM highway on the next eight consecutive positive transition of DCLKX. Similarly, which FSR (Frame synchronizing receive pulse) is high, data rate from the PCM highway are clocked into the codec on the next eight consecutive negative transitions of DCL KR.

Figure 9.42 shows the block diagram for a two channel-PCM-TDM system using a combo- chip in the variable-data-rate mode with a master clock frequency of 1.536 MHz, a sample rate of 8 kHz, and a transmit and receive data rate is 128 kbps.

If the sample rate is 8 kHz the frame time is 125 μs. Therefore one-eight bit PCM word for each channel is transmitted and/or received during each 125 μs frames. Now for 16 bits to occur in 125 μs, a 128 kHz transmit and receive data clock is required.

$$t_b = \left(\frac{1 \text{ channel}}{8 \text{ bits}}\right) \times \left(\frac{1 \text{ frame}}{2 \text{ channels}}\right) \times \left(\frac{125\ \mu\text{s}}{\text{Frame}}\right) = \frac{125\ \mu\text{s}}{16 \text{ bits}}$$

$$t_b = 7.8125\ \mu\text{s/bit}$$

So,

$$\text{Bit-rate} = \frac{1}{t_b} = \frac{1}{7.8125\ \mu\text{s}}$$

$$\text{Bit-rate} = 128 \text{ kbps.}$$

9.20. LINE CODING

When digital data is to be transmitted from one point to another point (for example, one computer to another computer), the digital data is to convert in to digital signal, so that it can be transmitted. This process of converting digital data in to digital signal is known as **line encoding**.

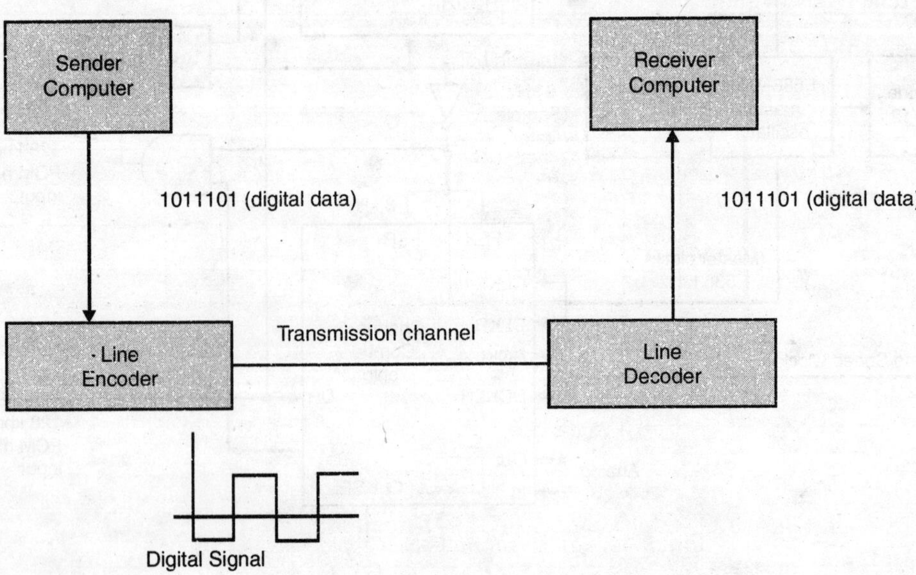

Fig. 9.43. *Block Diagram of Line Coding.*

Line encoding is the method used to represent the digital information on the media. A pattern, that uses either voltage or current, is used to represent the 1s and 0s of the digital signal on the transmission link.

Refer Fig. 9.43, from the sender computer the digital data is given to the block, Line encoders to convert the stream, of bits (0s and 1s) in to digital signal and then it is transmitted over the line or transmission media or channel (may be physically connected through wire or using wireless method). At the receiving end the digital signal is again converted in to digital data by using line decoder block and finally digital data is given to the receiver computer.

Important terms used in line encoding:

1. **Signal element :** A signal element carries data element and it is the shortest unit (time wire) of a digital signal.
2. **Data element (bit) :** Data element is the smallest entity that can represent a piece of information known a bit.

   "Data elements are what we need to send;

   Signal elements are what we can send.

   Data elements are being carried and signal

   Element are the carriers"
3. **Data rate or bit rate :** The **data rate** is defined the number of data element (or bits) sent in one second. Its unit is bps (Bits per second).
4. **Signal rate :** The **signal rate** is the number of signal element sent in one second. Its unit is baud. It is prefer in data communication to increase the data rate while decreasing the signal rate. Speed of transmission can be increase by increasing the data rate whereas decreasing the signal rate, decrease the bandwidth requirement.

We define a ratio '*r*' which is the number of data element carried by each digital signal element. It is the ratio of number of data elements to the number of signal elements. For example, a data element is represented by a signal element then the ratio '*r*' is '1'. Fig. 9.44(A, B, C and D) shows the different types of ratio's signals representation.

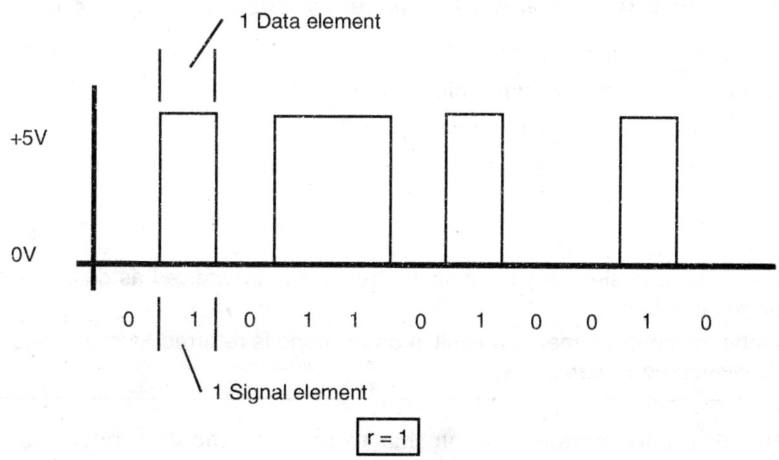

Fig. 9.44. *(a). One Data Element per signal (r = 1)*

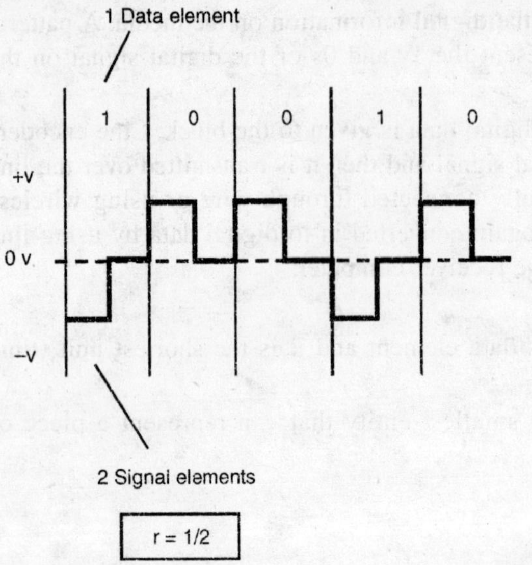

(b) *One data element per two signal element* ($r = 1/2$)

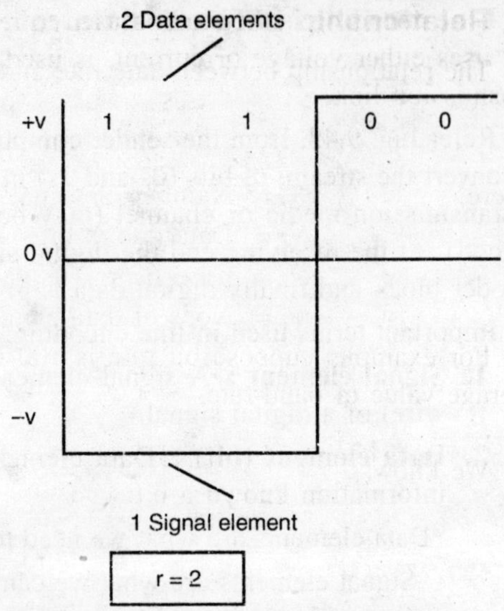

(c) *Two data elements per signal element* ($r = 2$)

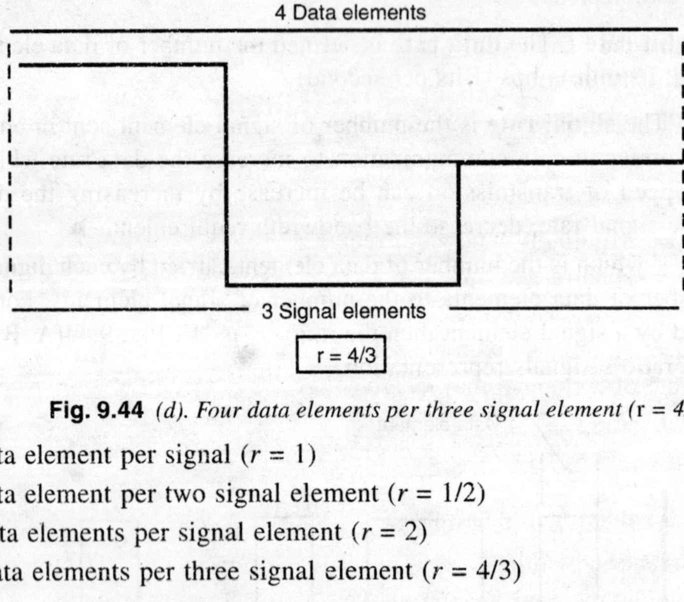

Fig. 9.44 (*d*). *Four data elements per three signal element* ($r = 4/3$)

A. One data element per signal ($r = 1$)

B. One data element per two signal element ($r = 1/2$)

C. Two data elements per signal element ($r = 2$)

D. Four data elements per three signal element ($r = 4/3$)

TIPS

- The number of data elements transmit in one second is referred as data and measured in *bits per second* (bps).
- The number of signal element transmit in one second is referred as signal rate or baud rate and it is measured in *baud.*

It is pretended in data communication that to increase the data rate which decreasing the signal rate. The requirement of bandwidth the data communication channel is also depend upon the ratio (r).

5. Relationship between data rate and signal rate:

The relationship between data rate and signal rate is given by,

$$S = C \times N \times \frac{1}{r} \quad \text{...(9.2)}$$

where, S = Number of signal elements
N = Data rate in bits.
C = Case factor (its values lies between 0 and 1)
r = No.'s of data element / each signal element

For example; suppose bit rate is 100 kbps, $r = 2$ and value of case factor (C) is 1. Find the average value of baud rate.

We know,

$$S = C \times N \times \frac{1}{r}$$

$$S = 1 \times 100000 \times \frac{1}{2}$$

$$S = 50 \text{ kbaud}$$

6. Bandwidth

Baud rate determine the bandwidth. Although the actual bandwidth of a digital signal is infinite but the effective bandwidth is finite.

By using following relation, the minimum bandwidth can be determined,

$$B_{\min} = C \times N \times \frac{1}{r} \quad \text{...(6.3)}$$

where, $B_{\min}$ = Minimum bandwidth required
N = Data rate in bps
C = Case factor
r = Number of data elements / each signal element

Maximum data rate of a channel may be given by

$$N_{\max} = \frac{1}{C} \times B \times r \quad \text{...(6.4)}$$

where, $N_{\max}$ = maximum data rate
B = Bandwidth

According to Nyquist theorem, maximum data of a channel is given as

$$N_{\max} = 2 \times B \times \log_2 (L) \quad \text{...(6.5)}$$

A signal with L levels actually can carry $\log_2 (L)$ bits per level, if $C = ½$ and $r = 1$;

$$N_{\max} = \frac{1}{C} \times B \times r = 2 \times B \times \log_2 (L)$$

7. DC Com ent

There are ; lines encoding schemes leaves residual **direct current (dc) components**. A dc signal has zero frequency so it cannot travel through a media and it is very difficult to handle dc components by channel.

The dc component is undesirable due to following two reasons :

(1) If the signal has to pass through a system, say transformer that does not allow dc component to pass through, so the signal is distorted and that may cause errors in the output signal.

(2) The dc component has extra energy lies on the line and it is waste full.

For example, we have two line encoding schemes as shown in Fig. 9.45 (*a*) and (*b*). In Fig. 9.45 (*a*), line coding has dc component because its average value is not zero, *i.e.*, the positive voltages are not cancelled by the negative voltages. This waveform does not pass through the transformer properly. In Fig. 9.45 (*b*) the average value is zero, than means there is no dc components in this case. This waveform can pass through the transformer properly.

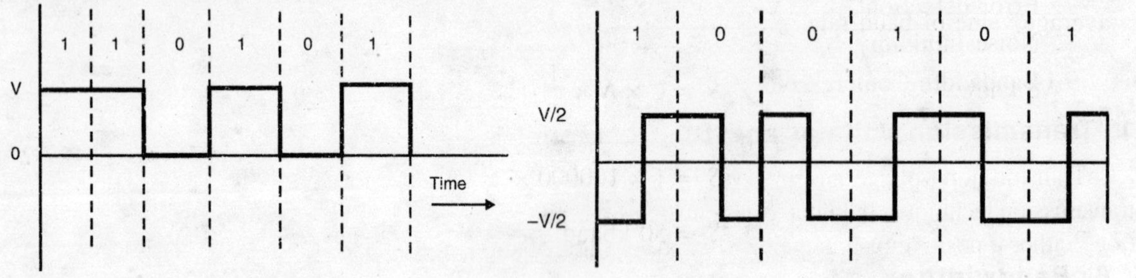

(*a*) *a signal with dc components* (*b*) *a signal without dc components*

Fig. 9.45.

8. Base-line Wandering

A long string of either logic 1's or logic 0's in NRZ produces a condition in which a receiver may loss its amplitude reference for optimum discrimination between received 1's and 0's. This state is known as **dc wandering or base line wandering**. This problem may also arise when there is a significant imbalance in the number of 1's and 0's transmitted. Figure 9.46 shows how dc wandering is produced from a long string of successive logic 1's. It is observed that after a long string of 1's; there is likely that 1 may be changed to 0 (1 to 0 errors). Similarly, when a long string of 0's increases the chances of that 0 may be changed to 1 (0 to 1 errors).

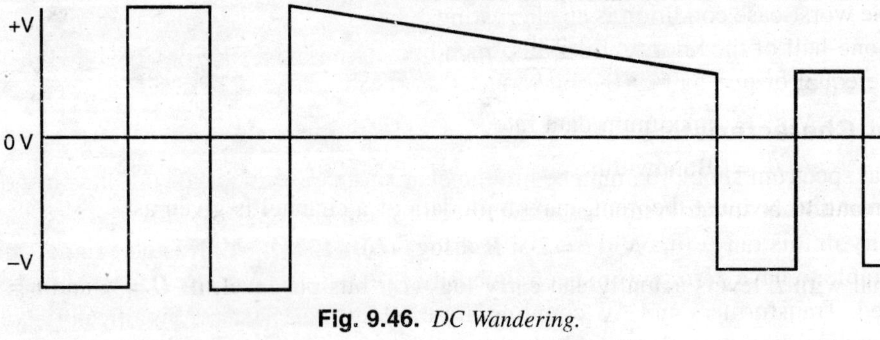

Fig. 9.46. *DC Wandering.*

DC components and DC wandering or base line wandering are short coming of few of the line encoding schemes.

9.21. FACTORS AFFECTING THE CHOICE OF LINE CODING SCHEME

Line coding is used to convert a sequence of bits in to a digital signal. There are several line codes that are used for the electrical representation of binary signal symbols 1 and 0. Each representation has its own advantages and disadvantages. The most important criteria for choosing a particular waveform are based on spectral characteristics, timing context, bandwidth, noise immunity etc.

These important factors are discussed as follows :

1. Transmission voltage and dc components
2. Bandwidth consideration
3. Spectral characteristics
4. Duty cycle
5. Clock and framing bit recovery
6. Power efficiency
7. Error detection
8. Noise Immunity
9. Bandwidth Compressive

1. Transmission Voltage and DC Components

Transmission voltage and or levels can be categories as being either unipolar or bipolar. In this transmission technique, the binary data involves the transmission of only a single non-zero voltage level (*e.g.*, either a positive or a negative voltage for logic '1' and zero voltage ground for logic 0).

In bipolar transmission, two non zero voltages are involved (for example, a positive voltage for a logic '1' and an equal magnitude negative voltage for a logic '0' or vice-versa. Considering an equal probability of the occurrence of logic 1 and 0, symmetrical bipolar voltage reduces the average power by 50%.

2. Bandwidth Consideration

To determine the maximum bandwidth required to propagate a line-encoded digital signal. We must determine the highest fundamental frequency in the signal. The highest fundamental frequency may be determined from the worst-case (*i.e.*, where fastest transitions occur) binary bit sequences. For example, with unipolar non-return to zero, the worst case condition is an alternating frequency 1 and 0 sequence, the period of the highest fundamental frequency is the time of two bits that is equal to one-half of the bit rate (bit rate/2). Similarity, in the case of bipolar non-return to zero, again the worst-case condition is an alternating 1 and 0 sequences, and the highest fundamental frequency is one-half of the bit rate. Like this, each line coding scheme will have highest frequency and it may be equal or one half of the bit rate.

3. Spectral Characteristics

The signal spectrum should be matched to the channel frequency response. Now, if the channel has high attenuation at lower frequencies the spectrum of the signal should have a small power spectral density in this range to avoid excessive distortion in the signal. It is also important to note that it is desirable to have zero power spectral density (PSD) at dc, *i.e.*, $\omega = 0$, unable the system to be AC coupled. Transformers and AC coupling are used at the repeaters. Significant lower in low frequency components causes dc wander in the pulse stream when AC coupling is used. The AC coupling is required since the DC path provide by the cable pairs between the repeaters sites are used to transmit the power required to operate the repeaters.

4. Duty Cycle

The duty cycle of a binary pulse may be used to categorize the type of transmission. If this binary pulse is maintained for the entire bit time, this is called non-return to zero. If the active time of binary pulse is less than 100% of the bit time, it is called return to zero (RZ).

5. Clock and Framing Bit Recovery

Bit synchronization is the most essential requirement in any digital communication system. To recover and maintain clock and framing synchronization from the received data, there should be enough transition in the data waveform. In unipolar NRZ and bipolar NRZ, encoding a long string of 1s and 0s generation a data signal void transition and therefore, is inadequate for clock recovery. In bipolar RZ, a transition occurs in each bit position regard less of whether the bit is a '0' or a '1'. This BPRZ is the best encoding scheme for clock recovery.

6. Power Efficiency

Power efficiency means the requirement of power for the transmission should be as small as possible, for a given bandwidth and a specified detection error probability.

7. Error Detection and Correction Capability

It is preferred that line encoding scheme should have some means to detect the error and preferably correct the detected errors. With unipolar NRZ, BPNRZ, unipolar RZ and Bipolar RZ encoding, there is no means to determine that the received data have the error or not. However, in BPRZ- AMI (Bipolar Return to zero- Alternate Mark Inversion) encoding, an error in any bit will cause a bipolar violation (*i.e.,* the reception of two or more consecutive logic 1s with the same polarity). Therefore, BPRZ-AMI has a build-in error detection mechanism.

8. Noise Immunity

The transmitted symbols (1s and 0s) should be protected against noise, in order to reduce the error in transmission. The bipolar signals are having more immunity against noise, them the unipolar signals. However more power is to be transmitted.

9. Bandwidth Compression

There are some line encoding scheme such as multilevel codes increase the efficiency of BW utilization by using more number of levels so that in a bit interval essentially 2 bit may be transmitted.

9.22. COMMON LINE CODING SCHEMES

There are following types of techniques or schemes used for Line Encoding (refer Fig. 9.47)

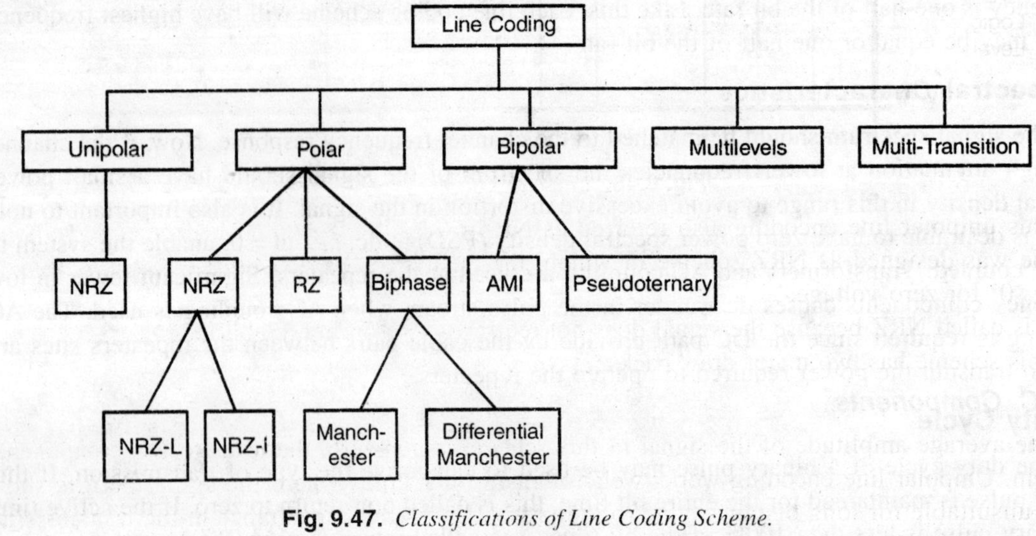

Fig. 9.47. *Classifications of Line Coding Scheme.*

1. Unipolar—NRZ (Non-return to zero)
2. Polar :
   - NRZ- L (Non-Return to zero-level)
   - NRZ-I (Non-Return to zero—Inversion)
   - R Z (Return to zero)
   - Biphase—Manchester and Differential Manchester
3. Bipolar :
   - AMI (Alternate Mark Inversion)
   - Pseudoternary
4. Multilevel
5. Multi-transition

9.23. LEVEL ENCODING

The simplest form of line coding uses a different signal level to encode each discrete symbol transmitted. Communications engineers commonly refer to the unbalanced code as a unipolar code and the balanced code as a polar code.

9.23.1. Unipolar (Non-Return to Zero)

- **Unipolar encoding** technique is the simplest technique to represent the digital signal with only one level of amplitude.
- One level of amplitude mean one voltage level (say V) stands for binary 1 and another level (say 0 V) stands for binary 0.
- In this scheme signal level is only one side of the reference level (0 V level) normally above the reference level that means having a single polarity only.

Figure 9.48 shows the representation of the digital data. A common example of Unipolar line encoding is the logic levels used in computers and digital logic. Logic High (1) is represented by +5V and a logic Low (0) is represented by 0 V.

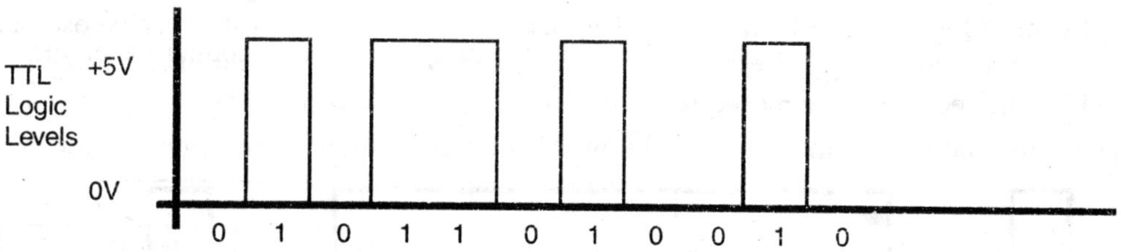

Fig. 9.48. *Unipolar Line Coding.*

This unipolar line encoding also referred as NRZ (Non-Return to Zero) scheme. A unipolar scheme was designed as NRZ scheme in which, the positive voltage defines the binary '1' and binary '0' for zero voltage.

It is called NRZ because the signal does not return to zero during the bit interval.

This scheme has two major draw backs.

1. DC Components

The average amplitude of the signal in this scheme is non-zero, the leads to DC components problem. Unipolar line encoding works well for inside machines where the signal path is short but is unsuitable for long distances due to the presence of stray capacitance in the transmission medium. On long transmission paths, the constant level shift from 0 volts to 5 volts causes the stray capacitance to charge up. There will be a "stray" capacitor effect between any two

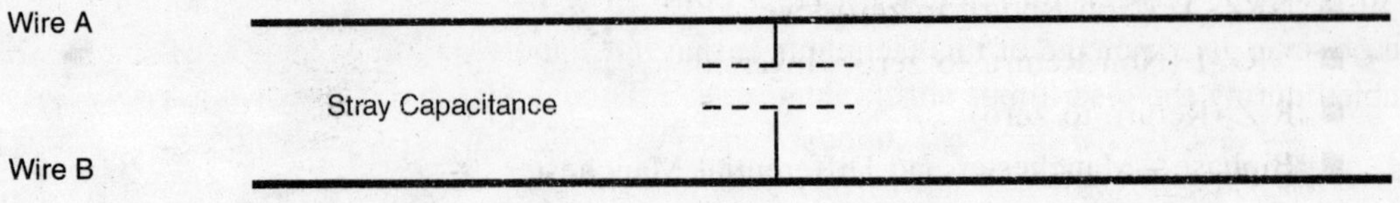

Fig. 9.49. *A "Stray" Capacitor Effect Between any Two Conductors.*

conductors that are in close proximity to each other. Parallel running cables or wires are very susceptible to stray capacitance.

If there is sufficient capacitance on the line and a sufficient stream of 1s, a DC voltage component will be added to the data stream. Instead of returning to 0 volts, it would only return to 2 or 3 volts. The receiving station may not recognize a digital low at voltage of 2 volts!

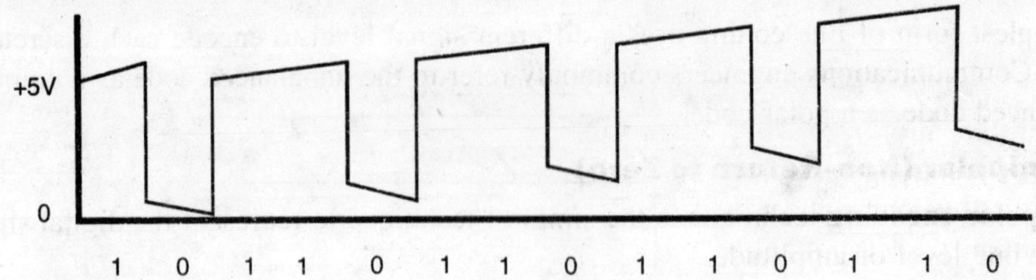

Fig. 9.50. *DC Voltage Component.*

2. Lack of Synchronization

If the data consists of a long sequence of 1's and 0's, and then there is no change in the signal level during this period, so that leads to synchronization problem. Unipolar line encoding can have synchronization problems between the transmitter and receiver's clock oscillator. The receiver's clock oscillator locks on to the transmitted signal's level shifts (logic changes from 0 to 1). If there is a long stream of logical 1s or 0s in a row than there is no level shift for the receive oscillator to lock to. The receive oscillator's frequency may drift and become unsynchronized. It could lose track of where the receiver is supposed to sample the transmitted data.

Unipolar line coding scheme is not used now days because of above mentioned draw backs.

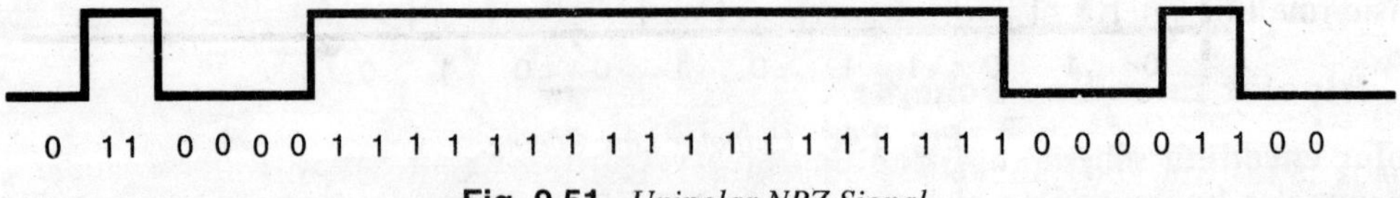

Fig. 9.51. *Unipolar NRZ Signal.*

The existence of low frequencies in a random data signal is the basic reason why modems are needed for data communications over the analog telephone network. Analog telephone circuits do not pass direct current. It is also the reason that NRZ coding is not often used for long-distance transmission. Direct-current wander is not unique to data transmission systems. It is a phenomenon that must be reconciled in television receivers, radar receivers, or radiation detectors.

One technique of offsetting dc wander is referred to as dc or baseline restoration. As illustrated in Figure 9.51, dc restoration involves passing received pulses through a capacitor, detecting them, and then removing the charge on the capacitor before the next pulse arrives. Charge on the capacitor is removed by driving the voltage to a specific threshold (0 V in Fig. 9.52) and then removing the driving voltage when threshold is reached. Since all charge on the capacitor is removed after each pulse, the baseline or decision reference level is constant at the beginning of a signal interval.

An obvious disadvantage of this technique is that the signal input must have zero amplitude or be disabled during the reset time.

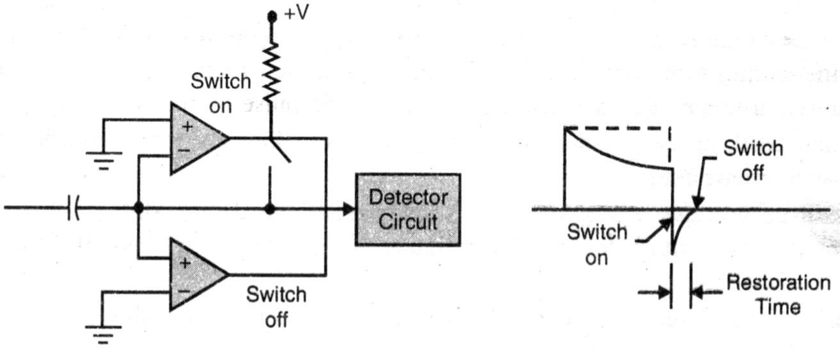

Fig. 9.52. *Direct-Current restoration for unipolar pulses.*

A generally more useful technique for overcoming baseline wander is to use decision feedback, also called quantized feedback equalization. In contrast to dc restoration, which drives the capacitor voltage to a constant, predetermined level, quantized feedback compensates for dc wander by locally generating the unreceived low-pass response and adding it to the received signal. To accomplish this, the original data stream is reconstructed. As shown in Fig. 9.53, the reconstructed data stream is passed through a low-pass filter that generates a pulse equal to the tail or droops characteristic of the channel. The feedback signal adds to the received signal to eliminate the droop (intersymbol interference). Using a frequency-domain analysis, the feedback response is complimentary to the channel response. Quantized feedback is used in ISDN basic rate line interfaces.

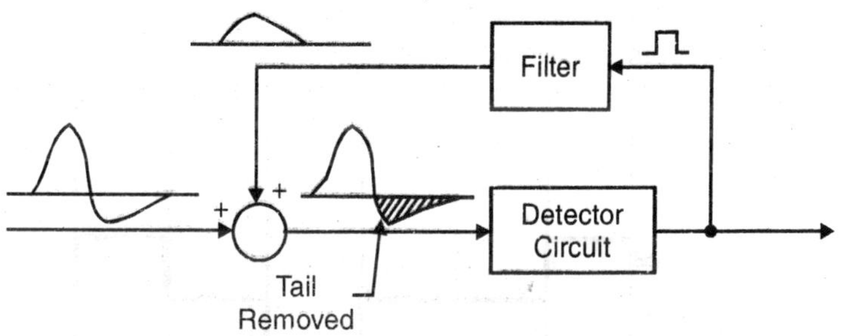

Fig. 9.53. *Decision feedback equalization.*

9.23.2. Bipolar Encoding Scheme

Bipolar encoding scheme utilizing the three voltage levels to represent digital data. These voltage levels are +v, 0v and –v. in this scheme, there is no transition at the middle of the bit interval. Bipolar coding solves the dc wander problem by using three levels to encode binary data. Specifically, logic 0 is encoded with zero voltage while logic 1 is alternately encoded with positive and negative voltages. Hence the average voltage level is maintained at zero-to eliminate dc components in the signal spectrum.

Since bipolar coding uses alternate polarity pulses for encoding logic 1's, it is also referred to as **alternate mark inversion (AMI)**.

The binary bit '0' bit is represented by zero level and binary bit '1' is represented by alternative positive and negative voltage level. There are three encoding schemes under bipolar encoding scheme.

(1) AMI (Alternate Mark Inversion)
(2) Pseudoternary (variation of AMI)
(3) Twinned binary codes

(1) Alternative Mark Inversion (AMI)

The most widely used of the multilevel binary codes is the Bipolar-AMI code. The Bipolar-AMI code was first used in the PCM based T1-carrier system by Bell Labs in 1962. Bipolar coding is the basic line-coding procedure used by T1 lines in the telephone network. Rather than using full-period pulses, however, T1 lines use a 50% duty cycle pulse to encode each logic 1.

The Bipolar-AMI suffers in that the receiver loses clocking information when a continuous stream of zeros is transmitted.

In AMI, three voltage levels are used to represent the digital data. A binary bit '0' is represented by the zero voltage level and alternative positive and negative voltage level represent the binary bit '1'.

Alternative Mark Inversion means alternately change in voltage level to represent mark (*i.e.,* binary bit '1' from telegraphy).

The drawbacks of AMI encoding schemes are :

- Problem of synchronization
- DC component occurs whenever there is a long string of binary bit '0'.

Figure 9.54 shown the representation of the digital data in AMI encoding scheme.

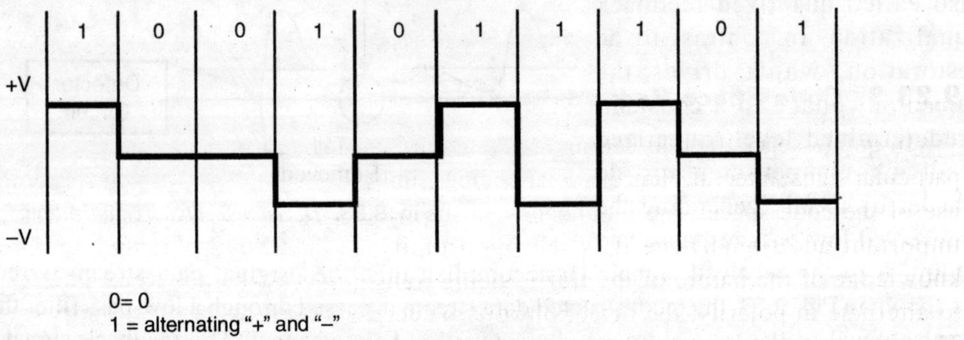

Fig. 9.54. *AMI Signal Representation.*

(2) Pseudoternary

This is a variation of the AMI, in this scheme binary bit '1' is represent with the zero level and binary bit '0' is represent with alternative +v and –v to voltage level.

It is used for long distance communication.

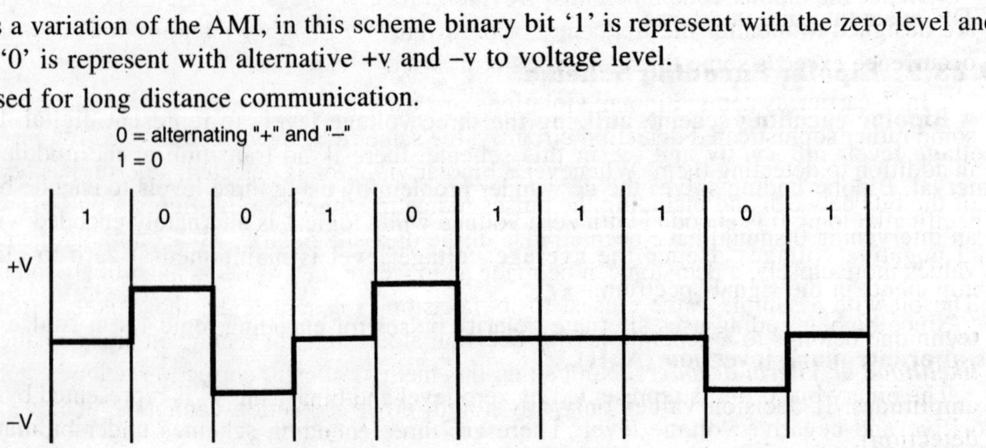

Fig. 9.55. *Pseudoternary Encoding Scheme.*

Figure 9.55 shown the representation of the digital data in Pseudoternary encoding scheme for the digital data 1001011101.

(3) Twinned Binary Codes

To eliminate the DC component from the spectrum and to keep from increasing the symbol rate, the **twinned binary codes** can be used. A twinned binary code increases the number of available symbols by increasing the number of signal levels.

The main disadvantage of twinned binary codes is a slight loss in noise immunity.

0 ⇒ 1 transmits "+"
1 ⇒ 0 transmits "–"
0 ⇒ 0 transmits "0"
1 ⇒ 1 transmits "0"

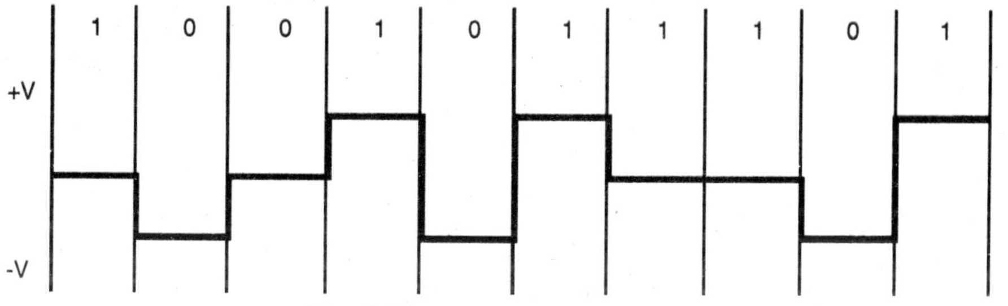

Fig. 9.56. *Twinned Binary Codes.*

9.23.3. Code Space Redundancy

In quintessence, bipolar coding uses a ternary code space but only two of the levels during any particular signal interval. Hence bipolar coding eliminates dc wander with an inefficient and redundant use of the code space. The redundancy in the waveform also provides other benefits. The most important additional benefit is the opportunity to monitor the quality of the line with no knowledge of the nature of the traffic being transmitted. Since pulses on the line are supposed to alternate in polarity, the detection of two successive pulses of one polarity implies an error. This error condition is known as a **bipolar violation**. No single error can occur without a bipolar violation.

Hence the bipolar code inherently provides a form of line code parity. The terminals of T1 lines are designed to monitor the frequency of occurrence of bipolar violations, and if the frequency of occurrence exceeds some threshold, an alarm is set.

In T-carrier systems, bipolar violations are used merely to detect channel errors. By adding some rather sophisticated detection circuitry, the same redundancy can be used for correcting errors in addition to detecting them. Whenever a bipolar violation is detected, an error has occurred in one of the bits between and including the pulses indicating the violation. Either a pulse should be a 0 or an intervening 0 should have been a pulse of the opposite polarity. By examining the actual sample values more closely, a decision can be made as to where the error was most likely to have occurred. The bit with a sample value closest to its decision threshold is the most likely bit in error. This technique belongs to a general class of decision algorithms for redundant signals called ***maximum likelihood*** *or* ***Viterbi decoders***. Notice that this method of error correction requires storage of pulse amplitudes. If decision values only are stored, error correction cannot be achieved (only error detection).

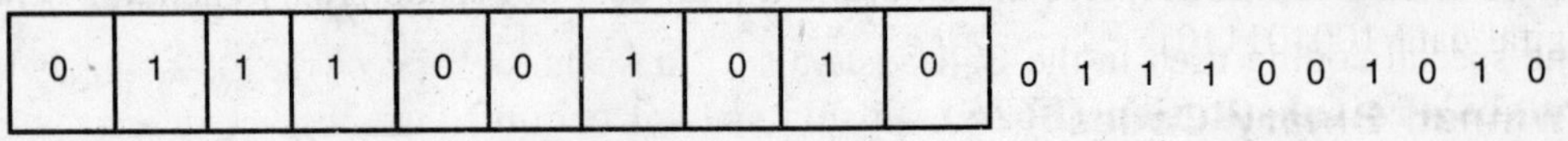

(*a*) Binary Code

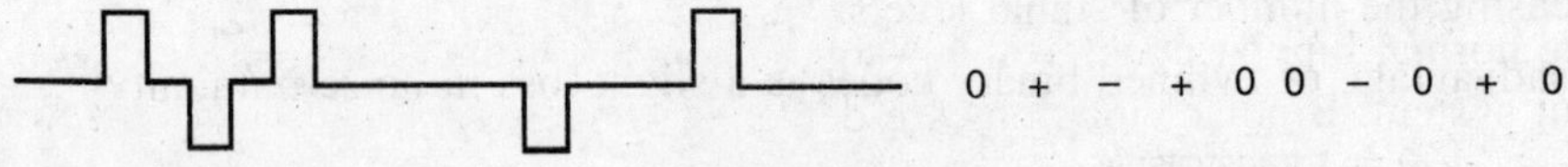

(*b*) Normal Bipolar

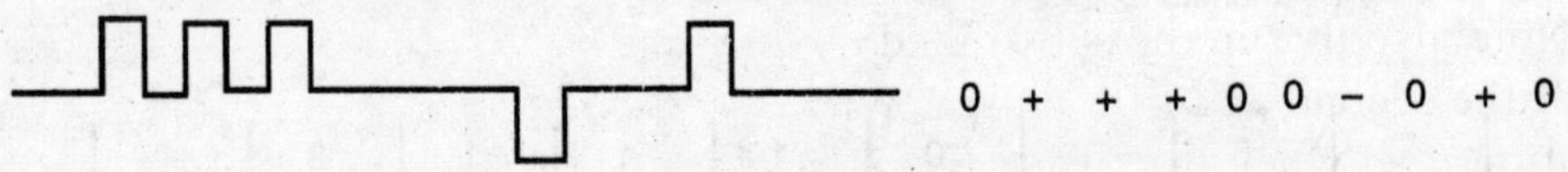

(*c*) Bipolar Violation

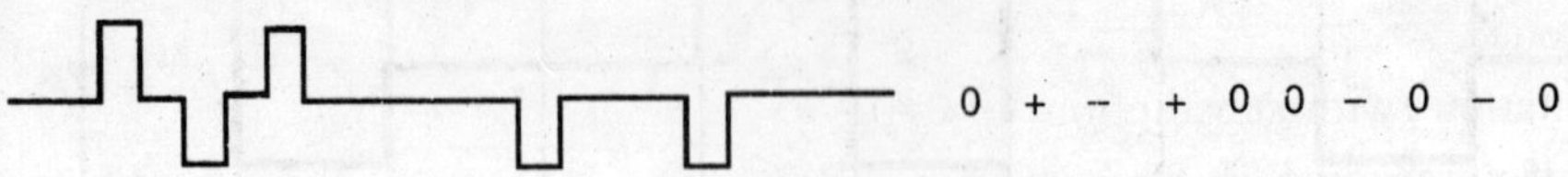

(*d*) Bipolar Violation

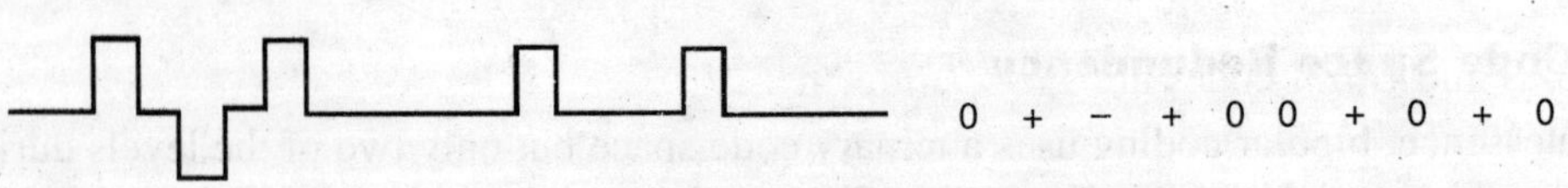

(*e*) Bipolar Violation

Fig. 9.57. *Code Representations of Binary Violation.*

An additional application of the unused code space in bipolar coding is to purposely insert bipolar violations to signify special situations such as time division multiplexing framing marks, alarm conditions, or special codes to increase the timing content of the line signals. Since bipolar violations are not normally part of the source data, these special situations are easily recognized. Of course, the ability to monitor the quality of the line is compromised when bipolar violations occur for reasons other than channel errors.

Figure 9.57 shows an example of binary violation.

9.23.4. N-Zero Substitution

The technique used to solve the pulse density problem is to use special coding to ensure that sufficient 1's are always present to provide accurate timing. The special code pattern is inserted into the transmitted bit stream. At the receiver, the presence of the special code is detected and the signal adjusted to the original information.

One way of ensuring that there are large numbers of 1's in any transmitted bit stream is to detect long strings of 0's and substitute for them code patterns containing mostly ones. At the receiver these code patterns are recognized and the inserted pattern is replaced with 0's. Such algorithms are called binary N-zero substitution schemes or BNZS schemes.

9.23.4.1. B6ZS Codes

One special coding used in the Bell System's T2 transmission lines (the DS2 signal) is called **binary six-zero substitution (B6ZS).** In this special coding, the receiver detects where the substitution has occurred by detecting a bipolar violation.

In a normal bipolar code, the pulses representing 1s are alternately positive and negative around zero. Zero is represented by 0 and 1 is represented by a + or – sign, which indicates the polarity of the pulse. For the normal bipolar code, + and – signs alternate in the code with 0s placed as they occur in the input system. Bipolar violations are detected when the alternating pattern of + and – signs is interrupted; that is if two + or two – signs occur in sequence.

In the B6ZS code, whenever a string of six 0s occurs, a special code, is substituted. The code pattern substituted depends on the polarity of the 1 pulse immediately preceding the string of six 0s. Note also that in the substituted code, a built- in bipolar violation is caused by the substitution of the second and fifth pulses. When the second pulse is substituted it causes a +0+ or –0– violation, and when the fifth pulse is substituted it causes a –0– or +0+ violation. Here are specific examples:

1. The 1 pulse preceding six zeros is a + :

   Binary code: 0101000000110000001001

   Representation (with substitution) 0-0+0+ –0– + – +0+ –0– + –00+

2. The 1 pulse preceding six zeros is a – :

   Binary code: 0111000000000000001101001

   Representation (with substitution): 0- + - 0 - + 0 + - 0 - + 0+ - + - + + 0 0 -

The B3ZS and B6ZS codes are specified for the DS3 and DS2 North American standard rates, respectively, and B8ZS is specified as an alternative to AMI for the DS1 and DS1C rates.

DO YOU KNOW

B6ZS Rules

1 Pulse Polarity	Special Code
+	0 + – 0 – +
–	0 – + 0 + –

9.23.4.2. HDB3

A somewhat different special code substitution is recommended by the CCITT in Europe. It is also a BNZS code called *high-density bipolar three-bit substitution* (HDB3). This technique replaces strings of four 0s with the substitute code shown in below :

DO YOU KNOW

HDB3 Rules

1 Pulse Polarity	Special Code Even	Special Code Odd
+	000 +	– 00– +
–	000 –	+ 00 +

Because four 0s are replaced (greater than 3), the code is called HDB3. Its substitution rules depend on how many 1s have occurred in the PCM bit stream since the last substitution as well as the polarity of the 1 pulse that is just ahead of the string of zeros.

For examples:

1. Plus, odd; minus, odd:

   Binary code: 010110000101010000

   Substitution : 0 + 0 – + 0 0 0 + – 0 + 0 – 0 0 0 –

9.23.4.3. Pair Selected Ternary

The BNZS substitution algorithms described in the preceding section are examples of selecting codes in the ternary codes pace to increase the timing content of a binary signal.

Pair selected ternary (PST) 20 is another example.

The PST coding process begins by pairing the binary input data to produce sequences of 2-bit codeword's. These codeword's are then translated in to two ternary digits for transmission. Since there are nine two-digit ternary codes but only four two-digit binary codes, there is considerable flexibility available in selecting the codes. The most useful of the possible coding formats is shown below. This particular format not only ensures a strong timing component but also prevents dc wander by switching modes to maintain a balance between positive and negative pulses. The codes are selected from one column until a single pulse is transmitted. At that time the encoder switches modes and selects codes from the other column until another single pulse (of the opposite polarity) is transmitted.

For example, encode the following binary data stream into a PST line code:

010 011 1010110 0

There are two possible solutions depending on whether the encoder is in the positive or negative mode at the beginning of the sequence:

	01	**00**	**11**	**10**	**10**	**11**	**00**
Case 1 (+ mode) :	0 +	– +	+ –	– 0	+ 0	+ –	– +
Case 2 (– mode) :	– +	– +	+ –	+ 0	– 0	+ –	– +

One potential drawback of the PST coding algorithm is that the binary data stream must be framed into pairs. Hence a PST decoder must recognize and maintain pair boundaries. Recognition of the boundaries is not difficult if random data are being transmitted since a pairwise misframe eventually produces unallowed codes (00, + +, – –). Furthermore, time division multiplex formats typically provide character and pairwise framing automatically.

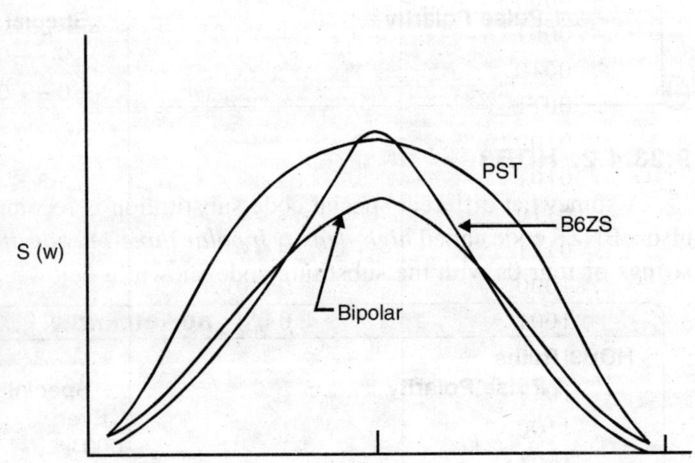

Fig. 9.58. *Spectrum of Bipolar, B3ZS, and PST Line Codes for Equally Likely 1's and 0's.*

The power spectrum of a PST line code with equal probabilities for 1's and 0's is plotted in Fig. 9.58. Also shown is the B6ZS power spectrum and the conventional bipolar power spectrum. An important point to notice in Fig. 9.57 is that bipolar coding and its extensions require equal bandwidths. Their only significant difference is that B6ZS and PST have higher energy levels as a result of greater pulse densities. The higher energy levels have the undesirable effect of increasing crosstalk interference in multipair cables.

However, the degradation from the increased crosstalk is somewhat offset by improved accuracy of the recovered sample clock (assuming all signals in the cable are using the same line code).

Pair Selected Ternary Encoding Rules :

Binary Input	+ Mode	– Mode
00	– +	– +
01	0 +	0 –
10	+ 0	– 0
11	+ –	+ –

9.23.5. Ternary Coding

Since bipolar and PST coding use a ternary code space to transmit binary data, they do not achieve as high an information rate as is possible with more efficient use of the code space. For example, an eight-element ternary code is capable of representing $3^8 = 6561$ different codes. In contrast, 8 bits of binary data produce only $2^8 = 256$ different codes. The previously described line codes do not take advantage of the higher information content of ternary codes; they select codes for their timing content and spectral properties.

Table 9.5

Binary Word	Ternary Word (Accumulated Disparity)		
	–	0	+
0000	– – –		+ + +
0001	– – 0		+ + 0
0010		– 0 –	+ 0 +
0011		0 – –	0 + +
0100		– – +	+ + –
0101	– + –		+ – +
0110	+ – –		– + +
0111	– 0 0		+ 0 0
1000	0 – 0		0 + 0
1001	0 0 –		0 0 +
1010		0 + –	
1011		0 – +	
1100		+ 0 –	
1101		– 0 +	
1110		+ – 0	
1111		– + 0	

One ternary encoding procedure involves mapping successive groups of 4 bits into three ternary (4B3T) digits. Since binary words of 4 bits require only 16 of the 27 possible three-digit ternary codewords, considerable flexibility exists in selecting the ternary codes. Table 6.1 presents one possible encoding procedure. Ternary words in the middle column are balanced in their dc content. Codewords from the first and third columns are selected alternately to maintain dc balance. If more positive pulses than negative pulses have been transmitted, column 1 is selected. When the disparity between positive and negative pulses changes, column 3 is chosen. Notice that the all '0's codeword is not used.

Hence a strong timing content is maintained. Because of the higher information efficiency, however, the ability to monitor performance is sacrificed, and framing is required on three-digit boundaries. 4B3T coding is used on the TI48 span line developed by ITT Telecommunication. This system provides T-carrier transmission for two DS1 signals (48 channels) using a bandwidth that is only 50% greater than a Tl bandwidth (carrying 24 channels).

9.24. POLAR LINE ENCODING

When the digital encoding is symmetrical around 0 Volts, it is called a **Polar Code.**

- Polar line encoding is also called Non-Return to Zero (NRZ).
- Polar line encoding is the simplest pattern that eliminates most of the residual DC problem.
- Polar line encoding scheme uses two voltage levels, one +ve and other is –ve. By using two voltage levels in this scheme reduces the average voltage level on line that means reducing the problem of dc component as observed in unipolar line encoding scheme.
- For example, the voltage level for '0' can be positive and voltage level for '1' can be negative.
- The RS-232D interface uses Polar line encoding.

Types of Polar Line Encoding Scheme

1. Polar NRZ
   - NRZ-L (Non-Return to Zero-Level)
   - NRZ-I (Non-Return to Zero-Inversion)
2. RZ (Return to zero)
3. Bi-phase
   - Manchester
   - Differential Manchester

9.24.1. Polar NRZ (Non-Return to Zero)

- In **polar NRZ** encoding scheme, we use two levels of voltages and the level is always positive or negative. In the scheme also signal level is not returning to zero during the bit interval as observed in unipolar line encoding scheme. There are two types of polar NRZ which are discussed here.

(i) NRZ-L (Non-Return to Zero-Level)

- In **NRZ-L (Non-return to Zero-Level)** the level of the voltage determines the value of the bit and depends upon the type of bit that it represents.
- Generally, it is a –ve logic encoding scheme, where positive level (+ve) represent the '0' binary and negative level (–ve) represent binary '1'. Therefore, the level of the signal depends upon the state of the bit.
- If there is a continuous string of same binary bit (say 1 or 0), if suffers the problem of synchronization.

- Since we use two voltage levels, that reduces the average voltage level on the line and hence problem of dc component is almost eliminated in this scheme.
- Figure 9.59 shows the representation of following digital data using NRZ-L encoding scheme.

For example, 110001100

Representation of binary bits:

To represent binary '1'—— negative voltage (–ve)

To represent binary '0'—— positive voltage (+ve)

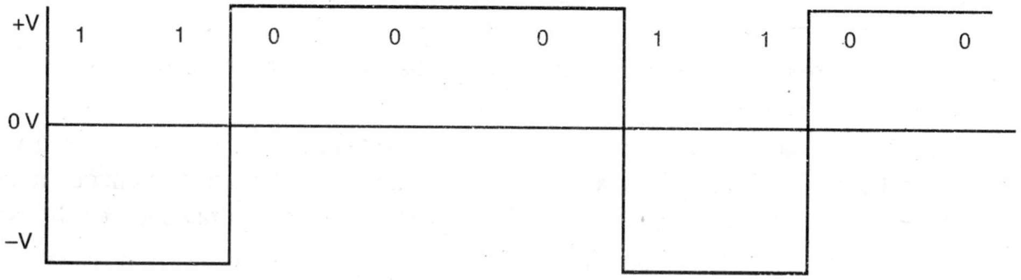

Fig. 9.59. *NRZ-L Encoding.*

(ii) NRZ-I (Non-Return to Zero–Inversion)

- In **NRZ-I (Non-Return to Zero–Inversion)** an inversion of the voltage level represents '1' bit and no change or non-inversion of the voltage level represents '0' bit.
- There is no representation of the bits with +ve or –ve voltage, as it was in polar NRZ-L, but in this scheme change of polarity voltage level will represent the binary bits.
- Synchronization problem is less in this scheme comparing to the NRZ-L, since there is a signal change for binary bit '1'. So, the existence of binary bit '1' in the data stream allows the receiver to synchronization its timer to actual arrival of the signal transmission.

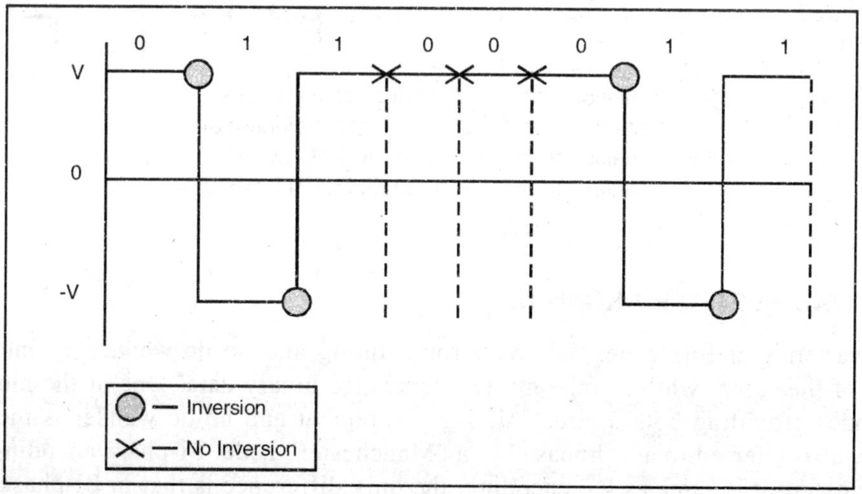

Fig. 9.60. *Shows the Representation of the Digital Data, 01100011.*

- Only when a long strings of '0' bit present in the data streams, the problem of synchronization will present, because at '0' bit there is no change of level.
- Figure 9.60 shows NRZ-I representation of the digital data, 01100011.

9.24.2. Return to Zero (RZ)

In Return to zero (RZ) line encoding scheme, we have main advantage of synchronization. In NRZ-L, if there is long string of '1' or '0' binary bits, the problem of synchronization occur and in NRZ-I, this problem occurs wherever we have a long string of '0' binary bit. But in RZ scheme, even if there is a long string of '0' binary bit; there is no problem of synchronization.

> **DO YOU KNOW**
>
> To avoid the synchronization problem, in this scheme the level of a bit change at the middle of bit interval, *i.e.*, at the middle of the bit duration the level comes back to the zero level, and that's why it is being called **Return to zero.**

- In this scheme, binary bit '1' is represented by positive to zero level and a '0' binary bit by negative to zero level.
- It may be noted that in this scheme, we use three voltage levels, +v, –v and zero voltage to represent the digital data. In this scheme, we need two signal change to encode a bit (either 1 or 0), and hence; takes more bandwidth. This is the primary drawback of this scheme.
- There is no problem of dc component and synchronization but it has few drawbacks that it required more bandwidth (twice of Polar NRZ) and system become more complex due to more signal levels.

The example of representation of digital data in RZ scheme is shown in **Fig. 9.61**; Let Binary bit '0' is represented from positive to zero level. Binary bit '1' is represented from negative to zero level.

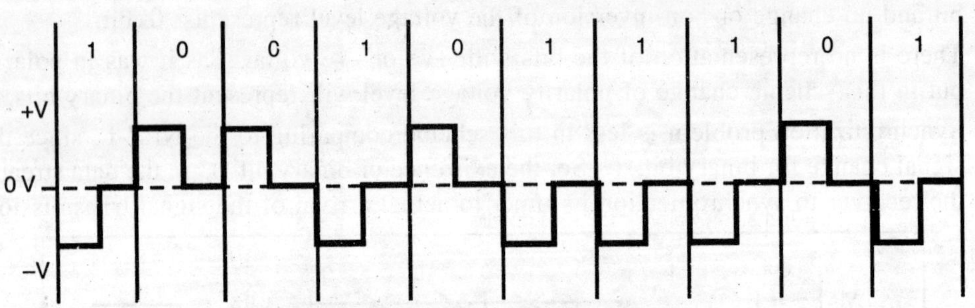

0 = transition from "0" to "+" at start of interval
transition from "+" to "0" at middle of interval
1 = transition from "0" to "–" at start of interval
transition from "–" to "0" at middle of interval

Fig. 9.61. *Return to Zero (RZ).*

9.25. DIGITAL BI-PHASE ENCODING

Many varieties of line codes achieve strong timing and no dc wander by increasing the bandwidth of the signal while using only two levels for binary data' One of the most common of these codes providing both a strong timing component and no dc wander is the digital bi-phase code, also referred to as "diphase" or a "Manchester" code. Bi-phase encoding scheme is up to some extent it is same as RZ encoding, the only difference is that in Bi-phase signalling, signal does not return to zero, instead it continues to opposite polarity, the middle transition of each bit is used for synchronization. So, there is no problem of synchronization in Bi-phase encoding scheme.

There are two types of bi-phase encoding schemes, which are as follows :

1. Manchester Encoding
2. Differential Manchester Encoding

9.25.1. Manchester Encoding

DO YOU KNOW

Manchester encoding scheme is the combination of RZ and NRZ-L encoding schemes.

In this encoding scheme, the duration of the bit is divided into two halves. The voltage level remains at one level during the first half and moves to the opposite voltage level in the second half.

- The transition at the halfway of the bit provides synchronization.
- In Manchester encoding scheme, a positive to negative transition represent a binary'0' and a negative to positive transition represent the binary '1'.
- In Manchester encoding scheme, the synchronization is achieved at the same level of RZ encoding scheme, with the two voltage levels (+v and –v) only whereas in RZ encoding to achieve synchronization, we have used three levels of voltages.
- Figure 9.62, shown the representation of digital data (1001011101) in Manchester encoding scheme.

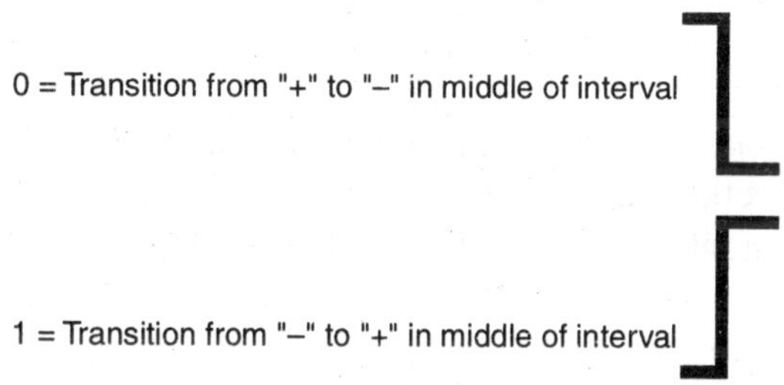

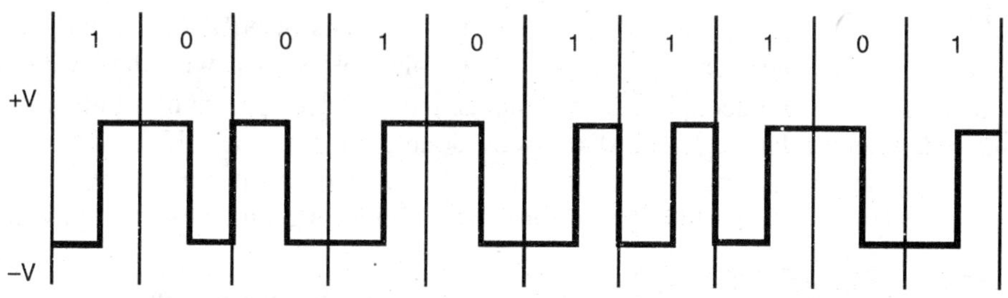

Fig. 9.62. *Manchester Encoding.*

Examination of the frequency spectrum in Figure 9.63 shows that the diphase spectrum is similar to an NRZ spectrum but translated so it is centered about 1/T instead of direct current. Hence digital biphase actually represents digital modulation of a square wave carrier with one cycle per signal interval. Logic 1's cause the square wave to be multiplied by +1 while logic 0's produce multiplication by –1. Diphase is primarily used on shorter links where terminal costs are more significant than bandwidth utilization. The "Ethernet" IEEE 802.3 local area data network uses digital biphase (Manchester) coding.

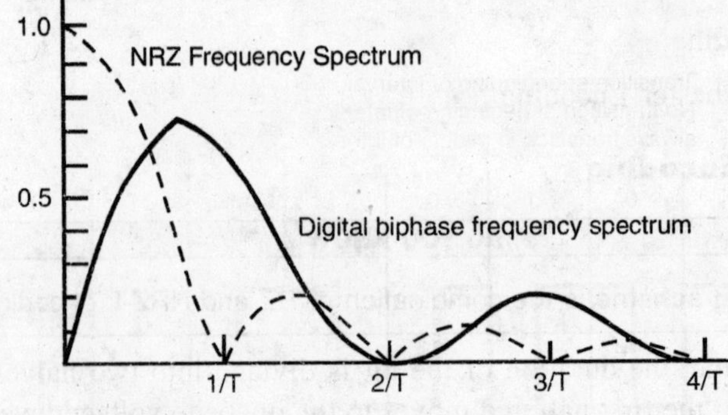

Fig. 9.63. *Power Spectral Density of Digital Biphase.*

9.25.2. Differential Manchester Encoding

Differential Manchester encoding continues the ideas of RZ and NRZ-I. We use the inversion at the middle of the each bit duration for synchronizing purpose but at the same time presence or absence of an additional transition at the beginning of the interval are used to determine the value of bit. One limitation of NRZ and digital biphase signals, as presented up to this point, is that the signal for a 1 is exactly the negative of a signal for a 0. On many transmission media, it may be impossible to determine an absolute polarity or an absolute phase reference.

Hence the decoder may decode all 1's as 0's and vice versa. A common remedy for this ambiguity is to use differential encoding that encodes a 1 as a change of state and encodes a 0 as no change in state. In this manner no absolute reference is necessary to decode the signal. The decoder merely detects the state of each signal interval and compares it to the state of the previous interval. If a change occurred, a 1 is decoded, otherwise a 0 is determined.

If the next bit is '0', transition takes place but if it is '1' than there is no transition of the level. That means to represent '0' bit we require two change in signal levels whereas to represent bit '1', there is only one change of signal level.

In this scheme also the transition at the middle of the bit interval used for synchronization.

As Manchester encoding scheme overcomes many problems of NRZ-L, similarly, Differential Manchester encoding scheme overcome the several problem associated with the NRZ-I.

Differential encoding and decoding do not change the encoded spectrum of purely random data (equally likely and uncorrelated 1's and 0's) but do double the error rate. If the detector makes an error in estimating the state of one interval, it also makes an error in the next interval. All signals of differentially encoded diphase retain a transition at the middle of an interval, but only the 0's have a transition at the beginning of an interval.

The advantages of Differential Manchester encoding schemes are as follows:

(1) Better synchronization

(2) No dc component because each bit has positive and negative voltage level continues.

(3) No base line wandering.

The only problem with this encoding scheme is the signal rate, which is double as compared to NRZ-L and NRZ-I and it required double Bandwidth than the NRZ-L and NRZ-I encoding scheme.

Figure 9.64 shows the representation of digital data (1001011101) in Differential Manchester encoding scheme.

1-No transition

0-Transition presents

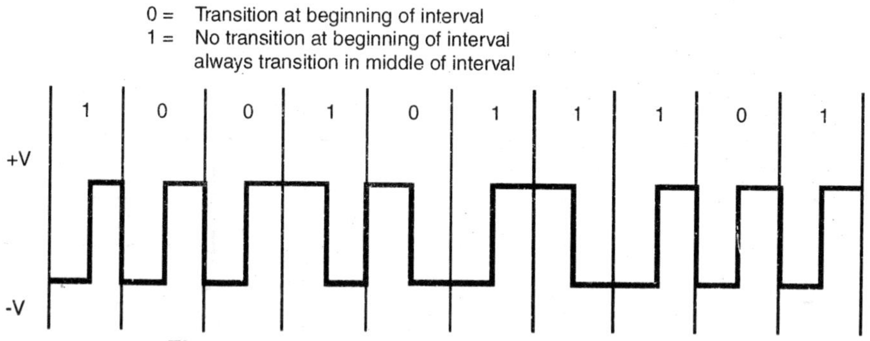

Fig. 9.64. *Differential Manchester Encoding Scheme.*

9.26. CODED MARK INVERSION

A variety of line codes have evolved similar to the digital biphase code described previously. One of these is referred to as coded mark inversion (CMI) in ITU recommendations. CMI encodes 1's (marks) as an NRZ level opposite to the level of the previous one and 0's as a half-cycle square wave of one particular phase. Figure 9.65 shows a sample encoding for CMI. There is no dc energy in the signal and an abundance of signal transitions exist as in diphase. Furthermore, there is no ambiguity between 1's and 0's. Elimination of the ambiguity actually leads to a major drawback of CMI coding: Its error performance is 3 dB worse than diphase. CMI is the specified interface code for the fourth-level (E4) ITU multiplex signal at 139.264M bps. CMI is also used for electrical interfaces of SONET - STS-3c signal.

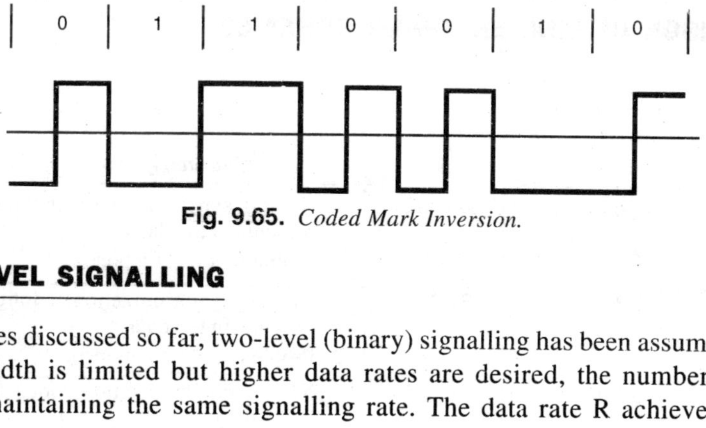

Fig. 9.65. *Coded Mark Inversion.*

9.27. MULTILEVEL SIGNALLING

In the line codes discussed so far, two-level (binary) signalling has been assumed. In applications where the bandwidth is limited but higher data rates are desired, the number of levels can be increased while maintaining the same signalling rate. The data rate R achieved by a multilevel system is given as

$$R = \log_2 (L)\left[\frac{1}{T}\right] \quad \text{...(9.1)}$$

where, L = number of levels that can be freely chosen during each interval

T = signaling interval

The signalling rate $1/T$ is often referred to as the symbol rate and is measured in bauds. Within the data communications industry it is common practice to use baud as being synonymous with bit rate. Strictly speaking, however, the bit rate is only equal to the baud rate if binary signalling (1 bit per signal interval) is used. Figure 9.66 show an example of an eight-level transmission format that achieves 3 bits per signal interval (*i.e.*, 3 bits per baud).

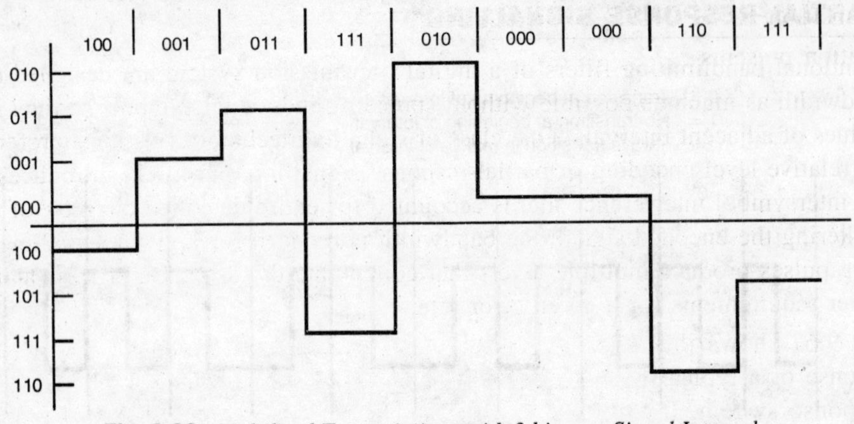

Fig. 9.66. *Multilevel Transmissions with 3 bits per Signal Interval.*

Multilevel transmission systems achieve greater data rates within a given bandwidth but require much greater signal-to-noise ratios for a given error rate. One aspect of wire line transmission that favors multilevel line coding is the lower baud rate for a given data rate, which in turn reduces the crosstalk. Hence, in crosstalk-limited systems the signal-to-noise ratio penalty of a multilevel line code is not as significant. The TlG T-carrier system developed by AT&T uses a four-level line code at the TIC baud rate (3.152 MHz) to double the capacity of a TlC system (from 48 channels to 96).

The error performance of CMI is 3 dB worse than diphase when bit-by-bit detection is used. The inefficiency rises because for one-half of an interval a 1 looks like a 0. Because CMI has redundancy, some of the inefficiency can be recovered with maximum likelihood (Viterbi) detection.

9.28. COMPARISON OF LINE ENCODING SCHEMES

S. No.	*Types of Encoding Scheme*	*Encoding Scheme*	*BW*	*Characteristics*
1	Unipolar	NRZ	B= N/2	1. No-self synchronization if long strings of 1 and 0 presents in the data. 2. Problem of dc component.
2	Polar	NRZ-L	B= N/2	1. No self-synchronization if long strings of 1 and 0 presents in the data. 2. Problem of dc component.
3	Polar	NRZ-I	B= N/2	1. No self-synchronization if long strings of 0's presents in the data. 2. Problem of dc component.
4	Polar	RZ	B= N	1. No synchronization problem. 2. Require more bandwidth. 3. More complex. 4. No dc component problem.
5	Biphase	Manchester and Differential manchester	B= N	1. Self-synchronization. 2. Require more bandwidth. 3. More complex. 4. No dc component problem.
6	Bipolar	AMI and Pseudoternary	B= N/2	1. Synchronization problem. 2. Less dc component problem

9.29. PARTIAL-RESPONSE SIGNALLING

Conventional bandlimiting filters of a digital transmission system are designed to restrict the signal bandwidth as much as possible without spreading individual symbols so they interfere with sample values of adjacent intervals. One class of signalling techniques variously referred to as duo binary, correlative level encoding or partial-response signalling, purposely introduces a prescribed amount of intersymbol interference that is accounted for entire detection circuitry of the receivers. By over filtering the encoded signal, the bandwidth is reduced for a given signalling rate, but the overlapping pulses produce multiple levels that complicate the detection process and increase the signal power requirements for a given error rate.

Figure 9.67 shows the pulse response of a typical partial-response system. If the channel is excited by a pulse of duration T, channel filters limit the spectrum such that the main part of the pulse extends across three signal intervals and contributes equally to two sample times. The reason for the term partial response is now apparent: The output only responds to one-half the amplitude of the input.

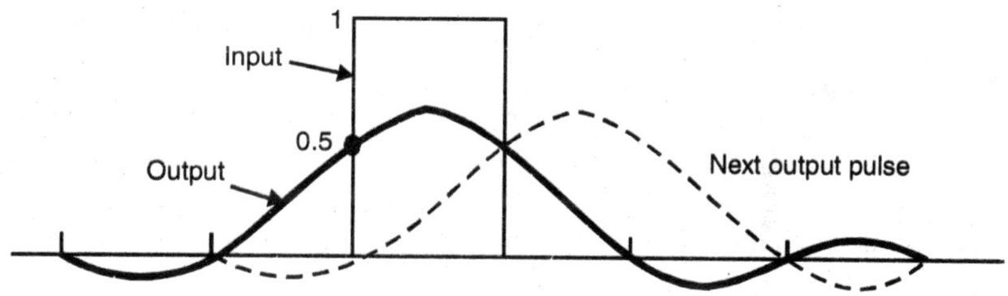

Fig. 9.67. *Output Pulse of Partial-Response Channel.*

If the input pulse of Fig. 9.68 is followed by another pulse of the same amplitude, the output will reach full amplitude by virtue of the overlap between pulses. However, if the next input pulse has a negative polarity, the overlap produces zero amplitude at the sample time. Thus, as shown in Figure 9.69, a partial-response system with two level inputs (+1, –1) produces an output with three levels (+1 ,0, –1). In similar fashion, Figure 6.69 shows a system with four input levels (+3, +1, –1, –3) and seven output levels (+ 3,+ 2,+ 1, 0, –1, –2, –3).

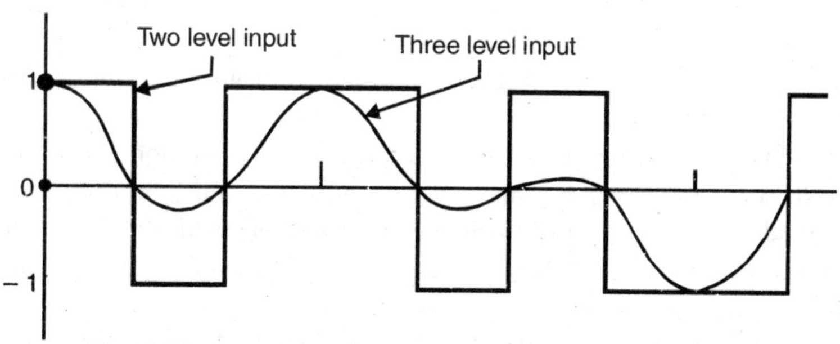

Fig. 9.68. *Three Level Partial-response Inputs and Outputs.*

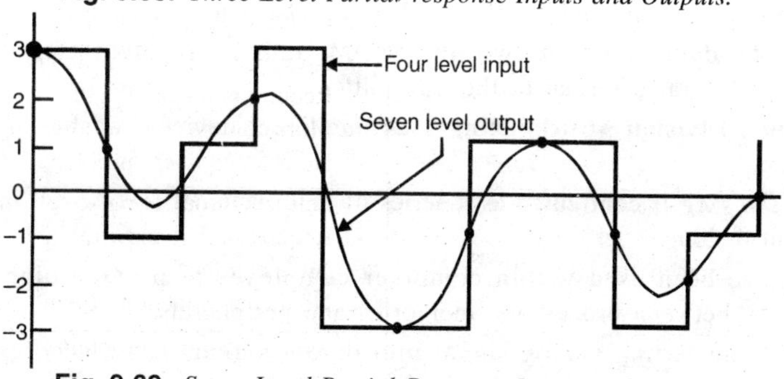

Fig. 9.69. *Seven Level Partial-Response Inputs and Outputs.*

Summary

- **Multiplexing** is an operation in which a number of independent signals are combined into a composite signal for transmission over a communication channel.
- The separation of the composite signal into the original signals is called **demultiplexing.**
- There are many domains in which multiplexing can be achieved. They are time, space, phase, frequency and wavelength.
- **Frequency Division Multiplexing (FDM)** is an analog technique where each communications channel is assigned a carrier frequency.
- The **message channel** is the basic building block of the FDM hierarchy.
- **Basic group :** A group is the next higher level in the analog hierarchy above the basic message channel and consequently is the first multiplexing step for combining message channels. 12 voice channels are multiplexed onto a higher bandwidth line to create a basic group. Twelve 4 kHz voice band channels occupy a combined bandwidth of 48 kHz.
- **Super groups :** The next higher level in the analog hierarchy is the super group, which is formed by FDM five groups containing 12 channels each. A super group has a bandwidth of 240 kHz and supports up to 60 voice channels.
- A master group must have 2.40 kHz of bandwidth, but the need for guard bands between the super groups increases the necessary bandwidth to 2.52 MHz Master groups support up to 600 voice channels.
- Master group can be further multiplexed in master group banks to create a jumbo group. A jumbo group must have 15.12 MHz (6 × 2.52MHz) but is augmented to 16.984 MHz to allow for guard bands between the master groups.
- Jumbo group can be further multiplexed, which is formed by 3 FDM Jumbo groups together. A Jumbo group mix support up to 10800 voice channels.
- A single message channel is baseband i.e the modulating signal in a communication system. A group, super group or signals prior to modulation of the final carrier is known as **composite baseband**.
- A radio channel includes either a single L 600 master group up to three U600 master groups (*i.e.*, 1800 voice-band channels).
- In space divisions multiplexing more than one physical transmission path are grouped together.
- TDM is the sharing of a common transmission medium in time. In TDM, the time available is divided into small slots, and each of them occupied by a piece of one of the signals to be sent.
- **Time-division multiplexing (TDM)** grants each user full channel capacity, but assigns time slots to each user.
- Time division multiplexing can be either bit by bit or byte by byte basis, known as bit interleaving and byte interleaving respectively. Bit interleaving also known as digit interleaving.
- To identify individual time slots within a TDM frame, a receiving terminal uses a counter synchronized to the frame format of the transmitter.
- **Statistical Time Division Multiplexing** uses intelligent devices capable of identifying when a terminal is idle.
- Basically, a *TDM ring* is configured as a series of unidirectional (two-wire) links arranged to form a closed circuit or loop.
- TDM loops have been used within computer complexes to provide high capacity and high interconnectivity between processors, memories and peripherals.
- The process of converting analog signal into digital signals is called **digitization**.

Review Questions

1. Explain in detail about CODECS? **[MDU Dec. 2008]**
2. What is meant by TDM ? Describe briefly about statistical and synchronous TDM. **[MDU May 2009]**
3. Explain CODECS and COMBO chips in detail. **[MDU May 2009]**
4. Explain AT and T's FDM in detail. **[MDU May 2009]**
5. Explain in detail about CCITT time division multiplexing. **[MDU Dec. 2009]**
6. Explain CODEC and COMBO chips in detail. **[MDU Dec. 2009]**
7. Assume a data stream is made of ten os (zeros). Encode this stream using following encoding schemes :
   (*i*) Polar NRZ-1
   (*ii*) Monchester
   (*iii*) Differential Monchester
   (*iv*) HDB3. **[MDU May 2010]**
8. Explain CCIT time division multiplexing in detail. **[MDU May 2010]**
9. Explain T-1 carrier system in detail. **[MDU May 2010]**
10. Draw and explain the AT&T's FDM hierarchy. Four digital data channels each transmitting 1 Mbps use a satellite channel of 1 MHz. Design an appropriate configuration using FDM. **[MDU Dec. 2010]**
11. In which type of multiplexing are CODECS and Combo chips used. What is the difference between the two of them ? Explain the transmitter and receiver sections of the CODECS with the help of block diagrams. **[MDU Dec. 2010]**
12. Write short note on T Carriers. **[MDU Dec. 2010]**
13. Describe the DS hierarchy. How are T lines related to DS service ? A DS-0 signal has a data rate of 64 Kbps. Where does this number come from ? **[MDU May 2011]**

❑❑❑

10

CHAPTER

Internet and TCP/IP

10.1. INTRODUCTION TO INTERNET

INSIDE THIS CHAPTER

The Internet is a worldwide collection of computer networks, cooperating with each other to exchange data using a common software standard. Through telephone wires and satellite links, Internet users can share information in a variety of forms. The size, scope and design of the Internet allow users to:

- connect easily through ordinary personal computers and local phone numbers;
- exchange electronic mail (E-mail) with friends and colleagues with accounts on the Internet;
- post information for others to access, and update it frequently;
- access multimedia information that includes sound, photographic images and even video; and
- Access diverse perspectives from around the world.

The Internet is a computer network made up of thousands of networks worldwide. No one knows exactly how many computers are connected to the Internet. It is certain, however, that these number in the millions and are growing.

No one is in charge of the Internet. There are organizations which develop technical aspects of this network and set standards for creating applications on it, but no governing body is in control. The Internet backbone, through which Internet traffic flows, is owned by private companies.

All computers on the Internet communicate with one another using the Transmission Control Protocol/Internet Protocol suite, abbreviated to TCP/IP. Computers on the Internet use client/server architecture. This means that the remote server machine provides files and services to the user's local client machine. Software can be installed on a client computer to take advantage of the latest access technology.

An Internet user has access to a wide variety of services: electronic mail, file transfer, vast information resources, interest group membership, interactive collaboration, multimedia displays, real-time broadcasting, breaking news, shopping opportunities, and much more.

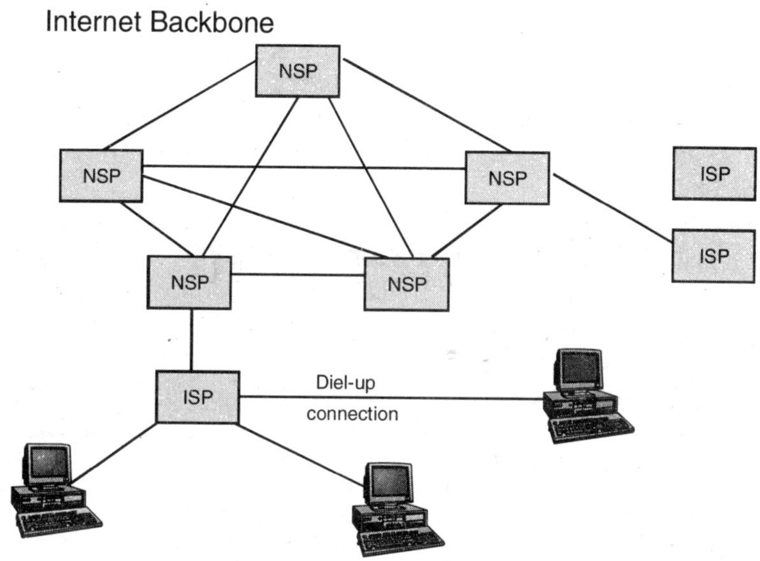

Fig. 10.1. *Typical Set-up of the Internet*

10.2. THE HISTORY OF THE INTERNET

Many people think that the Internet is a recent innovation, when in fact the essence of it has been around for over a quarter century. It's in the 60's, in the middle of the "cold war" between the east block and the western countries, that one finds the starting point of the Internet. The Internet began as ARPAnet, a U.S. Department of Defense project to create a nationwide computer network that would continue to function even if a large portion of it were destroyed in a nuclear war or natural disaster. The American army needed a way to coordinate their nuclear weapons in case of atomic war. They needed a system of communication that could "survive" even if several bases were destroyed.

During the next two decades, the network that evolved was used primarily by academic institutions, scientists and the government for research and communications. The appeal of the Internet to these bodies was obvious, as it allowed disparate institutions to connect to each others' computing systems and databases, as well as share data via E-mail.

A brief history of internet is as follows:

1964: (ARPANET project): The beginning of the project to create the first distributed network. The advantage of this technology is that it's decentralized. The system is not dependent on a main computer to coordinate communications. If one, two, even several computers go out of order, are overloaded or are unavailable, the system can by-pass these problems without any difficulty. This is the kind of network the military needed.

1969: The ARPANET is put in operation with the use of a protocol for communications called NCP (Network Control Protocol). It's the first time computers communicate among each other over long distances. The whole network is composed of super-powerful computers for the period. The network could even survive the devastation of an atomic war because every site is physically connected to four others. The communication and the coordination of the surviving armed forces in a nuclear attack were and still are assured. For reasons of national security, the armed forces refuse scientists the access to the military network and the use of the protocol of communication between sites.

1973: The scientists take their revenge by developing their own network through universities and by using an improved protocol of communication named TCP / IP (Transfer Control Protocol Protocol/Internet Protocol). The protocol IP makes sure that every computer connected with the network has a unique address and that each is capable of communicating with the others. This protocol is used on several platforms to allow a communication from these. The access became, in time, international. The foundations for the current Internet ware placed. Research started to increase the potential of the Internet with several other services.

Among these services, there is an E-mail. It allows people to send messages, "e-mail", to each other that have an e-mail address. There is also "gopher" who allows to show data, mostly text, and to make it available to all. It's in a sense the "grandpa" of the World Wide Web. There is also a FTP service, for File Transfer Protocol, which allows the transfer of files through the Internet to FTP servers. There are several other services that will be described further in this chapter.

1983: The management of ARPANET was handed over to Defense Communication Agency(DCA) which separated the military portion in to a separate **MILNET.**

1989: Tim Berners-Lee creates the bases of what will become World Wide Web.

1995: It's this year when the term "Internet" entered the vocabulary of most people. It's especially because of a "new" service: the World Wide Web (WWW or W3). Some people think, even today, that the WWW is the Internet.

The Internet explosion coincides with the advent of increasingly powerful yet reasonably priced personal computers with easy-to-use graphical operating systems. The result has been an attraction of recent computer "converts" to the network, and new possibilities for exploiting a wealth of multimedia capabilities.

10.3. USE OF INTERNET

In addition to text documents, the Internet makes available graphics files (digitized photographs and artwork), and even files that contain digitized sound and video. Through the Internet, you can download software, participate in interactive forums where users post and respond to public messages, and even join "chats," in which you and other users type (and, in some cases, speak) messages that are received by the chat participants instantly.

The Fig. 10.2 illustrated the various services on internet. Mainly we classify internet services in two categories such as communication services and information retrieval services.

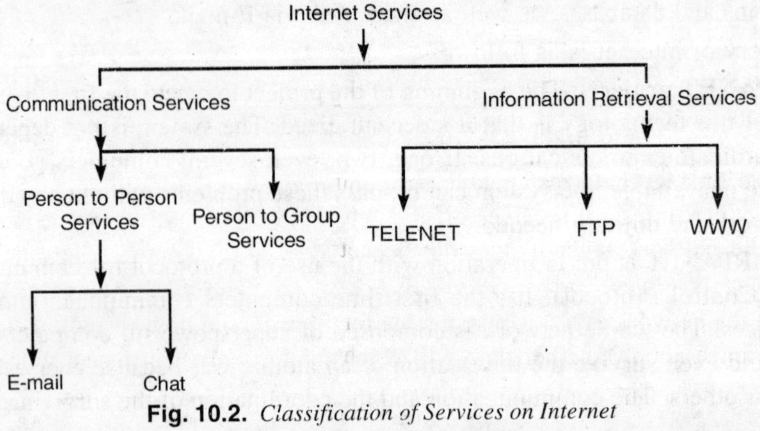

Fig. 10.2. *Classification of Services on Internet*

Among the ways that users like you are taking advantage of the Internet are:

- Sharing research and business data among colleagues and like-minded individuals.
- Communicating with others and transmitting files via E-mail.
- Requesting and providing assistance with problems and questions.
- Marketing and publicizing products and services.
- Gathering valuable feedback and suggestions from customers and business partners.

The Internet's potential is limited only by users' vision and creativity. And as the Internet grows, new and innovative uses will surely follow

10.4. TCP/IP REFERENCE MODEL

The aim of this article is to provide the reader with an introduction to TCP/IP, the protocols used on the Internet. We start with the typical overview of the protocol stack and then provide more detail on each layer. It provides a broad overview of the issues with few of the specifics.

10.4.1. The Protocol Stack

Before we examine each layer of TCP/IP, we should cover a few basic concepts. At its most basic level, a computer network is simply a series of connections between computers which allow them to communicate. The content, scope, size, speed, and reliability of the network vary depending on its protocols and implementation. Protocols are pre-established means of communication. The term TCP/IP (Transmission Control Protocol/Internet Protocol) actually refers to a whole family of protocols, of which TCP and IP are just two. Figure 10.2 contains the standard "stack" diagram of TCP/IP. Rather than make protocols monolithic (which would mean ftp, telnet, and gopher would each have a full network protocol implementation, including separate copies of kernel code for the devices each protocol uses), the designers of TCP/IP broke the job of a full network protocol suite into a number of tasks. Each layer corresponds to a different facet of communication. Conceptually, it is useful to envision TCP/IP as a stack. In implementations, programmers often blur the layers for increased performance.

Application	Telnet, FTP, RPC, etc.
Transport	TCP, UDP
Network	IP, ICMP, IGMP
Link	Network interface and device driver

Fig. 10.3. *The Layers of the TCP/IP Protocol Suite.*

The first, the link layer, is responsible for communicating with the actual network hardware (*e.g.*, the Ethernet card). Data it receives off the network wire it hands to the network layer; data it receives from the network layer it puts on the network wire. This is where device drivers for different interfaces reside.

The second, the network layer, is responsible for figuring out how to get data to its destination. Making no guarantee about whether data will reach its destination, it just decides where the data should be sent.

The third, the transport layer, provides data flows for the application layer. It is at the transport layer where guarantees of reliability may be made.

The fourth, the application layer, is where users typically interact with the network. This is where telnet, ftp, email, IRC, etc. reside.

Packets are the basic unit of transmission on the Internet. They contain both data and header information. Simply put, headers generally consist of some combination of checksums, protocol identifiers, destination and source addresses, and state information. Each layer may add its own header information, so it can interpret the data the lower layer is handing it. In Figure 3, we see a sample Ethernet frame. This is the product of a packet which has gone from that application layer all the way to the link layer. Each layer takes the previous layer's packet, viewing almost all of it as data, and puts its own header on it.

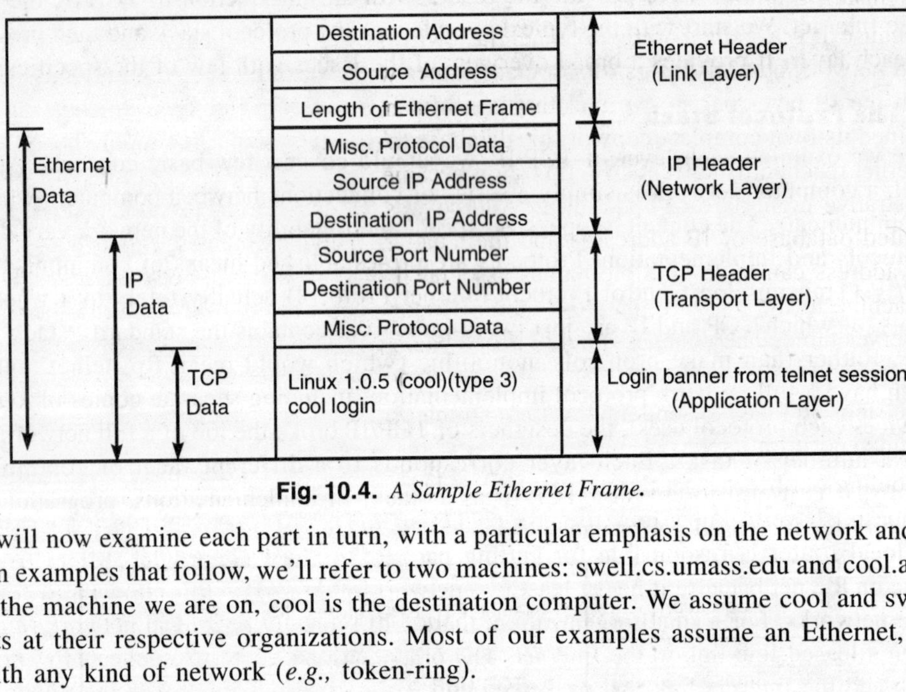

Fig. 10.4. *A Sample Ethernet Frame.*

We will now examine each part in turn, with a particular emphasis on the network and transport layers. In examples that follow, we'll refer to two machines: swell.cs.umass.edu and cool.alaska.edu. swell is the machine we are on, cool is the destination computer. We assume cool and swell are on Ethernets at their respective organizations. Most of our examples assume an Ethernet, but could work with any kind of network (*e.g.*, token-ring).

10.4.2. The Link Layer

The link layer is the simplest layer to understand. Composed of the network hardware and the device drivers, the link layer is the lowest level of the protocol stack. When receiving data from the network, it takes packets from the network wire, strips away any link layer header information, and hands it off to the network layer. When transmitting data onto the network, it takes packets from the network layer, sticks a link layer header on them, and send them out over the wire.

The benefit of separating out the hardware layer is that protocol implementers only have to write the network layer once. Then they provide a common interface to the network layer by writing different device drivers for each kind of network interface.

10.4.3. The Network Layer

This is where the Internet Protocol (IP) and the Internet Control Message Protocol (ICMP), among others, reside. ICMP is used both to provide network reliability information and by utilities like ping and trace route. IP is used for almost all other Internet communication. When sending packets, it is figures out how to get them to their destination; when receiving packets, it figures out where they belong. Because it does not worry about whether packets get to where they are going nor whether they arrive in the order sent, its job is greatly simplified. If a packet arrives with any problems (*e.g.*, corruption), IP silently discards it. Upper layers are responsible for insuring reliable reception of packets. We refer to IP's behavior as "stateless" or "connectionless" because the existence of previous or future packets is irrelevant when processing the current packet. We could unplug the network wire, wait a minute, plug it back in, and IP would never know the difference.

IP is able to get packets to their destinations because every network interface on the Internet has a unique, numeric address. Oddly enough, these numbers are called IP addresses. Notice, every interface has its own address. If a machine has multiple interfaces (as is the case with a router), each one has its own IP address. The Internic is responsible for assigning sets of addresses to organizations, thereby insuring uniqueness.

Because it's a pain to refer to machines with strings of numbers, the designers of TCP/IP allowed network administrators to associate names with IP addresses. Although this has nothing to do with the IP layer per se, we feel this is useful material. Originally, every host on the Internet maintained its own complete copy of this database (on Unix systems, it's in /etc/hosts). However, as the Internet reached its current size, this soon became unwieldy — both in terms of raw size and the administrative nightmare of updating it. And so was born the domain name system (DNS). It is a distributed database of IP addresses and their natural language names, called host names. In fact one IP address can have multiple names associated with it. When a network administrator adds a new machine to the network, network administrator is responsible for updating the organization's name server table. The changes quickly propagate. All communication with a machine is done via IP numeric addresses, so the hostname for a machine is only used at the beginning of a connection.

The steps IP takes to send a packet are simple: based on its IP address, figure out how to get it there and send it on its way.

Deciding out how to get the packet there, Alaska routing, is the critical task for IP. Fortunately, swell doesn't have to know how to get a packet all the way to Alaska; it just needs to figure out which local router is responsible for getting packets to Alaska. A router differs from a typical machine on the net because it has at least two network interfaces — this allows it to connect to two or more networks. For a small organization, there will typically be a local network (*e.g.*, Ethernet) and then a leased-line link to the Internet. The organization's router is connected to both the local network and the Internet link. All packets bound for the Internet are sent to the router, which then puts it on the leased line, bound for the next router.

Each router only needs to know about the routers to which it is connected. Those routers then know about all the routers to which they are connected. This allows swell's local router to say, "Well, all packets bound for the West go to MIT, so I'll just send it there and let MIT figure out what to do next." MIT puts it on a T3 line to Cleveland, from there it goes to Chicago, San Francisco, Seattle, and into Alaska, where it goes from the organization's router to the Ethernet interface on swell. The router at each hop is only concerned with where to send it next. It doesn't try to determine the full path which the packet will take.

To determine where a given packet will go next, machines on the Internet maintain routing tables. They consist of three major items: addresses of routers, addresses they can handle, and the interface to which they are connected. In the case of a machine on a local net (like cool.cs.umass.edu), it probably has three entries: one for the loopback interface (which allows a host to connect to itself), one for the local network and a default entry.

The local network entry lets IP know that the machine is directly connected to a certain set of IP addresses. Rather than try and route those packets, IP figures out the hardware address of the Ethernet interface to which the IP address corresponds and sends the packet there. With this entry, cool is essentially the router, the addresses it can handle are all the IP addresses on the local net, and the destination interface is an Ethernet card on the local net.

The default entry says "for all other addresses, send it to this router." Instead of trying to deliver a packet for cool.alaska.edu to a machine on the local net, cool sends it to the router's interface, saying, "Here, I don't know where this goes, you figure it out." The router then looks at its table, sees it doesn't have a direct connection to cool, and so sends it to its default destination, MIT. And so the process continues.

At this point, the reader should have a rough idea of how packets are transmitted on the Internet. When receiving data, IP takes the packet from the link layer, checks for any blatant corruption, and hands the packet to the proper process at the transport layer. If there is any problem with the packet, IP silently discards it because it doesn't have to worry about whether a packet reaches its destination.

We have left a huge amount out of this picture. Here are just some of the issues we're ignoring: packet fragmentation, netmasks and other routing tricks, network error handling, and the interactions between the network and transport layer.

10.4.4. The Transport Layer

There are two protocols at the transport layer: the transmission control protocol (TCP) and the user datagram protocol (UDP). TCP provides end-to-end reliable communication and UDP doesn't. UDP is as unreliable as IP, but allows people to write user level software that creates its own packet formats, which is particularly helpful if you want to write new protocols, don't have the kernel sources, and don't want the overhead of TCP.

TCP creates a "virtual circuit" between two processes. It insures that packets are received in the order they are sent and that lost packets are retransmitted. We won't go into the details of how it works, but interactive programs like ftp and telnet use it.

So far we have discussed addressing on the host level — how to identify a particular machine. But once at a machine, we need a way to identify a particular service (*e.g.,* mail). This is the function of ports — identification numbers included with every UDP or TCP packet. TCP/IP ports are not hardware-based. They are a just a way of labeling packets. A process on a machine "listens" on a particular port. When the transport layer receives a packet, it checks the port number and sends the data to the corresponding process. When a process starts up, it registers a port number with the TCP/IP stack. Only one process per protocol can listen on a given port. So while a process using UDP and one using TCP can both listen on port 111, two processes that both used TCP could not. There are a number of ports which are reserved for standard services. For example, SMTP, the mail protocol, is always on port 25, and telnetd is always on port 23. To see a list of the reserved ports on a Unix system, look at /etc/services.

We've examined how ports work on the server end — specific ports are reserved for set tasks. On the initiator end, port assignment is dynamic. When a telnet client on swell starts up, it gets a new port number (*e.g.,* 1066). This is the source port which swell's TCP layer puts on every packet. This allows the telnet daemon (telnetd) on cool to responds to the correct telnet process on swell. The combination of source/destination IP addresses and ports provides a unique conversation identifier. Each conversation is called a flow.

UDP is essentially IP with port numbers (flows). It gives the user access to IP-style datagrams. The network file system (NFS) and talk are two examples of UDP-based protocols.

This has been an extremely cursory exploration of TCP and UDP. At this point, you should have a decent understanding of how the network (IP) and transport (TCP/UDP) layers interact. We now turn to the final layer.

10.4.5. The Application Layer

This is where the user interacts with the network. All network programs like telnet, ftp, mail, news, and WWW clients are at the application layer. They then use either TCP or UDP to communicate with other machines. To provide a clearer picture, I'll examine telnet in a little detail.

Telnet is used for remote login. It removes the need for hardwired terminals. A user on swell types "telnet cool.alaska.edu" and he is rapidly connected to cool.alaska.edu, which asks him to login. He can then interact with cool. Here's a breakdown of the process:

The name address is turned into a numeric one (137.229.18.103), via a domain name server.

Telnet tells the transport layer it wants to start a TCP connection with 137.229.18.103, at port 23.

TCP initiates a conversation with cool. IP is used to route packets. The telnet process on swell gets a port number of, say, 1096. TCP places the source and destination port numbers in its packet header, IP the IP address.

Packets are now handed to the IP layer, which sends them to the link layer. They proceed from the user's machine to his organization's router and out onto the Internet. They make their way to cool, one router at a time.

TCP layer replies in a similar fashion.

The telnet daemon on cool and the telnet client on swell exchange terminal information and other parameters necessary for an interactive session. Control messages are sent in-band, as an escape byte of 255, followed by the control byte. Control messages include: echo, status, terminal type, terminal speed, flow control, line mode, and environmental variables.

The user sees a login prompt. After logging in, data is sent back and forth.

Once the task is broken into a number of steps, we see it's relatively simply. Because the transport layer provides a standard interface, network applications do not need to be rewritten or even recompiled if the transport layer code changes.

10.4.6. Internet Protocol

The Network Layer protocol for TCP/IP is the Internet Protocol (IP). It uses IP addresses and the subnet mask to determine whether the datagram is on the local or a remote network. If it is on the remote network, the datagram is forwarded to the default gateway which is a router that links to another network.

IP keeps track of the number of transverses through each router that the datagram goes through to reach its destination. Each transverse is called a hop. If the hop count exceeds 255 hops, the datagram is removed and the destination considered unreachable. IP's name for the hop count is called Time To Live (TTL).

10.5. IP ADDRESSES

IP addresses consist of a 32 bit number and are represented by the dot-decimal format. For example: 142.110.237.1 is an IP address. There are 4 decimal digits separated by three dots. Each digit is allowed the range of 0 to 255 which corresponds to 8 bits (one byte) of information.

A portion of an IP address represents the network address and the remaining portion the host address. For example: 142.110.237.1 is the IP address of a firewall. The network that the firewall resided on is 142.110.237.0 (Note: IP addresses that end in a 0 represent network addresses). The host address of the firewall is 0.0.0.1 (Note: the network portion of the IP address is represented by 0s). Each host on the network and Internet must have a unique IP address. There are ways around having each host a unique IP address and they are discussed under firewalls.

Obtaining an IP address

Originally, IP numbers were assigned by the Global Internet Registry (known as the InterNIC). Today, day-to-day responsibility for assignment of IP addresses, Autonomous System Numbers is given to Internet registries (IR):

1. RIPE-NCC for Europe;
2. APNIC for Asia and Pacific Area.

- RIPE has established local Internet Registries in countries in Europe
- RIPE-NCC has been delegated the following address ranges:

```
192.162.0.0 - 192.162.255.0
192.164.0.0 - 192.168.255.0
193.0.0.0   - 193.255.255.0
194.0.0.0   - 194.255.255.0
```

The Network Information Center (NIC) assigns network addresses to the Internet. You must apply to receive a IP network address. Depending on the class (more on this later) of the IP address, you can then assign as many hosts IP addresses as allowed.

10.5.1. Network Masking

The subnet mask is used to determine which portion of the IP address is the network address and which the host address is. This means that the portions of network to host in an IP address can change. The most common subnet mask is 255.255.255.0. The simple explanation is that wherever there is a 255, this indicates that it is the network portion. Wherever there is a 0, this indicates the host portion.

If we examine our IP address of 142.110.237.1, and use a subnet mask of 255.255.255.0. It can be seen that the network portion of the IP address is 142.110.237 and the host portion is 1. The network address is typically written 142.110.237.0 and the host is sometimes written 0.0.0.1.

Now if host **142.110.237**.1 wanted to send a datagram to **142.110.237**.21. It would look at the network portion of the IP address of the destination and determine that it is on the local network. It would then send the datagram out.

If host **142.110.237**.1 wanted to send a datagram to **142.110.150**.108. It would look at the network portion of the IP address of the destination and determine that it is not on the same network. It is on 142.110.150.0 network and it would send it to the default gateway. The default gateway is a router that knows how to reach the other networks.

10.5.2. Class Masking

Class A, B and C networks use masks and not subnet masks. Masks are similar to subnet masks except that usually they are used in routers and not workstations.

A Class A network has a mask of 255.0.0.0 which allows approximately 16.7 million host addresses. Also, a Class B network has a mask of 255.255.0.0 which allows approximately 65 thousand host addresses. Both classes of networks have too many hosts for one network to handle.

Imagine 65,000 users trying to access a network service at the same time. The network would be swamped with requests and slow down to a crawl.

The solution is to divide the network up into smaller workable networks called **subnets**. This is most commonly done by fooling the host machine into believing it is on a Class C network (only 254 hosts) by using a Class C mask: 255.255.255.0. This mask is called the **subnet mask**.

Thus for a Class A network using a subnet mask of 255.255.255.0, you can have roughly 65 thousand subnets of 254 hosts. On a Class B network using a subnet mask of 255.255.255.0, you can have roughly 254 subnets of 254 hosts

10.5.3. IP Address Classifications

There is a formal structure to the assignment of IP addresses. IP addresses are assigned by the Network Information Center (NIC) who is a central authority with the responsibility of assigning network addresses.

There are several classifications of IP addresses. They include network addresses and special purpose addresses.

Class A addresses

IP address range 1.0.0.0 to 127.0.0.0

Number of networks available: 125 (see special addresses below)

Number of hosts per network: 16,777,214

Net Mask: 255.0.0.0 (first 8 bits are ones)

Special Addresses: 10.0.0.0 is used for networks not connected to the Internet

127.0.0.0 is the loop back address for testing (ping)

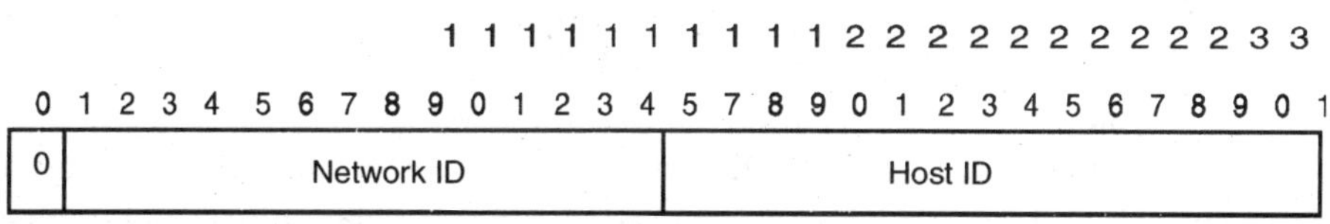

Class A addresses always have bit 0 set to 0, bits 1-7 are used as the network ID. Bits 8-31 are used as the host ID.

Class A networks are used by very large companies such as IBM, US Dept of Defense and AT&T.

Class B addresses

IP address range 128.0.0.0 to 191.0.0.0

Number of networks available: 16,382 (see special addresses below)

Number of hosts per network: 65,534

Net Mask: 255.255.0.0 (first 16 bits are ones)

Special Addresses: 172.16.0.0 to 172.31.0.0 are used for networks not connected to the Internet

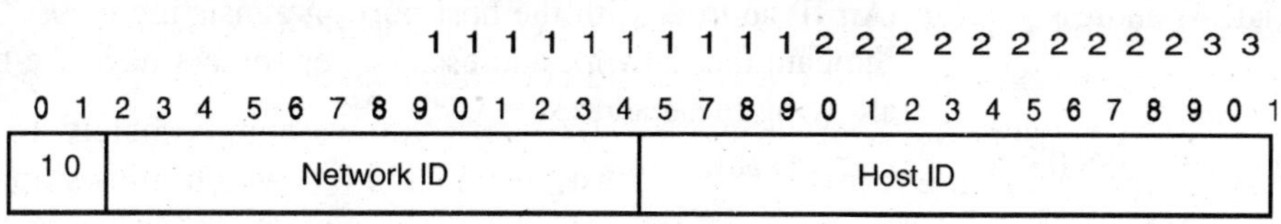

Class B addresses always have bit 0 and 1 set to 10, bits 2-15 are used as the network ID. Bits 16-31 are used as the host ID. Class B networks are assigned to large companies and universities.

Class C addresses

IP address range 192.0.0.0 to 223.0.0.0

Number of networks available: 2,097,150 (see special addresses below)

Number of hosts per network: 254

Net Mask: 255.255.255.0 (first 24 bits are ones)

Special Addresses: 192.168.1.0 to 192.168.255.0 are used for networks not connected to the Internet

1 1 1 1 1 1 1 1 1 1 2 2 2 2 2 2 2 2 2 2 3 3

0 1 2 3 4 5 6 7 8 9 0 1 2 3 4 5 6 7 8 9 0 1 2 3 4 5 6 7 8 9 0 1

1 1 0	Network ID	Host ID

Class C addresses always have bits 0-2 set to 110, bits 3-24 are used as the network ID. Bits 25-31 are used as the host ID. Class C network addresses are assigned to small companies and local Internet providers.

Class D Addresses

IP address range 224.0.0.0 to 239.0.0.0

Use: Multicasting addresses

1 1 1 1 1 1 1 1 1 1 2 2 2 2 2 2 2 2 2 2 3 3

0 1 2 3 4 5 6 7 8 9 0 1 2 3 4 5 6 7 8 9 0 1 2 3 4 5 6 7 8 9 0 1

1 1 1 0	Multicast Address

Class D addresses always have bits 0-3 set to 1110, bits 4-31 are used as the Multicast address.

Class D network addresses are used by multicasting. Multicasting is a method of reducing network traffic. Rather than send a separate datagram to each host if multiple host require the same information. A special multicast address can be used where one datagram is read by many hosts.

Class E Addresses

IP address range 240.0.0.0 to 255.0.0.0

Use : Reserved by the Internet for its own use.

If you try to ping a Class E address, you should get the error message that says that it is an invalid IP address.

10.5.4. Reserved IP Addresses

The following IP addresses are reserved:

127.0.0.0	Network addresses used for localhost mode (testing IP stack)
255.255.255.255	An IP address consisting of all 1s in binary (255).
Broadcast address x.x.x.0	An IP address with the host portion consisting of 0s. Used to Indicate the network address. Newer routers have the option of allowing these addresses.
224.0.0.0 - 255.0.0.0	Class D addresses.

10.5.5. Subnetting a network

Subnet masks can divide networks into smaller networks than the 254 hosts discussed previously. In order to understand this process, a discussion on binary to decimal number conversion is required.

The typical subnet mask 255.255.255.0 represents 4 bytes of data. Each number represents 1 byte and is displayed as a decimal number. One byte of information can represent a range of 0 - 255. One byte consists of 8 bits where 0000 0000 represents 0 in decimal and 1111 1111 represents 255 in decimal.

The convention for displaying bits is to group in nibbles (4 bits) to make it easier to read.

Each bit position has a weighting, where the weighting is equal to 2 to the power of the position starting at position 0 on the right. The easiest way to determine the decimal weighting is to start on the right with the number 1 (which is 2^0) and double it at each bit position. The weighting for each position is follows:

Decimal Weighting	128	64	32	16	8	4	2	1
Bit Position	2^7	2^6	2^5	2^4	2^3	2^2	2^1	2^0

Each position has its weighting multiplied by the binary bit value (0 or 1). For example, if bit position 2^3 had its bit set to 0, its decimal value would be 0 x 8 = 0. If bit position 2^5 had its bit set to 1, its decimal value would be 1 x 32 = 32.

To determine the decimal value of a binary number, add up all the decimal weighting values where ever there is a 1 in the binary number. For the following binary number 1111 1111, the decimal value would equal 255:

Decimal Weighting	128	64	32	16	8	4	2	1
Bit Position	1	1	1	1	1	1	1	1

128 + 64 + 32 + 16 + 8 + 4 + 2 + 1 = 255

For the binary number 0000 0000 the decimal value would equal 0 :

Decimal Weighting	128	64	32	16	8	4	2	1
Bit Position	0	0	0	0	0	0	0	0

0 + 0 + 0 + 0 + 0 + 0 + 0 + 0 = 0

For the binary number 1010 1001 the decimal value would equal 169:

Decimal Weighting	128	64	32	16	8	4	2	1
Bit Position	1	0	1	0	1	0	0	1

128 + 0 + 32 + 0 + 8 + 0 + 0 + 1 = 169

The significance of the decimal weighting to network routing becomes more evident when the method of rolling over the binary count is examined. For example, the decimal number 63 compared to 64 in binary produces an interesting observation:

Decimal 63 = 0011 1111

Decimal 64 = 0100 0000

The decimal number 63 is represented by all 1s in the first 6 bit locations. The decimal number 64 is represented by only bit 6 being a logical 1. If the count was further increased, similar relationships would occur at

Decimal 127 = 0111 1111

Decimal 128 = 1000 0000

Decimal 191 = 1011 1111
Decimal 192 = 1100 0000

Decimal 255 = 1111 1111
Decimal 0 = 0000 0000

Notice that bit 7 and 6 are the only bits that are changing from the initial example of 63 and 64. What this means is that the network can be subdivided into 4 logical networks of 64 hosts each. In actual fact the number is 62 hosts due to address 0 not being allowed (network address) and address 63 not being allowed (broadcast address).

In the introduction to subnetting, the portion of the IP address that corresponded to the network portion was easily identified as being the portion of the subnet mask that corresponded to the decimal number 255. This is really only for convenience for the dot decimal format of the IP address. In actual fact, the IP address is a 32 bit address and doesn't have byte "boundaries" as implied by the dot decimal notation. For example:

```
192.168.1.0  =  1100 0000 1010 1000 0000 0001 0000 0000
```

This means that the portion of the subnet mask that corresponds to the network address can be further broken down on the host bit positions.

A Class C network address of 192.168.1.0 has 254 hosts available to it. If your network consisted of 4 different physical locations each with a maximum of 50 hosts, then subnetting the network would be required. The locations could be different buildings or cities.

Bit 7 and bit 6 of the host portion can be used to describe the network portion of the subnet. The subnet masking would be:

```
1111 1111 1111 1111 1111 1111 1100 0000  =  255.255.255.192
```

The 4 subnets for Class C address 192.168.1.0 would be 192.168.1.0, 192.168.1.64, 192.168.1.128 and 192.168.1.192 with the following range of IP addresses:

Subnet	Host Range	Broadcast address
192.168.1.0	192.168.1.1 to 192.168.1.62	192.168.1.63
192.168.1.64	192.168.1.65 to 192.168.1.126	192.168.1.127
192.168.1.128	192.168.1.129 to 192.168.1.190	192.168.1.191
192.168.1.192	192.168.1.193 to 192.168.1.254	192.168.1.255

In this manner, a router with 4 interfaces could be configured with subnet masks of 255.255.255.192 to allow subdividing the Class C network into 4 smaller networks.

Theoretically, all of the host bits up to bit 1 and 0 can be used to make up to 64 subnets of 2 hosts each. In this case, 128 IP addresses would be lost to the network IP address and the broadcast IP address. The following table lists the number of hosts and networks that can implemented using subnet masking for a Class C network:

Subnet mask	Number of subnets	Number of hosts per subnet
255.255.255.128	2	126
255.255.255.192	4	62
255.255.255.224	8	30
255.255.255.240	16	14
255.255.255.248	32	6
255.255.255.252	64	2

The above example is based on subnetting a Class C network. Subnetting can get extremely complicated if you are subnetting assigned IP addresses that are in the middle of a Class C network such as when rented from an ISP.

10.6. DOMAIN NAME STRUCTURE

Domain names tend to follow a loose structure that gives a description of the network. For example, sait.ab.ca uses the Canada extension ca, preceded by the province of Alberta extension ab and then the abbreviation SAIT for the Southern Alberta Institute of Technology. This is a geographical designed domain name that follows the ISO-3166 country code structure. Example of country codes are:

In	India	ca	Canada
fi	Finland	gb	United Kingdom
na	Nambia	nz	New Zealand
tw	Taiwan	us	United States

There are top level domain (TLD) names extensions that attempt to describe the purpose of the domain. It is broken down into seven basic categories:

com - Commercial domains that are a business.
edu - Educational institutes
net - This is for computers of network providers such as Internet Service Providers
org - Standard organizations or non profit organizations
int - Organizations that have been established by international treaties.
gov - Municipal, federal, provincial, state governments.
mil - United States military

All in all, it is often quite difficult to establish where a domain is physically located or what it actually does from the domain name. But it makes remembering locations easier than attempting to remember an IP address.

10.6.1. Domain Name Hierarchy

For example, in Fig. 10.5, .edu is the top level domain, berkeley is the second level domain, and .cs is the sub-domain of berkeley. 'eos' is the host name. A DNS serve would store name reside in the tree.

DNS Hierarchy

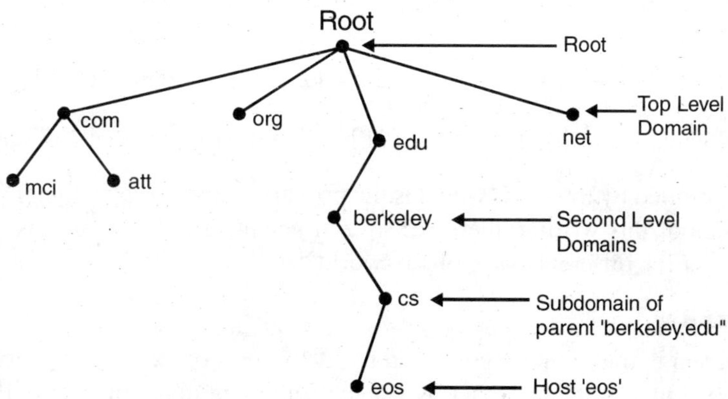

Fig. 10.5.

10.6.2. Domain Name Look-up Procedure

When a domain name is used, the IP stack doesn't understand domain names. It says "what is this? Is not an IP address!". The only thing the IP stack understands is IP addresses. The look-up order is as follows:

1. The IP stack checks the **hosts** file to see if there is a domain name match to IP address. If there is, the IP address is used.
2. If there is no match, the IP stack will look for a **NIS server** with its host file shared. If this service is not installed, the IP stack will jump to the next step.
3. If there is still no match, the IP stack will send out a request to the **domain name server** configured during the network configuration to see if it knows whose IP address belongs to the domain name.
4. If the domain name server doesn't know, it may make an enquiry to the next level up domain name server to see if it knows whose IP address belongs to the domain name and so on.

On the Internet, there are 13 top level root domain name servers. The current addresses and domain names are found at ftp://internic.net/domain/named.cache (also called named.ca and named.root)

Domain Name	IP Address	Description
A.ROOT-SERVERS.NET	198.41.0.4	formerly NS.INTERNIC.NET
B.ROOT-SERVERS.NET	128.9.0.107	formerly NS1.ISI.EDU
C.ROOT-SERVERS.NET	192.33.4.12	formerly C.PSI.NET
D.ROOT-SERVERS.NET	128.8.10.90	formerly TERP.UMD.EDU
E.ROOT-SERVERS.NET	192.203.230.10	formerly NS.NASA.GOV
F.ROOT-SERVERS.NET	192.5.5.241	formerly NS.ISC.ORG
G.ROOT-SERVERS.NET	192.112.36.4	formerly NS.NIC.DDN.MIL
H.ROOT-SERVERS.NET	128.63.2.53	formerly AOS.ARL.ARMY.MIL
I.ROOT-SERVERS.NET	192.36.148.17	formerly NIC.NORDU.NET
J.ROOT-SERVERS.NET	198.41.0.10	temporarily housed at NSI (InterNIC)
K.ROOT-SERVERS.NET	193.0.14.129	housed in LINX, operated by RIPE NCC
L.ROOT-SERVERS.NET	198.32.64.12	temporarily housed at ISI (IANA)
M.ROOT-SERVERS.NET	202.12.27.33	housed in Japan, operated by WIDE

These are controlled by InterNIC which is the primary agency responsible for registering domain names. At the time of this writing, there are several new agencies that are taking over the domain registration process for different parts of the world.

10.6.3. IP Header

The IP datagram is traditional represented by many rows of 32 bits (4 octets or bytes). Each 32 bit word is stacked on top of each other as per the following diagram for the IP header:

<table>
<tr><td>Version
(4 bits)</td><td>IHL
(4 bits)</td><td>Type pf Service
(8 bits)</td><td colspan="2">Total Length
(16 bits)</td></tr>
<tr><td colspan="3">Identification
(16 bits)</td><td>Flags
(3 bits)</td><td>Fragment Offset
(13 bits)</td></tr>
<tr><td colspan="2">Time to Live
(8 bits)</td><td>Protocol
(8 bits)</td><td colspan="2">Header Checksum
(16 bits)</td></tr>
<tr><td colspan="5">Source Address
(32 bits)</td></tr>
<tr><td colspan="5">Destination Address
(32 bits)</td></tr>
<tr><td colspan="5">Options and Padding
(multiples of 32 bits)</td></tr>
</table>

Fig. 10.6.

Field	**Description**	
Version (4 bits)	The IP version, currently it is version 4	
0	Reserved	
1-3	Unassigned	
4	IP	Internet Protocol
5	ST	ST Datagram Mode
6	SIP	Simple Internet Protocol
7	TP/IX	TP/IX: The Next Internet
8	PIP	The P Internet Protocol
9	TUBA	TUBA
10-14	Unassigned	
15	Reserved	

IHL (4 bits)Internet Header Length, the length of the IP header in 32 bit words.

Type of Service (8 bits) Flags to indicate precedence, delay, throughput and reliability parameters.

Bit 0-2: Precedence

000 - Routine

001 - Priority

010 - Immediate

011 - Flash

100 - Flash override

101 - CRITIC / ECP

110 - Internet control

111 - Network control

Bit 3: 0 = Normal delay, 1 = Low delay

Bit 4: 0 = Normal throughput, 1 = High throughput

Bit 5: 0 = Normal reliability, 1 = High reliability

Bit 6-7:	Reserved for future use.
Total Length (16 bits)	Total length in bytes of the IP datagram (IP header and data) Minimum length = 576, Maximum length = 65,535
Identification (16 bits)	Unique identifying number for this datagram

Flags (3 bits)Options that indicate if fragmentation s permitted and/or used

Bit 0:Reserved, always set to 0

Bit 1:0 = May fragment, 1 = Don't fragment

Bit 2:0 = Last fragment, 1 = More fragments

Fragment Offset (13 bits)	Indicates where in the entire datagram, this particular fragment belongs. Measured in 64 bit units from the beginning of the initial datagram.
Time to live (TTL) (8 bits)	Measured in hop counts or seconds. Every transverse through a router or gateway will decrement the hop count. When the TTL equals 0, the datagram is discarded. This stops datagrams from circulating the network forever. Starts at TTL = 255.
Protocol (8 bits)	Identifies the next protocol that follows the IP header.

Examples are:

Decimal	Protocol	Description
0		Reserved
1	ICMP	Internet Control Message
2	IGMP	Internet Group Management
6	TCP	Transmission Control
37	DDP	Datagram Delivery Protocol
46	RSVP	Reservation Protocol
93	AX.25	AX.25 Frames

Header Checksum (32 bit)	A 32 bit Cyclic Redundancy Check that may be checked at each gateway.
Source Address (32 bit)	An IP address indicating the sender. Ex. 142.110.237.1
Destination Address (32 bit)	An IP address indicating the recipient. Ex. 142.110.237.2
Options (variable)	Options from the sender such as route specifications.
Padding	Ensures that the IP header ends on a 32 bit boundary.

10.6.4. Domain Names

IP addresses are difficult for humans to remember, they're great for PCs! Domain names were invented to make it easier to navigate the Internet. A domain name is a vaguely descriptive name separated by dots. For example: www.healthyway.com as shown in Fig. 10.7.

Every machine that runs TCP/IP has a text file called **hosts**. It is a simple lookup table that the network stacks (IP) checks to see if it has a match between a domain name and an IP address. It is easily modified with a text editor and the contents look like the following:

127.0.0.1	localhost
142.110.237.1	e237-firewall.tech.el.sait.ab.ca
142.110.237.2	e237-bridge.tech.el.sait.ab.ca
142.110.237.3	ashley.tech.el.sait.ab.ca
142.110.237.4	mariah mariah.tech.el.sait.ab.ca

The IP address is listed on the left and the domain name is listed on the right. The actual registered domain name is sait.ab.ca (Southern Alberta Institute of Technology). The domain name el.sait.ab.ca (electronics dept.) is a subnet of sait.ab.ca. The domain name tech.el.sait.ab.ca (technical) is a subnet of el.sait.ab.ca.

The machine names are e237-firewall, e237-bridge, ashley and mariah. Mariah's entry is unique in that both the domain name mariah and mariah.tech.el.sait.ab.ca would be recognized by the IP stack as 142.110.237.4.

Fig. 10.7.

The problem with the hosts file is that each machine must have a current up to date copy of the network. For a small network (25 or less) not connected to the Internet this is not a problem to manage. If the network is larger, than problems can occur try to keep everyone updated.

Another solution is Unix's **Network Information Service** (NIS) (formerly called Yellow Pages until there was a copyright conflict with the Telcos). A central NIS server shares a master host's file to all the clients. In this way, only one file exists and is updated. This works well for a network not connected to the Internet.

If you are connected to the Internet then a **Domain Name Server** (DNS) is used. A DNS is a special server that communicates with other servers and keeps an up-to-date look-up table that matches IP addresses to domain names for the complete Internet. It is a hierarchical system where each DNS is authorities for the domain underneath it. This means that each server knows the domain name to IP address mapping of the network underneath it.

10.7. ADDRESS RESOLUTION PROTOCOL

Address Resolution Protocol (ARP) resides in the bottom half of the Network layer. It can be considered a mechanism for mapping addresses between the Network logical addresses and MAC (Media Access Control) layer physical addresses. For example: the Network layer protocol IP is not aware of 48 bit MAC addresses such as Ethernet. Likewise the MAC layer protocol such as Ethernet is not aware of 32 bit IP addresses. ARP provides the mechanism to map MAC addresses to IP addresses in a temporary memory space called the **ARP cache**.

The ARP cache is a dynamic cache and the information is stored only for 120 seconds. After which it is discarded. In this manner, the ARP cache remains small. The ARP cache can be viewed by using the "ARP -a" command at a command prompt. This should display the current ARP cache. If nothing is displayed, then most likely your computer hasn't communicated on the network for the past 120 seconds. Ping another device on the network and see if the ARP cache has changed.

The basic operation of ARP is as follows. When the IP layer wants to communicate with another device on the network, it checks the ARP cache to see if there is a match with an Ethernet address. If there is no matching entry in the ARP cache, an ARP broadcast datagram is sent out that basically says "Does anybody know whose Ethernet address belongs to this IP address?". The receiving station that has the IP address, responds with an ARP datagram that says "This is my IP address and here is my Ethernet address". The ARP cache is updated and the original IP layer information is then passed on to the MAC layer for processing.

10.8. REVERSE ADDRESS RESOLUTION PROTOCOL

Reverse Address Resolution Protocol (RARP) is the reverse of ARP. It is a mechanism to map MAC addresses to IP addresses. It is used mainly by diskless workstations upon boot-up to find out their IP addresses from a **BOOTP** server. The BOOTP server contains all of the boot-up configuration files that the workstation needs to boot-up.

On NICs (network interface cards) there is an empty DIP socket that is used for holding a **Boot PROM**. The Boot PROM holds a special software program that tells the workstation that upon powering up, to go and find a BOOTP server. One of the first tasks of the workstation is to find out its IP address. The MAC layer address is burnt into the NIC and is already known.

A RARP broadcast datagram is sent out that asks "Does any BOOTP server know what my IP address is?". The BOOTP server will reply with "Here's the IP address that belongs to your MAC address".

Once the IP address is known, then the rest of the configuration files can be downloaded and the diskless workstation booted up.

10.9. WORLD WIDE WEB

The **World Wide Web** (abbreviated as the Web or WWW) is a system of Internet servers that supports hypertext to access several Internet protocols on a single interface. The official definition of the WWW is *"wide-area hypermedia information retrieval initiative aiming to give universal access to a large universe of documents."*

The World Wide Web is just one of the many services that the Internet provides. Some other services provided by the Internet are email, FTP, gopher, telnet and usenet.

The World Wide Web was developed in 1989 by Tim Berners-Lee of the European Particle Physics Lab (CERN) in Switzerland. The initial purpose of the Web was to use networked hypertext to facilitate communication among its members, who were located in several countries. Word was soon spread beyond CERN, and a rapid growth in the number of both developers and users ensued. In addition to hypertext, the Web began to incorporate graphics, video, and sound. The use of the Web has reached global proportions and has become a defining aspect of human culture in an amazingly short period of time.

The World Wide Web provides a single interface for accessing all these protocols. This creates a convenient and user-friendly environment. Once upon a time, it was necessary to be conversant in these protocols within separate, command-level environments. The Web gathers these protocols together into a single system. Because of this feature, and because of the Web's ability to work with multimedia and advanced programming languages, the Web is by far the most popular component of the Internet.

It created Hypertext Markup Language (HTML) from another format used for documents levy the SGML. The WWW works in using the concept of hypertext. Inside a page, there are keywords or images that have links that, when you click above, bring you to another Web page.

All the popular software of word processor can now convert their documents in size (format) of Web page (HTML). But, there are also software specialized for the conception that is even more powerful.

The size (format) of the files of Web pages is called the HTML format for Hyper Text Markup Language. This size (format) consists of codes for the various available options. An instruction is always framed (supervised) among one <. The majority of the instructions work in peers to indicate the beginning and the end of an option. These instructions are almost identical without the command of the end has the character "/" in this one. For example, the instructions < P> and < / P > indicate the beginning and the end of a paragraph. So, the instructions < B > and < / B > indicates the beginning and the end of the stake in bold (bold) characters from these. There are also instructions that work only such as < BR > to end a line and begin on the line at once down.

10.10. THE ELEMENTS OF A WEB PAGE

Text

The most important on a Web page is always the *contents*. But, one can change the presentation of the text by changing font, by changing the color and the size of the text. You will use the same options as you would make it for a common document.

Images

The images are important constituents of a Web page. They draw the attention and supply with the *additional* data in the text. They should not however replace the contents of the text. It's also possible to align of the text next to the image.

On the World Wide Web, there are only three sizes (formats) of images that are supported (born).

1. The GIF (Graphic Interchange Format),
2. JPG (JPEG or Joined Photograph Experts Group) and
3. PNG (Portable Network Graphic)

The size (format) GIF is used to show drawings, diagrams or charts. This size (format) limits you to a palette that consists at most of 256 colors. Certain colors (40) are already reserved because they are used by the presentation of the windows of Windows (black, white, grey, blue...). So, you can choose 216 colors from the possible millions. The size (format) JPEG is especially used to show photos or diagrams having more than 256 colors. The size (format) PNG offers a better compression ratio than a size (format) GIF. So, for the same image, the image in size (format) PNG will have a smaller size of file and will transfer more quickly in the screen of the user. The only problem is that it's about a size (format) little known or used by the designers of Web pages.

10.10.1. Links

One of the basic characteristics of a Web page is that she should have at least a link on another page or another site. A Web page without at least a link is unthinkable! There is at least a link that returns to the address of the previous page or to the main page of the site. A link can be placed on the text or on the image.

The action that will arrive mostly when you'll click a link is to bring you to another Web page. But the system of addressing URL, for Uniform Resources Locator in English, allows reaching several services of the Internet. A link can have various types of actions.

By clicking a link, this one can:

(*i*) To bring you to another Web page.

(*ii*) To bring you to another place on *same* Web page.

(*iii*) Open a box of E-mail. If your Browser holds (detains) the data about your address of E-mail, the address of your server, your account number on this server and your password. Otherwise you'll have an error message.

(*iv*) Reach the other services on the Internet: Newsgroups, FTP, Gopher, Telnet, terminal emulator IBM 3270.

(*v*) To reach documents other than Web pages.

A link can have several forms. You can attach a link to the text, to the image and even to the part of an image. A link in the text is generally of a different color besides of the text and is often underlined. Unless the designer of the page changes the color of the link, this one is generally of the blue color.

10.10.2. Tables

Tables are practical to present of the data. In the HTML format, it's also an excellent tool to control the presentation in the screen. On a Web page, there are no tools as tabulations or columns to align of the text. However, you can put in the cells of various sizes of elements as text, images, options for a form etc. It's also possible to use to it to show the text on several columns. Nobody sees the difference if the border of cells is transparent.

10.10.3. The Navigation

There are several software to navigate, or "surfer", the World Wide Web. These are called the "pries" in French, the "browsers" in English. Both are more popular result from companies Netscape and Microsoft.

Netscape offers two pries: Navigator and Communicator. Navigator offers only the possibility of seeing Web pages, of answering addresses of E-mail if an address appears on a Web page and that you have to enter your coordinates (address of E-mail, name, name of the server where is your account of E-mail etc.). Communicator offer much more tools: see Web pages, access to Newsgroups, E-mail, pad of address of E-mail, conception of Web pages and more still.

Although Microsoft in taken time to react in the field of the pries and although it in taken some version to offer a Browser who could competition with what Netscape offered, the competition is now very wild between these two companies. Both you offer of excellent products that are free if to look to you for them directly on the Internet. Otherwise, there are loads (responsibilities) of purchase or delivery.

Although it has several pries available, and although each attempt to offer options that one does not find somewhere else, they have the same basic functioning. These include all a navigation bar, a bar of addresses and the other options.

10.10.4. The address bar

The bar of address is used to indicate the name of the service (http://, ftp://, news://...) and the name of the document. One can also use it to write the address of the Web page or the site that you want to visit. The Browser remembers last sites that you saw. He can so suggest you sites while you are spirit to write the address. This applies as much for Internet Explorer as for Netscape Navigator.

10.10.5. Major Protocols Accessible on the Web

Almost every protocol type available on the Internet is accessible on the Web. **Internet protocols** are sets of rules that allow for inter machine communication on the Internet. The following is a sample of major protocols accessible on the Web:

E-mail (Simple Mail Transport Protocol or SMTP)

Distributes electronic messages and files to one or more electronic mailboxes

Telnet (Telnet Protocol)

Facilitates login to a computer host to execute commands

FTP (File Transfer Protocol)

Transfers text or binary files between an FTP server and client

Usenet (Network News Transfer Protocol or NNTP)

Distributes Usenet news articles derived from topical discussions on newsgroups

HTTP (HyperText Transfer Protocol)

Transmits hyptertext over networks. This is the protocol of the Web.

Many other protocols are available on the Web. To name just one example, the Voice over Internet Protocol (VoIP) allows users to place a telephone call over the Web.

10.10.6. E-mail

Electronic mail, or e-mail, allows computer users locally and worldwide to exchange messages. Each user of e-mail has a mailbox address to which messages are sent. Messages sent through e-mail can arrive within a matter of seconds.

A powerful aspect of e-mail is the option to send electronic files to a person's e-mail address. Non-ASCII files, known as binary files, may be attached to e-mail messages. These files are referred to as MIME attachments. MIME stands for Multimedia Internet Mail Extension, and was developed to help e-mail software handle a variety of file types. For example, a document created in Microsoft Word can be attached to an e-mail message and retrieved by the recipient with the appropriate e-mail program. Many e-mail programs offer the ability to read files written in HTML, which is itself a MIME type.

10.10.7. Telnet

Telnet is a program that allows you to log into computers on the Internet and use online databases, library catalogs, chat services, and more. There are no graphics in Telnet sessions, just text. To Telnet to a computer, you must know its address. This can consist of words (locis.loc.gov) or numbers (140.147.254.3). Some services require you to connect to a specific port on the remote computer. In this case, type the port number after the Internet address. Example: telnet nri.reston.va.us 185.

Telnet is available on the World Wide Web. Probably the most common Web-based resources available through Telnet have been library catalogs, though most catalogs have since migrated to the Web. A link to a Telnet resource may look like any other link, but it will launch a Telnet session to make the connection. A Telnet program must be installed on your local computer and configured to your Web browser in order to work.

With the popularity of the Web, Telnet is less frequently used as a means of access to information on the Internet.

10.10.8. FTP

FTP stands for File Transfer Protocol. This is both a program and the method used to transfer files between computers. Anonymous FTP is an option that allows users to transfer files from thousands of host computers on the Internet to their personal computer account. FTP sites contain books, articles, software, games, images, sounds, multimedia, course work, data sets, and more.

If your computer is directly connected to the Internet via an Ethernet cable, you can use one of several PC software programs, such as WS_FTP for Windows, to conduct a file transfer.

FTP transfers can be performed on the World Wide Web without the need for special software. In this case, the Web browser will suffice. Whenever you download software from a Web site to your local machine, you are using FTP. You can also retrieve FTP files via search engines.

10.10.9. E-mail Discussion Groups

One of the benefits of the Internet is the opportunity it offers to people worldwide to communicate via e-mail. The Internet is home to a large community of individuals who carry out active discussions organized around topic-oriented forums distributed by e-mail. These are administered by various types of software programs.

A great variety of topics are covered by discussion groups. When you subscribe to a group, messages from other subscribers are automatically sent to your electronic mailbox. You subscribe by sending an e-mail message to the address of the group. You must have an e-mail account to participate in a listserv discussion group.

Listserv, majordomo and Listproc are among the programs that administer e-mail discussion groups. The commands for subscribing to and managing your list memberships are similar to those of listserv.

10.10.10. Usenet News

Usenet News is a global electronic bulletin board system in which millions of computer users exchange information on a vast range of topics. The major difference between Usenet News and e-mail discussion groups is the fact that Usenet messages are stored on central computers, and users must connect to these computers to read or download the messages posted to these groups. This is distinct from e-mail distribution, in which messages arrive in the electronic mailboxes of each list member.

Usenet itself is a set of machines that exchanges messages, or articles, from Usenet discussion forums, called newsgroups. Usenet administrators control their own sites, and decide which (if any) newsgroups to sponsor and which remote newsgroups to allow into the system.

There are thousands of Usenet newsgroups. These range from academic to recreational topics. Serious computer-related work takes place in Usenet discussions. A small number of e-mail discussion groups also exist as Usenet newsgroups.

The Usenet news feed can be read by a variety of newsreader software programs. For example, the Netscape suite comes with a newsreader program called Messenger. Newsreaders are also available as standalone products.

Usenet is not as popular nowadays as it once was. Blogs and RSS feeds are newer modes of communication that have caught the interest of Internet users.

10.10.11. Chat and Instant Messaging

Chat programs allow users on the Internet to communicate with each other by typing in real time. They are sometimes included as a feature of a Web site, where users can log into the "chat room" to exchange comments and information about the topics addressed on the site. Chat may take other, more wide-ranging forms. For example, America Online is well known for sponsoring a number of topical chat rooms.

Internet Relay Chat (IRC) is a service through which participants can communicate to each other on hundreds of channels. These channels are usually based on specific topics. While many topics are frivolous, substantive conversations are also taking place. To access IRC, you must use an IRC software program.

A variation of chat is the phenomenon of instant messaging. With instant messaging, a user on the Web can contact another user currently logged in and type a conversation. Most famous is America Online's Instant Messenger. ICQ, MSN and Yahoo also offer chat programs. Open Source chat programs include GAIM and Jabber.

10.10.12. Current Trends: Blogs and RSS

The Web is a welcoming medium for experimentation and user participation. It is becoming easier to post Web content and share comments with other users. The idea of the Web site is still very much alive, but Web participation is taking new forms and being driven by new technologies that foster social interaction. Here are two of the latest trends.

Blogs : A blog is an easy-to-create Web site that allows users to share their thoughts with the world managed by a lightweight content management system. The word "blog" comes from "Weblog" because a blog consists of a signed and dated log of individual postings. The topic of the blog can be anything, from the personal to the professional. A blog is what you make of it.

What is important about blogs is the content management system that manages the content. This system can offer a variety of features that can make the blog a useful tool. Examples include a calendar view of postings, organization of postings into categories, archived postings, options to send e-mail notification of new postings, and so on.

Blogging can be an interactive activity. Readers can add comments to a blogger's postings, other can respond, and a conversation ensues. Lately, bloggers have become well-known commentators on the political scene, but blogging can encompass any topic or no topic at all. If the blogging software allows it, bloggers can use RSS to distribute their postings.

RSS : RSS allows people to place news and other announcement-type items into a simple XML format that can then be pushed to RSS readers and Web pages. The initials RSS can stand for different things, including Rich Site Summary or Really Simple Syndication. Users can subscribe to the RSS newsfeeds of their choice, and then have access to the updated information as it comes in. RSS is used for all kinds of purposes, including the news itself and announcing new content on Web sites.

RSS content may be read by using an RSS reader, or aggregator. This is usually free software that you can install on your computer that posts new items and stores old ones in a graphical interface. An RSS reader similar to e-mail software in that it displays incoming items and can store content for offline reading. Subscribing to a newsfeed is usually as simple as entering the address of the RSS document.

It is also possible to subscribe to and read your own collection of RSS feeds on Web sites devoted to this purpose. Bloglines is one such example. The advantage here is that you can access your RSS feeds from any computer that is connected to the Web.

10.11. UNIFORM RESOURCE LOCATOR AND THE DOMAIN NAME SYSTEM (DNS)

URL stands for Uniform Resource Locator. The URL specifies the Internet address of a file stored on a host computer connected to the Internet. Every file on the Internet, no matter what its access protocol, has a unique URL. Web browsers use the URL to retrieve the file from the host computer and the specific directory in which it resides. This file is downloaded to the user's client computer and displayed on the monitor connected to the machine.

URLs are translated into numeric addresses using the **Domain Name System (DNS)**. The DNS is a worldwide system of servers that stores location pointers to Web sites. The numeric address, called the IP (Internet Protocol) address, is actually the "real" URL. Since numeric strings are difficult for humans to use, alphneumeric addresses are employed by end users. Once the translation is made by the DNS, the browser can contact the Web server and ask for a specific file located on its site.

Anatomy of a URL

This is the format of the URL:

protocol://host/path/filename

For example, this is a URL on the Web site of the U.S. House of Representatives:

http://www.house.gov/house/2004_House_Calendar.html

This URL is typical of addresses hosted in domains in the United States.

Structure of this URL

1. Protocol: **http**
2. Host computer name: **www**
3. Second-level domain name: **house**
4. Top-level domain name: **gov**
5. Directory name: **house**
6. File name: **2004_House_Calendar_html**

Note how much information about the content of the file is present in this well-constructed URL.

10.12. CONNECTION OPTIONS

Until recently, the two primary methods of accessing the Internet were through a **network connection**, allowing users of local area networks (LANs) to go online through their school or workplace systems, and **dial-up connections** through a modem and phone line. However, new connection options allow for greater speeds and flexibility, while keeping costs to a minimum.

The following are some of the newer connection options that you might want to investigate:

1. Cable Internet

These systems allow your computer to connect to the Internet through the same cable that carries your TV signal. Monthly service charges are usually not much more than standard modem connection costs, but you have to rent or purchase a "cable modem." Additionally, your computer will need an Ethernet card (a special circuit board that allows for network connections). Not all cable service operators offer this service; call your local operator for more information.

2. Satellite Connections

DirecPC allows you to download Internet files via a satellite connection. This is an efficient method for receiving large Web graphics and other items, but still need a modem connection for other features.

3. Integrated Services Digital Network (ISDN)

An ISDN line is a type of digital phone line that can transmit data many times faster than a conventional modem and phone line.

4. Wireless Connections

Pagers, cellular phones and personal digital assistants (PDAs) now allow varying levels of Internet access, from notification of E-mail to limited Web connections. Many of these services remain in the experimental stage.

5. WebTV

Introduced in late 1996, WebTV provides Web and E-mail access through ordinary television sets. The connection is made through a custom high-speed modem. You must purchase a special set-top unit for your TV, plus subscribe to the connection service. Recently, similar systems by other manufacturers have appeared on the market to compete with WebTV.

10.13. HOW TO ACCESS THE WWW: WEB BROWSERS

To access the World Wide Web, we must use a Web browser. A browser is a software program that allows users to access and navigate the World Wide Web. There are two types of browsers:

1. **Graphical :** Text, images, audio, and video are retrievable through a graphical software program such as Internet Explorer, Firefox, Netscape, Mozilla and Opera. These browsers are available for Windows, Apple, Linux and other operating systems. Navigation is accomplished by pointing and clicking with a mouse on highlighted words and graphics.
2. **Text :** Lynx is a browser that provides access to the Web in text-only mode. Navigation is accomplished by highlighting emphasized words in the screen with the arrow up and down keys, and then pressing the forward arrow (or Enter) key to follow the link. In these days of graphical browsers, it may be hard to believe that Lynx was once very popular.

10.14. INTRANET AND EXTRANET

Intranet

An Intranet is an organizations own internal network that uses the same services, protocols and technologies which are available on the Internet. An **intranet** is a private network utilizing Internet-type tools, but available only within that organization. With the advancements made in browser-based software for the Internet, many private organizations are implementing intranets. For large organizations, an intranet provides an easy access mode to corporate information for employees.

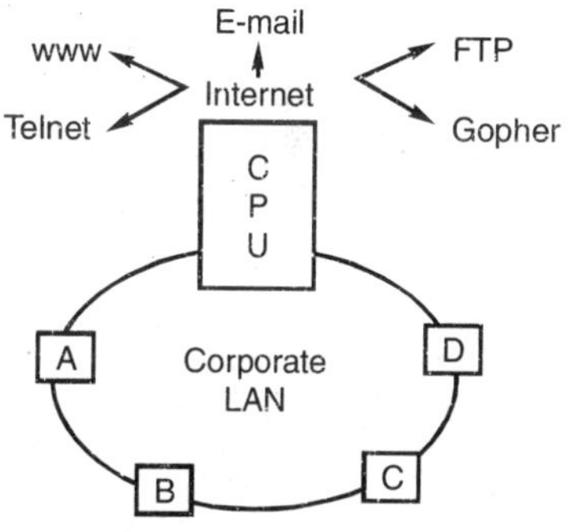

Fig. 10.8. *Intranet*

Users use the same software to utilize services of the Internet as well as Intranet. For example, more and more companies use internal documents (memos, company reports and policies) by using the same tools as on the Internet. However, these documents are only available from inside the company. The Intranet can provide all the services that the Internet provides, but only within the organization.

- *E-mail:* E-mail used for sending/receiving mails (SMTP, POP3).
- *PTP:* File transfer between internetwork nodes.
- *Gopher:* Gopher system used for organizing and displaying files on internet.
- *Telnet:* Telnet used for remote terminal emulation.
- *WWW:* World Wide Web, which allows users to create, edit and browse the hypertext documents.

Extranet

It's the former term to describe what is now called Business to Business or B2B. This consists in the purchase of services and materials by transactions on the Internet between companies. When an organization's, Intranet is made available to selected users outside its local area network, it is known as **extranet**.

It provides better communication and collaboration with customers.

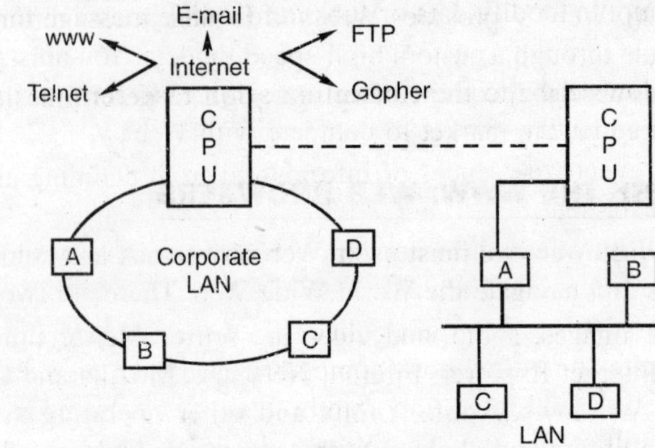

Fig. 10.9. *Extranet*

10.15. INTERNET SECURITY ISSUES

All communication over the Internet uses the Transmission Control Protocol/Internet Protocol (TCP/IP). TCP/IP allows information to be sent from one computer to another through a variety of intermediate computers and separate networks before it reaches its destination.

The great flexibility of TCP/IP has led to its worldwide acceptance as the basic Internet and intranet communications protocol. At the same time, the fact that TCP/IP allows information to pass through intermediate computers makes it possible for a third party to interfere with communications in the following ways:

(*i*) **Eavesdropping.** Information remains intact, but its privacy is compromised. For example, someone could learn your credit card number, record a sensitive conversation, or intercept classified information.

(*ii*) **Tampering.** Information in transit is changed or replaced and then sent on to the recipient. For example, someone could alter an order for goods or change a person's resume.

(*iii*) **Impersonation.** Information passes to a person who poses as the intended recipient. Impersonation can take two forms:

(*a*) **Spoofing.** A person can pretend to be someone else. For example, a person can pretend to have the email address jdoe@mozilla.com, or a computer can identify itself as a site called www.mozilla.com when it is not. This type of impersonation is known as spoofing.

(*b*) **Misrepresentation.** A person or organization can misrepresent itself. For example, suppose the site www.mozilla.com pretends to be a furniture store when it is really just a site that takes credit-card payments but never sends any goods.

Normally, users of the many cooperating computers that make up the Internet or other networks don't monitor or interfere with the network traffic that continuously passes through their machines. However, many sensitive personal and business communications over the Internet require precautions that address the threats listed above. Fortunately, a set of well-established techniques and standards known as **public-key cryptography** make it relatively easy to take such precautions.

Public-key cryptography facilitates the following tasks:

(*i*) **Encryption and decryption** allow two communicating parties to disguise information they send to each other. The sender encrypts, or scrambles, information before sending it. The receiver decrypts, or unscrambles, the information after receiving it. While in transit, the encrypted information is unintelligible to an intruder.

(*ii*) **Tamper detection** allows the recipient of information to verify that it has not been modified in transit. Any attempt to modify data or substitute a false message for a legitimate one will be detected.

(*iii*) **Authentication** allows the recipient of information to determine its origin—that is, to confirm the sender's identity.

(*iv*) **Nonrepudiation** prevents the sender of information from claiming at a later date that the information was never sent.

This article is a general introduction to network security issues and solutions in the Internet; emphasis is placed on route filters and firewalls.

http://www.acm.org/crossroads/xrds2-1/gfx/police.tifIn the last decade, the number of computers in use has exploded. For quite some time now, computers have been a crucial element in how we entertain and educate ourselves, and most importantly, how we do business. It seems obvious in retrospect that a natural result of the explosive growth in computer use would be an even more explosive (although delayed) growth in the desire and need for computers to talk with each other. The growth of this industry has been driven by two separate forces which until recently have had different goals and end products. The first factor has been research interests and laboratories; these groups have always needed to share files, email and other information across wide areas. The research labs developed several protocols and methods for this data transfer, most notably TCP/IP. Business interests are the second factor in network growth. For quite some time, businesses were primarily interested in sharing data within an office or campus environment, this led to the development of various protocols suited specifically to this task.

Within the last five years, businesses have begun to need to share data across wide areas. This has prompted efforts to convert principally LAN-based protocols into WAN-friendly protocols. The result has spawned an entire industry of consultants who know how to manipulate routers, gateways and networks to force principally broadcast protocols across point-to-point links (two very different methods of transmitting packets across networks). Recently (within the last 2 or 3 years) more and more companies have realized that they need to settle on a common networking protocol. Frequently the protocol of choice has been TCP/IP, which is also the primary protocol run on the Internet. The emerging ubiquitousness of TCP/IP allows companies to interconnect with each other via private networks as well as through public networks.

Security is probably the most well known of these problems. When businesses send private information across the net, they place a high value on it getting to its destination intact and without being intercepted by someone other than the intended recipient. Individuals sending private communications obviously desire secure communications. Finally, connecting a system to a network can open the system itself up to attacks. If a system is compromised, the risk of data loss is high.

It can be useful to break network security into two general classes:

- methods used to secure data as it transits a network
- methods which regulate what packets may transit the network

While both significantly affect the traffic going to and from a site, their objectives are quite different.

10.15.1. Transit Security

Currently, there are no systems in wide use that will keep data secure as it transits a public network. Several methods are available to encrypt traffic between a few coordinated sites. Unfortunately, none of the current solutions scale particularly well. Two general approaches dominate this area:

1. Virtual Private Networks
2. Packet Level Encryption

10.15.1.1. Virtual Private Networks

This is the concept of creating a private network by using TCP/IP to provide the lower levels of a second TCP/IP stack. This can be a confusing concept, and is best understood by comparing it to the way TCP/IP is normally implemented. In a nutshell, IP traffic is sent across various forms of physical networks. Each system that connects to the physical network implements a standard for sending IP messages across that link. Standards for IP transmission across various types of links exist, the most common are for Ethernet and Point to Point links (PPP and SLIP). Once an IP packet is received, it is passed up to higher layers of the TCP/IP stack as appropriate (UDP, TCP and eventually the application). When a virtual private network is implemented, the lowest levels of the TCP/IP protocol are implemented using an existing TCP/IP connection. There are a number of ways to accomplish this which tradeoff between abstraction and efficiency. The advantage this gives you in terms of secure data transfer is only a single step further away. Because a VPN gives you complete control over the physical layer, it is entirely within the network designer's power to encrypt the connection at the physical (virtual) layer. By doing this, all traffic of any sort over the VPN will be encrypted, whether it be at the application layer (such as Mail or News) or at the lowest layers of the stack (IP, ICMP).

The primary advantages of VPNs are:

(*i*) They allow private address space (you can have more machines on a network),

(*ii*) They allow the packet encryption/translation overhead to be done on dedicated systems, decreasing the load placed on production machines.

10.15.1.2. Packet Level Encryption

Another approach is to encrypt traffic at a higher layer in the TCP/IP stack. Several methods exist for the secure authentication and encryption of telnet and login sessions (Kerberos, S/Key and DESlogin) which are examples of encryption at the highest level of the stack (the application layer). The advantages to encrypting traffic at the higher layer are that the processor overhead of dealing with a VPN is eliminated, inter-operability with current applications is not affected, and it is much easier to compile a client program that supports application layer encryption than to build a VPN. It is possible to encrypt traffic at essentially any of the layers in the IP stack. Particularly promising is encryption that is done at the TCP level which provides fairly transparent encryption to most network applications.

It is important to note that both of these methods can have performance impacts on the hosts that implement the protocols, and on the networks which connect those hosts. The relatively simple act of encapsulating or converting a packet into a new form requires CPU-time and uses additional network capacity. Encryption can be a very CPU-intensive process and encrypted packets may need to be padded to uniform length to guarantee the robustness of some algorithms. Further, both methods have impacts on other areas (security related and otherwise- such as address allocation, fault tolerance and load balancing) that need to be considered before any choice is made as to which is best for a particular case.

10.15.1.3. Traffic Regulation

The most common form of network security on the Internet today is to closely regulate which types of packets can move between networks. If a packet which may do something malicious to a remote host never gets there, the remote host will be unaffected. Traffic regulation provides this screen between hosts and remote sites. This typically happens at three basic areas of the network: routers, firewalls and hosts. Each provides similar service at different points in the network. In fact the line between them is somewhat ill-defined and arbitrary. We will use the following definitions:

Router traffic regulation: Any traffic regulation that occurs on a router or terminal server (hosts whose primary purpose is to forward the packets of other hosts) and is based on packet characteristics. This does not include application gateways but does include address translation.

Firewall traffic regulation: Traffic regulation or filtering that is performed via application gateways or proxies.

Host traffic regulation: Traffic regulation that is performed at the destination of a packet. Hosts are playing a smaller and smaller role in traffic regulation with the advent of filtering routers and firewalls.

10.15.2. Filters and Access Lists

Regulating which packets can go between two sites is a fairly simple concept on the surface- it shouldn't be and isn't difficult for any router or firewall to decide simply not to forward all packets from a particular site. Unfortunately, the reason most people connect to the Internet is so that they may exchange packets with remote sites. Developing a plan that allows the right packets through at the right time and denies the malicious packets is a thorny task which is far beyond this article's scope. A few basic techniques are worth discussing, however.

- **Restricting access in, but not out**: Almost all packets (besides those at the lowest levels which deal with network reachability) are sent to destination sockets of either UDP or TCP. Typically, packets from remote hosts will attempt to reach one of what are known as the well known ports. These ports are monitored by applications which provide services such as Mail Transfer and Delivery, Usenet News, the time, Domain Name Service, and various login protocols. It is trivial for modern routers or firewalls only to allow these types of packets through to the specific machine that provides a given service. Attempts to send any other type of packet will not be forwarded. This protects the internal hosts, but still allows all packets to get out. Unfortunately this isn't the panacea that it might seem.
- **The problem of returning packets**: Let's pretend that you don't want to let remote users log into your systems unless they use a secure, encrypting application such as S/Key. However, you are willing to allow your users to attempt to connect to remote sites with telnet or ftp. At first glance, this looks simple: you merely restrict remote connections to one type of packet and allow any type of outgoing connection. Unfortunately, due to the nature of interactive protocols, they must negotiate a unique port number to use once a connection is established. If they didn't, at any given time, there could only be one of each type of interactive session between any given two machines. This result in a dilemma: all of a sudden, a remote site is going to try to send packets destined for a seemingly random port. Normally, these packets would be dropped. However, modern routers and firewalls now support the ability to dynamically open a small window for these packets to pass through if packets have been recently transmitted from an internal host to the external host on the same port. This allows connections that are initiated internally to connect, yet still denies external connection attempts unless they are desired.
- **Dynamic route filters**: A relatively recent technique is the ability to dynamically add entire sets of route filters for a remote site when a particular set of circumstances occur. With these techniques, it is possible to have a router automatically detect suspicious activity (such as ISS or SATAN) and deny a machine or entire site access for a short time. In many cases this will thwart any sort of automated attack on a site.

Filters and access lists are typically placed on all three types of systems, although they are most common on routers.

10.15.3. Address Translation

Another advancement has been to have a router modify outgoing packets to contain their own IP number. This prevents an external site from knowing any information about the internal network, it also allows for certain tricks to be played which provide for a tremendous number of additional internal hosts with a small allocated address space. The router maintains a table which maps an external IP number and socket with an internal number and socket. Whenever an internal packet is destined for the outside, it is simply forwarded with the routers IP number in the source field of the IP header. When an external packet arrives, it is analyzed for its destination port and re-mapped before it is sent on to the internal host. The procedure does have its pitfalls; checksums have to be recalculated because they are based in part on IP numbers, and some upper layer protocols encode/depend on the IP number. These protocols will not work through simple address translation routers.

10.15.4. Application Gateways and Proxies

The primary difference between firewalls and routers is that firewalls actually run applications. These applications frequently include mail daemons, ftp servers and web servers. Firewalls also usually run what are known as application gateways or proxies. These are best described as programs which understand a protocol's syntax, but do not implement any of the functionality of the protocol. Rather, after verifying that a message from an external site is appropriate, they send the message on to the real daemon which processes the data. This provides security for those applications that are particularly susceptible to interactive attacks. One advantage of using a firewall for these services is that it makes it very easy to monitor all activity, and very easy to quickly control what gets in and out of a network.

It is concluded that there are two basic types of network security, transit security and traffic regulation, which when combined can help guarantee that the right information is securely delivered to the right place. It should be apparent that there is also a need for ensuring that the hosts that receive the information will properly process it, this raises the entire specter of host security: a wide area which varies tremendously for each type of system. With the growth in business use of the Internet, network security is rapidly becoming crucial to the development of the Internet. Soon, security will be an integral part of our day to day use of the Internet and other networks.

10.16. INTERNET FIREWALL

10.16.1. What a Firewall Does ?

Computer networks are generally designed to do one thing above all others: allow any computer connected to the network to freely exchange information with any other computer also connected to the same network.

In an ideal world, this is a perfect way for a network to operate facilitating universal communications between connected systems. Individual computers are then free to decide who they want to communicate with, what information they want to allow access to and which services they will make available. This way of operating is called "host based security", because individual computers or hosts implement security mechanisms. The Internet is designed in this way, as is the network in your office.

In practice individual computers on say, an office network, are not terribly good at defining and securely enforcing a consistent security policy. They run very complex, and therefore by definition error prone software systems, and it is very difficult to ensure that they are consistently kept secure, much less that their users obey basic advice like choosing difficult to guess passwords etc.

This situation may be adequate where individual users on a network have a similar level of trust such that there is little chance or motive for a user to subvert host security, such as a small company network where everyone with physical access is trusted (*e.g.*, employee etc).

Once that network is connected to other networks where the trust relationships simply do not exist in the same way, and then other mechanisms need to be put in place to provide adequate security by protecting resources on the trusted network from potential access by attackers on the un-trusted part of the network.

The way this is done is by partially breaking connectivity at the network level so that nodes on the trusted and un-trusted parts of the network can no longer freely exchange information in an unfettered way. The device which does this is called a "**Firewall**", by reference to the analogue in American automobile engineering, where the Firewall is a thick steel plate barrier between engine and passenger compartments which prevents a fire in the former spreading to the latter.

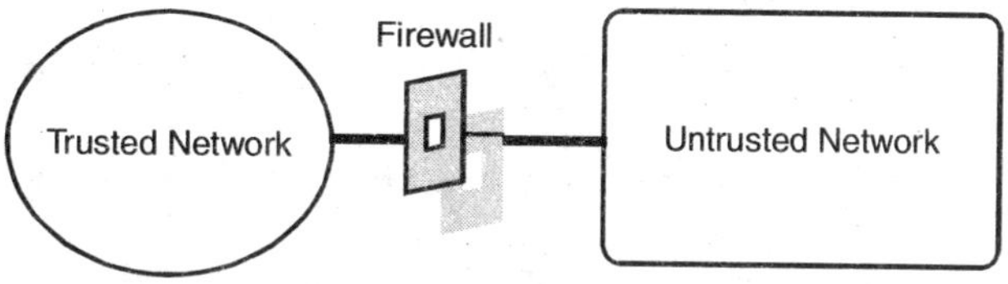

Fig. 10.10.

10.16.2. How a Firewall Works

A Firewall disrupts free communication between trusted and un-trusted networks, attempting to manage the information flow and restrict dangerous free access.

There are numerous mechanisms employed to do this, each one being somewhere between completely preventing packets flowing, which would be equivalent to completely disconnected networks, and allowing free exchange of data, which would be equivalent to having no Firewall.

If asked to write down a security policy that we would like our Firewall to implement in English, it would probably look something like:

"Allow internal users to access external www servers, but not allow external users to access our Intranet server".

In order to implement this policy, our Firewall needs to be able to examine packets and determine if they belong to either a conversation which should be allowed, or one which should be blocked.

To do this, it basically needs to know two things:

- The application being connected to.
- The direction of the conversation.

The first one of these can be guessed from the port number on the receiving end of the connection. For example, by convention, WWW servers run by default on port 80, e-mail servers run on port 25 etc.

Somewhat harder (and crucial to the above), is to determine the direction of conversation. Whilst each packet flowing through the Firewall is a self contained unit, by examining the sequence it is possible to see what the overall direction of the conversation is (*ie* who initiated it).

From the TCP transaction diagram it can be seen that the initial "Lets talk" TCP SYN packet is always seen coming from the originator of the connection, to the destination service.

Our Firewall then could implement the above security policy by translating to the following network level operations:

If packet is a TCP SYN from any inside address to any outside address, port 80, allow through.

If packet is a TCP SYN from any outside address to any inside address, port 80, block.

Allow through all other packets.

10.16.3. Types of Firewall

There are a number of different kinds of technique which may be employed by a Firewall in order to correctly identify a conversation and act on it.

The techniques used by a particular Firewall have an impact on the accuracy with which it can identify traffic, the level of sophistication of the checks it can implement, but also it's complexity and therefore cost and likelihood that it incorporates bugs.

Packet Filter

The network level operations corresponding to the security policy above were actually an example of a simple packet filter.

A Firewall implementing a packet filter looks at one packet at a time, and considers it in isolation in order to make a forwarding decision.

Because of the way that a packet filtering Firewall works, it can implement a restricted range of filtering decisions. The principal limitations of packet filtering are:

(*i*) TCP connections can be filtered on port and direction in order to implement simple directional traffic rules keyed on port number only.

(*ii*) It is not possible to completely filter TCP packets which aren't valid, or don't form part of an active connection.

It is not possible to fully filter UDP connections to ensure that they are part of a valid conversation.

The latter restriction is a fairly serious drawback of packet filtering. It means the Firewall implementer is left with the choice of either completely blocking UDP transactions, or accepting that packets may traverse the Firewall which should not be allowed through.

In the face of this, the only safe option is to block external to internal UDP transactions when using a packet filtering Firewall.

Although the above drawbacks may seem significant, there are also some quite strong advantages to a basic packet filtering Firewall:

(*i*) It is simple to implement, which means that it is much more unlikely that exploitable bugs exist in the Firewall code.

(*ii*) The same simplicity means that rule sets tend to be less complex, and again are less likely to contain unintentional access routes.

(*iii*) It can be implemented on relatively inexpensive hardware, meaning that simple, cheap boxes can do packet filtering for very large numbers of user connections.

Stateful inspection

Stateful inspection takes the basic principles of packet filtering and adds the concept of history, so that the Firewall considers the packets in the context of previous packets.

So for example, it records when it sees a TCP SYN packet in an internal table, and in many implementations will only allow TCP packets that match an existing conversation to be forwarded to the network.

This has a number of advantages over simpler packet filtering:

It is possible to build up Firewall rules for protocols which cannot be properly controlled by packet filtering (*e.g.,* UDP based protocols).

More complete control of traffic is possible.

Equally, there are some disadvantages to a stateful inspection solution, in that the implementation is necessarily more complex and therefore more likely to be buggy.

It also requires a device with more memory and a more powerful CPU etc for a given traffic load, as information has to be stored about each and every traffic flow seen over a period of time.

Application Proxies

Another mechanism for controlling risks when Internal servers must allow connections from the Internet is to use a technique called Application Proxies on a single external firewall.

These work by terminating the external connection at a special service within the firewall. As the name suggests, this service acts as a proxy for the real server, implementing the application protocol in the same way as the real server running on the internal network. It forms a connection to the internal server, only passing on application protocol elements that pass it's strict checks of correctness.

This way, most mechanisms for subverting the internal application server are blocked.

Using an application proxy is not without difficulty as their complexity tends to mean that they need to be implemented on firewalls which are significantly more powerful than the relatively simple systems used for basic packet filters. This, and the fact that such firewalls are typically sold to "Enterprise" customers mean that their cost is often uneconomic for small businesses.

Application proxy firewalls also tend to require frequent software updating to ensure that they are running latest versions of the proxy code. This occurs both when new exploits are identified which need to be blocked, but also when problems occur in interactions between the proxy and widely deployed applications (in other words when the proxy is actually breaking an otherwise working connection due to over strict or even erroneous checking).

10.17. DIGITAL SIGNATURES

Encryption and decryption address the problem of eavesdropping, one of the three Internet security issues mentioned at the beginning of this topic. But encryption and decryption, by themselves, do not address the other two problems mentioned in Internet Security Issues: tampering and impersonation.

This section describes how public-key cryptography addresses the problem of tampering. The sections that follow describe how it addresses the problem of impersonation.

Tamper detection and related authentication techniques rely on a mathematical function called a **one-way hash** (also called a **message digest**). A one-way hash is a number of fixed lengths with the following characteristics:

- The value of the hash is unique for the hashed data. Any change in the data, even deleting or altering a single character, results in a different value.
- The content of the hashed data cannot, for all practical purposes, be deduced from the hash—which is why it is called "one-way."

As mentioned in Public-Key Encryption, it's possible to use your private key for encryption and your public key for decryption. Although this is not desirable when you are encrypting sensitive information, it is a crucial part of digitally signing any data. Instead of encrypting the data itself the signing software creates a one-way hash of the data, and then uses your private key to encrypt the hash. The encrypted hash, along with other information, such as the hashing algorithm, is known as a **digital signature.**

Figure 10.11 shows a simplified view of the way a digital signature can be used to validate the integrity of signed data.

Figure 10.11 shows two items transferred to the recipient of some signed data: the original data and the digital signature, which is basically a one-way hash (of the original data) that has bee encrypted with the signer's private key. To validate the integrity of the data, the receiving software first uses the signer's public key to decrypt the hash. It then uses the same hashing algorithm that generated the original hash to generate a new one-way hash of the same data. (Information about the hashing algorithm used is sent with the digital signature, although this isn't shown in the figure.) Finally, the receiving software compares the new hash against the original hash. If the two hashes match, the data has not changed since it was signed. If they don't match, the data may have been tampered with since it was signed, or the signature may have been created with a private key that doesn't correspond to the public key presented by the signer.

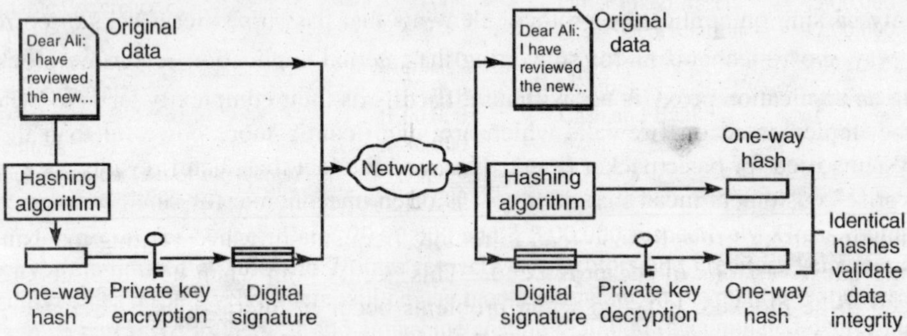

Fig. 10.11. *Using a digital signature to validate data integrity.*

If the two hashes match, the recipient can be certain that the public key used to decrypt the digital signature corresponds to the private key used to create the digital signature. Confirming the identity of the signer, however, also requires some way of confirming that the public key really belongs to a particular person or other entity. For a discussion of the way this works, see Certificates and Authentication.

The significance of a digital signature is comparable to the significance of a handwritten signature. Once you have signed some data, it is difficult to deny doing so later—assuming that the private key has not been compromised or out of the owner's control. This quality of digital signatures provides a high degree of no repudiation—that is, digital signatures make it difficult for the signer to deny having signed the data. In some situations, a digital signature may be as legally binding as a handwritten signature.

10.18. CERTIFICATES AND AUTHENTICATION

A certificate is an electronic document used to identify an individual, a server, a company, or some other entity and to associate that identity with a public key. Like a driver's license, a passport, or other commonly used personal IDs, a certificate provides generally recognized proof of a person's identity. Public-key cryptography uses certificates to address the problem of impersonation (see Internet Security Issues).

To get a driver's license, you typically apply to a government agency, such as the Department of Motor Vehicles, which verifies your identity, your ability to drive, your address, and other information before issuing the license. To get a student ID, you apply to a school or college, which performs different checks (such as whether you have paid your tuition) before issuing the ID. To get a library card, you may need to provide only your name and a utility bill with your address on it.

Certificates work much the same way as any of these familiar forms of identification. Certificate authorities (CAs) are entities that validate identities and issue certificates. They can be either independent third parties or organizations running their own certificate-issuing server software (such as Netscape Certificate Server). The methods used to validate an identity vary depending on the policies of a given CA—just as the methods to validate other forms of identification vary depending on who is issuing the ID and the purpose for which it will be used. In general, before issuing a certificate, the CA must use its published verification procedures for that type of certificate to ensure that an entity requesting a certificate is in fact who it claims to be.

The certificate issued by the CA binds a particular public key to the name of the entity the certificate identifies (such as the name of an employee or a server). Certificates help prevent the use of fake public keys for impersonation. Only the public key certified by the certificate will work with the corresponding private key possessed by the entity identified by the certificate.

In addition to a public key, a certificate always includes the name of the entity it identifies, an expiration date, the name of the CA that issued the certificate, a serial number, and other information. Most importantly, a certificate always includes the digital signature of the issuing CA. The CA's digital signature allows the certificate to function as a "letter of introduction" for users who know and trust the CA but don't know the entity identified by the certificate.

10.18.1. Authentication

Authentication is the process of confirming an identity. In the context of network interactions, authentication involves the confident identification of one party by another party. Authentication over networks can take many forms. Certificates are one way of supporting authentication.

Network interactions typically take place between a client, such as browser software running on a personal computer, and a server, such as the software and hardware used to host a Web site. Client authentication refers to the confident identification of a client by a server (that is, identification of the person assumed to be using the client software). Server authentication refers to the confident identification of a server by a client (that is, identification of the organization assumed to be responsible for the server at a particular network address).

Client and server authentication are not the only forms of authentication that certificates support. For example, the digital signature on an email message, combined with the certificate that identifies the sender, provide strong evidence that the person identified by that certificate did indeed send that message. Similarly, a digital signature on an HTML form, combined with a certificate that identifies the signer, can provide evidence, after the fact, that the person identified by that certificate did agree to the contents of the form. In addition to authentication, the digital signature in both cases ensures a degree of non repudiation—that is, a digital signature makes it difficult for the signer to claim later not to have sent the email or the form.

Client authentication is an essential element of network security within most intranets or extranets. The sections that follow contrast two forms of client authentication:

1. **Password-Based Authentication**. Almost all server software permits client authentication by means of a name and password. For example, a server might require a user to type a name and password before granting access to the server. The server maintains a list of names and passwords; if a particular name is on the list, and if the user types the correct password, the server grants access.
2. **Certificate-Based Authentication**. Client authentication based on certificates is part of the SSL protocol. The client digitally signs a randomly generated piece of data and sends both the certificate and the signed data across the network. The server uses techniques of public-key cryptography to validate the signature and confirm the validity of the certificate.

10.18.2. Password-Based Authentication

Figure 10.12 shows the basic steps involved in authenticating a client by means of a name and password. Figure 10.10 assumes the following:

(*i*) The user has already decided to trust the server, either without authentication or on the basis of server authentication via SSL.

(*ii*) The user has requested a resource controlled by the server.

(*iii*) The server requires client authentication before permitting access to the requested resource.

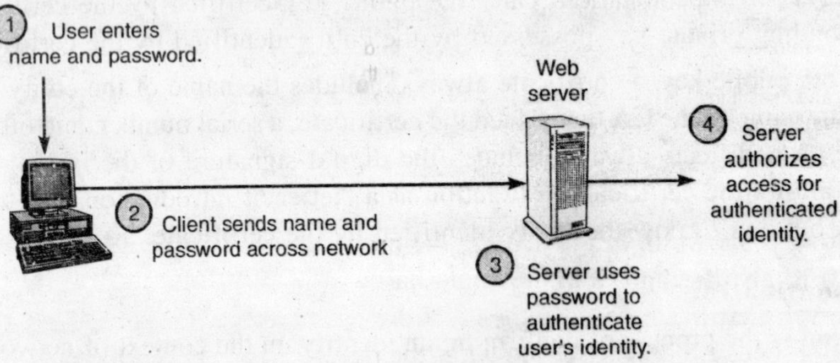

Fig. 10.12. *Using a password to authenticate a client to a server*

These are the steps shown in Fig. 10.10:

1. In response to an authentication request from the server, the client displays a dialog box requesting the user's name and password for that server. The user must supply a name and password separately for each new server the user wishes to use during a work session.
2. The client sends the name and password across the network, either in the clear or over an encrypted SSL connection.
3. The server looks up the name and password in its local password database and, if they match, accepts them as evidence authenticating the user's identity.
4. The server determines whether the identified user is permitted to access the requested resource, and if so allows the client to access it.

With this arrangement, the user must supply a new password for each server, and the administrator must keep track of the name and password for each user, typically on separate servers.

As shown in the next section, one of the advantages of certificate-based authentication is that it can be used to replace the first three steps in Figure 10.10 with a mechanism that allows the user to supply just one password (which is not sent across the network) and allows the administrator to control user authentication centrally.

10.18.3. Certificate-Based Authentication

Figure 10.10 shows how client authentication works using certificates and the SSL Protocol. To authenticate a user to a server, a client digitally signs a randomly generated piece of data and sends both the certificate and the signed data across the network. For the purposes of this discussion, the digital signature associated with some data can be thought of as evidence provided by the client to the server. The server authenticates the user's identity on the strength of this evidence.

Like Figure 10.10, Figure 10.11 assumes that the user has already decided to trust the server and has requested a resource, and that the server has requested client authentication in the process of evaluating whether to grant access to the requested resource.

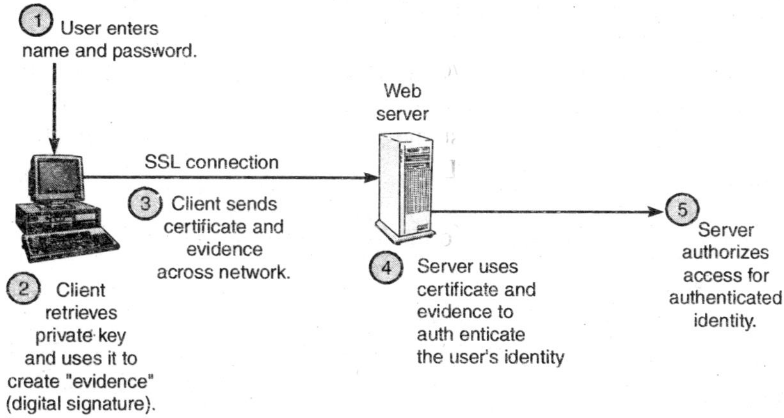

Fig. 10.13. *Using a certificate to authenticate a client to a server.*

Unlike the process shown in Figure 10.12, the process shown in Figure 10.13 requires the use of SSL. Figure 10.13 also assumes that the client has a valid certificate that can be used to identify the client to the server. Certificate-based authentication is generally considered preferable to password-based authentication because it is based on what the user has (the private key) as well as what the user knows (the password that protects the private key). However, it's important to note that these two assumptions are true only if unauthorized personnel have not gained access to the user's machine or password, the password for the client software's private key database has been set, and the software is set up to request the password at reasonably frequent intervals.

Important : Neither password-based authentication nor certificate-based authentication address security issues related to physical access to individual machines or passwords. Public- key cryptography can only verify that a private key used to sign some data corresponds to the public key in a certificate. It is the user's responsibility to protect a machine's physical security and to keep the private-key password secret.

These are the steps shown in Fig. 10.11 :

1. The client software, such as Communicator, maintains a database of the private keys that correspond to the public keys published in any certificates issued for that client. The client asks for the password to this database the first time the client needs to access it during a given session—for example, the first time the user attempts to access an SSL-enabled server that requires certificate-based client authentication. After entering this password once, the user doesn't need to enter it again for the rest of the session, even when accessing other SSL-enabled servers.
2. The client unlocks the private-key database, retrieves the private key for the user's certificate, and uses that private key to digitally sign some data that has been randomly generated for this purpose on the basis of input from both the client and the server. This data and the digital signature constitute "evidence" of the private key's validity. The digital signature can be created only with that private key and can be validated with the corresponding public key against the signed data, which is unique to the SSL session.
3. The client sends both the user's certificate and the evidence (the randomly generated piece of data that has been digitally signed) across the network.
4. The server uses the certificate and the evidence to authenticate the user's identity. At this point the server may optionally perform other authentication tasks, such as checking that the certificate presented by the client is stored in the user's entry in an LDAP directory. The server then

continues to evaluate whether the identified user is permitted to access the requested resource. This evaluation process can employ a variety of standard authorization mechanisms, potentially using additional information in an LDAP directory, company databases, and so on. If the result of the evaluation is positive, the server allows the client to access the requested resource.

As you can see by comparing Figure 10.11 to Figure 10.10, certificates replace the authentication portion of the interaction between the client and the server. Instead of requiring a user to send passwords across the network throughout the day, single sign-on requires the user to enter the private-key database password just once, without sending it across the network. For the rest of the session, the client presents the user's certificate to authenticate the user to each new server it encounters. Existing authorization mechanisms based on the authenticated user identity are not affected.

Summary

- The Internet is a computer network made up of thousands of networks worldwide. No one knows exactly how many computers are connected to the Internet. It is certain, however, that these number in the millions and are growing.
- The Internet makes available graphics files (digitized photographs and artwork), and even files that contain digitized sound and video. Through the Internet, you can download software, participate in interactive forums where users post and respond to public messages, and even join "chats," in which you and other users type (and, in some cases, speak) messages that are received by the chat participants instantly.
- IP addresses consist of a 32 bit number and are represented by the dot-decimal format. For example: 142.110.237.1 is an IP address. There are 4 decimal digits separated by three dots.
- Domain names tend to follow a loose structure that gives a description of the network. For example, sait.ab.ca uses the Canada extension ca, preceded by the province of Alberta extension ab and then the abbreviation SAIT for the Southern Alberta Institute of Technology.
- Address Resolution Protocol (ARP) resides in the bottom half of the Network layer. It can be considered a mechanism for mapping addresses between the Network logical addresses and MAC (Media Access Control) layer physical addresses.
- **Reverse Address Resolution Protocol (RARP)** is the reverse of ARP. It is a mechanism to map MAC addresses to IP addresses.
- The **World Wide Web** (abbreviated as the Web or WWW) is a system of Internet servers that supports hypertext to access several Internet protocols on a single interface.
- **URL stands for Uniform Resource Locator**. The URL specifies the Internet address of a file stored on a host computer connected to the Internet.
- The two primary methods of accessing the Internet were through a **network connection**, allowing users of local area networks (LANs) to go online through their school or workplace systems, and **dial-up connections** through a modem and phone line.
- To access the World Wide Web, we must use a Web browser. A browser is a software program that allows users to access and navigate the World Wide Web.
- Communication over the Internet uses the Transmission Control Protocol/Internet Protocol (TCP/IP).
- Computer networks are generally designed to do one thing above all others: allow any computer connected to the network to freely exchange information with any other computer also connected to the same network.
- There are a number of different kinds of technique which may be employed by a Firewall in order to correctly identify a conversation and act on it.
- The network level operations corresponding to the security policy above were actually an example of a simple packet filter.
- Encryption and decryption address the problem of eavesdropping, one of the three Internet security issues mentioned at the beginning of this topic.

Review Questions

1. Explain various classes of IP addressing. Where are class D and class E IP addressing used ? **[MDU Dec. 2008]**
2. Explain TCP/IP reference model in detail ? Explain various protocols used at each layer ? **[MDU Dec. 2008]**
3. Write short notes on any three :
   (*a*) DNS
   (*b*) WWW
   (*c*) Port Addressing
   (*d*) Security in Internet. **[MDU Dec. 2008]**
4. What is a Firewall ? Explain packet filtering firewall in detail. **[MDU May 2009]**
5. Write short notes on the following :
   (*a*) DNS
   (*b*) WWW
   (*c*) IP addressing. **[MDU May 2009]**
6. Explain in detail about packet filtering firewall. **[MDU Dec. 2009]**
7. Explain various classes of IP addressing. **[MDU Dec. 2009]**
8. Explain in detail about the various parts of world wide web. **[MDU Dec. 2009]**
9. Explain in detail about packet filtering firewalls. **[MDU May 2010]**
10. Explain TCP/IP reference modal in detail. Explain in detail about the various protocols used at each layer. **[MDU May 2010]**
11. Write short notes on any three :
    (*a*) Part Addressing
    (*b*) WWW
    (*c*) Uses of Internet
    (*d*) Physical Addressing **[MDU May 2010]**
12. How are internet addresses classified, explain with the help of its subfriend. What is the class of the following address
    10011101 10001111 11111100 11001111. **[MDU Dec. 2010]**
13. What are the four aspects of network security ? Explain the two types of firewalls with diagrams and the layers at which each of them is active. **[MDU Dec. 2010]**
14. Compare and contrast the three types of WWW documents. How are JAVA, HTML, and CGI related to these documents ? **[MDU Dec. 2010]**
15. Write notes on :
    (*a*) Intranet and Extranet
    (*b*) World Wide Web
    (*c*) Domain Name Service
    (*d*) Firewalls. **[MDU May 2011]**

❑❑❑

MODEL TEST PAPER

DATA COMMUNICATION

[Time : 3 Hours] *[Max. Marks: 100]*

Note : *Attempt any five questions. All questions carry equal marks.*

1. (*a*) Differentiate between ASK and PSK. What are the advantages of QAM over ASK and PSK ? 10

(*b*) For a signal, the bandwidth is 3 kHz and signal to noise ratio is Calculate the channel capacity. 15 If the BW is increased to 4 kHz and the signal is transmitted over the same channel, calculate the required SNR and the percentage change in the signal power. 10

2. (*a*) What is MODEM ? Differentiate between synchronous and asynchronous modems. 10

(*b*) Generate the hamming code for the codeword "1001101". 10

3. (*a*) Explain OSI model in detail. Differentiate between OSI and TCP/IP model. 10

(*b*) Compare and contrast the following :

(*i*) Serial and parallel transmission.

(*ii*) Connection-oriented and connection-less services. 10

4. (*a*) What is meant by TDM ? Describe briefly about statistical and synchronous TDM. 10

(*b*) Explain CODEC and COMBO chips in detail. 10

5. (*a*) Explain various classes of IP addressing. 10

(*b*) What is Firewall ? Explain packet filtering firewall in detail. 10

6. (*a*) Explain ISDN with the help of block diagram. 10

(*b*) Briefly explain the CCITT digital carrier system. 10

7. Write short notes on any of the three terms : (5 × 4)

(*i*) T-1 carrier system.

(*ii*) RS-232 serial interface.

(*iii*) LAN

(*iv*) Super frame & extended super frame.

(*v*) Error control

8. Explain TCP/IP reference model in detail. 20

❑❑❑

EXAMINATION PAPERS

B.E. SEVENTH SEMESTER (ECE) EXAMINATION MAY, 2012
DATA COMMUNICATION (EE-401-E)

[Time : 3 Hours] *[Max. Marks: 100]*

Note : *Attempt any five questions.*

1. (*a*) What is the number of bits per baud for following :
 (*i*) ASK with 8 different amplitude.
 (*ii*) QAM with a constellation of 28 points. (5)

(*b*) Define constellation diagram and its role in analog transmission. (5)

(*c*) Discuss in detail bit error rate and trellis encoding. (10)

2. (*a*) Discuss in detail various serial and parallel interfaces. (10)

(*b*) What is DDD network ? Discuss in brief private line service. (10)

3. (*a*) Define data encapsulation. Also give brief details of physical, iogical and service point addressing. (10)

(*b*) Discuss in detail various transmission modes and token pass ring. (10)

4. (*a*) What is CCITT time division multiplexed carrier system. Also discuss CODECS. (10)

(*b*) What are COMBO chips ? Discuss in detail synchronous time division multiplexing. (10)

5. (*a*) Discuss in detail various types of cryptography. (10)

(*b*) Differentiate intranet and extranet. (10)

6. What is the use of CIDR in Addressing ? Make the subnets using CIDR for IP address 192.168.1.1/27. (20)

7. Explain in detail synchronous and asynchronous protocols with suitable frame diagrams. (20)

8. Discuss the following : 20

(*a*) Clock Recovery

(*b*) LAN

(*c*) Firewall

(*d*) ISDN.

B.E. SEVENTH SEMESTER (ECE) EXAMINATION MAY, 2011
DATA COMMUNICATION (EE-401-E)

[Time : 3 Hours] *[Max. Marks: 100]*

Note : *Attempt any five questions. All questions carry equal marks.*

1. Why carrier recovery is needed in QAM and QPSK systems ? Draw and explain in detail all the methods of carrier recovery. 20

2. (*a*) Given the data word 1010011010 and the divisor 10111. Show the conversion of the codeword at the sender side using CRC and check the codeword at the receiver side. 10

(*b*) Which of the four digital to analog techniques is most susceptible to noise ? Defend your answer. Which is the best technique and why ? 10

3. List the names of all the synchronous and asynchronous protocols. What is the major drawback of byte oriented protocols that is overcome by bit oriented protocol ? Draw and explain the frame format of BSC protocol. 20

4. (*a*) Describe the DS hierarchy. How are T lines related to DS service ? A DS-0 signal has a data rate of 64 Kbps. Where does this number come from ? 14

(*b*) What is the difference between service point addressing, physical addressing and logical addressing in terms of the functions of the various layers of the OSI model ? 6

5. (*a*) What is the Shannon limit for information capacity for a standard telephone circuit with a signal to noise ratio fo 1000 and a bandwidth of 2.7 KHz. 5

(*b*) Draw and explain the electrical, mechanical and functional specification of the Centronics parallel interface. 15

6. What is the difference between a public telephone network and a public switched data network ? A message is broken up into three piece. Discuss the transmission of packets in a Public Data Network using : 20

(*a*) Datagram approach

(*b*) Permanent virtual circuit

(*c*) Switched virtual circuit

7. (*a*) Draw and discuss the frame formats of all the three types of frames in a token ring network. 14

(*b*) For an 8 PSK system operating at a bit rate of 24 Kbps determine baud, minimum bandwidth and bandwidth efficiency. 6

8. Write notes on : 20

(*a*) Intranet and Extranet

(*b*) World Wide Web

(*c*) Domain Name Service

(*d*) Firewalls.

B.E. SEVENTH SEMESTER (ECE) EXAMINATION DECEMBER, 2010

DATA COMMUNICATION (EE-401-E)

[Time : 3 Hours] *[Max. Marks: 100]*

Before answering the question, candidates should ensure that they have been supplied the correct, and complete question paper. No complaint in this regard, will be entertained after examination.

Note : *Attempt any five questions. All questions carry equal marks.*

1. (*a*) Draw and explain the DPSK transmitter and receiver. 15

   (*b*) In a 16 Mbps token ring network the length of token is 3 bytes. How long does it take for a station to produce a token ? 5

2. (*a*) Draw and explain the AT&T's FDM hierarchy. Four digital data channels each transmitting 1 Mbps use a satellite channel of 1 MHz. Design an appropriate configuration using FDM. 10

   (*b*) How are internet addresses classified, explain with the help of its subfriend. What is the class of the following address

   10011101 10001111 11111100 11001111. 10

3. Explain the architecture of the digital subscriber's loop used in the telephone network mentioning its seven major components. What is a DDD network ? 20

4. (*a*) Derive the relationship between minimum bandwidth, bit rate, mark and space frequency for an FSK system. Determine bandwidth and baud with a mark frequency of 32 KHz, space frequency of 24 KHz and bit rate of 4 Kbps. 10

   (*b*) What is a null modem connection ? Discuss RS 449 serial interface explaining its unbalanced and balanced modes and category I and category II pins. 10

5. What is the difference between In band and Out band signaling in context with ISDN. Explain all ISDN service and channels including its functional groupings ? 20

6. In which type of multiplexing are CODECS and Combo chips used. What is the difference between the two of them ? Explain the transmitter and receiver sections of the CODECS with the help of block diagrams. 20

7. (*a*) What are the four aspects of network security ? Explain the two types of firewalls with diagrams and the layers at which each of them is active. 10

   (*b*) Compare and contrast the three types of WWW documents. How are JAVA, HTML, and CGI related to these documents ? 10

8. Write notes on :

   (*a*) T Carriers

   (*b*) Ethernet

   (*c*) Data Modems

   (*d*) Trellis encoding 20

B.E. SEVENTH SEMESTER (ECE) EXAMINATION MAY, 2010
DATA COMMUNICATION (EE-401-E)

[Time : 3 Hours] *[Max. Marks: 100]*

Note : *Attempt any five questions.*

1. (*a*) Differentiate between ASK and PSK. What are the advantages of QAM over ASK and PSK ? 10

(*b*) Assume a data stream is made of ten os (zeros). Encode this stream using following encoding schemes :

(*i*) Polar NRZ-I

(*ii*) Monchester

(*iii*) Differential Monchester

(*iv*) HDB3 10

2. (*a*) A telephone line has a frequency range of 300 to 3300 kHz. The signal to noise ratio is 3162 (35 DB). For the channel calculate channel capacity.

(*b*) Construct the hamming code for one bit sequence 10011101. 10

(*c*) Explain RS-232 serial interface in detail. 5

3. (*a*) Differentiate between synchronous and asynchronous modems. Explain Modem synchronisation. 10

(*b*) Explain DDD networks in detail. 10

4. Explain OSI reference model in detail. Explain at which layer the following addressing is used :

(*a*) IP addressing

(*b*) Physical addressing

(*c*) Service point addressing. 20

5. (*a*) What is LAN ? Differentiate between star and mesh topology in detail. 10

(*b*) Explain CCIT time division multiplexing in detail. 10

6. (*a*) Explain T-I carrier system in detail. 10

(*b*) Explain in detail about packet filtering firewalls. 10

7. Explain TCP/IP reference modal in detail. Explain in detail about the various protocols used at each layer. 20

8. Write short notes on any three :

(*a*) Part addressing

(*b*) WWW

(*c*) Uses of internet

(*d*) Physical addressing 20

B.E. SEVENTH SEMESTER (ECE) EXAMINATION DECEMBER, 2009 DATA COMMUNICATION (EE-401-E)

[Time : 3 Hours] *[Max. Marks: 100]*

Note : *Attempt any five questions. All questions carry equal marks.*

1. (*a*) The SNR for a channel having 1-MHz bandwidth is 63. Calculate the following for this channel
(*i*) Bit Rate (*ii*) Signal Level

(*b*) Explain BPSK and how it is implemented.

(*c*) Define the term clock recovery and carrier recovery.

2. (*a*) Generate the hamming code for the codeword "1000".

(*b*) Differentiate between synchronous and asynchronous modems. Explain synchronous modem in detail.

3. Explain OSI model in detail. Give brief description about various layers.

4. (*a*) Explain in detail about CCITT time division multiplexing.

(*b*) Explain CODEC and COMBO chips in detail.

5. (*a*) Explain in detail about packet filtering firewall.

(*b*) Explain various classes of IP addressing.

6. Describe briefly about the following in context with the ISDN network. Explain its model ? Give

(*a*) BRI and PRI

(*b*) NT 1, NT 2

(*c*) TE 1, TE 2.

Brief description about various layers.

7. Explain the following in detail

(*a*) Frequency shift keying

(*b*) Phase shift keying.

8. (*a*) Describe in detail about the data communication circuits.

(*b*) Explain in detail about the various parts of world wide web.

B.E. SEVENTH SEMESTER (ECE) EXAMINATION MAY, 2009
DATA COMMUNICATION (EE-401-E)

[Time : 3 Hours] *[Max. Marks: 100]*

Note : *Attempt any five questions. All questions carry equal marks.*

1. (*a*) Explain in detail about the following :
   (*i*) Clock recovery
   (*ii*) Trellis encoding
   (*iii*) Carrier recovery.

(*b*) Differentiate between QAM and QPSK in detail ? What are the advantages of QAM over QPSK.

2. (*a*) Differentiate between serial and parallel interface ? Explain RS-449 serial interface in detail.

(*b*) Explain telephone networks in detail.

3. Given the data word "10110011" and one divisor 1001.

(*a*) Generate the codeword at the sender site.

(*b*) Generate the dataword from the codeword (assume no error and use binary division).

4. Explain OS-model in detail. Differentiate between OSI & TCP/IP model ? Explain in detail about the various layers of OSI model.

5. (*a*) Describe briefly about the token passing ring arrangement of LAN ?

(*b*) What is meant by TDM ? Describe briefly about statistical & synchronous TDM.

6. (*a*) Explain CODECS and COMBO chips in detail.

(*b*) Explain AT and T's FDM in detail.

7. (*a*) What is a Firewall ? Explain packet filtering firewall in detail.

(*b*) Describe briefly about the QAM.

8. Write short notes on the following :

(*a*) DNS

(*b*) WWW

(*c*) IP addressing.

B.E. SEVENTH SEMESTER (ECE) EXAMINATION DECEMBER, 2008 DATA COMMUNICATION (EE-401-E)

[Time : 3 Hours] *[Max. Marks: 100]*

Note : *Attempt any five questions. All questions carry equal marks.*

1. (*a*) Differentiate between QAM and QPSK in detail. What are the advantages of QAM over PSK?

(*b*) Explain frequency modulation in detail ? Compare FM bandwidth with AM bandwidth in terms of modulating signal.

2. (*a*) Given a remainder of "111", a data unit of "10110011", and a divisor of 1001, is there an error is data unit.

(*b*) Explain CCITT X.21 in detail.

(*c*) A telephone line has a frequency range of 300 to 3300 Hz. The signal to noise ratio is 3162 (35 dB). For this channel calculate the channel capacity.

3. (*a*) Explain DDD networks in detail.

(*b*) Explain RS-530 serial interface in detail.

4. Explain ISDN layer model in detail.

5. (*a*) What is meant by topology ? Differentiate between various topologies of LAN.

(*b*) Explain AT and T's FDM in detail.

6. (*a*) Explain in detail about CODECS?

(*b*) Explain various classes of IP addressing. Where are class D and class E IP addressing used ?

7. Explain TCP/IP reference model in detail ? Explain various protocols used at each layer ?

8. Write short notes on any three :

(*a*) DNS

(*b*) WWW

(*c*) Port Addressing

(*d*) Security in Internet.

❑❑❑

INDEX

D

E

F

❑❑❑

Win Prizes

ATTENTION READERS

Win Prizes

We request for your frank assessment regarding various aspects of the book as given below :

FUNDAMENTALS OF

DATA COMMUNICATION

By : K. K. SHARMA **(MDU)**

Please Fill up the Sheet in Neat and Clean Writing

(i) What topics of your syllabus, important for your examination point of view, are not covered in this book ?

....

....

....

....

(ii) What are the chapters wherein the treatment of the subject matter is not systematic or organised or updated ?

....

....

....

....

(iii) Have you come across misprints/mistakes/factual inaccuracies in the book ? Please specify the chapters and the page numbers.

....

....

....

....

(iv) What other books on the same subject have you found/heard better than the present book ? Please specify in terms of price and quality.

....

....

....

....

(v) Further suggestions and comments for the improvement of this book.

....

....

....

....

(PS : If need be please attach a separate sheet to write your views)

OTHER DETAILS

(i) Who recommended you this book ? (Please tick (✓) in the box near the option relevant to you.

☐ Teacher ☐ Friends ☐ Bookseller

....

....

....

....

(ii) Name of the recommending teacher, his designation and address:

....

....

....

....

(iii) Name and address of the bookseller you purchased the book from :

....

....

....

....

(iv) Name and address of your institution (Please mention the University or Board as the case may be)

....

....

....

....

(v) Your name and complete address :

....

....

....

....

(vi) Course you are enrolled in :

....

....

....

....

The best assessment will be awarded each month : The award will be in the form of our publications as decided by the Editorial Board.

Please mail the filled up coupon at the vearliest to :

S.K. KATARIA & SONS®

109, Prakash Mahal,
Dr. Subhash Bhargav Lane,
Opposite Delhi Medical Association,
Daryaganj, New Delhi-110002
Phone: 011-23243489, 23269324
TeleFax : 011-23243489
email : katariabooks@yahoo.com